Inequalities Involving Functions
and Their Integrals and Derivatives

Mathematics and Its Applications (*East European Series*)

Volume 53

Inequalities Involving Functions and Their Integrals and Derivatives

by

D. S. Mitrinović
University of Belgrade,
Yugoslavia

J. E. Pečarić
University of Zagreb,
Yugoslavia

and

A. M. Fink
Iowa State University,
Ames, U.S.A.

SPRINGER-SCIENCE+BUSINESS MEDIA, B.V.

Library of Congress Cataloging-in-Publication Data

Mitrinović, Dragoslav S.
Inequalities involving functions and their integrals and derivatives / by D.S. Mitrinović, J.E. Pečarić, and A.M. Fink.
p. cm. -- (Mathematics and its applications. East European series ; 53)
Includes index.
ISBN 978-94-010-5578-9 ISBN 978-94-011-3562-7 (eBook)
DOI 10.1007/978-94-011-3562-7
1. Integral inequalities. 2. Differential inequalities. I. Pečarić, J. E. II. Fink, A. M. (Arlington M.), 1932- . III. Title. IV. Series: Mathematics and its applications (Kluwer Academic Publishers). East European series ; v. 53.
QA295.M584 1991
515'.243--dc20 91-20780
CIP

ISBN 978-94-010-5578-9

Printed on acid-free paper

SERIES EDITOR'S PREFACE

'Et moi, ..., si j'avait su comment en revenir, je n'y serais point allé.'

Jules Verne

The series is divergent; therefore we may be able to do something with it.

O. Heaviside

One service mathematics has rendered the human race. It has put common sense back where it belongs, on the topmost shelf next to the dusty canister labelled 'discarded non-sense'.

Eric T. Bell

Mathematics is a tool for thought. A highly necessary tool in a world where both feedback and non-linearities abound. Similarly, all kinds of parts of mathematics serve as tools for other parts and for other sciences.

Applying a simple rewriting rule to the quote on the right above one finds such statements as: 'One service topology has rendered mathematical physics ...'; 'One service logic has rendered computer science ...'; 'One service category theory has rendered mathematics ...'. All arguably true. And all statements obtainable this way form part of the raison d'être of this series.

This series, *Mathematics and Its Applications*, started in 1977. Now that over one hundred volumes have appeared it seems opportune to reexamine its scope. At the time I wrote

> "Growing specialization and diversification have brought a host of monographs and textbooks on increasingly specialized topics. However, the 'tree' of knowledge of mathematics and related fields does not grow only by putting forth new branches. It also happens, quite often in fact, that branches which were thought to be completely disparate are suddenly seen to be related. Further, the kind and level of sophistication of mathematics applied in various sciences has changed drastically in recent years: measure theory is used (non-trivially) in regional and theoretical economics; algebraic geometry interacts with physics; the Minkowsky lemma, coding theory and the structure of water meet one another in packing and covering theory; quantum fields, crystal defects and mathematical programming profit from homotopy theory; Lie algebras are relevant to filtering; and prediction and electrical engineering can use Stein spaces. And in addition to this there are such new emerging subdisciplines as 'experimental mathematics', 'CFD', 'completely integrable systems', 'chaos, synergetics and large-scale order', which are almost impossible to fit into the existing classification schemes. They draw upon widely different sections of mathematics."

By and large, all this still applies today. It is still true that at first sight mathematics seems rather fragmented and that to find, see, and exploit the deeper underlying interrelations more effort is needed and so are books that can help mathematicians and scientists do so. Accordingly MIA will continue to try to make such books available.

If anything, the description I gave in 1977 is now an understatement. To the examples of interaction areas one should add string theory where Riemann surfaces, algebraic geometry, modular functions, knots, quantum field theory, Kac-Moody algebras, monstrous moonshine (and more) all come together. And to the examples of things which can be usefully applied let me add the topic 'finite geometry'; a combination of words which sounds like it might not even exist, let alone be applicable. And yet it is being applied: to statistics via designs, to radar/sonar detection arrays (via finite projective planes), and to bus connections of VLSI chips (via difference sets). There seems to be no part of (so-called pure) mathematics that is not in immediate danger of being applied. And, accordingly, the applied mathematician needs to be aware of much more. Besides analysis and numerics, the traditional workhorses, he may need all kinds of combinatorics, algebra, probability, and so on.

In addition, the applied scientist needs to cope increasingly with the nonlinear world and the extra mathematical sophistication that this requires. For that is where the rewards are. Linear models are honest and a bit sad and depressing: proportional efforts and results. It is in the non-linear world that infinitesimal inputs may result in macroscopic outputs (or vice versa). To appreciate what I am hinting at: if electronics were linear we would have no fun with transistors and computers; we would have no TV; in fact you would not be reading these lines.

There is also no safety in ignoring such outlandish things as nonstandard analysis, superspace and anticommuting integration, p-adic and ultrametric space. All three have applications in both electrical engineering and physics. Once, complex numbers were equally outlandish, but they frequently proved the shortest path between 'real' results. Similarly, the first two topics named have already provided a number of 'wormhole' paths. There is no telling where all this is leading - fortunately.

Thus the original scope of the series, which for various (sound) reasons now comprises five sub-series: white (Japan), yellow (China), red (USSR), blue (Eastern Europe), and green (everything else), still applies. It has been enlarged a bit to include books treating of the tools from one subdiscipline which are used in others. Thus the series still aims at books dealing with:

- a central concept which plays an important role in several different mathematical and/or scientific specialization areas;
- new applications of the results and ideas from one area of scientific endeavour into another;
- influences which the results, problems and concepts of one field of enquiry have, and have had, on the development of another.

Inequalities are everywhere. Whole series of conferences are devoted to them. Indeed in my more despondent moments, when struggling with one or another problem, I sometimes have the feeling that mathematics (especially analysis) is all inequalities. It is not that I dislike them or do not appreciate their usefulness; indeed the vast importance of inequalities is manifest and many, especially the more powerful ones, have great aesthetic appeal — for here, as in many other parts of mathematics and science, power and elegance go together. But, on the other side, there are so many inequalities and it is, by and large, such an unstructured mass of results. (This is, for that matter, an aspect that this book addresses with success.)

Since the time of the classical book by Hardy, Littlewood and Polya of 1934 (recently reprinted by Cambridge Univ. Press in their 'classics series'; a bargain!) the subject of this book, 'Differential and Integral Inequalities', has grown by about 800%. The senior author of this book, D.S. Mitrinovic, has worked in inequalities all his life and has collected results in the subsubject of inequalities between functions and their integrals and derivatives, i.e. differential and integral inequalities, since 1950. Other books by him and his coworkers on different topics within 'Inequality Theory' have also appeared in this series (P.S. Bullen, D.S. Mitrinovic, P.M. Vasic, Means and their Inequalities, 1987; D.S. Mitrinovic, J.E. Pecaric, V. Volenec, Recent Advances in Geometric Inequalities, 1989). The present volume is devoted to the particular topic of differential and integral inequalities. It systematizes and surveys the whole subject (thus making it possible for nonspecialists such as myself to find the results needed) and, additionally, it describes typical uses of these inequalities.

The shortest path between two truths in the real domain passes through the complex domain.

J. Hadamard

La physique ne nous donne pas seulement l'occasion de résoudre des problèmes ... elle nous fait pressentir la solution.

H. Poincaré

Never lend books, for no one ever returns them; the only books I have in my library are books that other folk have lent me.

Anatole France

The function of an expert is not to be more right than other people, but to be wrong for more sophisticated reasons.

David Butler

Bussum, June 1991

Michiel Hazewinkel

CONTENTS

CONTENTS

PREFACE

It is said that every mathematician loves an inequality. As Richard Bellman has so elegantly said, "There are three reasons for the study of inequalities: practical, theoretical, and aesthetic. In many practical investigations, it is necessary to bound one quantity by another. The classical inequalities are very useful for this purpose. From the theoretical point of view, very simple questions give rise to entire theories. For example, we may ask when the nonnegativity of one quantity implies that of another. This simple question leads to the theory of positive operators and the theory of differential inequalities. ... Another question which gives rise to much interesting research is that of finding equalities associated with inequalities. We use the principle that every inequality should come from an equality which makes the inequality obvious. Along these lines, we may also look for representations which make inequalities obvious. Often, these representations are maxima or minima of certain quantities. ... Finally, let us turn to the aesthetic aspects. As has been pointed out, beauty is in the eyes of the beholder. However, it is generally agreed that certain pieces of music, art, or mathematics are beautiful. There is an elegance to inequalities that makes them very attractive." (General Inequalities 2; Proc. Second International Conference on General Inequalities, Oberwolfach, July 30 - August 5, 1978, Ed. E. F. Beckenbach, Basel-Boston-Stuttgart, 1980).

This book gives an up to date report of inequalities that involve a relationship between a function and its derivatives or integrals. While the history of general inequalities surely goes back to K. F. Gauss (1777–1855) and A. L. Cauchy (1789–1857), the evolution of inequalities studied here is of more recent origin. The earliest such inequalities carry the names of P. L. Čebyšev (1821–1894), T. Gronwall (1877–1932), J. Hadamard (1865–1963), G. H. Hardy (1877–1947), D. Hilbert (1862-1943), E. Landau (1877–1938), A. A. Markov (1856–1922), H. Poincaré (1854 1912), and W. Wirtinger (1865–1945). A glance at the classical treatise on inequalities, "Inequalities" (1934,1952) by G. H. Hardy, J. E. Littlewood and G. Pólya, shows that two of the ten chapters are devoted to Hilbert's inequality and applications of the calculus of variations. These two chapters give a first systematic treatment of inequalities of the type studied in this monograph. It may be said that the growth of inequalities was surely stimulated by this book that was first published in 1934.

Since that time, several other books on inequalities have appeared. We can mention "Inequalities" (1961,1965,1971,1983) by E. F. Beckenbach and R. Bellman and "Analytic Inequalities" (1970) by D. S. Mitrinović. The latter book is a comprehensive study of inequalities of many types. The 386 pages of this monograph have short sections devoted to differential or integral inequalities numbering perhaps thirty pages. The book by Beckenbach and Bellman devotes the last fifty pages to differential inequalities.

Books of a different genre that might be mentioned are the books "Differential and Integral-Ungleichungen" by W. Walter, German edition, 1964, or the enlarged English edition, 1970, and "Differential and Integral Inequalities" by V. Lakshmikantham and S. Leela, vol I and II, 1969. The Eastern European literature is not always easily available to the West, so we should also mention the book "Theorems on inequalities", by Ja.D. Mamedov, S. Aširov, and S. Atdaev. The four volumes mentioned are devoted to the applications of certain differential and integral inequalities to the theory of differential or integral equations. Neither attempt to give comprehensive account of differential or integral inequalities.

Integral inequalities, especially Grownwall-Bellman type, are very important for applications in Theory of Motion Statibility. This can be seen from many books from Theory of Stability (the greater part of such books is in Russian).

Recently the following monograph was published: A. A. Martynyuk, V. Lakshmikanthan and S. Leela, "Motion of Stability: the Method of Integral Inequalities," Kiev 1989, 272 p. (Russian).

In this monograph wide applications of Grownwall-Bellman's inequality and related inequalities are given. Part of this book in which the Grownwall-Bellman and related inequalities are exposed is similar to our Chapters XII–XIV.

This book attempts to do both more and less than the above mentioned monographs. On the one hand we have undertaken the task of doing an exhaustive search of the literature on differential and integral inequalities. On the other hand in such an undertaking, the number of papers devoted to differential and integral inequalities is so large that space does not allow us to quote all the results or prove most of those that we do quote. We have given proofs only when they are relatively short and give insight into the general methods involved. We have mentioned some of the applications of these results, especially to the theory of differential equations.

Two books published as this manuscript was in final form are Hardy-type Inequalities by B. Opic and A. Kufner and the Proceedings of the Inequality Conference at University of Birmingham, edited by W. N. Everitt. The first named book is briefly mentioned in Chapter IV, Section 46. The latter was a conference entitled; "Inequalities: Fifty years on from Hardy, Littlewood and Pólya." Some of these papers have close connections with the text of our manuscript.

One of the authors, D.S. Mitrinović began the undertaking of a systematic search for papers in the subject areas of this book in 1950. In writing a systematic account we have taken the view that there are certain fundamental inequalities, perhaps very simple to state, that are the heart of the matter. The search for generalizations can lead to a determination of best (exact) constants that appear in inequalities, to some refinements, to a simpler form, a simpler proof, wider applicability, or the opposites of some of the above. It is generally recognized that the some generalizations are uninteresting inequalities, nevertheless they remain part of the lore of the subject, and some may become interesting if an appropriate application can be found. The above views are quite generally held in our correspondence with other mathematicians.

The organization of our book is relatively clear from a glance at the table of contents. We have attempted to classify each result as being closely related to a 'well-known classical result'. In this way certain inequalities, which from first appearance are quite different, when set side by side are seen as being of the same form. For example, the first chapter is devoted to inequalities related to ones proved by Landau in 1913 and Kolmogorov in 1938. These basic inequalities have been generalized in many directions, to more general differential operators than the differentiation operator, and even to abstract operators in a Banach space. It is beyond the scope of this book to prove these more general results, but we give the references and give abstracts of the results.

The succeeding chapters discuss inequalities related to those of Wirtinger and Poincaré, Opial, Hardy, Hilbert and Lyapunov. Some of the next chapters deal with unnamed inequalities or those whose names may not be well known. The next part of the monograph, Chapters XI–XIV are devoted to inequalities of Čaplygin and Gronwall type and there is some overlap in these chapters to the books by Walter and Laksmikantham and Leela, but the present exposition is more comprehensive. It is partially based on the monograph, "Diferencijalne i Integralne Nejednakosti" by D. S. Mitrinović and J. E. Pečarić (In Serbo-Croatian), published in Belgrade, 1988.

Chapter XVII outlines several methods of proof for inequalities. These are all illustrated with applications to Wirtinger's inequality and some of its generalizations.

The last chapter is not organized in the same manner as the first seventeen. It contains particular results which did not fit easily into one of the types of the earlier chapters.

Each chapter is written to be self-contained. There are occasional references to other chapters, mostly in the middle chapters, but each chapter can be read with full comprehension without reading the entire book. We hope this increases the value of the book to the occasional reader.

It is not possible to finish such an undertaking as this without making some errors. We have worked hard to eliminate all errors, but since we give the exact references, we hope that the reader will forgive the occasional lapse and feel free to consult the original reference.

The authors with to express their appreciation to Kluwer Academic Publishers, well known for their high quality productions, and particularly to the publisher, Dr. D. J. Larner, as well as the series editor, Professor Dr. M. Hazewinkel, for their most efficient handling of the publication of this monograph. A special thanks to Jan Nyhus who prepared the camera ready copy and to Deborah Fink for some work on the bibliography and suggestions on style.

This text was typed with AMSTeX.

Dragoslav S. Mitrinović
Smiljanićeva 38
11000 Belgrade, Yugoslavia
Josip E. Pečarić
Tehnološki Fakultet
41000 Zagreb, Yugoslavia
A. M. Fink
Mathematics Department
Iowa State University
Ames, Iowa 50011, USA

ORGANIZATION OF THE BOOK

We give a list of symbols common to all the chapters on page v.

Each chapter is divided into many sections, each of which contains results which are essentially the same or from the same paper. They are numbered consecutively in each chapter. References to sections from the same chapter are indicated by the number followed by a period, for example, if in Chapter III, section 10 we want to refer the reader to section 2 of Chapter III we merely say "see 2." If the section occurs in a different chapter we say "see 2. of Chapter III". Because we have organized the book by related inequalities, references between chapters are rare.

Each chapter begins with the inequality or inequalities that give the chapter its name. The bibliographic references are given at the end of each chapter.

We give a name index at the end of the book.

NOTATION

We give some of the notation used throughout the book.

R is the real numbers. N is the natural numbers.
I will be generic interval.
R^+ is $[0, \infty)$.
R^n is Euclidean n-space.

$x \leq y$, for x and y in R^n means that $x_i \leq y_i$ for $i = 1, \ldots, n$.

$C^n[a, b]$ means the n-times differential functions on $[a, b]$.
a.e. means almost everywhere with respect to the relevant measure

$\Gamma(x)$ is the gamma function. B is the beta function, $B(x, y) = \frac{\Gamma(x)\Gamma(y)}{\Gamma(x+y)}$.

$\|f\|_p = \left(\int |f(x)|^p dx\right)^{1/p}$ for $1 \leq p < \infty$, with $\|f\|_\infty = \text{ess sup}\,|f(x)|$, and in this context, $1/p + 1/p' = 1$ with the usual conventions that $1' = \infty$, and $\infty' = 1$.

$L^p[a, b] = \{f \mid \|f\|_p < \infty\}$, with $L = L^1$.

$\ell_p = \{\{a_n\} \mid \sum_1^\infty |a_n|^p < \infty\}$.

$AC[a, b]$ means the absolutely continuous functions on $[a, b]$.

$AC^n[ab]$ means $f, \ldots, f^{(n-1)}$ exist everywhere and $f^{(n-1)}$ is absolutely continuous.

$f(x)^2 = (f(x))^2 = f^2(x)$, and $\left(f^{(k)}(x)\right)^2 = \left(f^{(k)}(x)\right)^2 = \left(\frac{d^k}{dx^k} f(x)\right)^2$.

∇ is the gradient, i.e, $\nabla f(x) = \left(\frac{\partial f}{\partial x_1}, \ldots, \frac{\partial f}{\partial x_n}\right)^T$

Δ is the Laplacian, i.e. $\Delta f = \frac{\partial^2 f}{\partial x_1^2} + \cdots + \frac{\partial^2 f}{\partial x_n^2}$.

In R^n, $([a, b]) = [a, b] \times \cdots \times [a, b]$, ($n$ factors).

$p_+(x) = p^+(x) = \max\{p(x), 0\}$, $p_-(x) = p^-(x) = \min\{p(x), 0\}$.

References to other sections in the book are given in two different ways: references to sections from the same chapter are indicated by the number followed by a period, for example, if in Chapter III, section 10 we want to refer the reader to section 2 of Chapter III we merely say "see 2." If the section occurs in a different chapter we say "see 2. of Chapter III".

For reference to an inequality not discussed in this monograph, we will refer the reader to "Analytic Inequalities" by D. S. Mitrinović with the reference simply given as [**AI**, p. —].

CHAPTER I

LANDAU-KOLMOGOROV AND RELATED INEQUALITIES

1. Let f be a real function with n derivatives on an interval I of the real line. Define

$$M_k(p,I) = \|f^{(k)}\|_p, \qquad 0 \le k \le n.$$

We are interested in the problem of finding inequalities of the form

$$M_k(p,I) \le C_{nk}(p,I) M_0(p,I)^{1-\frac{k}{n}} M_n(p,I)^{\frac{k}{n}}, \tag{1.1}$$

where $C_{nk}(p,I)$ is to be the best possible constant. The interpretation of this inequality, the well-known Landau- Kolmogorov inequality, is that if M_0 and M_n are finite, then M_k is finite and the indicated inequality holds.

Here, we shall give history, variations and generalizations of this inequality and some related inequalities.

2. E. Landau in 1913 proved the following theorems:

THEOREM 1. *Let f be a real function defined on an interval whose length is not less than 2 and satisfies the conditions $|f(x)| \le 1$, and $|f''(x)| \le 1$. Then*

$$|f'(x)| \le 2, \tag{2.1}$$

where the constant 2 is the best possible.

THEOREM 2. *Let f be a real function on R such that $\sup_{x\in R} |f(x)| \le 1$ and $\sup_{x\in R} |f''(x)| \le 1$, then*

$$\sup_{x\in R} |f'(x)| \le \sqrt{2}. \tag{2.2}$$

where the constant $\sqrt{2}$ is the best possible.

In fact, from these results it follows that $C_{21}(+\infty, R^+) = 2$ and $C_{21}(+\infty, R) = \sqrt{2}$.

Landau [**1**] also proved that in the above results instead of $f''(x)$ we can put $|f'(y) - f'(x)|/(y-x)$, where $(x < y)$.

As we see, E. Landau obtained the inequality

$$M_1 \le 2M_0^{1/2} M_2^{1/2} \tag{2.3}$$

in the case $M_0 = 1 = M_2$. Inequality (2.3) is formulated by J. Hadamard [2] in 1914. He also proved that if f is a real function defined on an interval I, whose length is a, then

$$M_1(+\infty, I) \le 2a^{-1} M_0(+\infty, I) + \frac{a}{2} M_2(+\infty, I). \tag{2.4}$$

Hadamard showed that

$$C_{n1}(+\infty, R) \le 2^{n-1}.$$

G. H. Hardy and J. E. Littlewood [3] at about the same time showed that the constants $C_{nk}(+\infty, R)$ and $C_{nk}(+\infty, R^+)$ exist but did not get good estimates.

3. In [4] E. Landau gave some estimates similar to his Theorems from 2. but for a fixed interval $[0,1]$. (see also [7].)

4. Suppose that $f(x)$ is defined for $x > 0$ (this involves no loss of generality) and that $f''(x)$ exists for $x > 0$. If, as $x \to +\infty$, $g(x)$ and $h(x)$ are positive functions of x, both increasing or decreasing steadily, and $|f(x)| < g(x)$, $|f''(x)| < h(x)$, then, as $x \to +\infty$,

$$|f'(x)| < (2+\varepsilon)\,[g(x)h(x)]^{1/2}, \tag{4.1}$$

where $\varepsilon > 0$ and $\varepsilon \to 0$. In particular,

$$\begin{aligned}
&\text{if}\quad f(x) = O(g(x)),\ f''(x) = O(h(x)),\quad \text{then}\quad f'(x) = O(g(x)h(x))^{1/2},\\
&\text{if}\quad f(x) = o(g(x)),\ f''(x) = O(h(x)),\quad \text{then}\quad f'(x) = o(g(x)h(x))^{1/2},\\
&\text{if}\quad f(x) = O(g(x)),\ f''(x) = o(h(x)),\quad \text{then}\quad f'(x) = o(g(x)h(x))^{1/2}.
\end{aligned}$$

The last three results were given by Hardy and Littlewood [3], when g and h are increasing functions of x, and were proved by geometrical considerations. They were under the impression that the case when g and h are both decreasing function of x was rather more delicate and required additional restrictions. But this is not true as was shown in [5] where the above result is proved.

5. L. Neder [6] proved the inequality, $(L > 0)$,

$$\begin{aligned}
M_k(+\infty, [a,\ a+L]) \le{}& (2n)^{2n} L^{-k} M_0(+\infty, [a,\ a+L])\\
&+ L^{n-k} M_n(+\infty, [a,\ a+L])
\end{aligned} \tag{5.1}$$

He also gave generalizations of results from 4.

6. G. H. Hardy and J. E. Littlewood [8] in 1932 showed that $C_{21}(2,R) = 1$ and $C_{21}(2,R^+) = \sqrt{2}$ and three years later Hardy, Littlewood and Landau [10] showed $C_{21}(p,R^+) \le 2$ with $\lim_{p\to+\infty} C_{21}(p,R^+) = 2$.

Results from [8] are exposed in the book [9]. Here, we shall give results from this important book. (These results are also exposed in [188] and we use this exposition.)

THEOREM 1. *If f' is locally absolutely continuous and $f, f'' \in L_2(R^+)$, then*

$$\left(\int_0^\infty f'^2\,dx\right)^2 \le 4\left(\int_0^\infty f^2\,dx\right)\left(\int_0^\infty f''^2\,dx\right), \tag{6.1}$$

$$\int_0^\infty f'^2\,dx \le \int_0^\infty f^2\,dx + \int_0^\infty f''^2\,dx. \tag{6.2}$$

Equality holds in (6.2) if and only if $f(x) = Ae^{-x/2}\sin\left(\frac{\sqrt{3}}{2}x - \frac{\pi}{3}\right) \equiv f_A(x)$, and in (6.1) if and only if $f(x) = f_A(Bx)$, where A, B are arbitrary constants.

PROOF: The implication (6.1) implies (6.2) is obvious (by GA inequality). Now, we show that (6.2) implies (6.1). If (6.2) holds (with equality only as asserted), then for $t > 0$ and $F(x) = f(tx)$ we have

$$t^2\int_0^\infty f'(tx)^2\,dx \le \int_0^\infty f(tx)^2\,dx + t^4\int_0^\infty f''(tx)^2\,dx,$$

or, setting $tx = u$ and dividing by t,

$$\int_0^\infty f'(u)^2\,du \le t^{-2}\int_0^\infty f(u)^2\,du + t^2\int_0^\infty f''(u)^2\,du, \quad (t > 0).$$

Setting $t^2 = \left[\int_0^\infty f^2\,du \Big/ \int_0^\infty f''^2\,du\right]^{1/2} = \beta^2$, this inequality reduces to (6.1). Moreover, equality holds only if $f(\beta x) = f_A(x)$ for arbitrary B since it is easy to verify that if equality holds in (6.1) for a given f it also holds for $g(x) = f(kx)$ with arbitrary k.

To prove (6.2) note that for all $X > 0$

$$\int_0^X \left[f^2 - f'^2 + f''^2 - (f + f' + f'')^2\right] dx$$

$$= -2 \int_0^X \left[f'^2 + ff' + ff'' + f'f''\right] dx$$

or

$$\int_0^X \left[f^2 - f'^2 + f''^2 - (f + f' + f'')^2\right] dx \tag{6.3}$$

$$= [f(0) + f'(0)]^2 - [f(X) + f'(X)]^2 .$$

Also,

$$\int_0^X f'^2 \, dx = f(X)f'(X) - f(0)f'(0) - \int_0^X ff'' \, dx. \tag{6.4}$$

Since $f, f'' \in L_2$, it follows from Hölder's inequality that $\lim_{X\to\infty} \int_0^X ff'' \, dx = \int_0^\infty ff'' \, dx$ exists (finite). Hence also $\int_0^\infty f'^2 \, dx < +\infty$ since otherwise it would follow from (6.4) that $f(X)f'(X) \to +\infty$ and so also $\int_0^X f \, f' \, dx = \frac{1}{2} \left[f^2(X) - f^2(0)\right] \to +\infty$. This is impossible since $f \in L_2$. Thus both integrals in (6.4) converge to a finite limit as $X \to +\infty$, and so also $\lim_{x\to\infty} f(x)f'(x)$ exists. This limit must be zero since otherwise, as before, we could not have $f \in L_2$. But now, as $X \to \infty$ the integrals on the left side of (6.3) all exist (use Hölder's inequality for $\int ff' \, dx$ and $\int f'f'' \, dx$), and so

$$\lim_{X\to\infty} \left[f^2(X) + f'^2(X)\right] = L$$

exists, since $ff' \to 0$. If $L \neq 0$, either $\int_0^\infty f^2 \, dx$ or $\int_0^\infty f''^2 \, dx$ would be infinite. Hence $L = 0$ and

$$\int_0^\infty \left[f^2 - f'^2 + f''^2\right] \, dx \tag{6.5}$$

$$= [f(0) + f'(0)]^2 + \int_0^\infty (f + f' + f'')^2 \, dx.$$

The inequality (6.2) folows from (6.5), with strict inequality unless f satisfies $f + f' + f'' = 0$ and $f(0) + f'(0) = 0$, that is unless $f(x) = f_A(x)$ as specified.

The corresponding theorem for the interval R is also due to Hardy, Littlewood and Pólya [**9**, Th. 261], and is much easier to prove.

THEOREM 2. *If f' is locally absolutely continuous and $f, f'' \in L_2(R)$, then*

$$\left(\int_{-\infty}^{+\infty} f'^2\,dx\right)^2 \le \left(\int_{-\infty}^{+\infty} f^2\,dx\right)\left(\int_{-\infty}^{+\infty} f''^2\,dx\right), \tag{6.6}$$

with equality if and only if $f = 0$. The (unit) constant is best possible.

PROOF: As in the proof of Theorem 1, $f(X)f'(X) \to 0$ as $X \to \pm\infty$. Hence by Hölder's inequality,

$$\left(\int_{-\infty}^{+\infty} f'^2\,dx\right)^2 = \left(-\int_{-\infty}^{+\infty} ff''\,dx\right)^2 \le \left(\int_{-\infty}^{+\infty} f^2\,dx\right)\left(\int_{-\infty}^{+\infty} f''^2\,dx\right),$$

with strict inequality unless $f'' = kf$, that is unless $f(x) = Ae^{\alpha x} + Be^{-\alpha x}$ or $f(x) = A\sin(\alpha x + \beta)$ for some constants A, B, α, β. The only such f for which $\int f^2\,dx$ is finite is $f \equiv 0$.

To prove that the constant is best possible, note that both sides of (6.5) have the value 1 for the (non-admissible) function $f_0(x) = e^{-|x|}$. Given any $n \ge 1$ we may smooth the corner of the graph of f_0 at $x = 0$ to obtain a function f_n such that f_n'' exists for all x and

$$\int f_n'^2\,dx \ge 1 - \frac{1}{n}, \quad \int f_n^2\,dx \le 1 + \frac{1}{n}, \quad \int f_n''^2\,dx \le 1 + \frac{1}{n}.$$

Hence, given $\varepsilon > 0$ it follows that for all sufficiently large n,

$$\int f_n'^2\,dx > (1-\varepsilon)\left(\int f_n^2\,dx\right)\left(\int f_n''\,dx\right).$$

REMARK 1: The above proofs are those of Hardy, Littlewood and Pólya [**9**] who also gave two other proofs of (6.2) one of which was variational in character.

REMARK 2: By the same technique used in Theorem 2, one can show that *if f' is locally absolutely continuous and $f \in L_p(\mathbf{R})$, $f'' \in L_{p'}(\mathbf{R})$, where $p > 1$, $p' = p/(p-1)$, then*

$$\left(\int_{-\infty}^{+\infty} f'^2\,dx\right) < \left(\int_{-\infty}^{+\infty} |f|^p\,dx\right)^{1/p}\left(\int_{-\infty}^{+\infty} |f''|^{p'}\,dx\right)^{1/p'} \tag{6.7}$$

unless $f \equiv 0$. (cf. [**9**, Th. 269].)

REMARK 3: From [**9**, Theorems 272 and 276]: $f \in D$ if $f\colon R^+$ (or R) $\to R$, $f \in AC_{\text{loc}}(R^+)$ (or R), xf and f' both $\in L^2(R^+)$ (or L^2R); then $f \in L^2(R)$ (or L^2R) and

$$(6.8)\qquad \left(\int_0^\infty f(x)^2\,dx\right)^2 \le 4\int_0^\infty x^2 f(x)^2\,dx \int_0^\infty f'(x)^2\,dx \quad (f \in D),$$

with equality only when for some $A \in R$ and some $\rho > 0$

$$f(x) = A\exp[-\rho x^2] \qquad (x \in R^+)$$

(or similarly on R).

7. The first relevant results for functions of two variables are given by T. Tchou-Yun Tcheng [**11**].

Let function $u(x,y)$ have continuous second partial derivatives on R^2, and let m_0, m_1 and m_2 denote respectively supremum of

$$|u|,\ \left(u_x^2 + u_y^2\right)^{1/2},\ |u_{xx} + u_{yy}|.$$

Then

$$(7.1)\qquad m_1 \le \frac{8}{\pi\sqrt{3}}\sqrt{m_0 m_2}.$$

Now, let M_0, M_1 and M_2 denote respectively supremum of

$$|u|,\ |u_x + u_y|,\ |u_{xx} + 2u_{xy} + u_{yy}|,$$

then

$$(7.2)\qquad M_1 \le \sqrt{2}\sqrt{M_0 M_2}.$$

8. An attack on the general problem for $p = +\infty$ was begun by G. E. Šilov [**12**] in 1937, who found

$$C_{31}(+\infty, R) = (9/8)^{1/3},\ C_{32}(+\infty, R) = 3^{1/3},\ C_{41}(+\infty, R) = (512/375)^{1/4},$$
$$C_{42}(+\infty, R) = (6/5)^{1/4} \text{and } C_{52}(+\infty, R) = (125/72)^{1/5}.$$

9. A. Gorny, unifying his results published in [**13**], [**14**] and [**15**], but in a somewhat modified form, proved in [**16**] the following result:

If f is an n-times differentiable function on a closed interval I of length δ, and if $|f(x)| \leq M_0$ and $|f^{(n)}(x)| \leq M_n$, then for $x \in I$ and $0 < k < n$,

$$|f^{(k)}(x)| \leq 4e^{2k} \binom{n}{k}^k M_0^{1-(k/n)} {M'_n}^{k/n},$$

while, at the midpoint of I,

$$|f^{(k)}(x)| \leq 16(2e)^k M_0^{1-(k/n)} {M'_n}^{k/n},$$

where

$$M'_n = \max\left(M_n, M_0 n! \delta^{-n}\right).$$

If $I = (0, +\infty)$ or $(-\infty, +\infty)$, then $M'_n = M_n$ and for $(-\infty, +\infty)$ any point can be considered as the midpoint. He thus proved that

$$C_{nk}(+\infty, R) \leq 16(2e)^k.$$

The same results were independently proved by H. Cartan [**17**].

10. The complete solution of finding $C_{nk}(+\infty, R)$ was given by A. Kolmogorov in 1938 (see [**18**] and [**19**]). He showed that

$$C_{nk}(+\infty, R) = \frac{K_{n-k}}{K_n^{(k-n)/n}}, \tag{10.1}$$

where

$$K_n = \frac{4}{\pi} \sum_{\nu=0}^{+\infty} \left(\frac{(-1)^\nu}{2\nu+1}\right)^{n+1} \tag{10.2}$$

are the Favard constants ($K_0 = 1$, $K_1 = \frac{\pi}{2}$, $K_2 = \frac{\pi^2}{8}$, $K_3 = \frac{\pi^3}{24}, \ldots$).

Inequality (1.1) (with $p = +\infty$) becomes an inequality if $f(x) = k^{-n}\phi_n(kx)$, where

$$\phi_n(x) = \frac{4}{\pi} \sum_{\nu=0}^{+\infty} \frac{\sin\left((2\nu+1)x - \frac{2\pi}{2}\right)}{(2\nu+1)^{n+1}} \tag{10.3}$$

is the nth periodic integral, with zero mean value over the period of the function $\phi_0(x) = \operatorname{sign}(\sin x)$.

He also showed that $1 < C_{nk}(+\infty, R) < \pi/2$ for all n and k, and that $C_{n,n-1}(+\infty, R) \to \pi/2$ and $C_{n1}(+\infty, R) \to 1$ both as $n \to +\infty$.

11. The following integral inequality of A. Ostrowski [**20**] is related to Landau's theorems from 2.

Let f be a differentiable function on (a, b) and let, on (a, b), $|f'(x)| \leq M$. Then for every $x \in (a, b)$,

$$\left| f(x) - \frac{1}{b-a} \int_a^b f(x)\, dx \right| \leq \left(\frac{1}{4} + \frac{\left(x - \frac{a+b}{2}\right)^2}{(b-a)^2} \right) (b-a)M. \tag{11.1}$$

12. One of the earliest inequalities related to (6.2) is due to I. Halperin and H. Pitt [**21**] (here we used exposition of this result from [**188**]).

The inequality in question is

$$\int_0^X f'^2\, dx \leq K(\varepsilon) \int_0^X f^2\, dx + \varepsilon \int_0^X f''^2\, dx \tag{12.1}$$

valid for all $\varepsilon > 0$ and locally absolutely continuous f with $f'' \in L_2(0, X)$. Here

$$K(\varepsilon) = \frac{P}{\varepsilon} + \frac{Q}{X^2} \qquad \text{with } P = 1,\ Q = 12.$$

The constants $P = 1$, $Q = 12$ are best possible in the sense that if $P < 1$, or $Q < 12$, then for some $\varepsilon > 0$ and admissible f, (12.1) does not hold.

Note that if $X \to \infty$ and the norms on the right of (12.1) are finite we have for $\varepsilon > 0$

$$\varepsilon^2 \|f''\|^2 - \varepsilon \|f'\|^2 + \|f\|^2 \geq 0.$$

But this is also clearly true for $\varepsilon \leq 0$. Hence the discriminant of this quadratic in ε is less than or equal to zero. This gives $C_{21}(2, R^+) = \sqrt{2}$.

13. Let $f(x)$ be a complex function of a real variable, defined over the whole real line, which possesses n derivatives (the nth at least almost everywhere) and is such that $f^{(n-1)}(x) = \int^x f^{(n)}(t)\, dt$. Then, if k is any integer for which $0 < k < n$, Kolmogorov's inequality (1.1) for $p = +\infty$, is valid and $C_{nk}(+\infty, R)$ is defined as in 10. The constant C_{nk} is the best possible, i.e. there is a (real) function for which equality holds (see [**22**]).

14. B. Sz.-Nagy [**23**] proved the following:

Let $1 \le p < +\infty$, $1 < q \le +\infty$, $I = (-\infty, +\infty)$. If f is locally absolutely continuous with $f \in L$, $f' \in L_{p'}$, then

$$\|f\|_{L_q} \le Q \|f\|_L^u \|f'\|_{L_p}^v, \tag{14.1}$$

where $u = 1 - v$, $v = (1 - q^{-1})/(2 - p^{-1})$, $Q = \left(\frac{2p-1}{2p} H\left(\frac{2p-1}{p(q-1)}, \frac{p-1}{p}\right)\right)^v$,

$$H(x,y) = \frac{z(x+y)}{z(x)z(y)}, \; z(x) = x^{-x}\Gamma(1+x) \text{ for } x > 0 \text{ and } z(0) = 1.$$

The constant Q is the best possible. (B.Sz.-Nagy formulated functions for which equality in (14.1) is valid).

15. A. M. Rodov (see [**24**] and [**25**]) gave a solution of the following general problem: Given a finite set of integers $i_1, \dots, i_n$ such that

$$0 < i_1 < \cdots < i_n = n$$

under what conditions imposed on the positive fixed numbers

$$a_0, a_{i_1}, \dots, a_{i_n} = a_n,$$

is there a function f bounded, together with its n first derivative on R, such that

$$M_0 = a_0, \; M_{i_m} = a_{i_m} \quad \text{for } m = 1 \dots, k.$$

16. Some related results in the case when $I = [-1, 1]$ are given in [**26**].

17. The theorem of E. Landau from 2. was generalized by V. G. Avakumović and S. Aljančić [**27**] who gave geometric arguments that if f is a real function such that $|f''(x)| \le 1$ for $0 \le x \le 1$ then, for $0 \le x \le 1$,

$$|f'(x) - f(1) + f(0)| < \frac{1}{2} - x + x^2, \tag{17.1}$$

and that this bound cannot be improved.

They also proved that the conditions

$$|f''(x)| \le 1 \quad \text{for } 0 \le x \le 1, \quad \text{and } f'(0) = f'(1)$$

imply

$$|f'(x) - f(1) + f(0)| \le \frac{1}{4} \quad \text{for } 0 \le x \le 1. \tag{17.2}$$

This bound cannot be improved.

Note that (17.1) is equivalent to (11.1).

18. Inequalities

$$\|y^{(m)}\|_p^p \leq \varepsilon\|y^{(n)}\|_p^p + K(\varepsilon)\|y\|_p^p \quad \text{for } 1 \leq p < +\infty, \tag{18.1}$$

and

$$\|y^{(m)}\| \leq \varepsilon\|y^{(n)}\| + K(\varepsilon)\|y\| \qquad \text{for } p = +\infty \tag{18.2}$$

have been studied by L. Nirenberg in [**29**].

19. As we saw, the constant $C_{nk}(+\infty, R)$ is explicitly obtained. But the related problem for $C_{nk}(+\infty, R^+)$ seems more difficult. In 1955, A. P. Matorin showed that ([**30**])

$$C_{nk}(+\infty, R^+) \leq \frac{T_n^{(k)}(1)}{n!^{k/n} 2^{(1-\frac{1}{n})k}}, \tag{19.1}$$

where T_n is the nth Čebyšev polynomial .

He also proved the following result: If

$$\|f\| \leq 1 \quad \text{and} \quad \|f'''\| \leq 24,$$

then

$$\|f'\| \leq 9 \quad \text{and} \quad \|f''\| \leq 24. \tag{19.2}$$

In (19.2) the constants 9 and 24 are the best possible. From this result we can easily obtain $C_{31}(+\infty, R^+)$ and $C_{32}(+\infty, R^+)$.

20. E. M. Stein [**31**] in 1957 used a convolution technique to show that in general

$$C_{nk}(p, I) \leq C_{nk}(+\infty, I), \tag{20.1}$$

while Ju. Ljubič in 1960 had shown that ([**32**])

$$C_{nk}(2, R) = C_{n,n-k}(2, R). \tag{20.2}$$

Ju. Ljubič also gives an algorithm for finding $C_{nk}(2, R)$ in terms of computing the zero of an explicit polynomial.

21. J. Dieudonné [**33**] gave some results related to results from 17.

22. W. Müller [**34**] considered the inequality (12.1) with $P = 48$, $Q = 24$. The best possible result, given in 12., is proved by R. M. Redheffer in [**35**].

23. S. B. Stečkin [**36**] proved that in the case when $f(x) \geq 0$, we have $C_{21}(1, R) = 1$ and $C_{21}(1, R^+) = \sqrt{2}$. The same constants for $p = +\infty$, are given by V. M. Olovyanišnikov [**28**] who also showed that $C_{32}(+\infty, R^+) = 2 \cdot 3^{1/3}$. Stečkin also proved that $C_{31}(+\infty, R^+) = \frac{1}{2}3^{5/3}$. Moreover, he gave a result related to the Kolmogorov inequality in Banach space.

24. Let $K(M_0, M_2)$ be a class of functions $f(x)$ such that,

$$\sup_{-\infty<x<+\infty} |f(x)| = \|f\| \leq M_0,$$

$$\sup_{\substack{-\infty<x<y<\infty \\ |x-y|<\delta}} \frac{|f'(x) - f'(y)|}{|x-y|} \leq M_2, \quad 0 < \delta < +\infty.$$

For r, $0 < r < 1$, the derivative of order r is defined by

$$f^{(r)}(x) = \frac{\Gamma(r+1)\sin r\pi}{\pi} \int_0^\infty \frac{f(x) - f(x-s)}{s^{1+r}}\, ds,$$

i.e.

$$f^{(r)}(x) = \frac{\Gamma(r+1)\sin r\pi}{r\pi} \int_{-\infty}^0 \frac{f'(x+s)}{|s|^r}\, ds.$$

If $f \in K(M_0, M_2)$, $0 < r < 1$, then [**37**]:

$$\|f^{(r)}\| \leq C_r M_0^{1-(r/2)} M_2^{r/2}, \tag{24.1}$$

where

$$C_r = \frac{\Gamma(r+1)\sin r\pi}{r(1-r)(2-r)\pi} 2^{2-(r/2)} \left(2^{1/(1-r)} - 1\right)^{r-1}.$$

The constant C_r is the best possible.

25. A. M. Pfeffer [**38**] considered inequality (18.1) for $p = 2$ where $1 \le m < n$, $\varepsilon > 0$, $f \in C^n[a,b]$, $f^{(i)}(a) = f^{(i)}(b)$ $(0 \le i \le n-1)$. The proof of the existence of a constant $K(\varepsilon)$ in (18.1) is given in [**39**] by S. Goldberg.

26. Some related results in which we have linear differential operators instead of derivatives are considered in [**40**]–[**42**].

27. S. B. Stečkin [**43**] in 1967 gave estimates for $C_{nk}(+\infty, R^+)$:

$$a\left(\frac{n}{k}\right)^k \le C_{nk}(+\infty, R^+) \le A\left(\frac{e^2 n}{4k}\right)^k, \ (1 \le k \le \frac{n}{2},$$

$$a(n-k)^{-1/2}\left(\frac{n}{n-k}\right)^{n-k} \le C_{nk}(+\infty, R^+) \le A\left(\frac{2n}{n-k}\right)^{n-k}, \ \frac{n}{2} \le k < n,$$

where a and A are positive constants.

These inequalities present improvements and simplifications of inequalities obtained by V. M. Olovyanišnikov and A. P. Matorin which are discussed in [**43**].

28. An elementary proof of (6.2) was also indicated by D. C. Benson [**44**] who applied the same method to other inequalities involving $\int f^2\, dx$, $\int f'^2\, dx$, and $\int f''^2\, dx$.

29. An extremal problem related to Kolmogorov's inequality for bounded functions is considered in [**45**].

30. We gave the definition of a derivative of order r for $0 < r < 1$ in 24. For $r > 1$, if we put $D_r(f,x) = f^{(r)}(x)$, we can extend this definition by

$$D_r(f,x) = D_{r-[r]}\left(D_{[r]}(f,x)\right).$$

A generalization of the Kolmogorov inequality in the case when $k, n \in R^+$ is given in [**46**].

31. Let $1 \le p, r \le +\infty$ and let J be a finite or infinite interval of the real line. For n a positive integer $W^n_{p,r}(J)$ denotes the set all functions y and J to the real numbers such that $y \in L^p(J)$, $y^{(n-1)}$ is absolutely continuous on compact subintervals of J and $y^{(n)} \in L^r(J)$.

The following result is proved by V. N. Gabušin [**47**]:

Let $1 \le p, q, r \le +\infty$, $J = R$ or $J = R^+$, and let n, k be positive integers with $0 \le k < n$. Let

$$u = \frac{n-k-r^{-1}+q^{-1}}{n-r^{-1}+p^{-1}}, \qquad v = 1-u.$$

Suppose

$$(n-k)p^{-1}+kr^{-1}\geq nq^{-1}.$$

Then there exists a constant K, depending only on p,q,r,k,n and J, such that

$$\|y^{(k)}\|_q \leq K\|y\|_p^u\|y^{(n)}\|_r^v \tag{31.1}$$

holds for all $y\in W_{p,r}^n(J)$.

The inequality (31.1) is considered in papers [**48**]– [**53**].

For a special case $n=1$, $k=0$, $q\geq p$, $p>0$, $r\geq 1$, see 14. As we saw, the obtained constant is the best possible in 14. Similarly, in some other special cases of (31.1) there exist the best possible constants K. This problem in the case $0\leq k<n$, n arbitrary, $q=+\infty$, $r=p=2$ is solved in [**50**] for $J=R$ and in [**53**] for $J=R^+$, and in the case $n=2$, $k=0,1$; $q=+\infty$, $p\geq 1$, $r=+\infty$, $J=R$ in [**50**].

32. In 1970, I. J. Schoenberg and A. Cavaretta solved the problem of finding $C_{nk}(+\infty,R^+)$ by constructing the extremals. No closed form is given for the constants (see [**54**]).

33. V. I. Berdyšev [**55**] found that $C_{21}(1,R)=2$ and $C_{21}(1,R^+)=5/2$.

34. R. Kallman and G. C. Rota [**56**] have generalized the inequality (2.3) to a particular class of semi-groups. Let X be a B-space over the complex field, $\{T(s)\}$ a one-parameter semi-group of linear bounded transformations from X to itself such that $\|T(s)\|\leq 1$ for all $s\geq 0$. Let A be the infinitesimal generator of the semi-group, $D[A]\subset X$ its domain of definition. Then for any $f\in D[A^2]$,

$$\|Af\|^2\leq 4\|f\|\|A^2f\|. \tag{34.1}$$

PROOF: For $f\in D[A^2]$ we have

$$T(s)f=f+sAf+\int_0^s(s-u)T(u)A^2f\,du.$$

Whence

$$\begin{aligned}s\|Af\| &\leq \|f\|+\|T(s)f\|+\int_0^s(s-u)\|T(u)A^2f\|du\\ &\leq 2\|f\|+\frac{1}{2}s^2\|A^2f\|\end{aligned}$$

since $\|T(s)\| \leq 1$. Hence

$$\|Af\| \leq \frac{2}{s}\|f\| + \frac{1}{2}s\|A^2 f\|.$$

If $A^2 f \neq 0$, the minimum of the right hand side is reached for

$$s = 2\left(\frac{\|f\|}{\|A^2 f\|}\right)^{1/2}$$

and the minimum value is

$$2\left(\|f\|\ \|A^2 f\|\right)^{1/2}$$

which gives (34.1).

E. Hille [57] remarked that other unimodular semi-groups are generated by higher derivatives. So, for any positive integer n we get

$$\|f^{(n)}\|^2 \leq 4\|f\|\ \|f^{(2n)}\|. \tag{34.2}$$

The space is $C(R)$ or $L_p(R)$.

If $\|T(s)\| \leq M$, $s > 0$, then (S. Kurepa [58], S. Kurepa and H. Kraljević [59])

$$\|Af\|^2 \leq 2M(M+1)\|f\|\ \|A^2 f\|. \tag{34.3}$$

Kurepa and Kraljević [58],[59] also proved that if A is the infinitesimal generator of a strongly continuous cosine function $C(s)$ on a Banach space for $s > 0$ and if $M = \sup|C(s)| < \infty$, then for $f \in D(A^2)$ and $A^2 f \neq 0$,

$$\|Af\|^2 \leq \frac{2}{3}M(1+M)\|f\|\ \|A^2 f\|. \tag{34.4}$$

A special case of this result is

$$\|f''\|^2 \leq \frac{4}{3}\|f\|\ \|f^{(iv)}\|. \tag{34.5}$$

This is considerably better than (34.2) for $n = 2$.

Using the method from [59] E. Hille [60] gave the following inequalities

$$\|A^2 f\|^3 \leq 3\|f\|\ \|A^3 f\|^2, \tag{34.6}$$

$$\|Af\|^2 \leq 2\|f\|\ \|A^2 f\|, \quad \text{and} \tag{34.7}$$

$$\|Af\|^3 \leq \frac{9}{8}\|f\|^2\|A^3 f\|. \tag{34.8}$$

Hille also proved the following result: If A is the infinitesimal generator of a strongly continuous contraction semi-group, if $f \in D[A^n]$ and $1 \leq k < n$, then there exist constants $C_{n,k}$ independent of A, so that

$$\|A^k f\|^n \leq C_{n,k}^n \|f\|^{n-k} \|A^n f\|^k. \tag{34.9}$$

A direct proof of (34.4) and some applications of the previous results are given in Hille's paper **[61]**.

T. Kato **[62]** showed that (34.7) is valid whenever X is a Hilbert space. This suggests that this inequality can be generalized for any space in which the geometry is sufficiently nearly Euclidean (in some appropriate sense). So, J. A. R. Holbrook **[63]** proved:

Let $(X, \|\cdot\|)$ is a real or complex normed linear spacc, and let

$$a(X) = \sup\left\{\frac{\|x+y\|^2 + \|x-y\|^2}{2\|x\|^2 + 2\|y\|^2} : x, y \in X\right\},$$

$$b(X) = \sup\left\{\frac{\|x+y\|}{\|x-y\|} : x, y \in X, \text{ and } y \text{ is at an obtuse angle from } x\right\}.$$

If A is a dissipative linear operator with domain $D(A)$ in a normed space X, then

$$\|Ax\|^2 \leq a(X)\left(1 + b(X)^2\right)\|x\|\;\|A^2 x\|, \tag{34.10}$$

whenever $x, AX \in D(A)$. If X is C-Euclidean,

$$\|Ax\|^2 \leq C\left(1 + \lambda^2\left(2\lambda - 1 + 2(\lambda^2 - \lambda)^{1/2}\right)\right)\|x\|\;\|A^2 x\|, \tag{34.11}$$

where $\lambda = 1 + 19(C-1)$.

Note that if X is an inner product space, then $a(X) = b(X) = 1$, and we obtain Kato's inequality (i.e. (34.7)). In the same paper it was proved that L^p-spaces are $2^{|1-2/p|}$-Euclidean.

Some related results with fractional powers have been studied by W. Trcbcls and U. Westphal **[64]**. A special case with the Laplacian is also given.

(Note that some results of this kind, i.e. for Banach-spaces, are given in **[32]** ten years earlier.)

35. I. B. Lacković and M. S. Stanković **[65]** gave a generalization of (17.1) for functions with several variables. Further generalization of their result is given by D. D. Adamović in the same paper.

Some other results for functions with several variables are given by V. A. Solonnikov **[66]** and O. V. Besov **[67]**.

Results from [**65**] are generalized for Fréchet- differentiable function $F: X \to Y$ (X and Y are Banach spaces) by R. Ž. Djordjević and G. V. Milovanović in[**68**].

G. V. Milovanović [**69**] also gave a generalization of the Ostrowski inequality (11.1) for functions with several variables.

36. C. J. F. Upton [**70**] extended result from 13. to the cases when the sup norm on R is replaced by one of the following: (1) L_p norm on R, (2) L_p norm on R^+, (3) Stepanov norm on almost periodic functions on R, (4) Weyl norm on almost periodic functions on R, (5) Besicovitch norm on almost periodic function on R, (6) Love's variation norm on almost periodic functions on R and (7) $\|f\|^* = \sum_{r=0}^{k-1} \|f^{(r)}\|_\infty + \|f^{(k)}\|_\infty$ on almost periodic functions on R, where $\|\cdot\|_G$ is interpreted to be one of the norms (3)–(6).

(In fact, Upton also proved (20.1).)

37. V. V. Arestov [**71**] also considered the problem from 31. He used the result from 14. in the proof of (31.1) for $n = 2$, $p = +\infty$, $r \geq 1$, $q \geq 2r$, $u = 1 - v$, $v = (1 - q^{-1})/(2 - r^{-1})$. The best possible constant is

$$K = \begin{cases} 2Q, & \text{if } J = [0, +\infty), \\ 2^v Q, & \text{if } J = (-\infty, +\infty). \end{cases}$$

He also considered the case when $n = 3$, $q = p = +\infty$, $1 \leq r < +\infty$ for R and R^+. (Q is defined as in 14.).

Let $K^n_{p,r}$ be subclass of $W^n_{p,r}$ such that $\|f^{(n)}\|_{L_r} \leq 1$, and $\Sigma(N)$ be a set of linear bounded operators $S: L_p \to L_q$ with a norm $\|S\| = \|S\|^{L_q}_{L_p} \leq N$. Define

$$\rho(S) = \sup_{f \in K^n_{p,r}} \|f^{(k)}(x) - S(x, f)\|_{L_q}, \quad E(N) = \inf_{S \in \Sigma(N)} \rho(S).$$

Arestov [**71**] also proved the following inequality:

$$\text{(37.1)} \qquad D := \sup_{f \in W^n_{p,r}} \|f^{(k)}\|_{L_q} \|f\|^{-u}_{L_p} \|f^{(n)}\|^{-v}_{L_r} \leq \left(\frac{1}{v} E(N)\right)^v \left(\frac{1}{u} N\right)^u .$$

For $p = q = r$ we have a result from [**36**] and [**43**].

An inequality of type (31.1) is considered in [**72**] by W. N. Everitt and M. Giertz. Namely, they examined the inequality

$$\text{(37.2)} \qquad \|f'\|_2^2 \leq K \|f\|_p \|f''\|_q ,$$

for $p, q \in (1, +\infty)$, $p^{-1} + q^{-1} = 1$, $f \in L^p(R^+)$, f' absolutely continuous on all compact subintervals of $[R^+)$, $f'' \in L^q(R^+)$ and K a positive number independent of f but dependent of choice of p and q. Of course, (37.2) is an extension of (6.1). The authors show that the best constant K in (37.2) is finite for all (p, q), and they gave a characterization of this constant in terms of an equivalent problem from the calculus of variations.

Everitt and Giertz **[72]** also obtained bounds for $|f(0)|$ and $|f'(0)|$ in terms of $\|f\|_p$ and $\|f''\|_q$.

38. The following consideration is given in **[188]**.

In a series of papers (see e.g. **[73]**,**[75]**,**[76]**) W. N. Everitt has considered generalizations of (6.1) and (6.2) which involve the ordinary linear differential operator

$$M(f) = -(pf')' + qf, \tag{38.1}$$

and the inequality considered in **[73]**,**[75]** is

$$\left| \int_0^\infty \left(pf'^2 + qf^2\right) dx \right|^2 \le K(p,q) \left(\int_0^\infty f^2 \, dx \right) \left(\int_0^\infty M(f)^2 dx \right). \tag{38.2}$$

Let $\Delta(M)$ be the class of all $f \in L_2(R^+)$ such that f' is locally absolutely continuous and $M(f) \in L_2(R^+)$.

Suppose that the functions p, q satisfy the conditions (i) $p' \in C(R^+)$ *with* $p(x) > 0$ *for* $x \ge 0$, *and* (ii) $q \in C(R^+)$ *with* $q(x) \ge -d$ *for* $x \ge 0$ *and some* $d \ge 0$. In analogy with the relation between the inequalities (6.1), (6.2), it is shown in **[73]** that a necessary and sufficient condition for the existence of a positive constant $K = K(p, q)$ such that (38.2) holds for all $f \in \Delta(M)$ is that there exists a constant $k > 0$ such that

$$J_{p,k}(f) \equiv \int_0^\infty \left[\rho^2 f^2 - k\rho(pf'^2 + qf^2) + M(f)^2\right] dx \ge 0 \tag{38.3}$$

for all $f \in \Delta(M)$ and all $\rho \in R \setminus \{0\}$. Moreover, in this case $K(p,q) = 4k^{-2}$. A complete discussion of the possibility of such inequalities (38.2), (38.3) depends on a spectral analysis of the second order equation $M(y) = \lambda y$ and the fourth-order equation $(M - \beta)^2 y = \Lambda y$. (Additional differentiability assumptions on p, q are also required.) Here we only mention that when an inequality (38.2) holds for all $f \in \Delta(M)$, then $K \ge 4$, and that if $q(x) \ge 0$ but $q(x) \not\equiv 0$ on $[0, \infty)$, then $K(p,q) = +\infty$.

In the same way that the inequality (6.2) is easier than (6.1), the inequality

$$(38.4)\qquad \left\{\int_{-\infty}^{\infty}(pf'^2+qf^2)dx\right\}^2 \le K_0(p,q)\left(\int_{-\infty}^{\infty} f^2dx\right)\left(\int_{-\infty}^{\infty} M(f)^2\,dx\right)$$

is much easiler than (38.2). Indeed we shall follow [73] and show that *if* $\Delta(M)$, *p, and q are as above but with* R^+ *replaced by R, then* (38.4) *holds for all* $f \in \Delta(M)$, *with* $K_0(p,q) = 1$. *Moreover, with* $K_0 = 1$, *inequality holds in* (38.4) *if and only if f is a solution of the eigenvalue problem* $M(f) = \lambda f$ *such that* $f \in \Delta(M)$.

PROOF: The proof depends on the fact that

$$(38.5)\qquad \lim_{x\to\pm\infty} p(x)f(x)f'(x) = 0.$$

Assuming (38.5) for now, we have for all $X > 0$ and $f \in \Delta(M)$,

$$\int_{-X}^{X}\left[pf'^2+qf^2\right]\,dx = pf'f|_{-X}^{X} + \int_{-X}^{X} fM(f)\,dx.$$

Now let $X \to \infty$ and note that $fM(f) \in L(R)$ since f and $M(f)$ are in $L_2(R)$. Hence

$$\int_{-\infty}^{\infty}\left[pf'^2+qf^2\right]\,dx = \int_{-\infty}^{\infty} fM(f)\,dx$$

follows. But now (38.4), with $K_0 = 1$ and equality as asserted, follows from Hölder's inequality.

To prove (38.5) we deal with the case that $X \to +\infty$; the other limit will follow in the same way. As above, we have

$$\int_{0}^{X}\left[pf'^2+(q+d)f^2\right]\,dx = p(X)f'(X)f(X) - p(0)f'(0)f(0)$$
$$+\int_{0}^{X}\left[fM(f)+df^2\right]\,dx.$$

As $X \to \infty$, the integral on the right side converges since $fM(f) \in L(R)$ and $f \in L_2(R)$. However, so also does the integral on the left side coverage because if not then, since $q + d \geq 0$, we would have both the integral diverging to $+\infty$, and $p(X)f(X)f'(X) \to +\infty$. The latter is impossible since it would imply that

$f(X)$, $f'(X)$ have the same sign for sufficiently large X and this is impossible for $f \in L_2(R^+)$. It follows from this discussion that we necessarily have

(a) $p^{1/2}f' \in L_2(R^+)$, $|q|^{1/2}f \in L_2(R^+)$, and
(b) $\lim_{x\to\infty} p(X)f'(X)f(X) = \alpha$ exists (finite).

Clearly $\alpha \leq 0$ since $\alpha > 0$ would again imply that $f(X)f'(X) > 0$ for all sufficiently large X. We shall assume that $\alpha < 0$ and obtain a contradiction. Without loss of generality we may assume that $f(x) > 0$, $f'(x) < 0$ and $p(x)f(x)f'(x) < \alpha/2$ for $x \geq X_0$ say. Hence for $X > X_0$,

$$\int_{X_0}^{X} p^{1/2} f' f \, dx = \int_{X_0}^{X} p^{-1/2}(pff')dx < \frac{\alpha}{2}\int_{X_0}^{X} p^{-1/2}\, dx,$$

or

$$0 < \int_{X_0}^{X} p^{-1/2}\, dx < \frac{2}{\alpha}\int_{X_0}^{X} (p^{1/2}f')f\, dx, \quad (X > X_0).$$

Since $f \in L_2$ and $p^{1/2}f' \in L_2$, it follows that $\int_{X_0}^{+\infty} p^{-1/2}\, dx < \infty$. Hence

(c) $P(x) = \int_0^x p^{-1/2}\, dt \to R \quad$ as $x \to \infty$.

Let $Q(x) = \int_0^x (q + d + 1)dt$. Since $q + d \geq 0$, we have

(d) $Q(x) \geq x \quad for\ x \geq 0$.

In addition to (a), (c), (d) we shall need the following general result which is easy to prove (or see [**9**, Th. 223]):

(e) If $g \in L_2(R^+)$, then $\lim_{x\to\infty} x^{-1/2}\int_0^x g\, dt = 0$.

We have

$$\int_0^X p^{1/2}f'\, dx = \int_0^X p^{-1/2}pf'\, dx = P(X)p(X)f'(X) - \int_0^X P(pf')'\, dx$$

$$= P(X)p(X)f'(X) + \int_0^X PM(f)\, dx - \int_0^X Pqf\, dx.$$

We shall now show that the product of $[Q(X)]^{-1/2}$ with each of the three integrals in the last equation tends to zero as $X \to \infty$; since $P(X) \to R > 0$, it will follow that

$$\lim_{x\to\infty} [Q(x)]^{-1/2} p(x)f'(x) = 0. \tag{38.6}$$

Using (d) and (e) with $g = p^{1/2}q'$, we have

$$\left|[Q(X)]^{-1/2} \int_0^X PM(f)\, dx\right| \le X^{-1/2} \left|\int_0^X p^{1/2} f'\, dx\right| \to 0.$$

Similary, using (e) with $g = M(f)$,

$$\left|[Q(X)]^{-1/2} \int_0^X PM(f)\, dx\right| \le X^{-1/2} R \int_0^X |M(f)| dx \to 0.$$

For the third integral, we write

$$\int_0^X Pqf\, dx = \int_0^X P(q+d+1)f\, dx - (d+1) \int_0^X Pf\, dx,$$

and note that $Pf \in L_2(R^+)$ so that as above,

$$[Q(X)]^{-1/2} \int_0^X Pf\, dx \to 0.$$

For $0 < X_1 < X$,

$$\begin{aligned} 0 &< \int_0^X P(q+d+1)f\, dx \\ &= \int_0^{X_1} P(q+d+1)f\, dx + \int_{X_1}^X P(q+d+q)f\, dx, \\ &\le F(X_1) + \left(\int_{X_1}^X P^2(q+d+1)\, dx\right)^{1/2} \left(\int_{X_1}^X (q+d+1)f^2\, dx\right)^{1/2} \\ &\le F(X_1) + R[Q(X) - Q(X_1)]^{1/2} \left(\int_{X_1}^X (q+d+1)f^2\, dx\right)^{1/2}. \end{aligned}$$

Given $\varepsilon > 0$, there exists X_1 such that the last integral is $\leq \varepsilon^2 R^{-2}$ for all $X > X_1$. Hence

$$0 < [Q(X)]^{-1/2} \int_0^X P(q+d+1)f\, dx \leq X^{-1/2} F(X_1) + \varepsilon \quad \text{for } X > X_1,$$

and since ε is arbitrary, it follows that

$$[Q(X)]^{-1/2} \int_0^X P(q+d+1)f\, dx \to 0 \quad \text{as } X \to \infty.$$

This completes the proof of (38.6).

Next we consider $[Q(x)]^{1/2} f(x)$ as $x \to \infty$. Suppose that

$$\varliminf_{x\to\infty} [Q(x)]^{1/2} f(x) = \beta > 0.$$

Then there exists $X_0 > 0$ such that $f^2(x) \geq \beta^2/Q(x)$ for $x \geq X_0$, and so for $x \geq X_0$,

$$\int_{X_0}^X (q+d+1)f^2\, dx \geq \beta^2 \int_{X_0}^X \frac{q+d+1}{Q(x)} dx$$

$$\geq \beta^2 \log\left[\frac{X}{Q(X_0)}\right] \to \infty \quad \textit{as } X \to \infty.$$

This is impossible since $f \in L_2$ and $|q|^{1/2} f \in L_2$. Hence $\beta = 0$, so there must exist a sequence $\{x_n\}$ with $\lim x_n = \infty$ such that

$$\lim\, [Q(x_n)]^{1/2} f(x_n) = 0.$$

Combining this with (38.6), we have

$$\lim p(x_n) f'(x_n) f(x_n) = 0,$$

and this, together with the existence of the limit in (b), implies that $\alpha = 0$. This contradicts the assumption $\alpha < 0$, and proves that $\alpha = 0$ must hold.

REMARK 4: Note that for the original inequality (38.2), the above method of proof shows that under the given hypotheses on p, q, (a) and (38.5) hold, and

$$\int_0^\infty \left[pf'^2 + qf^2\right]\, dx \leq -p(0)f(0)f'(0) + \left(\int_0^\infty f^2\, dx\right)^{1/2} \left(\int_0^\infty M(f)^2\, dx\right)^{1/2}.$$

Hence a result of the form (38.2) would follow if one could obtain bounds of the form $|p(0)f(0)f'(0)| \leq \|f\| \, \|M(f)\|$. Unfortunately, as can be seen in [**72**], even in the simple case $M(f) = -f''$ such inequalities are both difficult to obtain and do not lead to sharp values of $K(p,q)$.

REMARK 5: A few specific examples of (38.2) are given in [**73**, §17], and in [**74**, pp. 13–14]. In particular, we state the following two results from [**74**]: if $f, f'' \in L_2(R^+)$, then

$$\left(\int_0^\infty \left[f'^2 + \mu f^2\right] \, dx\right)^2 \leq K(\mu) \int_0^\infty f^2 \, dx \int_0^\infty [f'' - \mu f]^2 \, dx, \tag{38.7}$$

where $K(\mu) = 4$ for $-\infty < \mu \leq 0$, $K(\mu) = +\infty$ for $\mu > 0$. If $a > 0$, $f \in L_2(a,\infty)$, $(x^\tau f')' \in L_2(a,\infty)$ then

$$\left(\int_a^\infty x^\tau f'^2 \, dx\right)^2 \leq K(\tau) \int_a^\infty f^2 \, dx \int_a^\infty \left[(x^\tau f'(x))'\right]^2 \, dx, \tag{38.8}$$

where $K(\tau) = 4$ for $-\infty < \tau \leq 0$, $K(\tau) = \left[\cos\left[(3-\tau)^{-1}\pi\right]\right]^{-2}$ for $0 \leq \tau < 1$, and $K(\tau) = +\infty$ for $\tau \geq 1$.

(For (38.7) see also [**77**].)

W. N. Everitt and M. Giertz [**76**] also considered inequalities of the form

$$A\|(pf')'\|^2 + B\|(pq)^{1/2} f'\|^2 + c\|qf\|^2 \leq D\|f\|^2 + \|M(f)\|^2$$

in the Hilbert function space $L_2(a,b)$ for arbitrary open intervals (a,b) with the differential operator M defined by (38.1).

39. The number $C_{42}(2,R^+)$ was discovered by J. S. Bradley and W. N. Everitt [**79**] in 1974. Namely, $C_{42}(2,R^+) = 2k_1^{-1}$, where k_1 is the smaller of the two positive roots of the equation

$$k^4 - 2k^2 - 4k + 1 = 0.$$

Note that $8.87 < C_{42}(2,R^+) < 8.88$.

N. P. Kupcov [**80**] in 1975 succeeded in reducing the problem of finding $C_{nk}(2,R^+)$ to finding roots of explicit polynomials. In this case inequality $\|f^{(k)}\|_2 \leq M_{n,k}\|f\|_2^{(n-k)/n}\|f^{(n)}\|_2^{k/n}$, where $M_{n,k} = C_{nk}(2,R^+)$, is equivalent to

$$\|f^{(k)}\|_2^2 \leq \frac{1}{\gamma_{n,k}}\left(\|f\|_2^2 + \|f^{(n)}\|_2^2\right), \tag{39.1}$$

where

$$M_{n,k}^2 = \frac{1}{\gamma_{n,k}}\left\{\left(\frac{n-k}{n}\right)^{k/n} + \left(\frac{k}{n-k}\right)^{(n-k)/n}\right\}.$$

In this way he obtained the results:

1^0 $0 < \gamma_{n,k} < \frac{n}{k^{k/n}(n-k)^{(n-k)/n}}$;

2^0 $\gamma_{n,n-k} = \gamma_{n,k}$;

3^0 $\gamma_{n,k}$ is the smallest positive root of the equation

$$D(\gamma) = \begin{vmatrix} \Delta_{n-k+1} & \Delta_{n-k+2} & \cdots & \Delta_n \\ \Delta_{n-k+2} & \Delta_{n-k+3} & & \Delta_{n+1} \\ \vdots & & & \\ \Delta_n & \Delta_{n+1} & & \Delta_{n+k-1} \end{vmatrix} = 0,$$

where

$$\Delta_{n-j} = \sum_{m=1}^{2k} \varepsilon_m^{-j} \exp\left(\frac{n}{i\varepsilon_m \pi k}\int_0^{\gamma} t^{\frac{1}{2k}-1} F_m(t)\, dt\right),$$

$$F_m(t) = \int_0^{+\infty} \frac{1 - s^{2k}t}{(s^2 t^{1/k} - \bar{\varepsilon}_m^2)(s^{2n} - ts^{2k} + 1)}\, ds,$$

$$\varepsilon_m = \exp\frac{m\pi i}{k} \quad (m = 1, \ldots, 2k);$$

4^0 $\frac{1}{\gamma_{n1}} = \frac{16n^2}{\pi^4} + \frac{4}{2\pi^2} + O\left(\frac{1}{n}\right) \qquad (n \to +\infty).$

The value of some constants (see [**80**]) are:

$$\gamma_{21} = 1, \ \gamma_{31} = \gamma_{32} = \sqrt[3]{3 - 2\sqrt{2}} \approxeq 0.555669.$$

The constant γ_{41} is characterized as the smallest positive zero of the polynomial $\gamma^8 - 6\gamma^4 - 8\gamma^2 + 1$, and γ_{42} is the smallest positive zero of the polynomial $\gamma^4 - 2\gamma^2 - 4\gamma + 1$ (as in [**79**]).

40. Some new proofs of results of Kolmogorov- Landau's type are given by I. J. Schoenberg [**81**] and A. S. Cavaretta [**82**]. Note that Cavaretta gave a surprisingly easy proof of the Kolmogorov inequality.

41. Z. Ditzian [**83**] proved the following result:

Let $T(s)$ be a group of operators on a Banach space X, $\|T(s)\| = 1$ and A the infinitesimal generator; then for $x \in D(A^2)$ inequality (34.7) is valid.

He also showed that $C_{nk}(1, R) = C_{nk}(+\infty, R)$.

42. Some related results are given by G. V. Kirsanova [**84**] and V. K. Dzjadyk and V. A. Dubovik [**85**], [**86**].

Results from [**85**] and [**86**] give generalizations of results of A. M. Rodov (see 15.).

43. Some integral inequalities related to that from 6. are considered by G. Talenti [**87**].

44. A. Russell [**88**] gave generlizations of results from [**73**].

45. Let f and f'' be continuous on the unit interval $I = [0, 1]$ and let $\|f\|_I :=$ ess sup$|f(t)| \leq 1$ and $\|f''\|_I \leq A$. Then

$$\|f'\|_I \leq \begin{cases} (4 + A)/2 & \text{if } 0 < A \leq 4, \\ 2\sqrt{A} & \text{if } A > 4. \end{cases} \tag{45.1}$$

Moreover, these inequalities are best possible.

This result is due to Ch. K. Chui and P. W. Smith [**89**].

Some related results are proved by A. Sharma and J. Tzimbalario [**90**]. They first proved (45.1) with $|f'(0)|$ instead of $\|f'\|_I$ on the left hand side. Further,

$$|f'(1/2)| \leq \begin{cases} 2 + \frac{A}{4} & \text{if } 0 \leq A \leq 8, \\ \sqrt{2A} & \text{if } A > 8. \end{cases} \tag{45.2}$$

The bounds are sharp.

Sharma and Tzimbalario also obtained best bounds for $|f'(0)|$, $|f''(0)|$, $|f'(1/2)|$, and $|f''(1/2)|$ in terms of $\|f\|_I$ and $\|f'''\|_I$.

46. Some inequalities with bounded derivatives of higher order are given by H. Kallioniemi in [**91**],[**92**], [**93**].

47. Let A be the infinitesimal generator of a one parameter semigroup of linear operators which are contractions simultaneously on $L^p(M, \mathcal{A}, \mu)$, $1 < p < +\infty$. Then, in [**95**],[**96**], the inequality

$$\|Af\|_p^2 \leq c_p \|A^2 f\|_p \|f\|_p \tag{47.1}$$

is shown to hold for all $f \in L^p(M, \mathcal{A}, u)$ for which the right-hand side is finite, where $c_p = 2^{2/p}$ for $1 < p \le 2$ and $c_p = 2^{2-2/p}$ for $2 \le p < +\infty$. Specializing A to be d/dx acting on $L^p[0, +\infty)$, the inequality becomes

$$\|f'\|_p^2 \le c_p \|f''\|_p \|f\|_p. \tag{47.2}$$

This is the classical inequality of Hardy, Landau and Littlewood, but with substantially improved constants.

J. A. Goldstein also proved in [**97**] the following upper bounds on $C_p(J)$, where $C_p(J)$ denote the smallest constant c_p for $f \in L^p(J)$ with second derivative $f'' \in L^p(J)$ (in (47.2) of course):

$$\begin{aligned} C_p(R^+) &\le 2(5/4)^{2/p-1} && \text{for } 1 < p < 2; \\ C_p(R) &\le 2^{2/p-1} && \text{for } 1 < p < 2; \\ C_p(R) &\le 2^{1-2/p} && \text{for } 2 < p < +\infty. \end{aligned}$$

48. A. S. Cavaretta [**98**] extended earlier work of Schoenberg and Cavaretta [**54**] (see 32.) concerning best bounds on intermediate derivatives for functions defined and n- times differentiable on the half-line. From the two hypotheses $-1 \le f(x) \le 1$ and $f^{(n)}(x) \le 1$ for all $x \le 0$, there follows best possible lower bounds on $f^{(j)}(x)$, $1 \le j \le n-1$, if $n+j$ is odd, and best possible upper bounds on $f^{(j)}(x)$ if $n+j$ is even.

49. G. G. Magaril-Il'Jaev [**99**] considered the inequality (31.1) for $J = R^+$, $q = +\infty$, $r = +\infty$, $u = (2-k)/(2+p^{-1})$, $v = (k+p^{-1})/(2+p^{-1})$, $n = 2$, $k = 0$ or 1, $1 \le p \le +\infty$ (if $k = 0$, the case $p = +\infty$ is excluded); as well as the case when $J = R^+$, $q = +\infty$, $p = 1$, $u = (2-k+r^{-1})/(3-r^{-1})$, $v = (k+1)/(3-r^{-1})$, $n = 2$, $1 < r < \infty$ and $k = 0$ or 1.

He gave some expressions for the best constants, and extremal functions for which these inequalities become equalities.

50. In paper [**100**] of A. A. Ligun some inequalities related to Kolmogorov's inequality are established. For example, the following result is valid:

Let $F(u)$ for $u \ge 0$ be an arbitrary N-function, i.e. it can be represented as

$$F(u) = \int_0^u p(t)\, dt,$$

where $p(t)$ is a nondecreasing function that is continuous from the right and satisfies the following conditions: $p(t) > 0$ $(t > 0)$, $p(0) = 0$, $\lim_{t \to +\infty} p(t) = +\infty$.

C is the space of all continuous 2π-periodic function with the norm $\|f\|_C = \max|f(x)|$.

For the elements $f \in C^{(r+1)}$ $(r = 0, 1, \dots)$ we have the following sharp inequality

$$(50.1) \qquad \int_0^{2\pi} F\left(\frac{|f'(x)|}{\|f^{(r+1)}\|_C}\right) dx \leq \int_0^{2\pi} F\left(\left(\frac{\|f\|_C}{K_{r+1}\|f^{r+1}\|_C}\right)^{r/(r+1)} |\phi_r(x)|\right) dx,$$

where K_r and ϕ_r are defined as in 10.

51. I. J. Schoenberg [**101**] proved that if $\|f\| \leq A$ and $\|f'' + f\| \leq 1$ ($\|\cdot\|$ denote the uniform norm on R) then

$$(51.1) \qquad \|f'\| \leq \sqrt{A(2 + A)};$$

and if $\|f\| \leq B < 1$ and $\|f'' - f\| \leq 1$, then

$$(51.2) \qquad \|f'\| \leq \sqrt{B(2 - B)}.$$

Furthermore, these results are sharp.

As we see, instead of D^2 $(D = d/dx)$ he considered differential operator $D^2 \pm 1$. In [**102**], he considered differential operator $D^2 - \lambda^2$ $(\lambda > 0)$. A. Sharma and J. Tzimbalario [**103**] considered the linear differential operator $(D - \lambda_1) \dots (D - \lambda_\nu)$, $\nu = 1, 2, \dots, n$.

52. Let $f \colon R \to R$ be n (> 1) times differentiable function such that $|f^{(n)}(x)| \leq M$ (for $x \in (a, b)$. Then, for every $x \in [a, b]$

$$(52.1) \qquad \left|\frac{1}{n}\left(f(x) + \sum_{k=1}^{n-1} F_k\right) - \frac{1}{b-a}\int_a^b f(t)dt\right| \quad \frac{M}{n(n+1)!} \cdot \frac{(x-a)^{n+1} + (b-x)^{n+1}}{b-a}$$

where F_k is defined by

$$F_k := F_k(f; n; x, a, b) := \frac{n-k}{k!} \cdot \frac{f^{(k-1)}(a)(x-a)^k - f^{(k-1)}(b)(x-b)^k}{b-a}.$$

This generalization of (11.1) is due to G. V. Milovanović and J. E. Pečarić [**104**]. For $n = 2$, we get: If $|f''(x)| \le M$ (on $x \in (a,b)$, then

$$\left|\frac{1}{2}\left(f(x)+\frac{(x-a)f(a)+(b-x)f(b)}{b-a}\right)-\frac{1}{b-a}\int_a^b f(t)\,dt\right| \le \frac{M}{4}(b-a)^2\left(\frac{1}{12}+\frac{\left(x-\frac{a+b}{2}\right)^2}{(b-a)^2}\right)$$

for every $x \in [a,b]$.

As a consequence of this inequality we can obtain for functions defined on $[a,+\infty)$ the following inequality

$$\|f'+f'(a)\|_\infty^3 \le \frac{9}{2}\|f\|_\infty^2\|f'''\|_\infty.$$

Another inequality related to (11.1) is given in [**105**] by A. Lupas.

Let f be a differentiable function such that $\int_a^b f'(x)^2 dx < +\infty$. Then

$$\left|f(x)-\frac{1}{b-a}\int_a^b f(x)\,dt\right| \le \frac{2}{\pi\sqrt{b-a}}\|f'\|_2 \quad \text{for } x \in (a,b). \tag{52.2}$$

See Chapter XV for more on these inequalities.

53. Denote by X a real Banach space with norm $\|\cdot\|$. If A is the generator of a strongly continuous contraction semigroup and x is in the domain of A^n, some integer $n \ge 2$, then

$$\|A^k x\|^n \le C_{nk}^n(+\infty, R^+)\|x\|^{n-k}\|A^n x\|^k \tag{53.1}$$

for integer k, $1 < k \le n-1$.

If A is the generator of a strongly continuous group of isometries and x is in the domain of A^n, some integer $n \ge 2$, then

$$\|A^k x\|^n \le C_{nk}^n(+\infty, R)\|x\|^{n-k}\|A^n x\|^k \tag{53.2}$$

for integer k, $1 \le k \le n-1$.

These results are due to M. W. Certain and T. G. Kurtz [**106**].

Related results for dissipative and Hermitian operators on a Banach space are given by B. Bollobás in [**107**].

54. E. T. Copson [**108**] proved the following theorem.

Let f, f', f'' be real and Lebesgue square integrable over R^+, and set $I_0 = \int_0^\infty f^2\,dx$, $I_1 = \int_0^\infty f'^2\,dx$, $I_2 = \int_0^\infty f''^2 dx$, $J = \int_0^\infty ff''\,dx$. Then

$$J^2 \le I_1\left(2\sqrt{I_0, I_2} - I_1\right).$$

His elementary proof is based on Gram's inequality (see for example, AI, pp. 45–47). As a corollary the inequality (6.1) is obtained.

In [**109**], Copson dealt with inequality (38.7) and with an analogous inequality over $(-\infty, +\infty)$.

55. In [**57**] E. Hille conjectured that for $L_p(R)$, $\|f^{(n)}\|_p^2 \le 4\|f\|_p\|f^{(2n)}\|_p$. Z. Ditzian [**110**] showed that

$$\|f^{(n)}\|_p^2 \le 2\|f\|_p\|f^{(n)}\|_p$$

always holds and actually

$$\|f^{(n)}\|_p^2 \le \left(\frac{4}{\pi} + \delta_n\right)\|f\|_p\|f^{(2n)}\|_p,$$

where $\delta_n = o(1)$ as $n \to +\infty$.

56. Further references of interest are [**111**], [**112**], and [**113**], where relationships with inequalities of this type are related to approximation problems.

57. Let r be fixed in $[0, 1]$ and let u be the solution of the equation $r = 4\cos\frac{1}{2}u/(3 + \cos u)$, $u \in [0, \pi]$. If f is complex-valued with

$$r \le |f(t)| \le 1,\ t \in R, \text{ and } \|f''\| \le 16\sin^2\frac{1}{2}u/(3 + \cos u)$$

then

$$\|f'\| \le 8\sin\frac{1}{2}u/(3 + \cos u). \tag{57.1}$$

Moreover, the constant on the right-hand side of (57.1) is the best possible.

This result is due to I. J. Schoenberg [**114**].

58. An expository account of the current development of Landau-Kolmogorov's type inequalities is given by T. N. T. Goodman and S. L. Lee [**115**]. It is not as complete as our discussion.

59. Let S be a closed and connected set in a finite dimensional Euclidean space, and consider the set $(S) = \{f(t)\colon f(t) \in S \text{ for all } t,\ f(t) \neq \text{constant}\}$, where $f \in C^1(R)$, $f''(t)$ is piecewise continuous with discontinuities only of first kind. Now consider the functional $F(t) = \|f'\|/\sqrt{\|f''\|}$, where $\|\cdot\|$ is the sup norm. The Landau problem for the set S is to determine the constant $L(S) = \sup_{f\in(S)} F(f)$. $L(S)$ is called the Landau constant of the set S. I. J. Schoenberg **[116]** showed that for any closed bounded and convex set S, $L(S) = (\text{diameter } S)^{1/2}$. If S is the solid shell $b \le (x^2+y^2+z^2)^{1/2} \le a$, $0 < b < a$, then $L(S) = \left(a + (a^2-b^2)^{1/2}\right)^{1/2}$.

60. Let $f(x)$ be a continuous differentiable function on $I = [0,1]$ with piecewise continuously differentiable derivative $f'(x)$, and let $\|\cdot\|_I$ be a sup norm. A. Sharma and J. Tzimbalario **[117]** establish the following sharp inequalities. If $\|f\|_I \le 1$, $\|f'' - \lambda^2 f\|_I \le A$, $\lambda > 0$, then

$$\|f' \pm \lambda f\|_I \le \begin{cases} \lambda + \lambda \coth\frac{\lambda}{2} + \frac{A}{\lambda}\tanh\frac{\lambda}{2}, & \text{for } 0 \le A \le \lambda^2\coth^2\frac{\lambda}{2}, \\ \lambda + 2\sqrt{A}, & \text{for } \lambda^2 \coth\frac{\lambda}{2} < A. \end{cases}$$

Sharp estimates are also obtained for $\|f'\|_I$ if $\|f\|_I \le 1$ and $\|f'' + \lambda^2 f\|_I \le A$.

61. Some integral inequalities are obtained by C. Bennewitz and W. N. Everitt **[118]** and P. R. Beesack **[119]**.

62. Some sharp inequalities for the best uniform approximations of periodic functions are established by V. V. Žuk **[120]**. For some previous results of the same author see references from **[120]**.

63. If $f, f^{(3,0)}, f^{(0,3)}$ (partial derivatives of f) are bounded on R^2, then

$$\|f^{(1,1)}\| \le \sqrt[3]{3}\|f\|^{1/3}\|f^{(3,0)}\|^{1/3}\|f^{(0,3)}\|^{1/3}, \tag{63.1}$$

$$\|f_e^{(2)}\| \le \sqrt[3]{3}\|f\|^{1/3}\left(\|f^{(3,0)}\|^{1/3}|e_1| + \|f^{(0,3)}\|^{1/3}|e_2|\right)^2, \tag{63.2}$$

where $\|g\|$ denotes the upper bound of $|g|$ on R^2 and $f_e^{(2)}$ the second derivative in the direction $e = (e_1, e_2)$.

This is a result from **[121]** (V. N. Konovalov).

64. R. Redheffer and W. Walter **[122]** considered an inequality of type (1.1) for $p = +\infty$, but with $M_n^*(+\infty, I) = \max(|I|^{-n}M_0(+\infty, I), M_n(+\infty, I))$ instead of

$M_n(+\infty, I)$, where $|I|$ is the length of interval I. Generalizations for functions of several variables are also given.

65. Inequalities for fractional derivatives on the half-line are considered in [**123**] by V. V. Arestov.

66. G. V. Milovanović and I. Ž. Milovanović [**124**] gave a weighted version of (52.2).

67. B. Neta [**125**] has determined the numerical values $\gamma_{41} = 0.339246$ and $\gamma_{42} = 0.225270$, for (39.1). In the same article B. Neta has given lower and upper bounds for the best possible constants $\gamma_{n,k}$ and $M_{n,k}$ for $n \le 10$.

An additive inequality of the form $\|f^{(n-1)}\|_p \le K\|f\|_p + \|f^{(n)}\|_p$ for $f \in L_p(0,1)$ is considered by V. I. Burenkov [**126**]. The best possible constant K is obtained.

68. Here, we shall give several results from [**127**] of M. K. Kwong and A. Zettl.

Let us consider the inequality (31.1), with conditions given in 31.

This inequality is equivalent to each of the following statements.

1. There exists a constant K, depending only on p, q, r, n, k and such that for all $\lambda > 0$ and $y \in W^n_{p,r}(J)$ we have

$$\|y^{(k)}\|_q \le K\left(\lambda^a\|y\|_p + \lambda^b\|y^{(n)}\|_r\right),$$

   where $a = u^{-1}$ and $b = -v^{-1}$.

2. There exists a constant K, depending only on p, q, r, n, k and such that for all $y \in W^n_{p,r}(J)$,

$$\|y^{(k)}\|_q \le K\left(\|y\|_p + \|y^{(n)}\|_r\right),$$

3. If $y \in W^n_{p,r}(J)$, then $y^{(k)} \in L^q(J)$.

In the following results in place of $W^n_{p,p}(J)$ we write $W^n_p(J)$, and $K(p,J)$ is the best, i.e. smallest, constant in

$$\|y'\|_p^2 \le K(p,J)\|y\|\;\|y''\|.$$

Let us consider the following set of functions:

$$\begin{aligned}
M_1 &= \{y \in W_p^2([0,1]) : y(0) = y(1) = 0\},\\
M_2 &= \{y \in M_1 : y(t) > 0 \text{ for } 0 < t < 1\},\\
M_3 &= \{y \in W_p^2([0,1]) : y(0) = y'(1) = 0,\ y(t) > 0 \text{ for } 0 < t < 1\},\\
M_4 &= \{y \in M_3 : y'(t) \geq 0 \text{ for } 0 < t < 1\},\\
M_5 &= \{y \in W_p^2([0,1]) : y'(0) = y'(1) = 0\},\\
M_6 &= \{y \in W_p^2(R^+) : y(0) = 0\},\\
M_7 &= \{y \in W_p^2(R^+) : y'(0) = 0\}.
\end{aligned}$$

For any p satisfying $1 \leq p \leq +\infty$ we have $K(p, M_i) = K(p, R)$, $i = 1,\dots,7$. Furthermore, the interval $[0,1]$ in the definition of M_i, for $i = 1,\dots,5$, can be replaced by any compact interval $[a,b]$.

We also have $K(p, M_8) = K(p, R)$, where $M_8 = \{y \in W_p^2(R^+) : y'(0)y(0) \geq 0\}$.

Further, let

$$\begin{aligned}
N_1 &= \{y \in W_p^2([0,1]) : y(1) = y'(1) = 0\},\\
N_2 &= \big\{y \in W_p^2([0,1]) : \text{ either } y(0) = y(1) = 0 \text{ or } y'(0)/y(0) = \\
&\qquad y'(1)/y(1),\ y \text{ vanishes exactly once in } [0,1]\big\}.
\end{aligned}$$

For any p, $1 \leq p \leq +\infty$ we have $K(p, R^+) = K(p, N_1) = K(p, N_2)$.

If $N_3 = \{f \in W_p^2(R^+),\ f \geq 0,\ f' \leq 0\}$, then $K(p, R^+) > K(p, N_3)$, and if $N_4 = \{f \in N_1,\ f \geq 0,\ f' \leq 0\}$, $1 < p < +\infty$, then $K(p, N_4) = K(p, N_3)$.

For $1 < p < +\infty$, we have that $K(p, R) < K(p, R^+)$. Further, $K(p, J)$ ($J = R$ or R^+) is a continuous function of p for $1 \leq p \leq +\infty$.

For any p, $1 \leq p < +\infty$, we have

$$2^{p+1}\Gamma\left(p+1+2^{-1}\right) / \left(\pi^{1/2}(p+1)\Gamma(p+2)\right) \leq K^p(p, R), \tag{68.1}$$

and if $3/2 < p < 2$, then

$$K(p, R) \leq \frac{1}{p-1}; \tag{68.2}$$

if $p \neq 2$, $1 \leq p \leq +\infty$, then

$$K(p, R) > 1; \tag{68.3}$$

if $1 \leq p \leq +\infty$, then

$$K(p, R) < K(p, R^+). \tag{68.4}$$

Note that H. A. Gindler and J. A. Goldstein [128] have proved

$$K(p,R) \leq p-1, \quad 2 < p < +\infty. \tag{68.5}$$

In view of $K(p,J) \leq K(+\infty,J)$, $1 \leq p \leq +\infty$, and $K(+\infty,R) = 2$, the inequality (68.5) is interesting only for $2 < p < 3$. In [130] Evans and Zettl independently found (68.5) and showed that it holds for a wider class of inequalities.

(The above results for $M_1, \dots, M_4$ are previously obtained in [130] by the same authors.)

69. Of course, related weighted inequalities are also of interest (see for example (6.8)). The following problems are considered:

(69.1) There is a constant K such that for all y in a class D of functions with domain I,

$$\int_I N|y^{(j)}|^p \leq K\left(\int_I W|y|^p + \int_I P|y^{(n)}|^p\right).$$

(69.2) For each $\varepsilon > 0$ there is a number $K(\varepsilon)$ such that for all $y \in D$,

$$\int_I N|y^{(j)}|^p \leq K(\varepsilon)\int_I W|y|^p + \varepsilon\int_I P|y^{(n)}|^p.$$

(69.3) There exists $\xi \geq 0$ ($\xi > 0$ for $j \neq 0$), $\eta > 0$, $K > 0$, and a set Γ of positive numbers such that for all $\varepsilon \in \Gamma$ and $y \in D$,

$$\int_I N|y^{(j)}|^p \leq K\left(\varepsilon^{-\xi}\int_I W|y|^p + \varepsilon^{\eta}\int_I P|y^{(n)}|^p\right).$$

In these inequalities I is an interval of the real line, $1 \leq p < +\infty$, $0 \leq j < n$, and N, W, P are positive measureable functions (which satisfy additional conditions).

(69.4) There are positive numbers u, v and K_1 with $u + v = 1$ such that for all $y \in D$,

$$\int_I N|y^{(j)}|^p \leq K_1\left(\int_I W|y|^p\right)^u \left(\int_I P|y^{(n)}|^p\right)^v.$$

Inequalities of type (69.2) have been given by Evans and Zettl [**130**] and M. Kwong and A. Zettl [**132**],[**131**], new results for (69.4) have been given by Kwong and Zettl [**127**],[**134**],[**133**] (see also R. Amos and W. N. Everitt [**135**] and W. N. Everitt [**136**]); for additional references see [**137**].

Recent work on product type inequalities involving powers of an operator may be found in M. Kwong and Z. Zettl [**134**] and [**138**] and V. Phóng [**139**].

70. Let $\varepsilon > 0$; S_ε a net on R (real line) with ε step; $W^n(S_\varepsilon, R)$ – set of all functions defined on R with locally absolutely continuous derivatives $f^{(n-1)}(x)$, such that $\|f\|_{S_\varepsilon} = \sup\limits_{x \in S_\varepsilon} |f((x)| < +\infty$, $\|f^{(n)}\|_R = \text{ess sup}|f^{(n)}(x)| < +\infty$.

If $f \in W^n(S_\varepsilon, R)$, then for all m, $0 \leq m < n$, we have

$$\|f^m\|_R \leq C_n \inf_{\delta \geq \varepsilon} \left(\delta^{-m}\|f\|_{S_\varepsilon} + \varepsilon^{n-m}\|f^{(n)}\|_R\right),$$

where the C_n have values depending only on n. This result is due to V. N. Konovalov [**140**].

71. G. G. Magaril-Il'jaev and V. Tihomirov [**141**] have given a generalization of a result from [**53**] for fractional derivatives (on the half-line).

72. The result from 59. is extended by the same author in [**142**].

73. In [**143**] H. A. Gindler and J. A. Goldstein investigated a Landau type inequality for infinite and biinfinite sequence $\{x_n\}$ in l_p or l_∞ and proved

$$\|\{\Delta x_n\}\| \leq a\|\{\Delta^2 x_n\}\|^{1/2}\|\{x_n\}\|^{1/2}, \tag{73.1}$$

where $\Delta x_n = x_{n+1} - x_n$, $\Delta^2 x_n = \Delta(\Delta x_n)$ and where the norm was that of l_p or l_∞, and a is a constant.

In [**143**] best constants for $l_2(N)$ and $l_2(Z)$ (where $N = \{1, 2, \dots\}$ and $Z = (\dots, -1, 0, 1, 2, \dots)$ were determined for (72.1) and estimates were given for the constant in the l_p case. This result also generalized an earlier result of E. T. Copoon [**144**].

74. Another class of inequalities of Landau-Kolmogorov's type, where explicit constants $\widehat{C}_{nk}$ are available, is given by A. M. Fink [**145**].

If f is a completely monotone function on R^+, then for any T, $0 < T \leq +\infty$,

$$\widehat{C}_{nk}(+\infty, [0, T]) = 1, \text{ and}$$
$$\widehat{C}_{nk}(p, [a, b]) = 1 \text{ for any } \{a, b\} \subset (0, +\infty)$$
$$\text{and } 1 \leq p \leq +\infty.$$

If f satisfies $f^{(j)} \geq 0$ for $j = 0, 1, \ldots, n+1$ and $f^{(j)}(0) = 0$ for $j = 0, \ldots, N-1$, then on $[0, T]$ we have

$$\widehat{C}_{nk}(+\infty, [0, T]) = \frac{(N-k)!}{(N!)^{1-k/N}}.$$

75. M. Sato and R. Sato [**146**] gave a simple proof of the Chui-Smith theorem from 45.

76. For a function f defined on R^+ we denote:

$$m(f) = \inf\{f(x) \colon x \in R^+\}, \quad M(f) = \sup\{f(x) \colon x \in R^+\}.$$

Let $f \in C^2(R^+)$, $f^{(3)}$ exist in R^+ and f be bounded and non-negative there. If $m(f') \leq 0$, $M\left(f^{(3)}\right) < +\infty$ and $f''(0) \leq 0$, then $M\left(f^{(3)}\right) \geq 0$ and:

$$-m(f')^3 \leq \left(\frac{9}{8}\right) \|f\|^2 M\left(f^{(3)}\right). \tag{76.1}$$

But, if $m\left(f^{(3)}\right) > -\infty$, $M(f') \geq 0$, $f''(0) \geq 0$, then $-m\left(f^{(3)}\right) \geq 0$ and

$$M(f')^3 \leq \left(\frac{9}{8}\right) \|f\|^2 \left(-m\left(f^{(3)}\right)\right). \tag{76.2}$$

These results are due to K. N. Boyadziev [**147**]. He also proved (76.1) and (76.2) under slightly different assumptions.

77. A. I. Zvyagintsev [**148**] shows that if

$$T^4 M_0^{-1}(+\infty, [0, T])\, M_4(+\infty, [0, T]) \leq 48\left(17 + 12\sqrt{2}\right),$$

then $T^k M_0^{-1}(+\infty, [0, T])\, M_k(+\infty, [0, T]) \leq P_k$, $k = 1, 2, 3$, the upper bounds P_k are exactly described. The inequality is best possible.

A. I. Zvyagintsev and A. Ya. Lepin [**149**] show that if

$$T^3 M_0^{-1}(+\infty, [0, T])\, M_3(+\infty, [0, T]) \leq 81,$$

then

$$T M_0^{-1}(+\infty, [0, T])\, M_1(+\infty, [0, T]) \leq (4y^{-1} - 6)(1 - y)^{-2},$$
$$T^2 M_0^{-1}(+\infty, [0, T])\, M_2(+\infty, [0, T]) \leq (4y^{-2} - 12)(1 - y)^{-2},$$

where y is uniquely determined from $[1/3, 1/2]$ by the relation

$$T^3 M_0^{-1}(+\infty, [0, T]) M_3(+\infty, [0, T]) y^2(1-y)^2 = 12 - 24y.$$

The inequality is the best possible.

78. A weighted version of (38.2) is also considered in many papers:

$$\left(\int_a^b (pf'^2 + qf^2)dx\right)^2 \leq K \int_a^b wf^2 dx \int_a^b w|w^{-1}M(f)|^2 dx. \tag{78.1}$$

For such results and for many new results see paper of W. D. Evans and W. N. Everitt [**150**].

79. I. J. Schoenberg [**151**] proved the following result (see also [**28**]):
Let g be k-convex for $k = 0, \ldots, n+1$ on $(-\infty, 0]$, and assume that g satisfies

$$g(0) \leq M_0 \text{ and } g^{(n)}(0)/n! \leq M_n.$$

Then

$$g^{(j)}(x)/j! \leq \binom{n}{j} M_0^{(n-j)/n} M_n^{j/n} \quad (j = 1, \ldots, n-2). \tag{79.1}$$

Equality holds in (79.1) for some $1 \leq j \leq n-1$ if and only if

$$g(x) = M_n(\lambda + x)^n \text{ on } [-\lambda, 0] \text{ and } g(x) = 0 \text{ in } (-\infty, -\lambda),$$

where $\lambda^n = M_0/M_n$.

A generalization of this result is given by R. Farwig and D. Zwick [**152**] and J. E. Pečarić, S. S. Dragomir and B. Crstici [**153**].

80. Z. Ditzian [**154**],[**155**] gave some further results about Kolmogorov inequalities. For example, in [**155**] he proved for a wide variety of Banach spaces of infinite or biinfinite sequences the analogue of (1.1), i.e. generalizations of (72.1), given by

$$\|\{\Delta^k x_m\}\|_B \leq c(n, k, B) \|\{\Delta^n x_m\}\|_B^{k/n} \|\{x_m\}\|_B^{1-k/n}, \tag{80.1}$$

where $\Delta x_m := x_{m+1} - x_m$, $\Delta^j x_m = \Delta(\Delta^{j-1} x_m)$ and where $c(n, k, B)$ is the best constant for B, n and k.

81. Let $L_{2,r}$ be a space of measurable functions on $[-1,1]$ such that

$$\|f\|_{2,r} = \left(\int_{-1}^{1} (1-x^2)^r |f(x)|^2 dx \right)^{1/2} < +\infty.$$

We shall write $L_{2,0} = L_2$, $\|f\|_{2,0} = \|f\|_2$. Let $\{X_k(x)\}_0^\infty$ be the system of normalized orthogonal polynomials of Legendre, and $c_k(f) = \int_{-1}^{1} f(x)X_k(x)dx$ $(k = 0,1,\dots)$ coefficients of Fourier-Legendre of the function f. Let $r \in N$, $s \geq 0$ an integer, $0 \leq x \leq r-1$.

Suppose that a function f is defined on $[-1,1]$ satisfying the following conditions:

1) $f^{(r-1)}$ exists in $(-1,1)$, and it is absolutely continuous on every interval $[-1+\varepsilon, 1-\varepsilon]$, $0 < \varepsilon < 1$;
2) $f^{(r)} \in L_{2,r}$;
3) $c_k(f) = 0$ $(k = s, s+1, \dots, r-1)$.

Then the following inequality holds (S. Z. Rafal'son [**156**])

$$\|f^{(s)}\|_{2,s} \leq \frac{\Gamma^{1/2}(r+s+1)}{\Gamma^{1/2}(r-s+1)\Gamma^{s/2r}(2r+1)} \|f\|_2^{1-s/r} \|f^{(r)}\|_{2,r}^{s/r},$$

with equality for $f = X_r$.

82. An application of Kolmogorov's inequality is given by A. A. Ligun in [**157**].

83. Some further remarks about results from 11., 17., 25., and 52. are given in [**158**], [**159**] by J. E. Pečarić and B. Savić. For example, a generalization of Ostrowski's inequality (11.1) in the case when f satisfies Lipschitz's condition of order α is given.

84. Let us consider the classical weighted Banach space $X_p = L^p(J,k)$ with the usual norm

$$\|f\|_p = \left(\int_J |f(x)|^p k(x) dx \right)^{1/p}, \quad 1 \leq p < +\infty.$$

Let $A = d/dx$ be the differentiation operator on X_p, and A the infinitesimal generator of a (C_0) semigroup $\{T(t)\colon t \geq 0\}$ given by

$$T(t)f(x) = f(x+t), \quad x \in J, \quad t \geq 0.$$

Suppose the weight function k satisfies

$$\sup_{t\subset J,\, s\geq 0} \frac{k(t)}{k(t+s)} = M < +\infty, \tag{84.1}$$

and $J = (a, +\infty)$ where $-\infty \leq a < +\infty$. Then

$$\|T(t)\| \leq M^{1/p}, \quad t \geq 0. \tag{84.2}$$

Now, as a simple consequence of (34.3) we get the following inequality of J. A. Goldstein, M. K. Kwong and A. Zettl [**160**]:

$$\|f'\|_p^2 \leq 2M^{1/p}\left(M^{1/p}+1\right)\|f\|_p\|f''\|_p. \tag{84.3}$$

Note that Goldstein, Kwong and Zettl did not note that (84.3) is a consequence of Kurepa's inequality (34.3). They also noted that if A generates a (C_0) group $\{T(t)\colon\; -\infty < t < +\infty\}$ on $(X, \|\cdot\|)$ satisfying $N = \sup\limits_{t\in R} \|T(t)\| < +\infty$, then

$$\|Af\|^2 \leq |Af|^2 \leq 2|A^2 f|\;|f| \leq 2N^2\|A^2 f\|\;\|f\|, \tag{84.4}$$

where $|f| = \sup\limits_{t\in R} \|T(t)f\|$.

Some other results similar to (84.3) are also obtained in [**160**].

85. Let $0 \leq a < 1$, $K(a, R) = 2/(1-a)$, $K(a, R^+) = 4/(1-a)$. Let $J = R$ or R^+. Suppose $y \in L^\infty(J)$, y' exists and is absolutely continuous on compact subintervals of J, and $y''|y|^a \in L^\infty(J)$. Then, $y' \in L^\infty(J)$ and

$$\|y'\|_\infty^2 \leq K(a, J)\|y\|_\infty^{1-a}\|y''|y|^a\|_\infty. \tag{85.1}$$

This inequality of M. K. Kwong and A. Zettl [**161**] is an extension of Landau's inequalities since $y''|y|^a$ might be bounded without y'' being bounded.

86. V. G. Timofeev [**162**],[**163**] proved for $f \in C(R^d)$ and $\Delta f \in L_\infty(R^d)$, where

$$\Delta f = \sum_{i=1}^{d}\left(\frac{\partial}{\partial x_i}\right)^2 f$$

is the Laplacian of f taken in the Sobolev (i.e. weak) sense, that

$$\left\|\frac{\partial}{\partial \varepsilon} f\right\| \leq \sqrt{2}\|f\|^{1/2}\|\Delta f\|^{1/2} \tag{86.1}$$

where $(\partial/\partial\varepsilon)f$ is a directional derivative and $\|\cdot\|$ is the $L_\infty(R^d)$ norm. In fact, in **[163]**, only $d \geq 3$ is dealt with and for $d = 2$, (86.1) was proved in **[162]**. Timofeev **[163]** also proved (86.1) in the case when $\|\cdot\|$ is $L_1(R^d)$ norm.

87. J. E. Pečarić **[164]** gave some remarks about results from **[145]**.

88. The inequality (78.1) is considered in **[165]** by C. Bennewitz and a special case of (78.1), i.e.

$$\left(\int_0^\infty \left(|f'(x)|^2 - x|f(x)|^2\right) dx\right)^2 \leq K \int_0^\infty |f(x)|^2\, dx \int_0^\infty |f''(x) + xf(x)|^2\, dx \tag{88.1}$$

in **[166]** (the integral on the left may be only conditionally convergent). In fact, W. D. Evans and W. N. Everitt **[166]** gave a complete analysis of the integral inequality (88.1) for $f \in \Delta$, where $\Delta = \{f\colon [0,+\infty) \to C \mid f'$ locally absolutely continuous on $[0,+\infty)$ and $f, f'' + xf \in L^2(0,+\infty)\}$.

A nonweighted case of (78.1), i.e. the inequality (38.2) is considered in C. Bennewitz **[167]**.

B. G. Pachpatte **[168]** proved the following theorem:

Let p, q be real valued continuous functions defined on $I = [a,b]$; p' exists and is continuous on I. Let f, g be real valued continuous functions defined on I, each twice continuously differentiable on I and $f(a) = f(b) = g(a) = g(b) = 0$. Then

$$\left(\int_a^b [p(t)f'(t)g'(t) + q(t)f(t)g(t)]\, dt\right)^2 \leq \leq \frac{1}{2}\left[\left(\int_a^b M[f(t)]^2 dt\right)\left(\int_a^b g^2(t) dt\right) + \left(\int_a^b M[g(t)]^2 dt\right)\left(\int_a^b f^2(t) dt\right)\right], \tag{88.2}$$

where $M(f)$ is defined by (38.1). Equality holds in (88.2) if either f is null or g is null on I.

By the same method the following stronger inequality is proved by D. S. Mitrinović and J. E. Pečarić [**169**]

$$\left(\int_a^b (pf'g' + qfg)\right)^2 \leq \left(\left(\int_a^b M(f)^2\right)\left(\int_a^b g^2\right)\left(\int_a^b M(g)^2\right)\left(\int_a^b f^2\right)\right)^{1/2}.$$

They also prove a discrete analogue.

89. Inequalities (69.1)–(69.4) were considered by R. C. Brown and D. B. Hinton in [**170**]. In the proofs the following result was used:

Let $J = [a, b] \subset I$ be a compact interval. Then for $y \in \mathcal{A}_n$, where $\mathcal{A}_n = \{y\colon y^{(n-1)} \in AC_{\text{loc}}(I)\}$, $t \in J$, and $j = 0, \ldots, n-1$,

$$|y^{(j)}(t)| \leq A_{nj} L^{-j-1} \int_J |y| + L^{n-j-1} \int_J |y^{(n)}|,$$

where $L = b - a$, $A_{n,n-1} = (n-1)!(2n-1)^n$ and the other A_{nj} are given inductively by $A_{n+1,j} = A_{nj} + A_{n+1,n}$, for $j = 1, \ldots, n-1$.

Some further considerations are given by the same authors in [**171**],[**172**].

Weighted multidimensional inequalities of these types were considered by L. Caffarelli, R. Kohn and L. Nirenberg [**173**], C. S. Lin [**174**], and R. C. Brown and D. B. Hinton [**175**].

90. A. I. Zvyagintsev [**176**], [**177**] gave extensions of results from [**148**] (see 77.). He also proved inequalities of type (1.1) for $n = 4$, $p = +\infty$, and $I = [0, T]$. He got the following constants in (1.1).

$$c_{41} = 2^{5/2} 3^{-1/4}, \quad c_{42} = 3^{-1/2} \cdot 10, \quad c_{43} = 2^{3/2} 3^{1/4} \tag{90.1}$$

in the case when $T^4 M_0 \left(+\infty[0,T]\right)^{-1} M_4 \left(+\infty, [0,T]\right) \geq 48 \left(17 + 12\sqrt{2}\right)$.

91. A generalization of Timofeev's result from 86. for the case of Kolmogorov-type inequality

$$\left\| \frac{\partial}{\partial \varepsilon_1} \cdots \frac{\partial}{\partial \varepsilon_k} f \right\| \leq c(n,k) \|f\|^{1-(k/2n)} \|\Delta^n f\|^{k/2n}$$

where $\partial/\partial\varepsilon_i$ is the derivative in the e_i direction, and the nth iterate of the Laplacian $\Delta^n f$ belongs to $L_\infty(R^d)$, $0 < k, 2n$; is proved by Z. Ditzian in [**179**]. This is an improvement of a result of K. N. Boyadzhiev [**178**]. A generalization for Banach space of function on R^d is also given in [**179**].

92. Expositions of Kolmogorov-type inequalities are given by V. M. Tihomirov in [**180**],[**181**], but not as comprehensive as the present exposition.

93. Various extensions of Fink's result from [**145**] (see 74.) are given by D. S. Mitrinović and J. E. Pečarić [**182**],[**183**],[**184**]. For example, the following result is valid:

Let $i, j, am, bm, an, bn \in N_0 = \{0, 1, 2, \dots\}$, where n and m are positive numbers such that $n \le m$. If f is a two-place completely monotone function on $[0, +\infty)^2$, i.e. let its partial derivatives of all orders exist and satisfy $(-1)^{n+m} D_1^n D_2^m (f) \ge 0$, $n, m = 0, 1, 2, \dots$, then

$$\left((-1)^{i+j+m(a+b)} D_1^{i+am} D_2^{j+bm} f(x,y)\right)^{\frac{1}{m}} \left((-1)^{i+j} D_1^i D_2^j f(x,y)\right)^{\frac{1}{n}-\frac{1}{m}}$$
$$\ge \left((-1)^{i+j+n(a+b)} D_1^{i+an} D_2^{j+bn} f(x,y)\right)^{\frac{1}{n}}.$$

94. The following results are obtained by H. Kraljević and J. E. Pečarić [**185**]

94.1. Semigroups.

In this section $T(t)$ is a strongly continuous semigroup on X which is uniformly bounded:

(94.1) $$\|T(t)\| \le M < +\infty, \quad t \ge 0$$

Let A be its infinitesimal generator.

LEMMA 1. *For every $t, s \in \underline{R}$, $0 < t < s$, and for every $x \in D(A^3)$ the following inequalities hold*

(94.2) $$\|Ax\| \le \left[\frac{M(t^2+s^2)}{ts(s-t)} + \frac{t+s}{ts}\right] \|x\| + \frac{Mts}{6}\|A^3x\|,$$

(94.3) $$\|A^2x\| \le 2\left[\frac{M(t+s)}{ts(s-t)} + \frac{1}{ts}\right] \|x\| + \frac{M(t+s)}{3}\|A^3x\|.$$

PROOF: By iterating the formula

(94.4) $$T(t)x = x + \int_0^t T(u)Ax\, du$$

we obtain for $x \in D(A^3)$:

(94.5) $$T(t)x = x + tAx + \frac{t^2}{2}A^2x + \frac{1}{2}\int_0^t (t-u)^2 T(u)A^3x\, du$$

Thus

$$(94.7)\qquad \begin{aligned} tAx + \frac{t^2}{2}A^2x &= T(x)x - x - \frac{1}{2}\int_0^t (t-u)^2 T(u)A^3x\,du,\\ sAx + \frac{s^2}{2} &= T(s)x - x - \frac{1}{2}\int_0^s (s-u)^2 T(u)A^3x\,du. \end{aligned}$$

By solving this system we obtain

$$(94.8)\qquad \begin{aligned} Ax = \frac{s^2T(t)x - t^2T(s)x}{ts(s-t)} - \frac{s+t}{ts}x\\ + \int_0^s K(t,s;u)T(u)A^3x\,du, \end{aligned}$$

$$(94.9)\qquad A^2x = \frac{tT(s)x - sT(t)x}{ts(s-t)} + \frac{2x}{ts} - \int_0^s L(t,s;u)(u)A^3x\,du,$$

where

$$K(t,s;u) = \begin{cases} \frac{t(s-u)^2}{2s(s-t)} - \frac{s(t-u)^2}{2t(s-t)} & 0 \le u \le t,\\ \frac{t(s-u)^2}{2s(s-t)} & t \le u \le s, \end{cases}$$

$$L(t,s;u) = \begin{cases} \frac{(s-u)^2}{s(s-t)} - \frac{(t-u)^2}{t(s-t)} & 0 \le u \le t\\ \frac{(s-u)^2}{s(s-t)} & t \le u \le s. \end{cases}$$

Note that $(K(t,s;u) \ge 0$ and $L(t,s;u) \ge 0$ for every $u \in [0,s]$ $(0 < t < s)$. Since

$$\int_0^s L(t,s;u)du = \frac{ts}{6}, \quad \int_0^s L(t,s;u)du = \frac{t+s}{3},$$

we obtain (94.2) and (94.3) immediately from (94.8) and (94.9) by the triangle inequality and using (94.1).

Now, we shall minimize the right-hand sides of the inequaliteis (94.2) and (94.3) over the domain $0 < t < s$. Fix $x \in D(A^3)$ and put $s = \alpha t$, $\alpha > 1$, $t > 0$. We obtain from (94.2) and (94.3)

$$(94.10)\qquad \|Ax\| \le \frac{a(\alpha)}{t} + b(\alpha)t^2, \qquad \|A^2x\| \le \frac{c(\alpha)}{t^2} + d(\alpha)t,$$

where

$$\text{(94.11)}\qquad \begin{aligned} a(\alpha) &= \left[\frac{M(1+\alpha^2)}{\alpha(\alpha-1)} + \frac{1+\alpha}{\alpha}\right]\|x\|, & b(\alpha) &= \frac{M\alpha}{6}\|A^3x\|,\\ c(\alpha) &= 2\left[\frac{M(1+\alpha)}{\alpha(\alpha-1)} + \frac{1}{\alpha}\right]\|x\|, & d(\alpha) &= \frac{M(1+\alpha)}{3}\|A^3x\|. \end{aligned}$$

For fixed $\alpha > 1$ the minimal values of the right hand side of the first (resp. second) inequality in (94.10) is attained at $t = 2^{-1/3}a(\alpha)^{1/3}b(\alpha)^{-1/3}$ (resp. $t = 2^{1/3}c(\alpha)^{1/3}d(\alpha)^{-1/3}$). Taking this minimal value we conclude that the inequalities

$$\text{(94.12)}\qquad \|Ax\|^3 \le \frac{9}{8}M\alpha\left[M\frac{1+\alpha^2}{\alpha(\alpha-1)} + \frac{1+\alpha}{\alpha}\right]^2\|x\|^2\|A^3x\|,$$

$$\text{(94.13)}\qquad \|A^2x\|^3 \le \frac{3}{2}M^2(1+\alpha)^2\left[M\frac{1+\alpha}{\alpha(\alpha-1)} + \frac{1}{\alpha}\right]\|x\|\,\|A^3x\|^2,$$

hold true for every $\alpha > 1$.

By differentiating the right hand sides of (94.12) and (94.13) we find that the minimal values of both functions are attained at some root of the cubic polynomial

$$F_M(\alpha) = (\alpha-1)^3 - \frac{6M}{M+1}(\alpha-1) - \frac{4M}{M+1}.$$

Now $F_M(1) < 0$ and $F_M'(1) < 0$. Furthermore, $F_M'(\alpha)$ has exactly one zero larger than 1. From this we see that F_M has exactly one root which is greater than 1. This root α can be written down explicitly by Cardano's formula and it is given by

$$\text{(94.14)}\qquad \alpha = 1 + 2\sqrt{\frac{2M}{M+1}}\cos\frac{\varphi}{3}, \quad \tan\varphi = \sqrt{\frac{M-1}{M+1}}, \quad \varphi \in \left[0, \frac{\pi}{4}\right).$$

Thus, we have proved:

THEOREM 1. *Let A be the infinitesimal generator of a uniformly bounded strongly continuous semigroup $T(t)$ $(t \geq 0)$ of linear operators on a Banach space. If $M \geq \|T(t)\|$ $(t \geq 0)$ and if $x \in D(A^3)$, the following inequalities hold:*

(94.15)
$$\|Ax\|^3 \leq \frac{9}{8} M\alpha \left[M \frac{1+\alpha^2}{\alpha(\alpha-1)} + \frac{1+\alpha}{\alpha} \right]^2 \|x\|^2 \|A^3 x\|,$$

(94.16)
$$\|A^2 x\|^3 \leq \frac{3}{2} M^2 (1+\alpha)^2 \left[M \frac{1+\alpha}{\alpha(\alpha-1)} + \frac{1}{\alpha} \right] \|x\| \cdot \|A^3 x\|^2,$$

where α is given by (94.10).

REMARK: By considering the matrix of the second partial derivatives of f and g it can be shown that $(t_1, \alpha t_1)$ and $(t_2, \alpha t_2)$ are actually the minimum points, respectively.

Specializing Theorem 1 for $M = 1$, and thus $\alpha = 3$, we obtain:

COROLLARY 1. *Let A be the infinitesimal generator of a strongly continuous contraction semigroup $T(t)$ $(\|T(t)\| \leq 1)$. For every $x \in D(A^3)$ the following inequalities hold true:*

$$\|Ax\|^3 \leq \frac{243}{8} \|x\|^2 \cdot \|A^3 x\|, \tag{94.17}$$

$$\|A^2 x\|^3 \leq 24 \|x\| \cdot \|A^3 x\|^2. \tag{94.18}$$

94.2. Groups.

Let $T(t)$ $(t \in R)$ be a uniformly bounded strongly continuous group of linear operators on X:

$$\|T(t)\| \leq M, \quad t \in R,$$

and let A be its infinitesimal generator. Then we have for $t > 0$:

(94.19)
$$T(t)x = x + tAx + \frac{t^2}{2} Ax + \frac{1}{2} \int_0^t (t-u)^2 T(u) A^3 x \, du, \quad \text{and}$$

(94.20)
$$T(-t)x = x - tAx + \frac{t^2}{2} Ax = \frac{1}{2} \int_0^t (t-u)^2 T(-u) A^3 x \, du.$$

Similar calculation as in Subsection 94.1 leads to:

LEMMA 2. *For every $x \in D(A^3)$ and every $t > 0$ the following inequalities hold:*

$$\|Ax\| \le \frac{M}{t}\|x\| + \frac{Mt^2}{6}\|A^3x\|, \quad \text{and} \tag{94.21}$$

$$\|A^2x\| \le \frac{2(M+1)}{t^2}\|x\| + \frac{Mt}{3}\|A^3x\|. \tag{94.22}$$

By minimizing the right-hand sides of (94.21) and (94.22) we obtain:

THEOREM 2. *Let A be the infinitesimal generator of a uniformly bounded group $T(t)$ $(t \in R)$ of linear operators $(\|T(t)\| \le M,\ t \in R)$. For every $x \in D(A^3)$ the following inequalities hold:*

$$\|Ax\|^3 \le \frac{9}{8}M^3\|x\|^2\|A^3x\|, \tag{94.23}$$

$$\|A^2x\|^3 \le \frac{3}{2}M^2(M+1)\|x\| \cdot \|A^3x\|^2. \tag{94.24}$$

COROLLARY 2. *Let A be the infinitesimal generator of a contraction group $T(t)$ $(\|T(t)\| \le 1,\ t \in R)$. Then for every $x \in D(A^3)$,*

$$\|Ax\|^3 \le \frac{9}{8}\|x\|^2\|A^3x\|, \quad \text{and} \tag{94.25}$$

$$\|A^2x\|^3 \le 3\|x\| \cdot \|A^3x\|^2. \tag{94.26}$$

REMARK: Using formulas (94.7) and (94.8) for any $t, s \in R$ (instead of choosing $s = -t$) and minimizing over the domain $t \ne s$ leads to the same inequalities.

94.3. Cosine Functions.

In this section $T(t)$ $(t \ge 0)$ will denote a uniformly bounded $(\|T(t)\| \le M)$ strongly continuous cosine function of linear operators on X and A its infinitesimal generator. Now, we have

$$T(t)x = x + \int_0^t (t-u)T(u)Ax\,du,$$

and by iterating this equality we obtain

$$T(t)x = x + \frac{t^2}{2}Ax + \frac{t^4}{24}A^2x + \frac{1}{120}\int_0^t (t-u)^5 C(u)A^3x\,du. \tag{94.27}$$

The similar calculation as in Subsection 94.1 gives:

LEMMA 3. *For every $x \in D(A^3)$ and for $0 < t < s$ the following inequalities hold*

(94.28)
$$\|Ax\| \le 2\left[M\frac{t^4+s^4}{t^2s^2(s^2-t^2)}+\frac{t^2+s^2}{t^2s^2}\right]\|x\| + \frac{Mt^2s^2}{360}\|A^3x\|, \quad \text{and}$$

(94.29)
$$\|A^2x\| \le 24\left[M\frac{t^2+s^2}{t^2s^2(s^2-t^2)}+\frac{1}{t^2s^2}\right]\|x\| + \frac{M(t^2+s^2)}{30}\|A^3x\|.$$

Note, that these inequalities can be obtained from (94.2) and (94.3) by substitutions $t \to t^2$, $s \to s^2$, $\|Ax\| \to \frac{1}{2}\|Ax\|$, $\|A^2x\| \to \frac{1}{12}\|A^2x\|$, $\|A^3x\| \to \frac{1}{120}\|A^3x\|$. Thus, we get immediately from Theorem 1 and its corollaries:

THEOREM 3. *Let A be the infinitesimal generator of a strongly continuous cosine function $T(t)$ ($\|T(t)\| \le M$, $t \ge 0$) on a Banach space X. For every $x \in D(A^3)$ we have*

(94.30)
$$\|Ax\|^3 \le \frac{3M\alpha}{40}\left[M\frac{1+\alpha^2}{\alpha(\alpha-1)}+\frac{1+\alpha}{\alpha}\right]^2\|x\|^2\|A^3x\|,$$

(94.31)
$$\|A^2x\|^3 \le \frac{9}{50}M^2(1+\alpha)\left[M\frac{1+\alpha}{\alpha(\alpha-1)}+\frac{1}{\alpha}\right]\|x\|\cdot\|A^3x\|^2,$$

where α is given by (94.10).

COROLLARY 3. *If A is the infinitesimal generator of a contraction cosine function ($\|T(t)\| \le 1$, $t \ge 0$), then for every $x \in D(A^3)$:*

$$\|Ax\|^3 \le \frac{81}{40}\|x\|^2\|A^3x\|, \tag{94.32}$$

$$\|A^2x\|^3 \le \frac{72}{25}\|x\|\cdot\|A^3x\|^2. \tag{94.33}$$

Kraljević and Pečarić [**186**] also gave some similar results related to Ostrowski's inequality (11.1).

95. D. S. Mitrinović, J. E. Pečarić and H. Kraljević [**187**] proved the following generalizations of results from [**68**] (see 35.).

THEOREM 1. *Let D be a convex set such that $a, b \in \overline{D}$ (the closure of D) and let $F\colon X \to Y$ be twice Frechet-differentiable on $\overline{D}$ and satisfy the conditions*

$$\|F(x)\| \le M \qquad \text{on } \overline{D} \tag{95.1}$$

and

$$\|F''_{(a)}(h,h)\| \le H(h) \quad \text{for all } h \in X \text{ and all } a \in D, \tag{95.2}$$

where H is a function from X to R^+. Then for all $x \in \overline{D}$

$$\|F'_{(x)}(b-a)\| \le 2M + \frac{1}{2}\left(H(a-x) + H(b-x)\right). \tag{95.3}$$

PROOF: Let $x \in \overline{D}$ and $x + th \in D$ $(0 < t < 1)$. Taylor's formula, namely

$$F(x+h) = F(x) + F'_{(x)}(h) + W(x,h) \qquad \left(W(x,h) = \frac{1}{2}F''_{(x+th)}(h,h)\right),$$

for $h = a - x$ and $h = b - x$, becomes in turn

$$F(a) = F(x) + F'_{(x)}(a-x) + W(x,a-x),$$
$$F(b) = F(x) + F'_{(x)}(b-x) + W(x,b-x).$$

From these equations it follows that

$$F'_{(x)}(b-a) = F(b) - F(a) + W(x,a-x) - W(x,b-x). \tag{95.4}$$

From (95.2) it immediately follows that

$$\|F'_{(x)}(b-a)\| \le \|F(b)\| + \|F(a)\| + \|W(x,a-x)\| + \|W(x,b-x)\|,$$

hence, using (95.1), we obtain (95.3).

REMARK: Theorem 1 holds if the first condition in (95.1) is substituted by the weaker condition $\|F(b) - F(a)\| \le 2M$.

REMARK: For $H(h) = N\|h\|^2$ we get Theorem 1 from [**68**], i.e. the first inequality in this theorem.

By using (95.4) we can also obtain the following theorem:

THEOREM 2. *Let in Theorem 1 only the second condition (95.2) be valid. Then*

$$\|F'_{(x)}(b-a) - F(b) + F(a)\| \le \frac{1}{2}\left(H(a-x) + H(b-x)\right).$$

Let $H(-h) = H(h)$. Of special interest is the case when $a = x - h$, $b = x + h$. Then (95.3) becomes

$$\|F'_{(x)}(2h)\| \le 2M + H(h). \tag{95.5}$$

COROLLARY 1. *Let* $X = Y = R$, $F(x) = \frac{1}{h}\int_x^{x+h} f(t)\,dt$, $H(h) = Nh^2$, $a < b$, $h > 0$. *Then we have for a differentiable function* $f(x)$ *on* $[a, b+h]$ *such that* $|f(x)| \le M$, $x \in [a, b+h]$ *and* $|\Delta_h f'(x)| \le N$, $x \in (a,b)$, *where* $\Delta_h g(x) = \frac{1}{h}(g(x+h) - g(x))$,

$$(95.6) \qquad (b-a)|\Delta_h f(x)| \le 2M + \frac{N}{2}\left((x-a)^2 + (b-x)^2\right).$$

This is a generalization of Theorem 6 from **[1]**. Namely from (95.6) we get

$$(95.7) \qquad \|\Delta_h f(x)\| \le \frac{2M}{b-a} + \frac{b-a}{2}N,$$

and if $b - a \ge 2\sqrt{M/N}$, we have

$$(95.8) \qquad \|\Delta_h f(x)| \le 2\sqrt{MN},$$

since the function $y = 2Mx^{-1} + \frac{N}{2}x$ has a minimum $2\sqrt{MN}$ for $x = 2\sqrt{M/N}$.

But, from (95.6) we can also obtain the following result. Let $f(x)$ be a differentiable function on R such that $|f(x)| \le M$ and $|\Delta_h f'(x)| \le N$, $x \in R$, $h > 0$. Then (95.5) becomes (i.e. (95.6) for $a \to x - H$, $b \to x + H$),

$$(95.9) \qquad |\Delta_h f(x)| \le \frac{M}{H} + \frac{HN}{2}.$$

The function $y = Mx^{-1} + \frac{1}{2}Nx$ has a minimum $\sqrt{2MN}$ for $x = \sqrt{2M/N}$, so for $H \ge \sqrt{2M/N}$ we get from (95.9)

$$(95.10) \qquad |\Delta_h f(x)| \le \sqrt{2MN}.$$

COROLLARY 2. *For* $X = R^n$, $Y = R$, $F(x) = f(x) = f(x_1, \ldots, x_n)$, *let*

$$(95.11) \qquad \left|\frac{\partial^2 f}{\partial x_i \partial x_j}\right| \le N_{ij}, \; x \in D = \{x = (x_1, \ldots, x_n) \,|\, s_i < x_i < b_i\},$$

$$\|x - y\| = \sum_{i=1}^{n} |x_i - y_i| \quad (x, y \in X),$$

and

$$\|F''_{(x)}(h,h)\| = \left|\sum_{i,j} \frac{\partial^2 f}{\partial x_i \partial x_j} h_i h_j\right| \le \sum_{i,j} \left|\frac{\partial^2 f}{\partial x_i \partial x_j}\right| \cdot |h_i| \cdot |h_j|$$
$$\le \sum_{i,j} N_{ij} |h_i| \cdot |h_j|.$$

Therefore, we have ($a = (a_1, \ldots, a_n)$, $b = (b_1, \ldots, b_n)$),

(95.12)
$$\begin{aligned}&\left|\sum_{i=1}^{n}(b_i - a_i)\frac{\partial f}{\partial x_i} - f(a) + f(b)\right| \\ &\qquad \leq \frac{1}{2}\sum_{i,j} N_{ij}\left((x_i - a_i)(x_j - a_j) + (b_i - x_i)(b_j - x_j)\right)\end{aligned}$$

and, if $|f(x)| \leq M$ ($\forall x \in \overline{D}$),

(95.13)
$$\begin{aligned}&\left|\sum_{i=1}^{n}(b_i - s_i)\frac{\partial f}{\partial x_i}\right| \\ &\leq 2M + \frac{1}{2}\sum_{i,j} N_{ij}\left((x_i - a_i)(x_j - a_j) + (b_i - x_i)(b_j - x_j)\right)\end{aligned}$$

By using the following inequality for nonnegative numbers a, b, c, d,

$$ab + cd \leq (a + c)(b + d),$$

(95.12) and (95.13) give, respectively,

(95.14)
$$\left|\sum_{i=1}^{n}(b_i - a_i)\frac{\partial f}{\partial x_i} - f(a) + f(b)\right| \leq \frac{1}{2}\sum_{i,j} N_{ij}(b_i - a_i)(b_j - a_j)$$

(95.15)
$$\left|\sum_{i=1}^{n}(b_i - a_i)\frac{\partial f}{\partial x_i}\right| \leq 2M + \frac{1}{2}\sum_{i,j} N_{ij}(b_i - a_i)(b_j - a_j).$$

If $b_1 - a_1 = \cdots = b_n - a_n = h$ (95.15) becomes

(95.16)
$$\left|\sum_{i=1}^{n}\frac{\partial f}{\partial x_i}\right| \leq \frac{2M}{h} + \frac{h}{2}N,$$

where $N = \sum_{i,j} N_{ij}$.

Therefore, if $\min(b_i - a_i) \geq 2\sqrt{M/n}$, then

(95.17)
$$\left|\sum_{i=1}^{n}\frac{\partial f}{\partial x_i}\right| \leq 2\sqrt{MN}.$$

On the other hand, for $b_i = x_i + h_i$, $a_i = x_i - h_i$ $(h_i > 0)$, (95.15) becomes

$$\left|\sum_{i=1}^{n} h_i \frac{\partial f}{\partial x_i}\right| \leq M + \frac{1}{2}\sum_{i,j} N_{ij} h_i h_j, \tag{95.18}$$

and for $h_1 = \cdots = h_n = h$,

$$\left|\sum_{i=1}^{n} \frac{\partial f}{\partial x_i}\right| \leq \frac{M}{h} + \frac{hM}{2}. \tag{95.19}$$

If $D = R^n$, inequality (95.19) gives

$$\left|\sum_{i=1}^{n} \frac{\partial f}{\partial x_i}\right| \leq \sqrt{2MN}, \tag{95.20}$$

where N is given as above.

Using (95.16) and (94.17) we can formulate the following result

$$\left|\sum_{i=1}^{n} \frac{\partial f}{\partial x_i}\right| \leq \begin{cases} \frac{2M}{h} + \frac{hN}{2}, & \text{if } h < 2\sqrt{M/n}, \\ 2\sqrt{MN}, & \text{if } h \geq 2\sqrt{M/n}, \end{cases} \tag{95.21}$$

where $h = \min(b_i - a_i)$.

A simple consequence of (95.21) is the following generalization of a result from [**189**] (see 45., i.e. inequality (45.1)):

Let $X = [0,1]^n$, $|f(x)| \leq 1$, and let (95.11) be valid with $N = \sum_{i,j} N_{ij}$. Then

$$\left|\sum_{i=1}^{n} \frac{\partial f}{\partial x_i}\right| \leq \begin{cases} \frac{4+N}{2}, & \text{if } 0 < N \leq 4 \\ 2\sqrt{N}, & \text{if } N > 4. \end{cases} \tag{95.22}$$

Note that in the case when $h_1 = \cdots = h_n = h$, we have

$$\|F''_{(x)}(h,h)\| = h^2 \left|\sum_{i,j} \frac{\partial^2 f}{\partial x_i \partial x_j}\right| \leq h^2 A,$$

if

$$\left|\sum_{i,j} \frac{\partial^2 f}{\partial x_i \partial x_j}\right| \leq A.$$

In this case, instead of (95.18) we can get

$$\left|\sum_{i=1}^{n} \frac{\partial f}{\partial x_i}\right| \leq \frac{M}{h} + \frac{hA}{2}, \tag{95.23}$$

and instead of (95.20), we have

$$\left|\sum_{i=1}^{n} \frac{\partial f}{\partial x_i}\right| \leq \sqrt{2MA}. \tag{95.24}$$

It is a generalization of (7.2).

By using (95.23) and (95.24) we can obtain the following generalization of (45.2):

Let $X = [0,1]^n$, $|f(x)| \leq 1$ $(x \in X)$ and $\left|\sum\limits_{i,j} \frac{\partial^2 f}{\partial x_i \partial x_j}\right| \leq A$ $(x \in X)$. Then

$$\left|\sum_{i=1}^{n} \frac{\partial f}{\partial x_i}\left(\frac{1}{2}, \dots, \frac{1}{2}\right)\right| \leq \begin{cases} 2 + \frac{A}{4}, & \text{if } 0 < A \leq 8, \\ \sqrt{2A}, & \text{if } A > 8. \end{cases} \tag{95.25}$$

If instead $|f(x)| \leq M$ we have the condition

$$0 \leq f(x) \leq M, \tag{95.26}$$

we can obtain better bounds. This is a consequence of the first remark after Theorem 1, and these results can be obtained from previous by substitutions $M \to M/2$. For example the following generalizations of results of V. M. Olovyanišnikov (see 23.) are valid:

$$\left|\sum_{i=1}^{n} \frac{\partial f}{\partial x_i}\right| \leq \sqrt{2MN}, \quad \text{and} \tag{95.17'}$$

$$\left|\sum_{i=1}^{n} \frac{\partial f}{\partial x_i}\right| \leq \sqrt{MA}, \tag{95.24'}$$

where the other conditions for (95.17) and (95.24) are fulfilled, respectively.

96. G. Talenti [**190**] considered inequality

$$\left(\int_0^\infty x^{-1/2} u \, dx\right)^2 \leq 4G\left(\int_0^\infty u^2 \, dx + 2\left(\int_0^\infty x u^2 \, dx \cdot \int_0^\infty x u'^2 \, dx\right)^{1/2}\right),$$

and W. N. Everitt and A. Zettl [**191**] studied

$$\left(\int_0^\infty x^b f'^2(x)\,dx\right)^2 \le N(a,b)\int_0^\infty x^a f(x)^2\,dx \int_0^\infty x^{-a}\left(\left(x^b f'(x)\right)'\right)^2\,dx.$$

97. The Landau-Kolmogorov inequality theory has been generalized to Hermitian and dissipative operators in a Banach space. See for example, B. Bollobás [**193**], B. Bollobás and E. Galanis [**194**], J. R. Partington [**195**] M. Protter [**196**], P. Chernoff [**197**], and B. Bollobás and J. R. Partington [**198**].

98. T. Kato and I. Satake [**199**] developed an algebraic theory of Landau-Kolmogorov inequalities.

99. Zvyaginstev [**200**] continues his investigation from [**176**], [**177**] of Landau-Kolmogorov inequalities.

100. Landau-Kolmogorov type inequalities for functions on R^n to a Banach space are studied in Redheffer and Walter [**201**]. The domains of the functions must be restricted in order to get these types of inequalities.

101. If f is only defined on $[a,b]$ then one can get inequalities of the Landau type $\|f'\|^2 \le K\|f\|\|f''\|$ under further boundary conditions. For example, in Gillman, Kaper, and Kwong [**202**], the boundary condition that f' has a zero in $[a,b]$ leads to a precise calculation of the best constant.

102. R. C. Brown and D. B. Hinton [**203**] study sufficient conditions for an inequality of the form

(102.1)

$$\left(\int_I |x^\beta f^{(j)}(x)|^p dx\right)^{1/p} \le K\left(\int_I |x^\gamma f(x)|^q dx\right)^{\frac{1-\lambda}{q}}\left(\int_I |x^\alpha f^{(n)}(x)|^r dx\right)^{\lambda/r},$$

where $1 \le r,p,q \le \infty$, $0 \le j \le n-1$, and $f^{(n-1)}$ is absolutely continuous. If the right hand side is finite, one seeks the existence of a constant K. If $\lambda = n = 1$, then this is a weighted Hardy inequality, see Chapter IV, Section 45. For $\lambda \le 1$ and $n > 1$ they offer the following result.

THEOREM 1. *Define* p_0 *by* $\frac{n}{p_0} = \frac{n-j}{q} + \frac{j}{r}$, λ_0 *by* $\lambda_0 = \frac{j+q^{-1}-p^{-1}}{n-r^{-1}+q^{-1}}$ ($\lambda_0 = 0$ *if* $n = r = 1$ *and* $q = \infty$), *and* $J = \lambda(\alpha + r^{-1} - n + j) + ((-\lambda)(\gamma + q^{-1} + j))$.

Then assume $p \geq p_0$ *and* $\lambda_0 \leq \lambda \leq 1$. *Then an inequality of the form* (102.1) *holds if*

(i) $I = R^+$, $\alpha + r^{-1}$ *is not an integer less than* $n - j$, $p \geq r$, *and* $\beta + p^{-1} = J$;
or
(ii) $I = [b, \infty)$, $\alpha + r^{-1}$ *is not an integer less than* $n - j$, $p \geq r$, *and* $\beta + p^{-1} \leq J$;
or
(iii) $I = [b, \infty)$, $\{\alpha + r^{-1}$ *is an integer less than* $n - j$ *or* $p < r\}$, *and* $\beta + p^{-1} < J$, *or*
(iv) $I = (0, b]$, $\alpha + r^{-1}$ *is not an integer less than* $n - j$, $p \geq r$, *and* $\beta + r^{-1} \geq J$;
or
(v) $I = (0, b]$, $\{\alpha + r^{-1}$ *is an integer less than* $n - j$ *or* $p < r\}$, *and* $\beta + p^{-1} > J$.

103. Existence of inequalities of the form

$$\sum_j \int_G |P_j f|^p dx + \int_G |f|^p dx \geq c \int_G |D^m f|^p dx$$

for $p > 1$ and certain differential operators P_j is given by K. T. Smith [**204**]. Specifically $\{P_i\}$ are differential operators with continuous coefficients on $\overline{G}$, G a domain in R^n whose characteristic polynomials have no common zeros and $f \in C^m(G)$.

104. D. Huang and J. Wang, [**205**], consider Landau-Kolmogorov inequalities for nth order splines f and operators $L = D \prod_{i=1}^{m} (D^2 - a_i^2)$. They show that if $\|Lf\|_\infty \leq 1$ and $\|f\|_\infty \leq 1$ then $\|f^{(j)}\|_\infty \leq \gamma_{n,j}$, $j = 1, \ldots, n-1$.

105. E. Neuman and J. Pečarić [**206**] gave a generalization of the Zwick-Farwig paper [**152**], see Section 79.

106. C. D. Pagani in a private communication to D. S. Mitrinovic in 1972, noted that in his paper "On the equation $\operatorname{sgn}(x)|x|^p u_y - u_{xx} = 0$ and a related one" he makes use of the following two inequalities:

$$|v(0)| \le c_p \left(\int_0^\infty x^p |v|^2 dx\right)^{\frac{p+3}{4(p+2)}} \left(\int_0^\infty x^{-p} |v''|^2 dx\right)^{\frac{p+1}{4(p+2)}}$$

$$|v'(0)| \le c'_p \left(\int_0^\infty x^p |v|^2 dx\right)^{\frac{p+1}{4(p+2)}} \left(\int_0^\infty x^{-p} |v''|^2 dx\right)^{\frac{p+3}{4(p+2)}}$$

where $p > -1$, v is a complex valued function belonging to $C^2((0,\infty))$ and such that $x^{p/2}v$ and $x^{-p/2}v''$ belong to $L^2(0,\infty)$. The best possible constants c_p and c'_p are calculated and the functions for which the equality holds are given.

In studying problems related to the equation cited in the title of the paper, he has also found the following inequality:

$$\left(\int_0^\infty x^{-1/2} v'(x) dx\right)^4 \le c \int_0^\infty x^p |v|^2 dx \int_0^\infty x^{-p} |v''|^2 dx.$$

107. Additional references to Landau-Kolmogorov inequalities are by Žuk [**207**]–[**210**], Nikol'skiĭ [**211**], Ligun [**212**], Hoa [**213**], Zwyaginsev [**214**], [**215**], Kraljević and Pečarič [**216**], and Kudrjavcev [**217**].

References

1. LANDAU, E., *Einige Ungleichungen für zweimal differentierbaren Funktionen*, Proc. London Math. Soc. Ser 2 **13** (1913), 43–49.
2. HADAMARD, J., *Sur le module maximum d'une fonction et de ses dérivées*, Comptes Rendus des séances de la Société mathematique de France (1914), 68–72.
3. HARDY, G. H. and J. E. LITTLEWOOD, Proc. Lond. Math. **2** (1913), 411–478.
4. LANDAU, E., *Die Ungleichungen für zweimal differertierbare Funktionen*, Meddelelser Kobenhavn 6, Nr. 10, 49 pp., 1925.
5. MORDELL, L. J., *The magnitude of the derivative of a function*, J. London Math. Soc. **3** (1928), 119–121.
6. NEDER, L., *Abschätzungen für die Ableitungen einer reellen Funktion eines reellen Arguments*, Math. Z. **31** (1930), 356–365.
7. LANDAU, E., *Über einen Satz von Herrn Esclangon*, Math. Ann. **102** (1929), 177–188.

8. HARDY, G. H. and J. E. LITTLEWOOD, *Some integral inequalities connected with the calculus of variations*, Quart. J. Math. (Oxford) **3** (1932), 242–252.
9. HARDY, G. H., J. E. LITTLEWOOD and G. PÓLYA, "Inequalities." Cambridge, 1934.
10. HARDY, G. H., E. LANDAU and J. E. LITTLEWOOD, *Some inequalities satisfied by the integrals or derivatives or analytic function*, Math. Z. **39** (1935), 677–695.
11. TCHENG, Tchou-Yun, "Sur les inégalités différentielles." Paris, 1934 , 41 pp.
12. ŠILOV, G. E., (Yu. G. Bossè), *On inequalities between derivatives* (Russian). Sb. Rabot Student. Naučnyh Kružkov Moskov. Gos. Univ. (1937), 17–27.
13. GORNY, A., *Sur les fonctions indéfiniment dérivables sur tout l'axe réel*, C. R. Acad. Sci. Paris **206** (1938), 733–735.
14. __________, *Sur les fonctions indéfiniment dérivables*, C. R. Acad. Sci. Paris **206** (1938), 1872–1874.
15. __________, *Sur les maxima des modules d'une fonction et de ses dérivées*, C. R. Acad. Sci. Paris **206** (1938), 1245–1247.
16. __________, *Contribution à l'étude des fonctions dérivables d'une variable réelle*, Acta Math. **71** (1939), 317–358.
17. CARTAN, H., Sur Les classes de fonctions définies par des inégalités portant sur leurs dérivées successives. Paris, 1940, 33 pages.
18. KOLMOGOROV, A., *Une généralisation de l'inégalite de J. Hadamard entre les bornes supérieurs des dérivés successives d'une fonction*, C. R. Acad. Sci. Paris **207** (1938), 763–765.
19. __________, *On inequalities between upper bounds of consecutive derivatives of an arbitrary function defined on an infinite interval* (Russian). Učen. Zap. Modov. Gos. Univ. Mt. **30** (1939), 3–16.
20. OSTROWSKI, A. *Über die Absolutabweichung einer differentierbaren Funktion von ihrem Integralmittelwert*, Commt. Mth. Helv. **10** (1938), 226–227.
21. HALPERIN, I. and H. PITT, *Integral inequalities connected with differential operators*, Duke Math. J. **4** (1938), 613–625.
22. BANG, T., *Une inégalité de Kolmogoroff et les fonctions presque-périodiques*, Meddelelser Kjobenhavn **19** Nr. 4, 1941, 28pp.
23. SZ.-NAGY, B., *Über Integralungleichungen zwischen einer Funktion und ihrer Ableitung*, Acta Sci. Math. **10** (1941), 64–74.

24. RODOV, A. M., *Relations between upper bounds of derivatives of functions of a real variable* (Russian). Izvestia Akad. Nauk SSSR **10** (1946), 257–270. See also a translation in Amer. Math. Soc. Translation No. 14 (1950), 32pp.
25. ________, *Sufficient conditions for existence of a function of a real variable with given upper bounds of moduli of the function and its first five derivatives* (Russian). Belorussk. Gos. Univ. Uč. Zap. Ser. Fiz.-Mat. **19** (1954), 65–72.
26. KLOOSTERMAN, H. D., *Over afgeleiden en differenties*, Actual. Math. Centrum .Amsterdam ZW (1948), 015, 10 pp.
27. AVAKUMOVIĆ and S. ALJANČIĆ, *Sur la meilleure limite de la dérivée d'une fonction assujettie à des conditions supplementaires*, Acad. Serbe Sci. Publ. Inst. Math. **3** (1950), 235–242.
28. OLOVYANIŠNIKOV, V. M., *On inequalities between upper bounds of consecutive derivatives on a half-line* (Russian). Uspehi. Mat. Nauk (N.S.) **642** 2 (1951), 167–170.
29. NIRENBERG, L., *Remarks on strongly elliptic partial differential equations*, Appendix. Comm. Pure. Appl. Math. **8** (1955), 649–675.
30. MATORIN, A. P., *On inequalities between the maxima of the absolute values of a function and its derivatives on a half-line* (Russian). Ukrain. Mat. Ž. **7** (1955), 262–266. See also a translation in Amer. Math. Soc. Translations, Series 2, **8** (1958), 13–17.
31. STEIN, E. M., *Functions of exponential type*, Ann. of Math. **65** (1957), 582–592.
32. LJUBIČ, Ju. I., *On inequalities between the powers of a linear operator*, Izv. Akad. Nauk SSSR, Ser. Mat. **24** (1960), 125–264. Amer. Math. Soc. Transl. **40** 2 (1964), 39–84.
33. DIEUDONNÉ, J., "Foundations of Modern Analysis". New York–London, 1960, pp. 187–188.
34. MÜLLER, W., *Über eine Ungleichung zwischen den Normen von f, f' und f''*, Math. Z. **78** (1962), 420–422.
35. REDHEFFER, R. M., *Über eine beste Ungleichung zwischn den Normen*, Math. Zeitschr. **80** (1963), 390–397.
36. STEČKIN, S. B., *Inequalities between the norms of derivatives for arbitrary functions* (Russian). Acta Sc. Math. Szeged **26** (1965), 225–240.
37. GEĬSBERG, S. P., *A generalization of an inequality of Hadamard* (Russian), Sb., in "Isledovania po nekotorim problemom konstruktivnoj teorii funkcií," Trudi Mat. Meh instituta No 50, 1965, pp. 42–54.
38. PFEFFER, A. M., *On certain discrete inequalities and their continuous analogs*, J. Res. Nat. Bur. Standards Sect. B. **70** (1966), 221–231.

39. GOLDBERG, S., "Unbounded linear operators". New York, 1966.
40. GALJAUTDINOV, I. G. and V. R. FRIDLENDER, *Some generalizations and applications of inequalities of Kolmogorov and Gorny-Cartan*, Izv. Vysš. Uč. Zaved. Matematika No. 3 **58** (1967), 20–26. (Russian).
41. __________, *K voprosu ob oboščenii neravenstv Kolmogorova i Gornogo-Kartana*, Dokl. Tret'eĭ sibirskoi konf. po mat. i. meh., Izd-vo Tomskogo un- ta, Tomsk (1964), 94–95.
42. GALJAUTDINOV, I. G., *Ob odnom obobščenii i primenenii neravenstv Kolmogorova i Gornogo-Kartana dlja verhnih graneĭ proizvodnyh*, Tezisy Kratkih naučnyh soobščeniĭ meždunarodnogo kongressa matematiov, Moskva 1966, sekcija 4: Klassičeskii analiz., str. 45.
43. STEČKIN, S. B., *Inequalities between the upper bounds of the derivatives of an arbitrary function on the half-line* (Russian). Mat. Zametki **1** (1967), 665–674.
44. BENSON, D. C., *Inequalities involving integrals of function and their derivatives*, J. Math. Anal. Appl. **17** (1967), 292–308.
45. DONAR, Y., *An extremal problem related to Kolmogoroff's inequality for bounded functions*, Ark. Mat. **7** (1969), 441–443.
46. GEĬSBERG, S. P., *Drobnye proizvodnye ograničennyh na osi funkciĭ*, Izv. Vysš. uč.zav Matematika No 11 **78** (1968), 51–69.
47. GABUŠIN, V. N., *Inequalities for norms of a function and its derivatives in L_p metrics* (Russian). Mat. Zametki **1** (1967), 291–289.
48. ARESTOV, V. V., *On nailučšem približenie lineĭnyh operatorov*, Mat. Zametki **1** (1967), 149- -154.
49. GABUŠIN, V. N., *Precise constants in inequalities between the norm of derivatives of a function* (Russian). Mat. Zametki **4** (1968), 221–232.
50. TAĬKOV, L. V., *Inequalities of Kolmogorov type and best formulas for numerical differentiation* (Russian). Mat. Zametki **4** (1968), 233–238.
51. SUBBOTIN, Ju. N. and L. V. TAĬKOV, *Nailuščee približenie operators differencirovanija v prostranstve L_2*, Mat. Zametki **3** (1968), 157–164.
52. ARESTOV, V. V., *O nailučšem ravnomernom približenii operatorov differencirovanija*, Mat. Zametki **5** (1969), 273–284.
53. GABUŠIN, V. N., *The best approximation for differentiation operators on the half-line*, Mat. Zametki **6** (1969), 573–582. (Russian).
54. SCHOENBERG, I. J. and A. CAVARETTA, *Solutions of Landau's problem concerning higher derivatives on the half line*, Proc. Int. Conf. of Constr. Func. Th., Varna (1970), 19–25.
55. BERDYŠEV, V. I., *The best approximation in $L(0,\infty)$ to the differentiation operator* (Russian). Mat. Zametki **5** (1971), 477–481.

56. KALLMAN, R. R. and G. C. ROTA, *On the inequality* $\|f'\|^2 \le 4\|f\|\cdot\|f''\|$, in "Inequalities II," (O. Shisha, Ed.), New York/London, 1970, pp. 187–192.
57. HILLE, E., *Remark on the Landau-Kallman-Rota inequality*, Aeq. Math. **4** (1970), 239–240.
58. KUREPA, S., *Remark on the Landau inequality*, Aeq. Math. **4** (1970), 240–241.
59. KRALJEVIĆ, H. and S. KUREPA, *Semigroups on Banach spaces*, Glasnik Mat. **5** (1970), 109–117.
60. HILLE, E., *Generalizations of Landau's inequality to linear operators*, in "Linear Operators and Approximations," (P. L. Butzer, J.-P. Kahane and B. Sz.- Nagy, eds.), Basel/Stuttgart, 1971, pp. 20–32.
61. __________, *On the Landau-Kallman-Rota inequality*, J. Approx. Th. **6** (1972), 117–122.
62. KATO, T., *On an inequality of Hardy, Littlewood and Pólya*, Advances in Math. **7** (1971), 217–218.
63. HOLBROOK, J. A. R., *A Kallman-Rota Inequality for Nearly Euclidean Spaces*, Advances in Math. **14** (1974), 335–345.
64. TREBELS, W. and U. WESTPHAL, *A note on the Landau-Kallman-Rotta-Hille Inequality*, in "Linear Operators and Approximation," (P. L. Butzer, J.-P. Kahane and B.Sz.-Nagy, Eds.), Basel/Stuttgart, 1971, pp. 115–119.
65. LACKOVIĆ, I. B. and M. S. STANKOVIĆ, *A note on a theorm of E. Landau*, Univ. Beograd. Publ. Elektrotehn. Fak. Ser. Mat. Fiz. No. 381–409 (1972), 113–116.
66. SOLONNIKOV, V. A., *O nekotoryh neravenstvah dlja funkciĭ iz klassov* $W_p(R^n)$, Zap. naučn. seminarov LOMI **27** (1972), 194–210.
67. BESOV, O. V., *Mul'tiplikativnye ocenki dlja integral'nyh norm differenciruemyh funkciĭ mnogih peremennyh*, Tr. matem. in-ta AN SSR, **131** (1974), 3–15.
68. DJORDJEVIĆ, R. Ž. and G. V. MILOVANOVIĆ, *A generalization of E. Landau's theorem*, Univ. Beograd. Publ. Elektrotehn. Fak. Ser. Mat. Fiz. No. 498–541 (1975), 97- -106.
69. MILOVANOVIĆ, G. V., *On some integral inequalities*, Univ. Beograd. Publ. Elektrotehn. Fak. Ser. Mat. Fiz. No. 498–541 (1975), 119–124.
70. UPTON, C. J. F., *On extensions of inequaities of Kolmogoroff and others and some applications to almost periodic functions*, Glasgow Math. J. **13** (1972), 1–16.
71. ARESTOV, V. V., *O točnyh neravenstvah meždu normami funkciĭ i ih proizvodnyh*, Acta. Sci. Math. Szeged **33** (1972), 243–257.

72. EVERITT, W. N. and M. GIERTZ, *On the integro-differential inequality* $\|f'\|_2^2 \leq \|f\|_p \|f''\|_q$, J. Math. Anal. Appl. **45** (1974), 639–653.
73. EVERITT, W. N., *On an extension to an integro-differential inequality of Hardy, Littlewood and Pólya*, Proc. Roy. Soc. Edinburgh (A) **69** (1971–72), 295–333.
74. __________, *Report on certain integral inequalities*, Sci Res. Council Symp. on Spectral Th. and Diff. Eq. Univ. of Dundee, 1974.
75. __________, *Integral inequalities and the Liouville transformation.*
76. EVERITT, W. N. and M. GIERTZ, *Some inequalities associated with certain ordinary differential operators*, Math. Zeit. **126** (1972), 308–326.
77. BRODLIE, R. W. and W. N. EVERITT, *On an inequality of Hardy and Littlewood*, Proc. Roy. Soc. Edinburgh **72** A (1973-74), 179–186.
78. EVERITT, W. N., *Integral inequalities and spectral theory*, in "Spectral Theory and Differential Equations," (A. Dold and B. Eckmann, Eds.), Lect. Notes in Math. **558**, Berlin/Heidelberg/New York, 1975, pp. 148–156.
79. BRADLEY, J. S. and W. N. EVERITT, *On the inequality* $\|f''\|^2 \leq K\|f\| \, \|f^{(4)}\|$, Quarterly J. Math. **25**, (1974), 241–252.
80. KUPCOV, N. P., *Kolmogorov estimates for derivatives in* $L_2(0,\infty)$, Proc. Steklov Inst. Math. **138** (1975), 94–117. (Eng. Transl.) 101–125.
81. SCHOENBERG, I. J., *The elementary cases of Landau's problem inequalities between derivatives*, Amer. Math. Monthly **80** (1973), 121–158.
82. CAVARETTA, A. S., *An elementary proof of Kolmogorov's theorem*, Amer. Math. Monthly **81** (1974), 480- -486.
83. DITZIAN, Z., *Some Remarks on Inequalities of Landau and Kolmogorov*, Aeq. Math. **12** (1975), 145–151.
84. KIRSANOVA, G. V., *Sootnošenija meždu verhnimi granjami module funkciĭ i ih starših proizvodnyh*, Mat. Zametki **14**, (1973), 329.
85. DZJADYK, V. K. and V. A. DUBOVIK, *K probleme A. N. Kolmogorova o zavisimostjah meždu verhnimi granjami proizvodnyh veščestve funkciĭ, zadannyh na vseĭ osi*, Ukrain. Mat. Ž. **26** 3 (1974), 246–259.
86. __________, *K neravenstvam A. N. Kolmogorov zavisimostjah meždu verhnimi granjami proizvodnyh veščestve funkciĭ, zadannyh na vseĭ osi*, Ukrain. Mat. Ž. **27** 3 (1975), 291–299.
87. TALENTI, G., *Une disuguaglianze fra u, u', u''*, Boll. Un. Mat. Ital. (4), **11**, Suppl. fasc. **3** (1975), 375–388.
88. RUSSELL, A., *Integral inequalities and spectral theory of certain ordinary differential operators*, Ph.D. thesis, University of Dundee, Scotland, U.K.
89. CHUI, C. K. and P. W. SMITH, *note on Landau's problem for bounded intervals*, Amer. Math. Monthly **82** (1975), 927–929.

90. SHARMA, A. and J. TZIMBALARIO, *Some inequalities between derivatives on bounded intervals*, Delta **6** (1976), 78–91.
91. KALLIONIEMI, H., *On representation formulas for intermediate derivatives*, Math. Scand. **39** (1976), 315–326.
92. ________, *On bounds for derivatives of a complex-valued function on a compact interval*, Math. Scand. **39** (1976), 295–314.
93. ________, *Uniform bounds for intermediate derivatives of a complex-valued function on a compact interval.*
94. ________, *On inequalities between the uniform bounds of derivatives of a complex-valued function of a real variable*, Inaugural dissertation, Uppsala Univ. (Sweden), 1970.
95. GOLDSTEIN, J. A., *On the Hardy-Landau- Littlewood Inequality*, Notices Amer. Math. Soc. **23** (1976), A–162.
96. ________, *On the Hardy-Landau-Littlewood inequality*, Trans. Amer. Math. Soc. (to appear).
97. ________, *Improvements in the Hardy-Landau-Littlewood Inequality*, Notices Amer. Math. Soc. **23** (1976), A–486–487.
98. CAVARETTA, A. S., *One-side inequalities for the successive derivatives of a function*, Bull. Amer. Math. Soc. **82** (1976), 303–305.
99. MAGARIL-IL'JAEV, G. G., *Kolmogorov inequalities on the half-line*, Vestnik Moskov. Univ. Ser. I. Mat. Meh. **31** 5 (1976), 33–41.
100. LIGUN, A. A., *Inequalities for upper bounds of functionals*, Analysis Mathematica **2** (1976), 11–40.
101. SCHOENBERG, I. J., *Notes on spline functions VI. Extremum problems of the Landau-type for the differential operators* $D^2 \pm 1$, in "Stud. Spline Funct. Approx. Theory," New York, 1976, pp. 353–369.
102. ________, *The Landau problem for the differential operator* $D^2 - a^2$ *in a circular ring*, Colloquia Math. Soc. János Bolyai **19**, Fourier Anal. and Approx. Th., Budapest, 1976, 713–723.
103. SHARMA, A. and J. TZIMBALARIO, *Landau-type inequalities for some linear differential operators*, Illinois J. Math. **20** (1976), 443–455.
104. MILOVANOVIĆ, G. V. and J. E. PEČARIĆ, *On generalization of the inequality of A. Ostrowski and some related applications*, Univ. Beograd. Publ. Elektrotehn. Fak. Ser. Mat. Fiz. No. 544–576 (1976), 155–158.
105. LUPAS, A., *Problem 313*, Mat. Vesnik **12** 27 (1975), 122–123.
106. CERTAIN, M. W. and T. G. KURTZ, *Landau-Kolmogorov inequalities for semi-groups and groups*, Proc. Amer. Math. Soc. (1977), 226–230.
107. BOLLOBÁS, B., *The spatial numerical range and powers of an operator*, J. London Math. Soc. **7** 2 (1973), 435–440.

108. COPSON, E. T., *On two integral inequalities*, Proc. Roy. Soc. Edinburgh Sect. A **77** (1977), 325–328.
109. _________, *On two inequalities of Brodlie and Everitt*, Proc. Roy. Soc. Edinburgh Sect. A **77** (1977), 329–333.
110. DITZIAN, Z., *Note on Hille's question*, Aeq. Math. **15** (1977), 143–144 .
111. FINK, A. M., *Best possible approximation constants*, Trans. Amer. Math. Soc. **226** (1977), 243–255.
112. CAVARETTA, A. S., *A refinement of Kolmogorov's inequality*, J. Approx. Theory **27** (1979), 45–60. (See also Notices Amer. Math. Soc. **25**, Nr. 5 (1978), A–530).
113. SCHOENBERG, I. J., "On Nicchelli's Theory of Cardinal $\mathcal{L}$-Splines," New York, 1976, pp. 251–276.
114. _________, *The Landau problem for motion in a ring and in bounded continua*, Mathematics Research Center Technical Summary Report #1563, Madison, Wisconsin, and in somewhat abbreviated form in Amer. Math. Monthly **84** (1977), 1–12.
115. GOODMAN, T. N. T. and S. L. LEE, *Landau and Hardy-Littlewood-Pólya inequalities involving derivatives of functions*, SEA Bull. Math. **2**, (1978), 82–100.
116. SCHOENBERG, I. J., *The Landau problem, I: The case of motions on sets*, Nederl. Akad. Wet., Proc., Ser. A. **81** (1978), 276–286.
117. SHARMA, A. and J. TZIMBALARIO, *Landau-type inequalities for bounded intervals*, Period. Math. Hungar. **9** 3 (1978), 175–186.
118. BENNEWITZ, C. and W. N. EVERITT, *Some remarks on the Titchmarsh-Weyl m-coefficient*, Proc. of the Pleijel Conf., Uppsala, (1979), 49–108.
119. BEESACK, P. R., *A simpler proof of two inequalities of Brodlie and Everitt*, Proc. Roy. Soc. Edinburgh A **84** (1979), 259–261.
120. ŽUK, V. V., *Nekotorye točnye neravenstva meždu nailučšimi približenijami periodičeskih funkciĭ i moduljami nepreryvnosti vysših porjadkov*, Vestnik Leningradskogo Univ. No. **19** (1978), 35–42.
121. KONOVALOV, V. N., *Točnye neravenstva dlja norm funkciĭ tret'ih častnyh, vtoryh smešannyh ili kosyh proizvodnyh*, Mat. Zametki **23** (1978), 67–78.
122. REDHEFFER, R. and W. WALTER, *Inequalities involving derivatives*, Pacif. J. Math. **85**, (1979), 165- -178.
123. ARESTOV, V. V., *Inequalities for fractional derivatives on the half-line*, Approx. theory, Warszawa 1975, Banach Cent. Pub. **4** (1979), 19–34.
124. MILOVANOVIĆ, G. V. and I. Ž. MILOVANOVIĆ, *On a generalization of certain results of A. Ostrowski and A. Lupas*, Univ. Beograd. Publ. Elektrotehn. Fak. Ser. Mat. Fiz. No. **634–677** (1979), 62-69.
125. NETA, B., *On Determination of Best-Possible Constants in Integral Inequalities Involving Derivatives*, Math. of Comptuation **35** (1980), 1191–1193.

126. BURENKOV, V. I., *O točnyh postojannyh v neravenstvah dlja norm promežutočnyh proizvodnyh na konečnom intervale*, Tr. Matem. in-ta AN SSSR **156** (1980), 22–29.
127. KWONG, M. K. and A. ZETTL, *Ramifications of Landau's inequality*, Proc. Roy. Soc. Edinburgh A **86** (1980), 175–212.
128. GINDLER, H. A. and J. A. GOLDSTEIN, *Dissipative operator versions of some classical inequalities*, J. Analysis Math. **28**, (1975), 213–238.
129. KWONG, A. K. and A. ZETTL, *Remarks on best constants for norm inequalities among powers of an operator*, J. Approx. Theory **3** (1979), 249–258.
130. EVANS, W. D. and A. ZETTL, *Norm inequalities involving derivatives*, Proc. Roy. Soc. Edinburgh A **82** (1978), 51–70.
131. KWONG, M. K. and A. ZETTL, *Norm inequalities for derivatives*, in "Lecture Notes in Mathematics," No. **846**, Berlin, 1980.
132. ________, *Weighted norm inequalities of sum form involving derivatives*, Proc. Roy. Soc. Edinburgh A **88** (1981), 121–134.
133. ________, *An extension of Hardy-Littlewood inquality*, Proc. Amer. Math. Soc. **77** (1979), 175–212.
134. ________, *Norm inequalities of product form in weighted L^p spaces*, Proc. Roy. Soc. Edinburgh A **89** (1981), 293-307.
135. AMOS, R. J. and W. N. EVERITT, *On integral inequalities and compact embeddings associated with ordinary differential expressions*, Arch. Rational Mech. Anal. **71**, (1979), 15–40.
136. EVERITT, W. N., *An integral inequality with an application to ordinary differential operators*, Proc. Roy. Soc. Edinburgh A **80** (1978), 35–44.
137. KWONG, M. K. and A. ZETTL, *Discreteness conditions for the spectrum of ordinary differential operators*, J. Differential Equations **40** (1981), 53–70.
138. ________, *Norm inequalities for dissipative operators on inner product spaces*, Houston J. Math. **5** (1979), 543–557.
139. PHÓNG, V. Q., *On inequalities for powers of linear operators and for quadratic forms*, Proc. Roy. Soc. Edinburgh A **89** (1981), 25–50.
140. KONOVALOV, V. N., *Eine Ergänzung zu den Ungleichungen von A. M. Kolmogorov*, Mat. Zametki **27** (1980), 209–215. (Russisch).
141. MAGARIL-IL'JAEV, G. G. and V. M. TIHOMIROV, *On the Kolmogorov inequality for fractional derivatives on the half-line*, Analysis Math. **7** (1981), 37–47.
142. SCHOENBERG, I. J., *The Landau problem. II: The case of motion on curves*, Indagationes Math. **43** (1981), 315–335.
143. GINDLER, H. A. and J. A. GOLDSTEIN, *Dissipative operators and series inequalities*, Bull. Austral. Math. Soc. **23** (1981), 429–442.
144. COPSON, E. T., *Two series inequalities*, Proc. Roy. Soc. Edinburgh A **83** (1979), 109–114.

145. FINK, A. M., *Kolmogorov-Landau inqualities for monotone functions*, J. Math. Anal. Appl. **90** (1982), 251–258.
146. SATO, M. and R. SATO, *A simple proof of the Chui-Smith theorem for Landau's problem*, Colloquium Math. **47** (1982), 119–122.
147. BOYADZIEV, K. N., *An inequality for bounds of successive derivatives*, The Math. Student **50** (1982), 214- -218.
148. ZVYAGINTSTEV, A. I., *Kolmogorov's inequalities for* $n = 4$ (Russian), Latv. Mat. Ezheg. **26** (1982), 165- -175.
149. ZVYAGINTSTEV, A. I. and A. Ya. LEPIN, *On Kolmogorov's inequality between the upper bounds for derivatives of functions for* $n = 3$ (Russian), Latv. Mat. Ezheg. **26** (1982), 176–181.
150. EVANS, W. D. and W. N. EVERITT, *A return to the Hardy-Littlewood integral inequality*, Proc. Roy. Soc. London A **280** (1982), 447–453.
151. SCHOENBERG, I. J., *Two applications of approximate differential formulae: An extremum problem for multiply monotone function differentiation of asymptotic expansions*, J. Math. Anal. Appl. (1982), 251–261.
152. FARWIG, R. and D. ZWICK, *Bounds on the derivatives of a function via the theory of n convex functions*, J. Math. Anal. Appl. **118** (1986), 97–106.
153. PEČARIĆ, J. E., S. D. DRAGOMIR and B. CRSTICI, *On the Olovyanišnikov Schoenberg Theorem and some similar results*, Studia Math. Bab. Bolyai Univ. Cluj-Napoca **34** (1989), 34–49.
154. DITZIAN, Z., *On Lipschitz classes and derivative inequalities for various Banach spaces*, Proc. of Conf. on Funct. Analysis, Holomorphy and Approx., Rio de Janeiro, Brazil, 1981.
155. __________, *Discrete and shift Kolmogorov type inequalities*, Proc. Roy. Soc. Edinburgh A **93** (1983), 307–317.
156. RAFAL'SON, S. Z., *Odno neravenstvo meždu normami funkcii i ee proizvodnyh v integral'nyh metrikah*, Mat. Zametki **33** (1983), 77–82.
157. LIGUN, A.A., *O neravenstvah meždu normami proizvodnyh periodičeskih funkciǐ*, Mat. Zametki **33** (1983), 385–391.
158. PEČARIĆ, J. E. and B. SAVIĆ, *O novom postupku razvijanja funkcja u red i nekim primjenama*, Zbornik radova AKov (Beograd) (1983), 171–202.
159. __________, *O generalizacijama nekih rezulat Karamate i nekim primjenama*, Zbornik radov AKov (Beograd) (1984), 245–268.
160. GOLDSTEIN, J. A., M. K. KWONG and Z. ZETTL, *Weighted Landau Inequalities*, J. Math. Anal. Appl. **95** (1983), 20–28.
161. KWONG, M. K. and A. ZETTL, *Landau's inequality*, Proc. Roy. Soc. Edinburgh A **97** (1984), 161–163.

162. TIMOFEEV, V. G., *Kolmogorov inequalities with Laplace operator*, in "Theory of Functions and Approximation," Lecture Notes Saratov State University, 1983, pp. 83–92 (Russian).
163. _________, *Landau inequality for function of several variables* (Russian) , Mat. Zametki **37** (1985), 676–689.
164. PEČARIĆ, J. E., *Remarks on some inequalities of A. M. Fink*, J. Math. Anal. Appl. **104** (1984), 428–431.
165. BENNEWITZ, C., *A general version of the Hardy-Littlewood-Pólya-Everitt (HELP) inequality*, Proc. Roy. Soc. Edinburgh A **97** (1984), 9–20.
166. EVANS, W. D. and W. N. EVERITT, *On an inequality of Hardy-Littlewood type: I.*, Proc. Roy. Soc. Edinburgh A **101** (1985), 131–140.
167. BENNEWITZ, C., *The HELP inequality in regular case*, Internat. Ser.Numer. Math. **80** (1987), 337–346.
168. PACHPATTE, B. G., *A Note on Everitt Type Integral Inequality*, Tamkang J. Math. **18** (1987), 7–10.
169. MITRINOVIĆ, D. S. and J. E. PEČARIĆ, *Remarks on the paper "A note on Everitt type integral inequality" of B. G. Pachpatte*, Tamkang J. Math. **21** (1990), 169–170.
170. BROWN, R. C. and D. B. HINTON, *Sufficient Conditions for Weighted Inequalities of Sum Form*, J. Math. Anal. Appl. **112** (1985), 563–578.
171. _________, *Sufficient conditions for weighted Gabushin inequalities*, Časopis Pěst. Mat. **111** (1986), 113–122.
172. _________, *Necessary and sufficient conditions for one variable weighted interpolation inequalities*, J. London Math. Soc. **35** 2 (1987), 439–453.
173. CAFFARELLI, L. R. KOHN and L. NIRENBERG, *First order interpolation inequalities with weights*, Compositio Math. **53** (1984), 259–275.
174. LIN, C. S., *Interpolation inequalities with weights*, Comm. Partial Differential Equations **11** (1986), 1515–1538.
175. BROWN, R. C. and D. B. HINTON, *Weighted interpolation inequalities of sum and product form in R^n*, Proc. London Math. Soc. **56** 3 (1988), 261–280.
176. ZVYAGINSTEV, A. I., *Nekotorye ocenki dlja norm funkcii i ee proizvodnyh na konečnom intervale*, Latv. Mat. Ezheg. **29** (1985), 198–210.
177. _________, *Utočnenie konstant v neravenstvah Kolmogorova dlja $n = 4$*, Latv. Mat. Ezheg. **32** (1988), 187–207.
178. BOYADZHIEV, K. N., *A many-variable Landau-Kolmogorov inequality*, Math. Proc. Camb. Phil. Soc. **101** (1987), 123–129.
179. DITZIAN, Z., *Multivariate Landau-Kolmogorov-type inequality*, Math. Proc. Camb. Phil. Soc. **105** (1989), 335–350.
180. TIHOMIROV, V. M., "Some questions of approximation theory," (Russian), MGU, 1976.
181. _________, *A. N. Kolmogorov i teorija približeniĭ*, Uspehi Mat. Nauk 44m vtp **265** 1 (1989), 83–122.

182. MITRINOVIĆ, D. S. and J. E. PEČARIĆ, *On some inequalities for monotone functions*, Boll. Un. Mat. Ital. (to appear).
183. __________, *On two-place completely monotone functions*, Anzeiger Öster. Akad. Wiss. **126** (1989), 85–88.
184. __________, *On completely monotonic sequences*, (Manuscript).
185. KRALJEVIĆ, H. and J. E. PEČARIĆ, *Some Landau's type inequalities for infinitesimal generators*, Aeq. Math. **40** ((1990), 147–153.
186. __________, *Some Ostrowsky-Landau type inequalities for infinitesimal generators*, (Manuscript).
187. MITRINOVIĆ, D. S., J. E. PEČARIĆ and H. KRALJEVIĆ, *On Landau's theorems*, Glasnik Mat. (Zagreb), (to appear).
188. BEESACK, P. R., D. S. MITRINOVIĆ and P. M. VASIĆ, *Integral inequalities*, (the manuscript not achieved and not published).
189. BUSLEAV, P., *About an extremal problem associated with inequalities for derivatives* (Russian), Vestn. Mosk. un-ta. Ser. mat. meh. No. 3 (1978), 67–77.
190. TALENTI, G., *Una disuguaglianza concernente equazioni integrali di Wiener-Hopt a nucleo singolare*, Rend. Sem. mat. fis. Nilano **42** (1972), 121–141.
191. EVERTITT, W. N. and A. ZETTL, *On a class of integral inequalities*, J. London Math. Soc., II. Ser. **17** (1978), 291–303.
192. STEČKIN, S. B., *Inequalities between the upper bounds of the derivatives of an arbitrary function on the half-line* (Russian), Mat. Zametki **1** (1967), 665–674.
193. BOLLOBÁS, B., *A characterization of the circle and its applications to heritian operators*, Proc. Camb. Phil. Soc. **74** (1973), 67–72.
194. BOLLOBÁS, B. and E. GALANIS, *An inequality involving a hermitian operator and its square in a normal space*, Bull. Soc. Math. Grèce **17** (1976), 44–49.
195. PARTINGTON, J. R., *Constants relating to hermitian operator and its square*, Proc. Camb. Phil. Soc. **85** (1979), 325–333.
196. PROTTER, M., *Inequalities for powers of unbounded operators*, Manuscripta Math. **78** (1979), 71–79.
197. CHERNOFF, P., *Optimal Landau-Kolmogorov inequalities for dissipative operators in Hilbert and Banach spaces*, Adv. in Math. **34** (1979), 137–144.
198. BOLLOBÁS, B. and J. R. PARTINGTON, *Inequalities for quadratic polynomials in Hermitian and dissipative operators*, Advances in Math. **31** (1984), 271- -280.
199. KATO, T. and J. SATAKE, *An algebraic theory of Landau-Kolomogorov inequalities*, Tôhuku Math. J. **33** (1981), 421–478.

200. ZVYAGINESTEV, A. I., *Estimates for an intermediate derivative of a function* (Russian), Latv. Mat. Ezhegodnik No. **32** (1988), 183-186.
201. REDHEFFER, R. and W. WALTER, *Inequalities involving derivatives*, Pac. J. Math. **85** (1979), 165–178.
202. GILLMAN, D., H. KAPER, and M. K. KWONG, *L^2-norm inequalities for the differentiation operator on compact intervals*, Proc. Roy. Soc. Edinburgh Sect. A. **110** (1988), 335–342.
203. BROWN, R. C. and D. B. HINTON, *Interpolation inequalities with power weights for functions of one variable*, preprint 1990.
204. SMITH, K. T., *Inequalities for formally positive integro-differential forms*, Bull. Amer. Math. Soc. **67** (1961), 368–370.
205. HUANG, D. and J. WANG, *The Kolmorogov-type inequalities related with selfconjugate ordinary differential operators* (Chinese), Acta Mech. **26** (1983), 715–722.
206. NEUMAN, E. and J. PEČARIĆ, *Inequalities involving multivariate convex function*, J. Math. Anal. Appl. **137** (1989), 541–459.
207. ŽUK, V. V., *O nekotoryh sootnšenijah meždu moduljami nepreryvnosti i funkcionalami, zadannymi na množestvah periodičeskih funkciĭ*, Izv. Vuzov., Matematika, No. **5** (1970), 24–33.
208. ________, *O nekotoryh točnyh neravenstvah meždu nailučšimi približenijami i moduljami nepreryvnosti*, Sib. matem. ž **12** (1971), 1283–1291.
209. ________, *Nekotorye točnye neravenstva meždu ravnomernymi približenijami periodičeskih funkciĭ*, Dokl. AN SSSR **201** (1972), 263–265.
210. ________, *Nekotorye točnye neravenstva meždu ravnomernymi približenijami periodičeskih funkciĭ*, Izv. vuzov, Matematika, No. **1** (1973), 51–56.
211. NIKOL'SKIĬ, S. M., *Približenie funkciĭ trigonometričeskimi polinomami v srednem*, Izv. AN SSSR, Ser. Matem., **10** (1946), 207–256.
212. LIGUN, A. A., *Točnye neravenstva dlja vernih graneĭ polunorm na klassah periodičeskih funkciĭ*, Mat. Zametki **13** (1973), 647–654.
213. HOA, N. T. T., *Kolmogorov's inequality for differential operators* (Russian), Vestn. Mosk. Un.-ta, Ser. 1, Matematika, Mehanika, 1988 No. 1, 17–21.
214. ZWYAGINSEV, A. I., *Ocenki dlja promežtočnoĭ proizvodnoĭ funkciĭ*, Latv. Mat. Ezheg. **32** (1988), 183–186.
215. ________, *Ob odnoĭ probleme A. N. Kolmogorova*, Latv. Mat. Ezheg. **33** (1989), 192–203.
216. KRALJEVIĆ, H. and J. PEČARIČ, *Some Landau's inequalities for infinitesimal generators*, Aeq. Math. **40** (1990), 147–153.
217. KUDRJAVCEV, L. D., *O postroenii posledovatel'nosti finitnyh funkciĭ, approksimiryjuščih funkciĭ vesoyh klassov*, Trudy Mat. Inst. AN SSSR, **156** (1980), 121–129.

CHAPTER II

AN INEQUALITY ASCRIBED TO WIRTINGER AND RELATED RESULTS

1. In this Chapter we shall consider inequalities of the form

$$\|f\|_p \le C(p,q,n)(b-a)^{n-\frac{1}{p}-\frac{1}{q}}\|f^{(n)}\|_q \tag{1.1}$$

on an interval $[a,b]$ under the assumptions that $f^{(n-1)}$ is absolutely continuous and $f^{(n)}$ is in L^q. The exponent of $(b-a)$ is selected to make the inequality invariant under translation of the interval and invariant under scale change, as it must be. We attempt as much as possible to get the best possible constant and reserve the notation $C(p,q,n)$ for the best possible constant. Now, such an inequality as (1.1) does not exist for all functions f since the kernel of the map from f to $f^{(n)}$ is n-dimensional. Hence in order to get such an inequality we must impose at least n independent conditions on f.

The most famous example of such an inequality is one usually ascribed to Wirtinger. Let f be a periodic function with period 2π with $\int_0^{2\pi} f(x)\,dx = 0$. Then

$$\int_0^{2\pi} (f(x))^2\,dx \le \int_0^{2\pi} (f'(x))^2\,dx \tag{1.2}$$

with equality if and only if $f(x) = A\cos x + B\sin x$ for A and B constants.

In the literature, (1.2) is known as Wirtinger's inequality (see [**1**] and [**2**]). We will also accept this name. The French and Italian mathematical literature do not mention the name of Wirtinger. It appears that the first published proof by W. Wirtinger was in the 1916 book of W. Blaschke [**3**]. In fact, however, an inequality of the form (1.2) with weaker conditions on the function f was given in 1905 by E. Almansi [**4**] who showed

$$\int_a^b f'(x)^2\,dx \ge \left(\frac{2\pi}{b-a}\right)^2 \int_a^b f(x)^2\,dx, \tag{1.3}$$

if f and f' are continuous, $f(a) = f(b)$, and $\int_a^b f(x)\,dx = 0$.

These conditions were weakened in turn by E. E. Levi [**5**] in 1911 and L. Tonelli [**6**] in 1914.

2. L. Scheeffer in 1885 [**7**] proved the following result:

If f and f' are continuous on the interval $[a,b]$ and if $f(a) = f(b) = 0$, then

$$\int_a^b f'(x)^2\,dx \geq \left(\frac{\pi}{b-a}\right)^2 \int_a^b f(x)^2\,dx. \tag{2.1}$$

Note that Scheeffer's result is also given in Kneser's article [**8**].

However, in 1896, E. Picard [**9**] considered the problem of finding the function f which maximizes the expression

$$\frac{\int_a^b p(x)f(x)^2\,dx}{\int_a^b f'(x)^2\,dx} \tag{2.2}$$

where f and f' are continuous functions, $f(a) = f(b)$ and where p is a positive continuous function on (a,b).

In papers [**10**],[**11**] W. Stekloff proved that (2.1) is valid if either $f(a) = f(b) = 0$ or $\int_a^b f(x)\,dx = 0$ is valid. A further generalization of (2.1) is given in Stekloff's paper [**12**].

3. The two dimensional version of the quotient (2.2) was studied by H. A. Schwarz [**13**] in 1885. He determined the maximum of the quotient (with appropriate boundary conditions)

$$Q = \frac{\iint_R p(x,y)f(x,y)^2\,dxdy}{\iint_R \left(\left(\frac{\partial f}{\partial x}\right)^2 + \left(\frac{\partial f}{\partial y}\right)\right)^2 dxdy}.$$

If R is convex and $\iint_R f(x,y)\,dxdy = 0$ and $p(x,y) \equiv 1$, the maximum value of this quotient Q is $7\ell^2/24$, where ℓ is the maximal chord of the region R. This is a result of H. Poincaré [**14**]. He also considered an analogous problem in three dimensions. For modern versions of this classical Poincaré inequality see 64. below.

The two dimensional quotient in the case $p(x,y) \equiv 1$ is considered by E. E. Levi [**15**]. He shows that if R is convex with respect to a point P whose minimum and maximum distance to the boundary of R are ℓ and L respectively, and $\iint_R f(x,y)\,dxdy = 0$, then the quotient Q is less that $1/\kappa$ where $\kappa = \min\left(\frac{3\ell}{2L^3}, \frac{1}{13L^2}\right)$.

In a private communication, H. Kalf gave the following comment concerning the priority of Wirtinger-type inequalities.

If one does want to dig for even earlier sources, one has to go back to Jacobi's treatment of the second variation (Crelle's Journal für die reine und angewandte Mathematik **17**, pp. 68–82, 1837; see also the papers cited on p. 60 of Bolza's "Vorlesungen über Variationsrechuung" in this context), but maybe this is carrying things too far.

4. Inequality (1.3) can be also found in the book **[16]** of J. Hadamard from 1910, but in that year Ja. D. Tamarkine **[17]** proved the following result:

If a continuous function f has two continuous derivatives in (a, b) and if $f(a) = f(b)$, $\int_a^b f(x)\,dx = 0$, then

$$\int_a^b f''(x)^2\,dx \geq \left(\frac{\pi}{b-a}\right)^4 \int_a^b f(x)^2\,dx. \tag{4.1}$$

A related results was given by N. M. Kriloff **[18]**:

Let $f(x)$ be a real function such that it has an absolutely continuous first derivative in (a, b) and that its Fourier's series is uniformly convergent on an interval (a, b). If $f(a) = f(b)$ and $\int_a^b f(x) = 0$, then

$$\int_a^b f''(x)^2\,dx \geq \left(\frac{2\pi}{b-a}\right)^4 \int_a^b f(x)^2\,dx. \tag{4.2}$$

For some related results see papers of M. Picone **[19]** and G. Cimmino **[20]**.

5. The form of the inequalitiy (1.2) was slightly changed by E. E. Levi **[21]** in order to accomodate other boundary conditions. For example, let f have a bounded derivative on (a, b) and let $f(a) = f(b) = 0$. If A and B are disjoint measurable sets such that $A \cup B = (a, b)$ then

$$\int_a^b f(x)^2\,dx \leq \frac{(b-a)^2}{8} \int_A f'(x)^2\,dx + \|f\|_\infty (b-a) \int_B |f'(x)|\,dx,$$

and

$$\int_a^b |f(x)f'(x)|\,dx \le \frac{(b-a)^2+8}{16}\int_A f'(x)^2\,dx + \|f\|_\infty\left(\frac{b-a}{2}+1\right)\int_B |f'(x)|\,dx.$$

6. A. Pleijel [**22**] proved that if f is periodic of period 2π with $f'' \in L^2(a,b)$ then

$$2\pi\frac{\int_0^{2\pi} f'(x)^2\,dx}{\int_0^{2\pi} f''(x)^2\,dx}\int_0^{2\pi} f'(x)^2\,dx \le 2\pi\int_0^{2\pi} f(x)^2\,dx - \left(\int_0^{2\pi} f(x)\,dx\right)^2$$
$$\le 2\pi\int_0^{2\pi} f'(x)^2\,dx.$$

The right hand inequality implies (1.3) but has stronger hypothesis. On the other hand it is a better inequality as it stands. Janet [**23**] had proved this part earlier as well as other results in [**24**].

As a follow-up of Janet's results, Cimmino [**25**] considered quotients of the form

$$I = \frac{\int_a^b f^{(n)}(x)^2\,dx}{\int_a^b f^{(p)}(x)^2\,dx} \qquad (p < n) \tag{6.1}$$

under the conditions that f have continuous derivatives of order $n-1$ and f have an n-fold zero at a and b. He showed that the minimum of I is $\left(\frac{v_{n,p}}{b-a}\right)^{2n-2p}$ where $v_{n,p}$ is the least positive zero of the Wronskian of n independent solutions of the differential equation

$$y^{(2n)} - (-1)^{n+p}y^{(2p)} = 0.$$

Janet in [**23**], [**24**] continued the study of finding the explicit minimum of I. The papers [**23**], [**24**] also contain the following geometric result. Let

$$C:\begin{cases}\left(\frac{t(1+\cos t)}{t+\sin t},\ t^2\frac{t-\sin t}{t+\sin t}\right), & -\pi \le t \le 0,\\ \left(\frac{t(1+\cosh t)}{t+\sinh t},\ t^2\frac{\sinh t - t}{t+\sinh t}\right), & 0 \le t.\end{cases}$$

The two branches of this curve descend from $(0, \pi^2)$ to $(1, 0)$ and then ascend to $(+\infty, +\infty)$. The result is that if f and f' are continuous on $(1, 1)$, then the point

$$\left(\frac{\frac{1}{2}\left(f(0)^2 + f(1)^2\right)}{\int_0^1 f(x)^2\,dx}, \frac{\int_0^1 f'(x)^2\,dx}{\int_0^1 f(x)^2\,dx} \right)$$

lies on or above C.

Other results related to the quotient I are in M. Janet **[28]**,**[29]**,**[30]**, G. Cimmino **[31]** and the doctoral dissertation **[32]** of Tcheng Tchou-Yun.

7. Several results of G. H. Hardy and J. E. Littlewood **[33]** are given in the well-known book of Hardy, Littlewood and Pólya **[1]**, as follows:

(i) (Th. 253) If y' belongs to $L^2(0, \infty)$, $y(0) = 0$ and y is not always zero, then

$$\int_0^\infty \left(4y'^2 - \frac{y^2}{x^2}\right) dx > 0.$$

(ii) (Th. 254) If $u > 4$, $y(0) = 0$, $y(1) = 1$, and y' is L^2, then

$$\int_0^1 \left(uy'^2 - \frac{y^2}{x^2}\right) dx \geq \frac{2}{1 - 2a},$$

where $a = \left(\frac{1}{4} - \frac{1}{u}\right)^{1/2}$. The only case of equality is that $y = x^{a+1/2}$.

(iii) (Th. 255) If $k > 1$, $u > \left(\frac{k}{k-1}\right)^k$, $y(0) = 0$, $y(1) = 1$, and y' is L^k, then

$$\int_0^1 \left(uy'^k - \frac{y^k}{x^k}\right) dx \geq \frac{1}{(k-1)(1-\lambda)},$$

where λ is the (unique) root of

$$u(k-1)\lambda^{k-1}(\lambda - 1) + 1 = 0$$

which lies between $(k-1)/k$ and 1.

(iv) (Th. 256) If $y(0) = 0$ and $2k$ is an even positive integer, then

$$\int_0^1 y^{2k}\,dx \leq C \int_0^1 y'^{2k}\,dx,$$

where

$$C = \frac{1}{2k-1}\left(\frac{2k}{\pi}\sin\frac{\pi}{2k}\right)^{2k}.$$

There is equality only for certain hyperelliptic curves.

(v) (Th. 262) If $y(0) = y(1) = 0$ and y' is L^2, then

$$\int_0^1 \frac{y^2}{x(1-x)}\, dx < \frac{1}{2}\int_0^1 y'^2\, dx,$$

unless $y = cx(1-x)$.

(vi) (Th. 220) If y' is L^2 then

$$\int_0^{+\infty} \frac{y^2}{x^3}\, dx < \frac{3}{2}\left(\int_0^{+\infty} y'^2\, dx\right)^2$$

unless $y = x/(ax+b)$, where a and b are positive, in which case there is equality. More generally, if $m > k > 1$, $r = \frac{m}{k} - 1$, and y' is positive and L^k, then

$$\int_0^{+\infty} \frac{y^m}{x^{m-r}}\, dx < K\left(\int_0^{+\infty} y'^k\, dx\right)^{m/k},$$

where

$$K = \frac{1}{m-r-1}\left(\frac{r\Gamma(m/r)}{\Gamma(1/r)\Gamma((m-1)/r)}\right),$$

unless $y = x/(ax^r+b)^{1/r}$.

Note that Theorems 257 and 258 and Scheeffer's and Wirtinger's inequalities, respectively, and that for (vi) is given the reference Bliss [**34**].

8. The Bohr inequality from 1935 (see [**35**]) for almost periodic functions is connected to the previous results. Namely, if f has no spectrum in $[-\lambda, \lambda]$, then

$$\|f\|_\infty \le \frac{\pi}{2\lambda}\|f'\|_\infty. \tag{8.1}$$

(For an interesting proof see Esseen [**36**].) L. Hörmander [**37**] has shown that (8.1) holds for functions in the Schwarz class φ', with spectrum defined in the distribution sense. He also showed that there is a more general theorem where only a one sided bound of the derivative is assumed.

A related and in a certain sense dual inequality of (8.1) is Bernstein's inequality (see [**58**, II, p.7]). If $f(x) = \sum_{\nu=-N}^{N} C_\nu e^{i\nu x}$ is a trigonometric polynomial of degree N, then

(8.2) $$\|f'\|_\infty \leq N\|f\|_\infty.$$

Bernstein's inequlaity also holds for entire functions of exponential type

$$\|f'\|_\infty \leq \tau\|f\|_\infty,$$

where τ is the type of f. (see, for example, Boas [**38**, p. 206].)

D. G. Northcott [**39**] proved the following result:

If A_k is the class of functions such that $f(x) = f(x+2\pi)$, $f(x), f'(x), \dots, f^{(k-1)}(x)$ are absolutely continuous, $\int_0^{2\pi} f(x)\,dx = 0$, then

(8.3) $$\|f\|_\infty \leq \frac{4}{\pi}\sum_{\nu=0}^{+\infty} \frac{(-1)^{\nu(k+1)}}{(2\nu+1)^{k+1}}\|f^{(k)}(x)\|_\infty.$$

Further, for each value of k, there is always a function $f(x)$ belonging to A_k and not identically zero, for which equality is valid in (8.3).

This is a known result of Favard (Zentralblatt für Mathematik **16**, 58–59).)

9. A very simple proof of Wirtinger's inequality is given by A. Dingas [40].

10. Generalizations of results from 7., i.e. of result from Chapter VII of [**33**] is given by V. I. Levin [**41**]. His paper contains the proof and some applications of three general inequalities:

(10.1) $$\int_0^1 f(x)y(x)^k\,dx \leq \int_0^1 y'(x)^k\,dx,$$

(10.2) $$\int_0^1 g(x)z(x)^{2q}\,dx \leq \int_0^1 z'(x)^{2q}\,dx,$$

and

(10.3) $$\int_0^\infty h(x)w(x)^k\,dx \leq \int_0^\infty w'(x)^k\,dx,$$

where k is a real number > 1; if k is an even integer, we put $k = 2q$.

In (10.1) $y(x)$ is the integral of y' ($y(0) = 0$), $y' \not\equiv 0$, $y' \geq 0$ if $k \neq 2q$ and $y' \in L^k(0,1)$; $f(x)$ is given by

$$f(x) = -(k-1)\frac{\phi'(x)^{k-2}\phi''(x)}{\phi(x)^{k-1}}, \tag{10.4}$$

where $\phi \in \phi_I^*$, i.e. a) $\phi(0) = 0$ b) $\phi'(1) = 0$, c) ϕ'' exists almost everywhere in $(0,1)$, $\phi'' \not\equiv 0$, $\phi'' \leq 0$, and d*) $(\phi^{1+\varepsilon})' \in L^k(0,1)$, where $\varepsilon > 0$ is arbitrarily small. If ϕ satisfies instead of d*) the condition d) $\phi' \in L^k(0,1)$, we say that $\phi \in \phi_I$. The sign of equality can occur in (10.1) only if $\phi \in \phi_I$, and then it occurs for $y = C\phi(x)$ ($C \geq 0$ if $k \neq 2q$); if $\phi \in \phi_I^* \setminus \phi_I$ we have in (10.1) strict inequality with the best constant.

In (10.2) $z(x)$ is the integral of z' ($z(0) = 0$), $z(1) = 0$, $z \not\equiv 0$ and $z' \in L^{2q}(0,1)$; $g(x)$ is given by (10.4) with $k = 2q$, where $\phi \in \phi_{II}^*$, i.e. ϕ satisfies a), b') $\phi(1) = 0$, c) and d*) with $k = 2q$. Accordingly, $\phi \in \phi_{II}$, if ϕ satisfies d) with $k = 2q$ instead of d*). The sign of equality can occur in (10.2) only if $\phi \in \phi_{II}$, and then it occurs for $z = C\phi(x)$; if $\phi \in \phi_{II}^* \setminus \phi_{II}$, we have in (10.2) strict inequality with the best constant.

Finally, in (10.3) $w(x)$ is the integral of w' ($w'(0) = 0$), $w' \not\equiv 0$, $w \geq 0$ if $k \neq 2q$ and $w' \in L^k(0,\infty)$; $h(x)$ is given by (10.4) ($f \to h$), where $\phi \in \phi_{III}^*$, i.e. ϕ satisfies a), c') ϕ'' exists almost everywhere in $(0,\infty)$, $\phi'' \not\equiv 0$, $\phi'' \leq 0$, and d'*) $(\phi^{1+\varepsilon})' \in L^k(0,\xi)$, $(\phi^{1-\varepsilon})' \in L^k(\xi,\infty)$ with a certain $0 < \xi < \infty$, where $\varepsilon > 0$ is arbitrarily small. We say that $\phi \in \phi_{III}$ if ϕ satisfies d') $\phi \in L^k(0,\infty)$ instead of d'*). The equality can occur on (10.3) only if $\phi \in \phi_{III}$ and then it occurs for $w = C\phi(x)$; if $\phi \in \phi_{III}^k \setminus \phi_{III}$, we have in (10.3) strict inequality with the best constant.

11. S. L. Sobolev [42],[43] considered the following inequalities:

$$\left(\int_B |f|^q \, dx\right)^{1/q} \leq C\left(\int_B \sum\left|\frac{\partial u}{\partial x_i}\right|^p dx\right)^{1/p}, \tag{11.1}$$

$$(q^{-1} = p^{-1} - n^{-1}, \; B \subset R^n)$$

and

(11.2)

$$\left(\int_B |f|^q\,dx\right)^{1/q} \le M\left(\int_B |f|^r dx\right)^{1/r} + N\left(\int_B \sum_{s_1,\ldots,s_n} \left|\frac{\partial^s f}{\partial x_1^{s_1}\ldots\partial x_n^{s_n}}\right|^p dx\right)^{1/p}$$

$(p \le n/s,\ q^{-1} = p^{-1} - s/n,\ 1 \le r \le q)$.

He only proved the existence of constants C, M and N.

12. A slightly different set of boundary conditions were considered by E. Schmidt [**44**]. Let f' be piecewise continuous on $[0,a]$ with $f(0) = f(a)$ and let $m = \min_{0\le x\le a} f(x)$ and $M = \max_{0\le x\le a} f(x)$ satisfy $m + M = 0$. Then for $p > 0$, $q \ge 1$

$$\left(\frac{1}{a}\int_0^a |f(t)|^p dt\right)^{1/p} \le \frac{1}{4}H\left(\frac{1}{p}, \frac{q-1}{q}\right)\left(a^{q-1}\int_0^a |f'(t)|^q dt\right)^{1/q} \tag{12.1}$$

where

$$H(u,v) = \frac{u^u v^v}{(u+v)^{u+v}}\,\frac{\Gamma(1+u+v)}{\Gamma(1+u)\Gamma(1+v)}.$$

If $m + M$ is not required to be zero and $p = q \ge 1$, then Schmidt showed that

$$\int_0^a |f(t) - \frac{1}{2}(m+M)|^p\,dt \le \frac{1}{p-1}\left(\frac{p}{4\pi}\sin\frac{\pi}{p}\right)^p a^p \int_0^a |f'(t)|^p\,dt. \tag{12.2}$$

Now if one takes $p = 2$ and $\int_0^a f(t)dt = 0$, we get

$$\int_0^a f(t)^2\,dt \le \frac{a^2}{4\pi^2}\int_0^a f'(t)^2\,dt - a\left(\frac{m+M}{2}\right)^2 \tag{12.3}$$

which improves (1.3). This inequality for $a = 2\pi$ appears in Benson [**45**]. Of course a scale change is not important in obtaining such results. For other results related to (12.1) see E. Schmidt [**46**], Bellman [**47**] and B. Sz. Nagy [**48**].

13. Northcott's result (8.3) was generalized by R. Bellman [**49**] who showed that if $f^{(n)} \in L^{2k}$, $f(x) = f(x + 2\pi)$ and $\int_{-\pi}^{\pi} f(x)dx = 0$, then

$$\int_{-\pi}^{\pi} f(x)^{2k}\, dx \le a_n^{2k} \int_{-\pi}^{\pi} f^{(n)}(x)^{2k}\, dx, \tag{13.1}$$

where k, n are integers and a_n are the numbers that appear on the right hand side of (8.3).

For $k = n = 1$ the inequality (13.1) is weaker than (1.2).

14. Wirtinger type inequalities have a close connection with differential equations. Beesack [**50**] established the following one.

THEOREM 1. *Let the differential equation*

$$y'' + py = 0 \tag{14.1}$$

be disconjugate (see Ch. 6) on $(-a, a)$, *i.e. let a solution* y_1 *exist such that* $y_1(x) > 0$ *on* $(-a, a)$. *Further assume* $\int_{-a}^{a} p(x)dx \ge 0$. *If* $f(\pm a) = 0$, $f' \in L^2$, *and* $\int_{-a}^{a} f(x)p(x)dx = 0$, *then*

$$\int_{-a}^{a} f'(x)^2\, dx \ge \int_{-a}^{a} p(x)f(x)^2\, dx. \tag{14.2}$$

Equality holds if and only if $f(x) = Ay_1(x)$ *where* $A = 0$ *if either* $y_1(a) \ne 0$ *or* $y_1(-a) \ne 0$.

Beesack also considered inequalities where $\int_{-a}^{a} f'(x)^2\, dx$ on the left in (14.2) is replaced by $\int_{-a}^{a} f''(x)^2\, dx$.

Other higher derivative versions are given by W. J. Kim [**51**] and Z. Nehari [**52**]. Kim's result reads as follows.

THEOREM 2. *Let* f *have* m *continuous derivatives with* $f^{(j)}(a) = f^{(j)}(b) = 0$, $j = 0, \dots, n-1$, *then*

$$\int_a^b f^{(m)}(x)^2 > \left(\frac{b-a}{2}\right)^{2m} \prod_{k=0}^{m-1} (2k+1)^2 \int_a^b \frac{f(x)^2}{(a-x)^{2m}(b-x)^{2m}} dx.$$

Beesack's [**53**] version of this is slightly different.

THEOREM 3. *Let* $f' \in L^{2k}$ *on* $[-\pi, \pi]$ *with* $f(\pi) = f(-\pi)$ *and* $\int_{-\pi}^{\pi} f(x)^{2k-1}\, dx = 0$ *then*

$$\int_{-\pi}^{\pi} f(x)^{2k}\, dx \leq \frac{1}{2k-1}\left(k \sin\frac{\pi}{2k}\right)^{2k} \int_{-\pi}^{\pi} f'(x)^{2k}\, dx.$$

See **[53]** for the cases of equality. Beesack also replaced the differential equation (14.1) by

$$(14.3) \qquad \left(r(x)\left(y'(x)^{p-1}\right)\right)' + s(x)y(x)^{p-1} = 0,$$

where r, r', and s are continuous on (a, b) with $r(x) > 0$ and $s(x) \geq 0$ where r may have a zero or a discontinuity at some single point.

Further let $p = 2k$ and $q = \frac{2k}{2k-1}$. Suppose there is a solution $y(x)$ of (14.3) on (a, b) which is absolutely continuous and $y(x) < 0$ on $(a, \bar{x})$ and $y(x) = 0$ on $(\bar{x}, b)$. Suppose y also satisfies

$$\frac{y'(x)}{y(x)} = O\left(\frac{1}{x-\bar{x}}\right), \quad \frac{y'(x)}{y(x)} = O\left(\frac{1}{x-a}\right), \quad \frac{y'(x)}{y(x)} = O\left(\frac{1}{b-x}\right)$$

for x near $\bar{x}$, a, b respectively.

The multiplier $r(x)$ must satisfy

$$r(x) = O\left((x-\bar{x})^{p-1}\right) \text{ or } r(x)^{q/p} \int_{\bar{x}}^{x} r(t)^{-q/p}\, dt = O(x-\bar{x}),$$

near $\bar{x}$;

$$r(x) = O\left((x-a)^{p-1}\right) \text{ and } r(x)^{q/p} \int_{x} r(t)^{-q/p}\, dt = O(x-a),$$

near a; and

$$r(x) = O\left((b-x)^{p-1}\right) \text{ and } r(x)^{q/p} \int^{x} r(t)^{-q/p}\, dt = O(b-x)$$

near b.

If f is absolutely continuous with $\int_a^b r(t)f'(t)^p\, dt < \infty$ and $\int_a^b s(t)f(t)^{p-1}\, dt < \infty$ then

$$(14.4) \qquad \int_a^b s(t)f(t)^p\, dt \leq \int_a^b r(t)f'(t)^p\, dt$$

with equality if and only if $f(x) = cy(x)$ where $c = 0$ unless $\int_a^b r(t)y'(x)^p < \infty$.

15. Various generalizations of (1.2) along the lines of Beesack are given by W. J. Coles in [**54**], [**55**]. One of them is given here. Let m be an integer and set $n = 2m$. Let k_i be numbers that are 0 and 1 so that $\sum_{i=0}^{m} k_i$ is even. Define $h_i = \sum_{j=0}^{i} k_i$ and

$$p_i = (-1)^{h_i},\ q_i = (-1)^i p_i,\ c_i = (1-k_i)a + k_i b,$$
$$d_i = k_{i+1}a + (1-k_{i+1})b, \text{ and } d_i' = a + b - d_i.$$

Let $p(x)$ be continuous and assume the differential equation

$$y^{(n)}(x) - p(x)y(x) = 0 \tag{15.1}$$

has a solution $y(x)$ such that $(-1)^m p(x)y(x) \geq 0$, with strict inequality somewhere, and satisfying the boundary conditions

$$p_i y^{(m-1)}(c_i) \geq 0,\ 1 \leq i \leq m;\ q_i y^{(m+i)}(d_i) \geq 0,\ 0 \leq i \leq m-1.$$

Then for functions f such that $f^{(m)} \in L^2$ and $f^{(i)}(d'_{m-i-1}) = 0 \quad 0 \leq i \leq m-1$ we have

$$(-1)^m \int_a^b p(x)f(x)^2\,dx \leq \int_a^b f^{(m)}(x)^2\,dx.$$

For cases of equality, see Coles' papers. He also discusses inequalities of the form

$$\sum_{i=1}^{n}(-1)^{n+i}\int_a^b p_i(x)f^{(i)}(x)^2 dx \geq 0.$$

16. Inequality (14.4) was also considered in Beesack [**56**], i.e.

$$\int_a^b s|u|^p\,dx \leq \int_a^b r|u'|^p\,dx \qquad (p < 0 \text{ or } p > 1) \tag{16.1}$$

or the same inequality with $\leq$ replaced by $\geq$ if $0 < p < 1$.

In (16.1) *we suppose that* r, s *are continuous and positive on* (a, b), *where* $-\infty \le a < b \le \infty$. The Euler differential equation corresponding to the minimum problem (16.1) is

$$(16.2) \qquad \left(ry'^{p-1}\right)' + sy^{p-1} = 0 \text{ or } \left[r(-y')^{p-1}\right]' - sy^{p-1} = 0$$

if we suppose $y > 0$, $y' > 0$ or $y > 0$, $y' < 0$ respectively. In (16.1) the *admissible functions* u *are those of class* $AC(a, b)$ *which satisfy* $u(x) = \int_a^x u'\,dt$ *or*

$$u(x) = \int_x^b u'\,dt \quad \textit{as specified below.}$$

Suppose the first (second) of differential equations (16.2) *has a solution* y *such that* $y > 0$ *and* $y' > 0$ $(y' < 0)$ *on* (a, b) *with* $y(x) = \int_a^x y'\,dt$ $\left(y(x) = -\int_x^b y'\,dt\right)$ *while as* $x \to a+$ $(x \to b-)$ *we have*

$$f\left(\frac{y'}{y}\right)^{p-1}\left(\int_a^x r^{-q/p}\,dt\right)^{p/q} = O(1),$$

$$\left(r\left(-\frac{y'}{y}\right)^{p-1}\left(\int_x^b r^{-q/p}\,dt\right)^{p/q} = O(1)\right),$$

where $p > 1$ *and* $q = p/(p-1)$. *Let* u *be any function such that* $\int_a^b r|u'|^p\,dx < \infty$ *and* $u(x) = \int_a^x u'\,dt$ $\left(u(x) = \int_x^b u'\,dt\right)$. *Then the inquality* (16.1) *holds. Moreover equality holds in* (16.1) *if and only if* $u = cy$ *where* $c = 0$ *unless* $\int_a^b r|y'|^p\,dx < \infty$ *and* $r\ y|y'|^{p-1}(x) = o(1)$ *as* $x \to a+$ *(as* $x \to b-$*).*

By taking $y(x) = x^{(k-1)/p}$ and $r(x) = \lambda x^{p-k}$ with $\lambda = (p/|k-1|)^p$ we get the case $p > 1$ of the well-known Hardy's inequality. Many other special cases of the inequality (16.1) are given in [**56**] of which we list only the following here: *If* $p > 1$, $u \in AC[0, \pi/2)$ *with* $u(0) = 0$, *then*

$$(16.3) \qquad \int_0^{\pi/2} u^p\,dx \le \frac{1}{(p-1)}\left(\frac{p}{2}\sin\frac{\pi}{p}\right)^p \int_0^{\pi/2} u'^p\,dx,$$

and equality holds only if $u = cy(x)$ *where* $y(x)$ *is the unique solution of the equation* $x = \frac{1}{2}p\sin(\pi/p)\int\limits_0^y(1-t^p)^{-1/p}\,dt$ $0 \le y \le 1$. (cf. E. Schmidt [44] and [**33**, Th. 256], i.e. (iv) from 7.)

17. A. Yu Levin [**57**] gave the following two results:

THEOREM 1. *Let* $x(t)$ *be* n *times continuously differentiable on the interval* $[a,b]$, *and satisfy the conditions*

$$x(s_1) = x'(s_2) = \cdots = x^{(n-1)}(s_n) = 0 \tag{17.1}$$

where $s_1, s_2, \ldots, s_n$ *are certain points in the interval* $[a,b]$. *Then the estimate*

$$|x(t)| \le C_n(b-a)^n \max_{a\le t\le b}|x^{(n)}(t)|, \qquad a \le t \le b, \tag{17.2}$$

is valid on $[a,b]$, *where the numbers* $C_1, C_2, \ldots$ *are defined by the expansion*

$$\tan t + \sec t = 1 + \sum_{k=1}^{\infty} C_k t^k \qquad \left(|t| < \frac{\pi}{2}\right). \tag{17.3}$$

With regard to (17.3) the values of C_k can be expressed as follows in terms of the Bernoulli numbers B_k and the Euler numbers E_k

$$C_{2n-1} = 2^{2n}(2^{2n}-1)B_n/(2n!), \quad C_{2n} = E_n/(2n!).$$

THEOREM 2. *Let* $x(t)$ *satisfy conditions* (17.1), *where we have one of the chains of inequalities*

$$(a\le)\ a_2 \le a_3 \le \cdots \le a_n\ (\le b)$$
$$(a\le)\ a_n \le a_{n-1} \le \cdots \le a_2\ (\le b).$$

Then the estimate

$$|x(t)| \le \frac{1}{n\left[\frac{n-1}{2}\right]!\left[\frac{n}{2}\right]!}(b-a)^n \max_{a\le t\le b}|x^{(n)}(t)|, \tag{17.4}$$

holds on $[a,b]$.

Levin himself noted that Theorem 1 had been obtained by S. N. Bernštein in 1910 ([**58**]).

M. Hukuhara [59] indicates that M. Tumura [50] has found inequalities in the form

$$\|f\|_\infty \le (n-1)^{n-1}(b-a)^n\|f^{(n)}\|_\infty/n!n^n \tag{17.5}$$

and

$$\|f^{(k)}\|_\infty \le k(b-a)^{n-k}\|f^{(n)}\|_\infty/n(n-k)!, \quad k=1,\dots,n-1,$$

for $f \in C^n[a,b]$ having n zeros in $[a,b]$ including zeros at a and b.

These results are also obtained by G. A. Bessmertnyh and A. Ju. Levin [61]. A proof of (17.4) is also given in Levin [62].

18. J. Brink [63] considered generalizations of these results replacing the infinity norm by any of the usual p-norms. In particular he has considered the inequality (1.1) with the hypotheses that f has n zeros on the interval $[a,b]$. If $1 < p \le \infty$ then he proves that one may assume that there are n zeros at the endpoints. He therefore considered the problem: Find the best possible constants $K(n,k,p,q)$ such that

$$\|f\|_p \le K(n,k,p,q)(b-a)^{n+\frac{1}{p}-\frac{1}{q}}\|f^{(n)}\|_q \tag{18.1}$$

where f has a k-fold zero at a and an $(n-k)$-fold zero at b, $(0 \le k \le n)$. Let $1/p+1/p'=1$ with the usual conventions when $p=1$. He shows that for $B(x,y)$ the beta function

(18.1′)

$$K(n,k,p,\infty) = \frac{B(pk+1\ pn-pk+1)}{n!}, \qquad 1 \le p < \infty,$$

with

$$K(n,k,\infty,\infty) = \frac{k^k(n-k)^{n-k}}{n!(n-k)!}.$$

In particular (17.5) is best possible. Furthermore he showed that

$$K(n,k,1,q) = K(n,k,q',\infty) \tag{18.2}$$

so that these numbers are known explicitly. Furthermore, the constants may be of interest to study in themselves. For example Brink has shown that:

(a) $K(n,k,\cdot,q)$ is increasing;
(b) $K(n,k,p,\cdot)$ is decreasing; and
(c) $K(n,k,\cdot,q)$ is log convex.

Furthermore he has derived a (generally nonlinear) differential equation for the extremal. In particular he can in this way show that $K(2,1,2,2) = 1/\pi^2$ which is a version of Wirtinger's inequality. Moreover it follows from a) and b) that

(18.3) $$K(n,k,1,\infty) \le K(n,k,p,q) \le K(n,k,\infty,1).$$

The left hand side is known by (18.2) and it is conjectured that the right hand side is less than $K(n,0,\infty,1)$ which in turn is known to be $(n-1)!$.

19. Fink [64],[65],[66] continued the study of these kinds of inequalities. In particular he found simple proofs of the properties a), b), c) of 18., and gave succinct arguments for the computation of extremals. Extremals exist in all cases for the problem (18.1) except for $(p,q) = (1,1)$,

In particular he showed that (18.2) can be extended to the general case:

(19.1) $$K(n,k,p,q) = K(n,k,q',p').$$

Thus the two inequalities are related. He called them conjugate inequalities. Generally, if one replaces the zero conditions of Brink's problem by some other boundary conditions, then the related problem has boundary conditions that are in some sense adjoint to the original ones. For example consider the problems:

(19.2) $$\|f\|_p \le D(n,p,q)(b-a)^{n+\frac{1}{p}-\frac{1}{p}}\|f^{(n)}\|_q$$
with f having n-fold zeros at a and at b;

(19.3) $$\|f\|_p \le F(n,p,q)(b-a)^{n+\frac{1}{p}-\frac{1}{q}}\|f^{(n)}\|_q$$
with $|f|^{p-1}\ \mathrm{sgn}(f) \perp \pi_{n-1}$;

where π_n is the set of polynomials of degree $\le n$. Here $g \perp f$ means $\int_a^b f(x)g(x)dx = 0$.

Then it is shown that for $p < \infty$ and $1 < q$

(19.4) $$F(n,p,q) = D(n,q',p').$$

The conditions are connected in that if f satisfies the boundary conditions of (19.2) then $f^{(n)} \perp \pi_{n-1}$ as well as the integration by parts formula, $\int_a^b f^{(n)}g = (-1)^n\int_a^b fg^{(n)}$ for g satisfying the boundary conditions for (19.3). Fink also shows that $F(n+1,p,q)$ is the best possible constant in the inequality

(19.5) $$\min_{p\in\pi_n}|f(x)-p(x)| \le B(n,p,q)(b-a)^{n+\frac{1}{p}-\frac{1}{q}}\|f^{(n)}\|_q,$$

i.e. $B(n,p,q) = F(n+q,p,q)$. Phillips **[218]** has shown that

$$B(n,p,\infty) = \frac{\delta(n-1,p)}{n!} \tag{19.6}$$

where $\delta(n,p) = \min_{q\in\pi_n} \|x - q(x)\|$.

The constants D and F satisfy conditions a), b) and c) of article 18. Some of the constants can be computed as approximation problems to certain splines. However one always has

$$\frac{1}{(n+1)!2^{2n+2}} = B(n,1,\infty) \le B(n,p,q) \le B(n,\infty,1) = \frac{1}{n!2^{2n}}. \tag{19.7}$$

Other results are $B(n,2,q) = K(2n+2,n+1,q',q)^{1/2}$ with $B(n,2,\infty) = ((n+1)!/(2n+2)!)\,(2n+3)^{-1/2}$. Also $B(n,2,1) = (n!2^{2n+1})^{-1}(2n+1)^{-1/2}$. The latter two give upper and lower bounds for $B(n,2,2)$.

Other boundary conditions are considered in **[64]** but we shall here mention those in **[66]**. The inequalities are of the form

$$\begin{gathered}\|f\|_p \le H(n,p,q)(b-a)^{n+\frac{1}{p}-\frac{1}{q}}\|f^{(n)}\|_q \\ \text{with } f^{(i)}(a_i) = 0, \text{ for } a \le a_0 \le a_1 \le \cdots \le a_{n-1} \le b.\end{gathered} \tag{19.8}$$

For $1 < p, q$ one may take all the a_i at the ends of the intervals. See the discussion in Chapter VI on disconjugacy for applications of these inequalities. Again the properties a), b), and c) of 18. hold. We particularly want to mention the following special cases:

$$H(n,1,1) = H(n,\infty,\infty) = \binom{n-1}{\left[\frac{n-1}{2}\right]} \Big/ n! \tag{19.9}$$

$$\begin{aligned}H(n,1,\infty) &= \binom{n}{\left[\frac{n}{2}\right]} \Big/ (n+1)!, \\ H(n,\infty,1) &= \binom{n-2}{\left[\frac{n-2}{n}\right]} \Big/ (n-1)!\end{aligned} \tag{19.10}$$

These numbers serve as crude upper and lower bounds for $H(n,p,q)$, however various other better upper and lower bounds are obtained. In particular this provides another proof of (17.4) and verifies that it is a best possible result.

20. H. Federer, W. Fleming and R. Rishel **[67]**, **[68]**,**[69]** proved the inequality:

$$\left(\int_{R^m} |u|^{m/(m-1)}\,dx\right)^{1-1/m} \le \frac{(\Gamma(1+m/2))^{1/m}}{\sqrt{\pi}\,m} \int_{R^m} |Du|\,dx \tag{20.1}$$

where $m > 1$, $|Du|$ is the length of the gradient Du of u, for every u belonging to a broad class of functions which vanish at infinity. The inequality (20.1) is sharp.

Note that the constant in the right-hand side of (20.1) is the isoperimetric constant.

In connection with this result are papers of M. Miranda [**70**] [**71**],[**72**]. Sobolev's inequalities are considered in his book [**73**].

21. A slightly different weighted version of (1.2) is given by Troesch [**74**]. Let h be a positive function with $-h$ convex and h' piecewise continuous on $[0,1]$ and $h'(0) \le 0$. If f' is piecewise continuous with $f(0) = 0$, then

$$\frac{\int_0^1 h(x) f'(x)^2\, dx}{\int_0^1 h(x)\, dx \int_0^1 f(x)^2\, dx} \ge \frac{\pi^2}{4} \tag{21.1}$$

with equality if and only if h is a constant and $f(x) = A\sin\left(\frac{\pi x}{2}\right)$. This is connected with the results in 3. and those in 67. Related results are given by Troesch in [**75**].

22. Inequalities which involve linear combination of values of f that are somewhat like (1.2) are given by J. B. Diaz and F. T. Metcalf [**76**].

THEOREM 1. *If f is continuously differentiable on (a,b) and suppose $f(t_1) = f(t_2)$ for $a \le t_1 \le t_2 \le b$, then*
(22.1)

$$\int_a^b [f(t) - f(t_1)]^2\, dt \le \frac{4}{\pi^2} \max\left((t_1 - a)^2, (b - t_2)^2, \left(\frac{t_2 - t_1}{2}\right)^2\right) \int_a^b f'(x)^2\, dx.$$

If moreover $f(a) = f(b)$ then the right hand side of (22.1) may be replaced by

$$\frac{1}{\pi^2} \max\left((t_2 - t_1)^2, (b - a - t_2 + t_1)^2\right) \int_a^b f'(x)^2\, dx.$$

If further, all of the above hypotheses and $(b-a) f(t_1)^2 \ge 2 f(t_1) \int_a^b f(t) dt$ holds, then

$$\int_a^b f(x)^2\, dx \le \frac{1}{\pi^2} \max\left((t_2 - t_1)^2, (b - a - t_2 + t_1)^2\right) \int_a^b f'(x)^2\, dx.$$

Cases of equality are also given.

In this paper [**76**] more similar results are also proved.

23. In another direction one can replace integrals by sums of integrals as in the following result, obtained in various ways by I. Halperin and H. Pitt [**77**], W. Müller [**78**], L. Nirenberg [**79**], and R. Redheffer [**80**]. If f'' is continuous on $[a,b]$ then for every $\varepsilon > 0$

$$(23.1) \qquad \int_a^b f'(x)^2\,dx \le \varepsilon \int_a^b f''(x)^2\,dx + K(\varepsilon)\int_a^b f(x)^2\,dx$$

where $K(\varepsilon) = \frac{P}{\varepsilon} + \frac{Q}{(b-a)^2}$ and $P = 1$, $Q = 12$. This is best possible in the sense that if $P < 1$ or $Q < 12$ then (23.1) is not correct for all f, and $\varepsilon > 0$.

A. M. Pfeffer [**81**] generalized (23.1) to

$$(23.2) \qquad \int_a^b f^{(k)}(x)^2\,dx \le \varepsilon \int_a^b f^{(m)}(x)^2\,dx + H_{k,m}(\varepsilon)\int_a^b f(x)^2\,dx,$$

where one assumes that $f^{(m)}$ is continuous $1 \le k < m$, $f^{(i)}(a) = f^{(i)}(b)$, $i = 0,\dots,m-1$ and $\varepsilon > 0$. The best possible inequalities hold when

$$H_{k,m}(\varepsilon) = \begin{cases} 0, \quad \text{if } \varepsilon > \left(\frac{b-a}{2\pi}\right)^{2m-2k}, \\ \left(\frac{k}{m\varepsilon}\right)^{\frac{k}{m-k}}\left(1-\frac{k}{m}\right), \text{ if } J \text{ is a positive integer,} \\ \max\left(\left(\frac{2J\pi}{b-a}\right)^{\varepsilon k} - \varepsilon\left(\frac{2J\pi}{b-a}\right)^{2m}, \left(\frac{2(J+1)\pi}{b-a}\right)^{2k} - \varepsilon\left(\frac{2(J+1)\pi}{b-a}\right)^{2m}\right), \text{ else,} \end{cases}$$

where $J = \frac{b-a}{2\pi}\left(\frac{k}{m\varepsilon}\right)^{\frac{1}{2m-2k}}$.

The exponent $\frac{k}{m-k}$ replaces the typographical error $\frac{k}{m} - k$ noticed by Beesack.

Note that inequalities of this type are connected to the well- known Kolmogorov inequalities. So many other results, related to the above, are given in Chapter I on Kolmogorov's inequalities.

24. H. D. Block [**82**] has formed a class of integral inequalities which contain inequalities of Wirtinger's type. Here, we shall give some special cases of Block's general results.

Let y be a real function, defined and continuous on $a \le x \le b$ and having a derivative which is sectionally continuous. Then for every $c \ne 0$

$$|y(t)|^2 \le \frac{\coth c(b-a)}{c}\int_a^b \left((y')^2 + c^2y^2\right)dx.$$

If $y(a) = 0$, then

$$|y(t)|^2 \leq \frac{\tanh c(b-a)}{c} \int_a^b ((y')^2 + c^2y^2)\, dx.$$

If $y(a) = y(b) = 0$, then

$$|y(t)|^2 \leq \frac{\tanh \frac{c(b-a)}{2}}{2c} \int_a^b ((y')^2 + c^2y^2)\, dx,$$

and (when $c \to 0$)

$$|y(t)|^2 \leq \frac{b-a}{4} \int_a^b (y')^2\, dx.$$

25. R. Redheffer [**83**] gave several interesting related results. For example he gave some comments about (21.1), an inequality of Block type (see 24.), etc. Here we shall give his inequality of Wirtinger type:

Let v and w be absolutely continuous and let $v > 0$, $v' > 0$, $w > 0$, $w' \leq 0$. Then $u(a+) = 0$ implies

$$\int_a^b \frac{u'}{r'} w\, dx \geq -\int_a^b \frac{w'}{v} u^2\, dx$$

and the inequality is strict unless v satisfies

$$\int_a^b v'w\, dx < \infty, \qquad \lim_{x\to a+} v(x) = 0, \qquad \lim_{x\to b-} v(x)w(x) = 0.$$

26. Some related results are also given in papers by Rybarski [**84**] and Krzywicki and Rybarski [**85**],[**86**]. For example, the following result is proved in [**86**]:

Let a real function $f(t)$, defined on the interval $[-1,1]$ and absolutely continous there, satisfy the boundary conditions $f(-1) = f(1) = 0$.

If $a \geq -1$ then

$$(1+a)\int_{-1}^{1} p^{a+4} f^2 \leq \int_{-1}^{1} p^a f'^2, \tag{26.1}$$

where $p = (1 - t^2)^{-1/2}$.

There is equality in (26.1) only if either $a > 0$ and $f = \text{const}\,(1 - t^2)^{(1+a)/2}$ or $a \le 0$ and $f = 0$ besides the trivial case when the integrals in (26.1) become infinite.

We also have, for $a > -2$,

$$(26.2) \qquad (2 + a) \int_{-1}^{1} p^{a+2} f^2 \le \int_{-1}^{1} p^2 f'^2.$$

There is equality in (26.2) only if $f = \text{const}(1 - t^2)^{1+a/2}$ besides the trivial case as above.

27. V. I. Levin and S. B. Stečkin [**87**] have indicated a number of interesting analogues of inequality (1.2). Related results for functions of several variables are given in papers of J. Calvert [**88**],[**89**], S. F. Morozov and V. I. Plotnikov [**90**] and D. Mangeron and A. Oprea [**91**]. For example, in his papers J. Calvert considered the following norms:

$$\|u\|_q = \left(\int_D \sum_{i=1}^{n} |D_i^j u|^2 + q|u|^2 \, dx \right)^{1/2}, \quad j = 1, 2,$$

where q is a function of special type.

28. A differently weighted version of (1.2) was studied by D. W. Boyd [**92**] namely inequalities of the form

$$(28.1) \qquad \int_a^b |f(x)|^p |f^{(n)}(x)|^q w(x) \, dx \le K \left(\int_a^b |f^{(n)}(x)|^r v(x) \, dx \right)^{\frac{p+q}{r}}.$$

He gave a method for discovering the best possible constants K under the assumptions that $f^{(n-1)}$ is absolutely continuous and f has n zeros at a. For sufficiently smooth weights v and w the extremal solves a boundary value problem for a differential equation. For $n = 1$ and $v \equiv w \equiv 1$ he obtained the best constants.

The cases $p = q = r = 2$, $n = 1$; $p = 4$, $q = 0$, $r = 2$, $n = 1$ for $v \equiv w \equiv 1$ are considered by O. Aramă and D. Ripianu [**93**] under the hypothesis that $f(b) = 0$ and they do not obtain the best possible constants.

The inequality generalizes both Wirtinger's inequality (1.2) and Opial's inequality which is considered in the next Chapter.

29. Discrete versions of Wirtinger's inequality are given by various people. K. Fan, O. Taussky, and J. Todd [**94**] proved that if $x_1 = 0$ then

$$\sum_{k=1}^{n-1} (x_k - x_{k-1})^2 \geq 4\sin^2 \frac{\pi}{2(2n-1)} \sum_{k=1}^{n} x_k^2 \tag{29.1}$$

with equality if and only if

$$x_k = \sin\frac{(k-1)\pi}{2n-1}, \quad k = 1, \ldots, n.$$

If instead $x_{n+1} = x_1$ and $\sum_{k=1}^{n} x_k = 0$ then

$$\sum_{k=1}^{n} (x_k - x_{k-1})^2 \geq 4\sin^2 \frac{\pi}{n} \sum_{k=1}^{n} x_k^2 \tag{29.2}$$

with equality if and only if

$$x_k = A\,\cos\frac{2k\pi}{n} + B\,\sin\frac{2k\pi}{n}, \quad k = 1, \ldots, n.$$

If in (29.2) one passes to the limit one recovers (1.2).

The paper [**94**] also considers discrete analogues of (6.1). Related results are given by H. D. Block [**95**]. A. M. Pfeffer [**81**] gave discrete analogues of (23.1) and (23.2). A. M. Fink [**96**] also gave discrete versions of (1.2) where he gave a theory parallel to that of 18. for the continuous case. Further results can be found in papers of R. H. Sadikova [**97**], J. Novotá [**98**], and G. V. Milovanović and I. Ž. Milovanović [**99**] that sharpens 29.2. Let $n = 2m$, $n \geq 4$ and suppose $\sum_{i=1}^{n} x_i = 0$ with $x_{n+i} = x_i$ $i = 1, \ldots, m$. Then

$$\sum_{i=1}^{n} (x_i - x_{i+1})^2 \geq \left(\sin^2\frac{2\pi}{n} - \sin^2\frac{\pi}{n}\right) \sum_{i=1}^{n} (x_i + x_{i+m})^2 + 4\sin^2\frac{\pi}{n}\sum_{i=1}^{n} x_i^2$$

with equality if and only if

$$x_i = A\,\cos\frac{2\pi i}{n} + B\,\sin\frac{2\pi i}{n} + C\,\cos\frac{4\pi i}{n} + D\,\sin\frac{4\pi i}{n}$$

$i = 1, \ldots, n$.

30. In the papers [**100**] of Mitrinović and Vasić, and [**101**] of Janet the history of inequality (1.2) and its relatives are given. The name of Wirtinger's inequality is not justified. In the paper [**100**] the topics are exposed chronologically as in this Chapter, where the text has been updated and enlarged.

31. Extensions of 7(i) are given by R. Redheffer [**102**]. Some results concerning Wirtinger's inequality can be found in papers of U. Richard [**103**] and A. Lupaş [**104**].

32. Generalizations of results from 14. and 16. are given by the same author in [**105**].

33. There exists a constant M such that

$$\|f\|_\infty \leq M \sup_{|x-y|\leq 1/\lambda} |f(x) - f(y)|$$

for every $f \in L^\infty$ with no spectrum in $[-\lambda, \lambda]$.

This is a variant of Bohr's inequality (see 8.). Inequalities of such type have been used by T. Ganelius [**106**],[**107**] in his work on Tauberian remainder theorems. Ganelius also suggested that this inequality could be obtained in a more general group context, and paper [**108**] of L. Landberg is devoted to a discussion of this matter.

34. If ϕ is a real-valued smooth (e.g. C^2) function defined on three-dimensional Euclidean space and $\|\cdot\|$ is the L^2 norm, then

$$\|\phi^3\| \leq c\|\nabla\phi\|^3. \tag{34.1}$$

This was shown by L. Nirenberg [**109**], and with the constant $c = 4/\pi$ by H. Fujita [**110**]. G. Rosen [**111**] gave the best possible constant

$$c = \frac{4}{3}/\sqrt[3]{\pi^2} = 0.078\ldots . \tag{34.2}$$

35. Some results about inequality (6.1) (for $p = 0$) is given by Z. Denkowski [**112**] as well as some discrete analogues.

36. I. J. Schoenberg [**113**] proved Theorem 1 from 17. but with less restrictive conditions for functions, i.e. $x(t) \in C^{n-1}(I)$, $x^{(n-1)}(t)$ satisfies a Lipshitz condition and $x^{(\nu)}(t)$ $(\nu = 0, 1, \ldots, n-1)$ vanishes at some points of I. He also noted that his result for entire functions from 1936 [**114**] is a consequence of these results.

37. A related result for functions of several result is given by V. G. Hryptun [**115**]. One-dimensional case of his result is proved by S. Mandelbrojt [**116**] in 1940.

38. Beesack's paper [**50**] gives a connection of Wirtinger's inequality to differential equations. Generalizations of his results with respect to partial differential equations are given by P. K. Wong [**117**],[**118**],[**119**]. (See also [**120**] of M. J. Winter and P. K. Wong.)

39. A paper [**121**] of J. Peetre is in connection to a result given in 24.

40. G. A. Šadrin [**122**] studied the best constants in some inequalities of the form

$$\int_0^\pi \rho u^2\, dx \le C \int_0^\pi \left(pu_x^2 + qu^2\right)\, dx \quad (\rho(x) > 0,\ p(x) > 0,\ q(x) > 0). \tag{40.1}$$

41. N. Anderson, A. M. Arthurs and R. R. Hall [**123**] proved the following result: if

$$y(0) = y'(0) = y(L) = y'(L) = 0 \tag{41.1}$$

then

$$\int_0^L y''(x)^2\, dx \ge \frac{\pi^2 K}{L^4} \int_0^L y(x)^2\, dx \tag{41.2}$$

where K is the smallest root of

$$\tanh \frac{\pi K^{1/4}}{2} + \tanh \frac{\pi K^{1/4}}{2} = 0. \tag{41.3}$$

It is shown in the appendix of [**123**] that $K \doteqdot 81/16$ and that the inequality is best possible.

42. A new proof of (29.1) is given by O. Shisha [**124**]. In connection to (26.1) is a paper [**125**] of B. Florkiewicz and A. Rybarski. Some inequalities for functions of two variables are considered by S. Easwaran [**126**].

43. Let f be a 2π-periodic function which is continuous and absolutely continuous over a period such that $f' \in L^p[0, 2\pi]$, where $1 < p \leq 2$. If there exists a natural number $m > 1$ such that the Fourier coefficients of f satisfy $a_k = 0$ for $k = 0, 1, 2, \ldots, m-1$ and $b_k = 0$ for $k = 1, 2, \ldots, m-1$, then

$$|f(x)| \leq 2^{-1/p} \|f'\|_p m^{-1/q} \quad \text{where } p^{-1} + q^{-1} = 1. \tag{43.1}$$

This inequality of A. A. Jagers is an improvement of an inequality of W. A. J. Luxemburg (see [**127**]).

44. For a result similar to (43.1), see Tsatsanis [**128**].

45. Let u be any real (or complex) valued function, defined on the whole m-dimensional Euclidean space R^m, sufficiently smooth and decaying fast enough at infinity. Moreover let p be any number such that: $1 < p < m$. Then

$$\left(\int_{R^m} |u|^q \, dx \right)^{1/q} \leq C \left(\int_{R^m} |Du|^p \, dx \right)^{1/p} \tag{45.1}$$

where: $|Du|$ is the length of the gradient Du of u, $q = mp/(m-p)$ and

$$C = \pi^{-1/2} m^{-1/p} \left(\frac{p-1}{m-p} \right)^{1-1/p} \left(\frac{\Gamma(1+m/2)\Gamma(m)}{\Gamma(m/p)\Gamma(1+m-m/p)} \right)^{1/m}. \tag{45.2}$$

The equality sign holds in (45.1) if u has the form

$$u(x) = \left(a + b|x|^{p/(p-1)} \right)^{1-m/p}, \tag{45.3}$$

where $\|x\| = \left(x_1^2 + \cdots + x_m^2 \right)^{1/2}$ and a, b are positive constants.

This result is due to G. Talenti [**129**]. If $p \to 1$, we get (22.1).

Talenti also gave an analogue of the Sobolev inequality in one dimensional case:

(45.4)

$$\left(\int_0^L |u|^q \, dx \right)^{1/q} \leq L^{1/q+1/p'} \frac{q(1+p'/q)^{1/p}}{2(1+q/p')^{1/q} B(1/q, 1/p')} \left(\int_0^L |u'|^p \, dx \right)^{1/p},$$

where u is any real valued absolutely continuous function on a bounded interval $[0, L]$ with $u(0) = u(L) = 0$ and p, p', q are numbers such that $1 \leq q < \infty$, $1 < p < \infty$, $p' = p/(p-1)$. The constant in (45.4) is the best.

46. Wirtinger (or Sobolev) type inequalities are also considered by N. C. Meyers [**130**].

47. Sobolev type inequalities are also considered in papers of G. Rosen [**131**], F. B. Weissler [**132**], T. Aubin [**133**] and A. Alvino [**134**].

48. Let $M(f) = -f'' + qf$ be a given differential operator. Inequality

$$a_0\|qf\|^2 + a_1\|qf^{1/2}f'\|^2 + a_2\|f''\|^2 \leq \|M(f)\|^2$$

is considered by W. N. Everitt and M. Giertz [**135**]. Related inequalities are considered by the same authors in [**135**].

Inequality

$$\int_a^b (p|f'|^2 + q|f|^2) \geq \lambda \int_a^b w|f|^2$$

was considered in papers [**137**], [**138**] of J. S. Bradley and W. N. Everitt and [**139**],[**140**] and [**141**] of R. J. Amos and W. N. Everitt. A similar inequality is established by W. N. Everitt [**142**].

49. If $f \in C^2(-\infty, \infty)$ and $\lambda > 0$, then

$$\|f\|_\infty \leq \frac{1}{\lambda}\|f'' - \lambda f\|_\infty, \quad \|f'\| \leq \frac{1}{\sqrt{\lambda}}\|f'' - \lambda f\|.$$

These results are due to J. S. Muldowney [**143**].

50. G. Talenti [**144**] proved: Let f be twice continuously differentiable on the positive half line and let $\int_0^\infty x f(x)^2\, dx$ and $\int_0^\infty x f''(x)^2\, dx$ be convergent. Then $\int_0^\infty x f'(x)^2\, dx$ is convergent too and

$$\int_0^\infty x f'(x)^2\, dx \leq C \int_0^\infty x\left(f(x)^2 + f''(x)^2\right) dx$$

where $C = 1/(2(2\nu - 1))$ is the best posible constant, ν being the solution between 1/2 and 1 of

$$\int_0^{\pi/2} (1 - 2\nu \sin^2 t)(1 - \nu \sin^2 t)^{-1/2}\, dt = 0.$$

51. The following extension of Wirtinger's inequality is given by W. N. Everitt [**145**]:

Let p be a given positive integer and let $f \in AC[0, 2\pi]$ with: $f' \in L^2(-2\pi)$, $f(0) = f(2\pi)$ and

$$\int_0^{2\pi} f(x) \cos nx\, dx = \int_0^{2\pi} f(x) \sin nx\, dx = 0 \quad \text{for } n = 0, 1, \ldots, p - 1.$$

Then the following inequality is valid

$$\|f'\| \geq p\|f\| \quad \text{i.e.} \quad \int_0^{2\pi} |f'|^2 \geq p^2 \int_0^{2\pi} |f|$$

with equality if and only if for some A and B

$$f(x) = A \cos px + B \sin px \ (x \in [0, 2\pi]).$$

52. Best constants in integral inequalities of the form (1.2) for functions with n zeros are considered by Ju. A. Melencova [**146**].

53. An interesting generalization of Wirtinger's inequality is given by J. M. Feinberg [**147**]:

Let f be a real-valued piecewise C^1 function defined on a finite closed interval $[a, b]$, with $f(a) = f(b)$. If $\nu \geq 1$ is the quotient of odd integers, and $\int_a^b f^\nu\, dt = 0$, $f \not\equiv 0$, then

$$\int_a^b |f'|^\nu\, dt \geq K(\nu) \cdot (4/(b - a))^\nu \cdot \int_a^b |f|^\nu\, dt$$

holds with $K(1) = 1$ and $K(\nu) = (\nu - 1)((\pi/\nu)\csc(\pi/\nu))^\nu$, $\nu > 1$. The constant $K(\nu)$ is best possible in each case, but is not attained for $\nu = 1$. The case $\nu = 1$

is extended to vector-valued functions, and a discrete analogue of the result in this case is given in the same paper.

54. Generalizations of Wirtinger's inequality are given in papers **[148]** of C. A. Swanson and **[149]** of K. Kreith and C. A. Swanson. For example, the general nth order type of inequality dealth with in **[149]** is

$$\int_a^b \sum_{k=0}^{n} p_k(x) \left(u^{(k)}(x)\right)^2 \, dx \geq \int_a^b \rho(x) u(x)^2 \, dx,$$

valid for admissible functions u, i.e. functions u such that $u^{(n-1)} \in AC[a,b]$, $u^{(n)} \in L^2(a,b)$ and which satisfy appropriate boundary conditions. Three main methods for obtaining such inequalities are given. This paper contains many examples of new Wirtinger type inequalities, as well as earlier results due to P. R. Beesack **[50]**, W. J. Coles **[54]**, K. Fan, O. Taussky and J. Todd **[94]**, and W. Leighton **[150]**.

55. Best constants in Sobolev's inequalities are considered in **[151]** by H. A. Levine, **[152]**, **[153]** by P. S. Crooke, **[154]** by L. Gross, **[155]** by R. A. Adams and F. H. Clarke and **[156]** by R. A. Adams.

So, H. A. Levine proved the following:

Let $H^1(R^2)$ denote the Sobolev space of all real valued functions, which together with their gradients, are square integrable in the plane. If $\phi \in H^1(R^2)$, then

$$\pi < \iint_{R^2} |\nabla \phi|^2 \, dxdy \cdot \left(\iint_{R^2} \phi^2 \, dxdy \Big/ \iint_{R^2} \phi^6 \, dxdy \right)^{1/2} < \pi^{4/3} + 5 \times 10^{-4}.$$

The logarithmic Sobolev inequality, published by L. Gross in 1976, states that for any continuously differentiable function u on Euclidean space R^n

$$\int_{R^n} |u(x)|^2 \ln |u(x)| d\mu_n(x) \leq \frac{1}{2} \|\nabla u\|^2 + \|u\|^2 + \|u\|^2 \ln \|u\|, \tag{55.1}$$

where μ_n denotes Gaussian measure,

$$d\mu_n(x) = \pi^{-n/2} \exp\left(-|x|^2\right) dx,$$

and $\|\cdot\|$ denotes the norm on the Hilbert space $L^2(R^n, d\mu_n)$,

$$\|\nu\|^2 = \int_{R^n} |\nu(x)|^2 \, d\mu_n(x).$$

A simple proof of (55.1) is given in [**155**]. A generalization of (55.1) is given in [**156**], and this result is also a generalization of a result of G. Rosen [**157**].

56. Inequality (16.1) was considered in papers [**158**] of B. Florkiewicz, and A. Rybarski, [**159**] of A. Kufner and H. Triebel, [**160**] of G. Tomaselli, [**161**], [**162**] of B. Florkiewicz, and [**163**] of D. T. Shum. For some related results see paper [**164**] of B. Florkiewicz.

Here we shall give some results from [**161**] (see also [**162**]).

Let r, φ be given functions which are absolutely continuous and positive, and define for $p > 1$ $s = -\phi^{1-p}\left(r|\varphi'|^{p-1}\sin\varphi'\right)$ and $v = \varphi^{1-p}\left(r|\varphi'|^{p-1}\sin\varphi'\right)$. Assume $r|\varphi'|^{p-1}\sin\varphi'$ is also absolutely continuous.

If f is absolutely continuous such that $\int_a^b r|f'|^p\,dt < \infty$, $\int_a^b s|f|^p\,dt > -\infty$, $\varliminf_{t\to a+} v|h|^p < \infty$ and $\varlimsup_{t\to b-} v|h|^p > -\infty$, then (limits are shown to exist)

$$(56.1) \qquad \int_a^b s|f|^p\,dt + \lim_{t\to b} v|h|^p \le \int_a^b r|h'|^p\,dt + \lim_{t\to a} v|h|^p.$$

If $\varliminf_{t\to a} v|h|^p \le \varlimsup_{t\to b} v|h|^p$ then the two limit terms in (56.1) may be dropped. Conditions for equality are given in [**161**].

57. L. Boček [**165**] gave multidimensional version of a results from [**166**] of Z. Nádeník, and R. C. Brown [**167**] gave Wirtinger's inequality for linear operators.

58. Bohr-Favard's inequality (see 8.) is considered by A. I. Kamzolov [**168**]. The following result is proved by Z. Ditzian [**169**]:

Let f be 2π-periodic, and let $f, \ldots, f^{(k-1)}f$ be absoutely continuous and such that $\int_0^{2\pi} f = 0$ and $f^{(k)} \in L_p(0, 2\pi)$ for some p, $1 \le p < \infty$. Let $a_k = (4/\pi)\sum_{i=1}^{\infty} i^{-k-1}$ (k odd), $a_k = (4/\pi)\sum_{i=1}^{\infty}(-1)^{i+1}i^{-k-1}$ (k even). Then

$$\|f\|_p \le a_k \|f^{(k)}\|_p.$$

In the case $p = 1$ the constant a_k is best possible (that is, equality can be achieved for a suitable f). The case $p = \infty$ is (8.3) while the case $p = 2r$ (r a positive integer) is (13.1).

Ditzian makes an extension from L_p to more general Banach spaces and also considered the discrete analogues with differences in place of derivatives.

59. Let f_i, $i = 1, 2, \ldots, n$, have absolutely continuous $(k-1)$th derivatives on $[a, b]$. If in addition

$$f_{i+}^{(r)}(a) = 0, \; i = 1, 2, \ldots, n, \; r = 0, 1, 2, \ldots, k-1,$$

where $f_+(t) = \max\{f(x), 0\}$, then

$$\int_a^b \left| \left(\prod_{j=1}^n f_j(t) \right)^{(k)} \right| dt \leq \frac{(b-a)^{(n-1)(k-1)}}{((k-1)!)^{n-1}} \binom{n(k-1)}{k-1} \prod_{j=1}^n \left(\int_a^b |f_j^{(k)}(t)| dt \right)$$

where the constant on the right-hand side is best possible.

This is a result of C. J. F. Upton [**170**]. For some previous results see papers [**171**] and [**172**] of A. M. Russell. Related results can be found in Saitah [**216**] and [**217**].

60. Several related results are given in papers [**173**], [**174**], [**175**] and [**176**] of B. G. Pachpatte. The first two papers gave generalizations of the weighted version of Wirtinger's inequality given by J. Traple [**177**]:

Let p be a real-valued nonnegative continuous function defined on $[0, b]$. If f, g are real-valued absolutely continuous functions defined on $[0, b]$ with $f(0) = f(b) = 0$, $g(0) = g(b) = 0$, then

$$\text{(60.1)} \qquad \int_0^b p(s)|f(s)|\,|g(s)| ds \leq \frac{b}{8} \left(\int_0^b p(s)\, ds \right) \int_0^b \left(|f'(s)|^2 + |g'(s)|^2 \right) ds.$$

Equality holds in (60.1) if $f(t) \equiv g(t) \equiv 0$ on $[0, b]$ or $p(t) = 0$ almost everywhere on $[0, b]$.

Papers [**175**] and [**176**] are connected to an inequality of J. Serrin [**178**], which is of Sobolev's type.

An inequality of Sobolev type is also considered in paper [**179**] of B. Bojarski.

61. D. S. Mitrinović and J. E. Pečarić [**180**] showed that the following result of N. Ozeki [**181**] is a generalization of (29.1):

If $x_1, \ldots, x_n$ are real numbers, then

$$\sum_{k=1}^{n-1} x_k x_{k-1} - x_n x_1 \le \left(\cos\frac{\pi}{n}\right) \sum_{i=1}^{n} x_i^2.$$

62. B. Florkiewitz [**182**], using a method from [**158**] and [**161**] examine certain integral inequalities of Block type.

63. Sobolev's inequalities are considered by J. H. Michael and L. M. Simon [**183**], and Wirtinger type inequalities by H. Horst [**184**].

64. Poincaré's Inequality and its Ramification.

H. Poincaré [**185**], [**186**] in trying to formulate and derive his equations in mathematical physics discovered a very significant inequality. It gives us very elegant estimates in the problems of global variational analysis and in the study of the existence of solutions of nonlinear elliptic partial differential equations. Work in this direction can be found in R. Courant and D. Hilbert [**187**], C. B. Morrey, Jr. [**188**], and Th. M. Rassias [**189**]–[**192**]. A related paper is B. G. Pachpatte [**193**].

Poincaré's Inequality: Consider Ω to be a bounded domain in the Euclidean plane R^2. The Poincaré inequality states: There exists a constant c depending on Ω alone such that

$$\iint_{\Omega} f^2 \, dxdy \le c \iint_{\Omega} |\nabla f|^2 \, dxdy, \tag{64.1}$$

where $\nabla f = \left(\frac{\partial f}{\partial x}, \frac{\partial f}{\partial y}\right)$, for all functions f with finite Dirichlet integral over Ω in some class of functions Σ.

It seems that two dimensional versions of Wirtinger's inequality should be given the name of Poincaré's inequality.

Poincaré proved that (64.1) is true if the function space Σ consists of real valued functions which are continuous with their derivatives on Ω and vanishing on $\partial\Omega$ (cf. C. b. Morrey, Jr. [**188**, p. 69]). It is also true that (64.1) holds in the case that the function space Σ consists of all real valued functions under the restriction that Ω be bounded and decomposable into a finite number of convex subdomains; and moreover (64.2) holds, where

$$\iint_{\Omega} f \, dxdy = 0. \tag{64.2}$$

(cf. C. B. Morrey [**188**], p. 83, Theorem 3.6.5. The requirements of continuity, etc., may be replaced by the requirement that f be in a proper Sobolev space). The restriction on the domain Ω to be decomposable into a finite number of convex subdomains became clear after Courant- Hilbert gave an example to explain that (64.1) is not true under the restriction (64.2) in the case where no restriction on the domain Ω is mentioned.

Themistocles M. Rassias [**189**],[**190**] has given an example to prove that (64.1) is not necessarily true under the restriction (64.2) in the case where the function space Σ is restricted to harmonic functions on a bounded domain. Th. M. Rassias [**189**] also proved that the inequality (64.1) is not always true for harmonic functions (bounded or unbounded) in an unbounded domain subject to the condition (64.2). See also 45.

65. There is a close connection between certain eigenvalue problems and Wirtinger type inequalities. The extrema of inequalities in 18. are solutions to (nonlinear) differential equations with the values of the constants being eigenvalues. Only in the case that $p = q = 2$ are these differential equations linear. So the Raleigh quotient

$$I = \frac{\int_a^b (f'(x))^2\,dx}{\int_a^b f(x)^2\,dx}$$

has a minimum λ under certain boundary conditions. Then λ is an eigenvalue of a boundary value problem and the inequality $I \geq \lambda$ leads to a Wirtinger inequality. The result in 41. is of this type where I is replaced by

$$J = \frac{\int_a^b f''(x)^2\,dx}{\int_a^b f(x)^2\,dx},$$

with the boundary conditions

$$f(a) = f'(a) = f(b) = f'(b) = 0. \tag{65.1}$$

The minimum of J is λ_0, the smallest (positive) eigenvalue of the boundary value problem

$$y^{(iv)} - \lambda y = 0 \tag{65.2}$$

with (65.1) as boundary condition.

See **[187]** for the validity of this procedure. It is an elementary exercise to compute the general solution of (65.2) and find the conditions for (65.1) to hold. It leads to $\lambda = \mu_0^4/(b-a)^2$ where μ_0 is the least positive root of $\cos\mu\ \cosh\mu = 1$.

Horgan **[194]** has computed $\mu_0 \doteq 4.7300$. He has also considered J with other boundary conditions and carried out the above program in some detail. We refer to his paper for the detailed resutls, as well as to the book Mikhlin **[195]**. A. M. Bresquar **[196]**,**[197]** gave further generalizations of Wirtinger's inequality also considering J.

We also may mention the paper of A. Weinstein **[198]** which is the two dimensional version. He studied the quotients I/H, I/D, D/H where

$$I = \iint\limits_R \left(\frac{\partial^2 f}{\partial x^2} + \frac{\partial^2 f}{\partial y^2}\right)^2 dx\,dy$$

$$H = \iint\limits_R f(x,y)^2\ dx\,dy$$

$$D = \iint\limits_R \left[\left(\frac{\partial f}{\partial x}\right)^2 + \left(\frac{\partial f}{\partial y}\right)^2\right] dx\,dy$$

with the idea of determining their minimum value. He mentions Janet **[28]** but not the work of Poincaré **[14]** or Levi **[15]**.

D. M. Mangeron **[199]** generalized the inequalities of Schwarz (see 3.) by finding

$$A_n = \min_f \frac{\int\cdots\int p(x_1,\ldots,x_n)\left(\frac{\partial^n f}{\partial x_1\ldots\partial x_n}\right)^2 dx_1\ldots dx_n}{\int\cdots\int q(x_1,\ldots,x_n) f(x_1,\ldots,x_n)^2\ dx_1\ldots dx_n}.$$

In the case $n = 2$ he considered instead the quotient $K = \min\left(\frac{D}{H}\right)^{1/2}$ and found that $K < \frac{\pi}{2}\,\frac{L}{A}$ where R is a simply connected domain with perimeter L and area A. That is is best possible was shown by Pólya [200]. If R is convex the constant $\pi/2$ can be replaced by $\pi/4$ according to Makai [201]. See also Pólya and Szegö **[202]** and Pólya **[203]** and the papers [201] and [202] for references.

In Boll. Un. Mat. Ital. **8** (1929), 113, 164–165, the inequality

$$\int\limits_R f(x_1,\ldots,dx_n)^2 dA \le \frac{1}{n}\left(\frac{D}{\pi}\right)^2 \int\limits_R \left[\left(\frac{\partial f}{\partial x_1}\right)^2 + \cdots + \left(\frac{\partial f}{\partial x_n}\right)^2\right] dA$$

was proved when R is bounded and f is zero on the boundary of R and D is the diameter of R, see **[203]**. Refinements of this result are given by I. D. Nečaev **[204]**,**[205]** as well as the number A_n.

Determination of eigenvalues is difficult but if a lower bound can be given, then a Wirtinger inequality is proved. The literature on this subject is vast and a comprehensive survey of these results is beyond the scope of this book. We are content to mention the book by V. G. Sigillito [**206**] and some of his results which are of the type discussed here. The inequalties are given below where the notation shows which constants are the proved to be the same. They are

$$\int_R u^2\,dx \le q^{-1} \oint_{\partial R} u^2\,dS, \qquad \Delta u = 0 \text{ on } R; \tag{65.3}$$

$$\oint_{\partial R} \left(\frac{\partial u}{\partial n}\right)^2 dS \le q^{-1} \int_R (\Delta u)^2\,dx, \quad u = 0 \text{ on } \partial R. \tag{65.3'}$$

$$\int_R u^2\,dx \le \lambda^{-1} \int_R u_{,i}\,u_{,i}\,dx, \quad u = 0 \text{ on } \partial R; \tag{65.4}$$

$$\int_R u^2\,dx \le \lambda^{-2} \int_R (\Delta u)^2\,dx, \quad u = 0 \text{ on } \partial R; \tag{65.4'}$$

$$\int_R u_{,i}\,u_{,i}\,dx \le \lambda^{-1} \int_R (\Delta u)^2\,dx, \quad u = 0 \text{ on } \partial R. \tag{65.4''}$$

(65.5)

$$\int_R u^2\,dx \le \mu^{-1} \int_R u_{,i}\,u_{,i}\,dx, \quad \int_R u\,dx = 0;$$

(65.5′)

$$\int_R u^2\,dx \le \mu^{-2} \int_R (\Delta u)^2\,dx, \quad \frac{\partial u}{\partial n} = 0 \text{ on } \partial R, \quad \int_R u\,dx = 0;$$

(65.5″)

$$\int_R u_{,i}\,u_{,i}\,dx \le \mu^{-1} \int_R (\Delta u)^2\,dx, \quad \frac{\partial u}{\partial n} = 0 \text{ on } \partial R.$$

(65.6)

$$\int_R u^2\,dx \le \xi^{-1} \oint_{\partial R} \left(\frac{\partial u}{\partial n}\right)^2 dS, \quad \Delta u = 0 \text{ on } R, \quad \int_R u\,dx = 0;$$

(65.6′)

$$\oint_R u^2\,dS \le \xi^{-1} \int_R (\Delta u)^2\,dx, \quad \frac{\partial u}{\partial n} = 0 \text{ on } \partial R, \quad \oint_{\partial R} u\,dS = 0.$$

(65.7)
$$\int_R u^2\,dx \le \Omega^{-1}\int_B (\Delta u)^2\,dx, \quad u = \frac{\partial u}{\partial n} = 0 \text{ on } \partial R;$$

(65.8)
$$\oint_{\partial R} u^2\,dS \le p^{-1}\int_R u_{,i}u_{,i}\,dx, \quad \oint_{\partial R} u\,dS = 0;$$

(65.8′)
$$\oint_{\partial B} u^2\,dS \le p^{-2}\oint_{\partial R}\left(\frac{\partial u}{\partial n}\right)^2 dS, \quad \Delta u = 0 \text{ on } R, \quad \oint_{\partial B} u\,dS = 0;$$

(65.8″)
$$\int_R u_{,i}u_{,i}\,dx \le p^{-1}\oint_{\partial R}\left(\frac{\partial u}{\partial n}\right)^2 dS, \quad \Delta u = 0 \text{ on } R.$$

(65.9)
$$\int_B u_{,i}u_{,i}\,dx \le \Lambda^{-1}\int_B (\Delta u)^2\,dx, \quad u = \frac{\partial u}{\partial n} = 0 \text{ on } \partial R;$$

The optimal values are given as eigenvalues (first unless noted) of the boundary value problems (with consistent notation)

(65.10) $\Delta u + \lambda u = 0,$ on R, $u = 0$ on ∂R,

(65.11) $\Delta u + \mu u = 0,$ on R, $\dfrac{\partial u}{\partial n} = 0$ on ∂R,
(second eigenvalue, $\mu_1 = 0$),

(65.12) $\Delta^2 u - \Omega u = 0$ on R, $u = \dfrac{\partial u}{\partial n} = 0$ on ∂R,

(65.13) $\Delta^2 u - \Lambda\Delta u = 0$ on R, $u = \dfrac{\partial u}{\partial n} = 0$ on ∂R,

(65.14) $\Delta u = 0$ on R, $\dfrac{\partial u}{\partial n} = pu$ on ∂R,
(second eigenvalue),

(65.15) $\Delta^2 u = 0$ on R, $\dfrac{\partial u}{\partial n} = \dfrac{\partial \Delta u}{\partial n} + \xi u = 0$ on ∂R,
(second eigenvalue)

(65.16) $\Delta^2 u = 0$ on R, $u = \Delta u - q\dfrac{\partial u}{\partial n} = 0$ on ∂R

He also gives some estimates for these eigenvalues for certain geometries. We refer to his book [**206**, Chapter 3] for details.

66. Variations of Wirtinger type inequalities which put boundary values into the inequality are given by A. P. Burton and P. Smith [**207**]. For example one of their inequalities is

$$\|x\|_2^2 \leq \lambda^{-2}\|x'\|_2^2 + \frac{2\alpha\beta - (\alpha^2+\beta^2)\cos\lambda(b-a)}{\lambda\sin\lambda(b-a)},$$

where the interval is $[a,b]$ with $x(a)=\alpha$, $x(b)=\beta$, and $0<\lambda<\pi/(b-a)$.

67. Pachpatte [**208**], [**209**] gives inequalities of the Sobolev type for two functions on R^n and other types. In [**210**], he gives discrete versions of Wirtinger inequalities.

68. Fontes [**211**] gives a condition that an inequality

$$\int_\Omega |\phi|^2 dx \leq k \int_\Omega |\nabla\phi|^2 dx$$

holds for some k and all functions for which $|\nabla\phi|^2$ is integrable, and Ω is a fixed open bounded set in R^n. The condition is that for every sequence $\{\phi_n\}$ in $H_0^1(\Omega)$, there is a subsequence that converges in $L^2(\Omega)$.

69. Kreith and Swanson [**212**] have given various higher order generalizations of Wirtinger's inequality. These are inequalities of the form

$$\int_a^b \sum_{n=0}^{n} p_k(x) f^{(k)}(x)^2 dx \geq \int_a^b \rho(x) f(x)^2 dx \tag{69.1}$$

for $f^{(n)} \in L^2(a,b)$, and appropriate boundary conditions. For example, (69.1) holds for all f with n zeros at a and n zeros at b, with $\rho\equiv 0$ provided the p_i's are continuous and $p_n(x)>0$ on (a,b) and $b<\eta_1(a)$, the first $n-n$ conjugate point of $Ly=0$ on (a,b). Here $Ly=\sum_{k=0}^{n} p_k(x)y^{(n)}(x)$.

70. Everitt and Jones [**213**] completed a study of an inequality showing that by choices of coefficients the best possible constants can be arbitrarily large.

71. Two additional references are Nečaev [**214**] and Viszus [**215**].

REFERENCES

1. HARDY, G. H., J. E. LITTLEWOOD and G. PÓLYA, "Inequalities," 2nd ed., Cambridge, 1952, pp. 184–187.
2. BECKENBACH, E. F. and R. BELLMAN, "Inequalities," 2nd ed., Berlin-Heidelberg-New York, 1965, pp. 177–180.
3. BLASCHKE, W. "Kreis and Kugel," Leipzig, 1916, pp. 105–106.
4. ALMANSI, E., *Sopra una delle esperienze di Plateau*, Ann. Mat. Pura Appl. **12** 3 (1905), 1-17.
5. LEVI, E. E., *Sulle condizioni sufficienti per il minimo nel calcolo delle variazioni (Gli integrali sotto forma non parametrica)*, Atti Accad. Naz. Lincei Ren. I. Cl. Sci. Fis. Mat. Natur. **22** 5 (1911), 525–431; or *Opere*, vol. 2, Bologna (1960), pp. 347–359.
6. TONELLI, L., *Su una proposizione deli'Almansi*, Rend. R. Accad. Lincei **23** (1914), 676–682; or *Opere scelte*, vol. 1, Roma (1960), pp. 236–242.
7. SCHEEFFER, L., *Über die Bedeutung der Begriffe "Maximum und Minimum" in der Variationsrechung*, Math. Ann. **26** (1885), 197–208. See, in particular, p. 207.
8. KNESER, A., *Variationsrechnung*, "Encyklopädie der mathematischen Wissenschaften" II A 8, Leipzig (Abgeschlossen in September 1900), pp. 551–625. See, in particular, p. 605.
9. PICARD, É., "Traité d'analyse," Vol. 3, Paris 1896, pp. 100–128.
10. STEKLOFF, W., *Sur le problème de refroidissement d'une barre hétérogène*, C. R. Sci. Paris **126** (1898), 215–218.
11. _________, *Problème de refroidissement d'une barre hétérogène*, Ann. Fac. Sci. Toulouse **3** 2 (1901), 281–313.
12. _________, *Sur certaines égalités remarquables*, C. R. Sci. Paris **135** (1902), 783–786.
13. SCHWARZ, H. A., *Über ein die Flächen kleinsten Flächeninhalts betreffendes Problem der Variationsrechnung*, Acta Soc. Sci. Fenn. **15** (1885), 315–362, or "Gesammelte Math. Abhandlungen von H. A. Schwarz," vol. 1, Berlin, 1890, pp. 223–268 and, in particular, pp. 257–259.
14. POINCARÉ, H., *Sur les équations de la physique mathématique*, Rend. Circ. Mat. Palermo **8** (1894), 57–156 ; or "Oeuvres de Henri Poincaré," Paris, 1954, pp. 123–196.
15. LEVI, E. E., *Su un lemma del Poincaré*, Atti Accad. Naz. Lincei Rend. Cl. Sci. Fis. Mat. Natur. **15** 5 (1906), 83–89 and 353–358; or *Opere*, vol. 2, Bologna, 1960, pp. 3–18.
16. HADAMARD, J., "Leçons sur le calcul des variations," vol. 1, Paris, 1910, pp. 334–336 and p. 467.

17. TAMARKINE, Ja. D., *Applications de la méthode des fonctions fondamentales*, Soobšćenija Har'kovskago Mat. Obščestva (1910).
18. KRILOFF, N. M., *Sur certaines inégalités trouvées dans l'exposition de la méthode de Schwarz- Poincaré-Stekloff et qu'on rencontre dans la résolution de nombreax problèmes de minimum* (Russian), Zapiski Gornago Instituta imeni Ekaterini 2, vol. 6, fasc. 1 (1915), 10–14.
19. PICONE, M., *Sulle autosoluzioni e sulle formole di maggiorazione ecc.* Math. Zeitschrift, Bd. 28 Heft 4, 1928, p. 552.
20. CIMMINO, G., *Problem 37*, Risposte. Boll. Un. Mat. Ital. **8** (1929), 225–226.
21. LEVI, E. E., *Sui criteri sufficienti per il massimo e per il minimo nel calcolo delle variazioni*, Ann. Mat. Pura Appl. **21** 3 (1913), 173–218; or *Opere*, vol. 2, Bologna, 1960, pp. 375–426.
22. PLEIJEL, A., *An inequality* (Swedish), Mat. Tidsskr. A (1949), 67–69.
23. JANET, M., *Sur le rapport des valeurs moyennes des carrés de deux dérivées d'ordre consécutifs*, C. R. Acad. Sci. Paris **188** (1929), 681--683.
24. ________, *Sur une suite de fonctions considérées par Hermite et son application à un problème de calcul des variations*, C. R. Acad. Sci. Paris **190** (1930), 32–34.
25. CIMMINO, G., *Su una questione di minimo*, Boll. Un. Mat. Ital **9** (1930), 1–6.
26. JANET, M., *Détermination explicite de certains minima dans des problèmes sans conditions aux limites*, C. R. Acad. Sci. Paris **194** (1932), 2109–2111.
27. ________, *Détermination explicite de certains minima*, Congrès Intern. Math. Zürich (1932).
28. ________, *Les valeurs moyennes des carrés de deux dérivées d'ordre consécutifs, et le développement en fraction continue de tan x*, Bull. Sci. Math. **55** 2 (1931), 1–13.
29. ________, *Sur la méthode de Legendre-Jacobi-Clebsch et quelques-unes de ses applications*, Bull. Sci. Math. **53** 2 (1929), 144–160.
30. ________, *Sur le minimum du rapport de certaines intégrales*, C. R. Acad. Sci. Paris **193** (1931), 977–979.
31. CIMMINO, G., See references 20 and 25.
32. TCHENG, Tchou-Yun, "Sur les inégalités différentielles," Paris, 1934, 39 pp.
33. HARDY, G. H. and J. E. LITTLEWOOD, *Some integral inequalities connected with the calculus of variations*, Quart. J. Math. **2** 2 (1932), 241–252.

34. BLISS, G. A., *An integral inequality*, J. Lond. Math. Soc. **5** (1930), 40–46.
35. BOHR, H., *Ein allgemeiner Satz über die Integration eines trigonometrschei Polynoms* , Prace Matem.-Fiz. **43** (1935), "Collected Math. Works" II C 36, 273–288.
36. ESSEEN, C. G., *Fourier analysis of distribution functions*, Acta Math. **77** (1945), 1–125.
37. HÖRMANDER, L., *A new proof and a generalization of an inequality of Bohr*, Math. Scand. **2**, (1954), 33–45.
38. BOAS, R. P., "Entire Functions," New York, 1954.
39. NORTHCOTT, D. G., *Some inequalities between periodic functions and their derivatives*, J. London Math. Soc. **14** (1939), 198–202.
40. DINGAS, A., *Beweis einer Ungleichung für konvexe Körper*, Abhandl. Preuss. Akad. Wissensch., Math.-naturwiss. Klasse **11** (1939), no. 11, 13 pp.
41. LEVIN, V., *Notes on inequalities II: On a class of integral inequalities*, Mat. Sbornik T. **4** 4, N. 2, (1938), 309–324.
42. SOBOLEV, S. L., *Sur un théorème de l'analyse fonctionnelle*, C. R. Acad. Sci. URSS **20** 2 (1936), 5–9.
43. __________, *Sur l'évaluation de quelques sommes pour une fonction définie sur un réseau*, Bull. Acad. Sci. URSS **4** (1940), 4–16.
44. SCHMIDT, E., *Über die Ungleichung, welche die Integrale über eine Potenz einer Funktion und über eine andere Potenz ihrer Ableitung verbindet*, Math. Ann. **117** (1940), 301–326.
45. BENSON, D. C., *Inequalities involving integrals of functions and their derivatives*, J. Math. Anal. Appl. **17** (1967), 292–308.
46. SCHMIDT, E., *Über die nebst ihren Ableitungen orthogonalen Polynomensysteme und zugehörige Extremum*, Math. Ann. **119** (1944), 165–204.
47. BELLMAN, R., *A note on an inequality of E. Schmidt*, Bull. Amer. Math. Soc. **50** (1944), 734–736.
48. SZ-NAGY, B. *Über Integralungleichungen zwischen einer Funktion und ihrer Ableitung*, Acta Sci. Math. (Szeged) **10** (1941), 64–74.
49. BELLMAN, R., *A note on periodic functions and their derivatives*, J. London Math. Soc. **18** (1943), 140- -142.
50. BEESACK, P. R., *Integral inequalities of the Wirtinger type*, Duke Math. J. **25** (1958), 477–498.
51. KIM, W. J., *On the zeros of solutions of* $y^{(n)} + py = 0$, J. Math. Anal. Appl. **25** (1969), 189–208.
52. NEHARI, Z., *The Schwarzian derivative and schlicht functions*, Bull. Amer. Math. Soc. **55** (1949), 545- -551.

53. BEESACK, P. R.,*Extensions of Wirtinger's inequality*, Trans. Roy. Soc. Canada Sect. III **53** 3 (1959), 21–30.
54. COLES, W. J., *A general Wirtinger-type inequality*, Duke Math. J. **27** (1960), 133–138.
55. _________, *Wirtinger- type integral inequalities*, Pacific J. Math. **11** (1960), 871–177.
56. BEESACK, P. R., *Hardy's inequality and its extension*, Pacif. J. Math. **11** (1961), 39–61.
57. LEVIN, A. Yu., *On some estimates of a differentiable function*, Soviet Mathematics **2** (1961), 523- -524.
58. BERNSTEIN, S. N., "Collected Works," Vol. II, Izdat Akad. Nauk. SSSR, Moscow, 1954, Article 100, p. 497.
59. HUKUHARA, K., *On the zeros of solutions of linear ordinary differential equations*, Sugaku **15** (1963), 108–109.
60. TUMURA, M., *Kōkai zyôbibunhôteisiki ni tuite*, Kansû Hôteisiki **30** (1941), 30–35.
61. BESSMERTNYH, G. A. and A. Ju. LEVIN, *Some inequalities satisfied by differentiable functions of one variable*, Dokl. Akad. Nauk SSSR **144** (1962), 471–474. Soviet Math. Dokl. **3** (1962), 737–740.
62. LEVIN, A. Yu., *A bound for a function with monotonely distributed zeros of successive derivatives*, Mat. Sb., (N.S.) **106** 64 (1964), 396–409, (Russian).
63. BRINK, J., *Inequalities involving* $\|f\|_p$ *and* $\|f^{(n)}\|_q$ *for* f *with* n *zeros*, Pacif. J. Math. **42** (1972), 289–311.
64. FINK, A. M., *Best possible approximation constants*, Trans. Amer. Math. Soc. **226** (1977), 243–255.
65. _________, *Conjugate inequalities for functions and their derivatives*, SIAM J. Math. Anal. **5** (1974), 399–411.
66. _________, *Differential inequalities and disconjugacy*, J. Math. Anal. Appl. **49** (1975), 758–772.
67. FEDERER, H. and W. FLEMING, *Normal and integral currents*, Ann. Math. **72** (1960), 458–520.
68. FLEMING, W., *Functions whose partial derivatives are measures*, Illinois J. Math. **4** (1960), 452–478.
69. FLEMING, W. and R. RISHEL, *An integral formula for total gradient variation*, Arch. Math. **11** (1960), 218–222.
70. MIRANDA, M., *Distribuzioni aventi derivate misure*, Ann. Scuola Norm. Sup. Pisa **18** (1964).
71. _________, *Sul minimo dell'integrale del gradiente di una funczione*, Ann. Scuola Norm. Sup. Pisa **19** (1965).

72. _________, *Disuguaglianze di Sobolev sulle ipersuperfici minimali*, Rend. Sem. Mat. Univ. Padova **38** (1967).
73. SOBOLEV, S. L., "Applications of Functional Analysis in Mathematical Physics," Amer. Math. Soc., 1963.
74. TROESCH, B. A., *Integral inequalities for two functions*, Arch. Rational Mech. Anal. **24** (1967), 128- -140.
75. _________, *An isoperimetric sloshing problem*, Comm. Pure and Applied Math. **18** 1/2 (1965), 319–338.
76. DIAZ, J. B. and F. T. METCALF, *Variations of Wirtinger's inequality*, in "Inequalities," edited by O. Shisha, New York, 1967, pp. 79–103.
77. HALPERIN, I. and H. PITT, *Integral inequalities connected with differential operators*, Duke Math. J. 4 (1938), 613–625.
78. MÜLLER, W., *Über eine Ungleichung zwischen den Normen von f, f' und f''*, Math. Z. **78**, (1962), 420–422.
79. NIRENBERG, L., *Remarks on strongly elliptic partial differential equations*, Appendix. Comm. Pure Appl. Math. **8** (1955), 649–675.
80. REDHEFFER, R., *Über eine beste Ungleichung zwischen den Normen von f, f', f''*, Math. Z. **80** (1963), 390–397.
81. PFEFFER, A. M., *On certain discrete inequalities and their continuous analogs*, J. Res. Nat. Bur. Standards **70** B (1966), 221–231.
82. BLOCK, H. D., *A class of inequalities*, Proc. Amer. Math. Soc. **8** (1957), 844–851.
83. REDHEFFER, R., *Inequalities with three functions*, J. Math. Anal. Appl. **16** (1966), 219–242.
84. RYBARSKI, A., *Angeäherte Schwingungfrequenzformeln für konservative Systeme* (2), Zastosowania Matematyki **7** (1964), 255–269.
85. KRZYWICKI, A. and A. RYBARSKI, *On a linearization of an equation of an elastic rod* (II), Zastosowania Matematyki **7** (1964), 383–390.
86. _________, *On some integral inequalities involving Chebyshev weight function*, Colloq. Math. **18**, (1967), 147–150.
87. LEVIN, V. I. and S. B. STEČKIN, *Inequalities*, Amer. Math. Soc. Transl. **14** 2 (1960), 1- -29. See, in particular, pp. 11–13.
88. CALVERT, J., *An integral inequality with applications to the Dirichlet problem*, Pacif. J. Math. **22** (1967), 19–29.
89. _________, *Integral inequalities involving second order derivatives*, Pacific J. Math. **27** (1968), 39–47.
90. MOROZOV, S. F. and V. I. PLOTNIKOV, *Integral'nye neravenstva tipa Virtingera*, Izv. Vysš. Uč. Zaved. Matematika **74** 7 (1968), 56–60.
91. MANGERON, D. and A. OPREA, *Inégalités dans le problème spectrale concernant les phénomènes polyvibrants*, Bul. Inst. Politehn. Iaşi (S.N.) **14** 18, fasc 3–4, (1968), 429–432.

92. BOYD, D. W., *Best constants in a class of integral inequalities*, Pacific J. Math. **30** (1969), 367-383.
93. ARAMĂ, O., *Sur un problème d'interpolation relatif aux solutions des équations différentielles linéaires du quatrième ordre*, Mathematica (Cluj) **10** 33 (1968), 5–15. See,in particular, pp. 7–8.
94. FAN, K., O., TAUSSKY and J. TODD, *Discrete analogues of inequalities of Wirtinger*, Monatsh. Math. **59** (1955), 73–90.
95. BLOCK, H. D., *Discrete analogues of certain integral inequalities*, Proc. Amer. Math. Soc. **8** (1957), 852–859.
96. FINK, A. M., *Discrete inequalities of generalized Wirtinger type*, Aequationes Math. **11** (1974), 31–39.
97. SADIKOVA, R. H., *Obobščenija diskretnyh neravenstv Tauski i ih kontinual'nyh analogov*, Volž, Mat. Sb. **7** (1969), 161–165.
98. NOVOTNÁ, J., *Variations of discrete analogues of Wirtinger's inequality*, Čas. pro. pěst. mat. **105**, (1980), 278–285, Praha.
99. MILOVANOVIĆ, G. V. and I. Ž. MILOVANOVIĆ, *On discrete inequalities of Wirtinger's type*, J. Math. Anal. Appl. **88** (1982), 378–387.
100. MITRINOVIĆ, D. S. and P. M. VASIĆ, *An inequality ascribed to Wirtinger and its variations and generalizations*, Univ. Beograd Publ. Elektrotehn. Fak. Ser. Mat. Fiz. No. **247–273** (1969), 157–170.
101. JANET, M., *Sur l'inégalité classique du problème isopérimétrique*, Univ. Beograd. Publ. Elektrotehn. Fak. Ser. Mat. Fiz. No. **274–301** (1969), 9- -10.
102. REDHEFFER, R., *Integral Inequalities with Boundary Terms*, in "Inequalities 2," New York, 1970, pp. 261- -291.
103. RICHARD, U., *Sur des inégalités du type de Wirtinger et leur application aux équations diff. ordinaires*, Proc. Colloquium Analysis Rio de Janeiro (1970).
104. LUPAŞ, A. *An interpolation problem for the solutions of a class of linear differential equations*, Mathematica 12 **35** (1970), 87–104.
105. BEESACK, P. R., *Integral inequalities involving a function and its derivative*, Amer. Math. Monthly, **78** (1971), 705–741.
106. GANELIUS, T., *Some applications of a lemma on Fourier series*, Acad. Serbe Sci. Publ. Inst. Math. **11** (1957), 9–18.
107. ________, *Tauberian remainder theorems*, (mineogr. lecture notes), Madras, 1970.
108. LANDBERG, L., *Some variants of Bohr's inequality*, Report No. 1971–17, Charmers Inst. of Technology and the University of Götenborg, 1971.
109. NIRENBERG, L., *On elliptic partial differential equations*, Ann. Scuola Norm. Sup. Pisa **13** 3 (1959), 115–162.

110. FUJITA, H., *On the existence and regularity of the steady-state solutions of the Navier-Stokes equation*, J. Fac. Sci. Univ. Tokyo Sect. I **9** (1961), 59–102.
111. ROSEN, G., *Minimum value for c in the Sobolev inequality* $\|\phi^3\| \leq c\|\nabla\phi\|^3$, SIAM J. Appl. Math. **21**, (1971), 30–32.
112. DENKOWSKI, Z., *Inequalities of Wirtinger's type and their discrete analogues*, Zesz. Nauk. Iniw. Jagiellońskiego Prace Matem. Z. **15** CCLII (1971), 27–37.
113. SCHOENBERG, I. J., *Norm inequalities for a certain class of* C^∞ *functions*, Isreal J. Math. **10** 3 (1971), 364–372.
114. __________, *On the zeros of successive derivatives of integral functions*, Trans. Amer. Math. Soc. (1936), 12–23.
115. HRYPTUN, V. G. *On a certain representation of infinitely differentiable functions*, Doklady Akad. Nauk SSSR **199** (1971), 282–284, (Russian).
116. MANDELBROJT, S., *Sur les fonctions indéfiniment dérivables*, Acta Math. **72** (1940), 15–29.
117. WONG, P. K., *Wirtinger type inequalities and elliptic differential inequalities*, Tôhoku Math. J. **23** (1971), 117–127.
118. __________, *Integral inequalities of Wirtinger-type and fourth-order elliptic differential inequalities*, Pacif. J. Math. **40** (1972), 739- -751.
119. __________, *A Sturmian theorem for first order partial differential equations*, Trans. Amer. Math. Soc. **166** (1972), 125–131.
120. WINTER, M. J. and P.-K. WONG, *Comparison and maximum theorems for systems of quasilinear elliptic differential equations*, J. Math. Anal. Appl. **40**, (1972).
121. PEETRE, J., *The best constant in some inequalities involving* L_q *norms of derivatives*, Ricerche di Matematica **21** (1972), 176–183.
122. ŠADRIN, G. A., *Ob ocenkah sverhu točnyh konstant v neravenstvah tipa Puankare*, Uč. zap. Mosk. gos. ped. in-ta im. V. I. Lenia **460** (1972), 33–37.
123. ANDERSON, N., A. M. ARTHURS and R. R. HALL, *Extremum principle for a nonlinear problem in magneto- elasticity*, Proc. Cambridge Philos. Soc. **72** (1972), 315- -318.
124. SHISHA, O., *On the discrete version of Wirtinger's inequality*, Amer. Math. Monthly **80** (1973), 755–760.
125. FLORKIEWICZ, B. and A. RYBARSKI, *On an integral inequality connected with Hardy's inequality* (III), Colloquium Math. **27** (1973), 293–296.
126. EASWARAN, S., *Fundamental inequalities for polyvibrating operators*, Bul. Inst. Politehn. din Iaşi **19** 23 (1973), 71–76.
127. *Problem No. 319*, Nieuw Archief voor Wiskunde **21** 3 (1973), 108–109.

128. *Problem 72–12*, Ontario Math. Bulletin **10** (1974).
129. TALENTI, G., *Best constant in Sobolev inequality*, Istit. Mat. Univ. Firenze **22** (1974–75), 1- -32.
130. MEYERS, N. C., *Wirtinger and Poincaré inequalities for BV Functions*, Notices AMS **22** (1975), A 403.
131. ROSEN, G., *Sobolev-type lower bounds on* $\|\nabla\Psi\|^2$ *for arbitrary regions in two-dimensional Euclidean space*, Quarterly Appl. Math. July (1976), 200–202.
132. WEISSLER, F. B. *Logarithmic Sobolev inequalities for the heat-diffusion semi-group on* R^n, Notices AMS **24** (1977), A–117.
133. AUBIN, T., *Problèmes isopérimétriques et espaces de Sobolev*, C. R. Acad. Sci. Paris, Sér. A–B **280** (1975), 279–281.
134. ALVINO, A., *Sulla diseguaglianza di Sobolev in sparzi di Lorentz*, Bollettino U. M. I. **14-A** 5 (1977), 148–156.
135. EVERITT, W. N. and M. GIERTZ, *Inequalities and separation for certain ordinary differential operators*, Proc. London Math. Soc., III. Ser. **28** (1974), 352–372.
136. ________, *On certain ordinary differential expressions and associated integral inequalities*, in "New Dev. Differ. Equat.," Proc. 2nd Scheveningen Conf. Differ. Equat. 1975, pp. 161–174.
137. BRADLEY, J. S. and W. N. EVERITT, *Inequalities associated with regular and singular problems in the calculus of variations*, Trans. Amer. Math. Soc. **182** (1973), 303–321.
138. ________, *A singular integral inequality on a bounded interval*, Proc. Amer. Math. Soc. **61** (1976), 29–35.
139. AMOS, R. J. and W. N. EVERITT, *On integral inequalities associated with ordinary regular differential expressions*, Differ. Equat. Appl. Proc. 3rd Scheveningen Conf., Math. Stud. **31** (1978), 237–255.
140. ________, *On a quadratic integral inequality*, Proc. roy. Soc. Edinburgh, Sect. A **78** (1978), 241–256.
141. ________, *On integral inequalities and compact embeddings associated with ordinary differential expressions*, Arch. Ration. Mech. Anal. **71** (1979), 15–40.
142. EVERITT, W. N., *An integral inequalities with an application to ordinary differential operators*, Proc. roy. Soc. Edinburgh, Sect. A **80** (1978), 35–44.
143. MULDOWNEY, J. S., *On an inequality of Caplygin and Pólya*, Proc. roy. Irish Acad., Sect. A **76** (1976), 85–100.
144. TALENTI, G., *Una disuguaglianza fra* u, u', u'', Boll. Un. mat. Ital. IV. Ser. 11, Suppl. Fasc. **3** (1975), 375–388.

145. EVERITT, W. N., *Spectral theory of Wirtinger inequality*, Ordin. part. Differ. Equat., Proc. 4th. Conf., Dundee 1976 , Lecture Notes Math. **564** (1976), 93- -105.
146. MELENCOVA, Ju. A., *Best constants in integral inequalities for functions with n zeros*, Ural. Gos. Univ. Mat. Zap. 10, no. 2. Issled. Sovremen. Mat. Anal. **217** (1977), 93–107.
147. FEINBERG, J. M., *Some Wirtinger-like inequalities*, SIAM J. Math. Anal. **10** (1979), 1258–1271.
148. SWANSON, C. A., *Wirtinger's ienquality*, SIAM J. Math. Anal. **9** (1978), 484–491.
149. KREITH, K. and C. A. SWANSON, *Higher order Wirtinger inequalities*, Proc. roy. Soc. Edinburg, Sect. A **85** (1980), 87–100.
150. LEIGHTON, W., *Quadratic functionals of second order*, Trans. Amer. Math. Soc. **151** (1970), 309–322.
151. LEVINE, H. A., *A numerical estimate for the best constant in a two dimensional Sobolev inequality with three integral norms*, Notices AMS **26** (1979), A–205.
152. CROOKE, P. S., *On two inequalities of Sobolev type*, Applicable Analysis **3** (1974), 345–358.
153. __________, *An isoperimetric bound for a Sobolev constant*, Colloquium Math. **38** (1979), 263–267.
154. GROSS, L., *Logarithmic Sobolev inequalities*, Amer. J. Math. **97** (1976), 1061–1083.
155. ADAMS, R. A. and F. H. CLARKE, *Gross's logarithmic Sobolev ienquality: A simple proof*, Amer. J. Math. **101** (1979), 1265–1269.
156. ADAMS, R. A., *General logarithmic Sobolev inequalities in Orlicz imbeddings*, Notices AMS **25** (1978), A–431.
157. ROSEN, G., *Sobolev inequalities for weight spaces and supercontractivity*, Trans. Amer. Math. Soc. **222** (1976), 367–376.
158. FLORKIEWICZ, B. and A. RYBARSKI,*Some integral inequalities of Sturm-Liouville type*, Colloquium Math. **36** (1976), 127–141.
159. KUFNER, A. and H. TRIEBEL, *Generalizations of Hardy's inequality*, Conf. Semin. Mat. Univ. Bari **156** (1978), 21 p.
160. TOMASELLI, G., *A class of inequalities*, Boll. Un. Mat. Ital **2** 4 (1969), 622–631.
161. FLORKIEWICZ, B., *Some integral inequalities of Hardy type*, Colloquium Math. **42** (1980), 321–330.
162. __________, *Integral inequalities of Hardy type and Weyl type*, Dissertation.
163. SHUM, D. T., *On integral inequalities related to Hardy's*, Canad. Math. Bull. **14** (1971), 225- -230.

164. FLORKIEWICZ, B., *On some integral inequalities of Weyl type*, Colloquium Math. **49** (1984), 91- -107.
165. BOČEK, L., *Eine Verschärfung der Poincaré-Ungleichung*, Čas. pro. Pěst. Mat. **108** (1983), Praha, 78–81.
166. NÁDENÍK, Z., *Die Verschärfung einer Ungleichung von Frobenius für den gemischten Flächeninhalt der konvexen ebenen Bereiche*, Čas. pro Pěst. Mat. **90** (1965), 220–255.
167. BROWN, R. C., *Wirtinger inequalities, Dirichlet functional inequalities, and the spectral theory of linear operators and relations*, in "Spectral Theory of Differential Operators," Proc. Conf., Birmingham USA 1981, North-Holland Math. Stud. **55** (1981), 69–79.
168. KAMZOLOV, A. I., *O neravenstve Bora-Favara dlja funkcií i kompaktnyh simmetričeskih prostranstvah ranga I*, Mat. Zametki **33** (1983), 187–193.
169. DITZIAN, Z., *On inequalities of periodic function and their derivatives*, Proc. Amer. Math. Soc. **87** (1983), 463–466.
170. UPTON, C. J. F., *An integral inequality associated with generalized variation*, Appl. Anal. **14** (1982), 1–9.
171. RUSSELL, A. M., *A commutative Banach algebra of functions of bounded variation*, Amer. Math. Monthly **87** (1980), 39–40.
172. ________, *Some inequalities arising from a Banach algebra norm*, J. Austral. Math. Soc. (Series A) **34** (1983), 199–202.
173. PACHPATTE, B. G., *A note on an inequality ascribed to Wirtinger*, Tamkang J. Math. **17** 1 (1986), 69- -73.
174. ________, *On Wirtinger-Opial type integral inequalities*, Tamkang J. Math. **17** 4 (1986), 1–6.
175. ________, *On some new multidimensional discrete inequalities*, Tamkang J. Math. **17** 3 (1986), 21–29.
176. ________, *On Sobolev type integral inequalities*, Proc. Roy. Soc. Edinburgh A **103** (1986), 1–14.
177. TRAPLE, J., *On a boundary value problem for systems of ordinary differential equations of second order*, Zeszyty Nauk, Uni. Jagiell. Prace Math. **15** (1971), 159- -168.
178. SERRIN, J., *The initial value problem for the Navier-Stokes equation*, in "Nonlinear Problems," Ed. R. E. Langer, The Univ. of Wisconsin Press, Madison, 1963, pp. 69–78.
179. BOJARSKI, B., *Remarks on Sobolev imbedding inequalities*, Lect. Notes Math. **1351** (1988), 52- -68.
180. MITRINOVIĆ, D. S. and J. E. PEČARIĆ, *On two lemmas of N. Ozeki*, J. Coll. Arts and Sci. Chiba Univ. **B–21** (1988), Nov. 30, 19–21.
181. OZEKI, N., *On P. Erdős' inequality to the triangle*, J. Coll. Arts and Sci. Chiba Univ. **2** (1957), 247–250.

182. FLORKIEWICZ, B., *On some integral inequalities of Block type*, Colloquium Math. **55** (1988), 179–189.
183. MICHAEL, J. H. and L. M. SIMON, *Sobolev and mean-value inequalities on generalized submanifolds of R^n*, Communications Pure and Appl. Math. **26** (1973), 361–379.
184. HORST, H., *Integralungleichungen mit Anwendungen auf lineare Differenialgleichungen*, Appl. Anal. **29** 1–2 (1988), 55–62.
185. POINCARÉ, H., "Oeuvres," Vol. 1, Paris, 1928.
186. __________, "Analysis Situs. Collected Works," Vol. 6, Paris, 1953.
187. COURANT, R. and D. HILBERT, "Methods of Mathematical Physics," Vol. I., New York, 1953.
188. MORREY, C. B., Jr., "Multiple Integrals in the Calculus of Variations," New York, 1966.
189. RASSIAS, Th. M., *Un contre-exemple à l'inégalité de Poincaré*, C. R. Acad. Sciences Paris **284** (1977), 409–412.
190. __________, *On some properties of harmonic mappings of the Poincaré inequality*, Proceedings of the Conference on Complex Analysis and Applications, Varna, Bulgaria (1981), Bulgarian Acad. Sciences (1984), 453–457.
191. __________, *Foundations of global nonlinear analysis*, Teubner–Texte zur Mathematik, Leipzig, Band **86** (1986).
192. __________, *Global Analysis–Analysis on Manifolds: (Dedicated to Marson Morse: 1892–1976)*, Teubner-Texte zur Mathematik, Leipzig, Band 57 (1983).
193. PACHPATTE, B. G., *A note on Poincaré and Sobolev type integral inequalities*, J. Math. **18** (1987), 1- -7.
194. HORGAN, C. O., *A note on a class of integral inequalities*, Proc. Cambridge Phil. Soc. **74** (1973), 127–131.
195. MIKHLIN, S. G., "The problem of the minimum of a quadratic functional," San Francisco, 1965.
196. BRESQUAR, A. M., *Sulla diseguaglianza di Wirtinger*, Rend. Sem. Mat. Univ. Padova **51** (1974), 758–772.
197. __________, *Ancora sulle diseguaglianze di Wirtinger concernenti la derivata seconda*, Rend. Sem. Mat. Univ. Pdova **53** (1975), 335–362.
198. WEINSTEIN, A., *Étude des spectres des équations aux dérivées partielles de la théorie des plaques élastiques*, Mémorial Sci. Math. No. 88, Paris (1937). See, in particular, pp. 8–14.
199. MANGERON, D., *Sur le rapport des valeurs moyennes des carrés de deux dérivées totales d'ordre consécutif*, C. R. Acad. Sci. Paris A **266** (1968), 1103–1106.

200. PÓLYA, G., *Two more inequaliteis between physical and geometrical quantities*, J. Indian Math. Soc. (N.S.) **24** (1960), 413–419.
201. MAKAI, E., *On the principal frequency of membrane and the torsional rigidity of a beam*, "Studies in Mathematical Analysis and Related Topics," Stanford, 1962, pp. 227–231.
202. PÓLYA, G. and G. SZEGÖ, "Isoperimetric Inequalities in Mathematical Physics," Princeton, 1951.
203. PÓLYA, G., *Circle, sphere, symmetrization and some classical physical problems*, in "Modern Mathematics for the Engineer," second series, New York-Toronto-London, 1961, pp. 420–442, (edited by E. F. Beckenbach).
204. NEČAEV, I. D., *Nekotorye obobščenija mnogomernogo neravenstva tipa Virtingera. I*, Uč. zap. Mosk. gos. ped. in-ta im. V. I. Lenina **460** (1972), 96- -101.
205. _________, *Nekotorye obobščenija mnogomernogo neravenstva tipa Virtingera. II*, Uč. zap. Mosk. gos. ped. in-ta im. V. I. Lenina **460** (1972).
206. SIGILITTO, V. G., "Explicit a priori inequalities with applications to boundary value problems," London/San Francisco/Melbourne, 1977.
207. BURTON, A. P. and P. SMITH, *An approach to Friedricks and Poincaré integral inequalities using convexity*, Rad. Mat. **5** (1989), 107–114.
208. PACHPATTE, B. G., *On two inequalities of the Sobolev type*, Chinese J. Math. **15** (1987), 247–252.
209. _________, *A note on Poincaré and Sobolev type integral inequalities*, Tamkang J. Math. **18** (1987), 1–7.
210. _________, *On Wirtinger like discrete inequalities*, Tamkang J. Math. **20** (1989), 211–219.
211. FONTES, F. G., *On Friedricks inequality and Rellichs Theorem*, J. Math. Anal. Appl. **145** (1990), 516- -533.
212. KREITH K. and C. A. SWANSON, *Higher order Wirtinger inequalities*, Proc. Roy. Soc. Edinburgh Sect. A. **85** (1980), 87–110.
213. EVERITT, W. N. and P. S. JONES, *On an integral inequality*, Proc. Roy. Soc. London Ser A **357** (1977), 271–288.
214. NEČAEV, I. D., *On an inequality of Steklov-Wirtinger Type* (Russian), Gos. Ped. Institut Barnaul, 11pp.
215. VISZUS, E., *A remark on type of Poincaré's inequality*, Acta Math. Univ. Comen. **56–57** (1990), 19–22.
216. SAITAH, S., *Hilbert spaces admitting reproducing kernels on the real line and related fundamental inequalities*, Riazi J. Kar. Math. Assoc. **0** (1984), 25–31.
217. _________, *Norm inequalities derived from the theory of reproducing kernels*, General Inequalities **6**, Oberwolfach, 1990.
218. PHILLIPS, G. M., *Error estimates for best polynomial approximations*, Approximation Theory (Proc. Sympos., Lancaster, 1969), London, 1970, 1–6.

CHAPTER III

OPIAL'S INEQUALITY

1. We devote this chapter to ideas surrounding the inequality of Z. Opial [**1**].

THEOREM 1. *If $f \in C^1[0,a]$ with $f(0) = f(a) = 0$ and $f(x) > 0$ on $(0,a)$, then*

$$\int_0^a |f(x)f'(x)|dx \leq \frac{a}{4}\int_0^a f'(x)^2\, dx. \tag{1.1}$$

The constant $a/4$ is best possible.

Proofs simpler than Opial's original ones are given in C. Olech [**2**], P. R. Beesack [**3**], N. Levinson [**4**], C. L. Mallows [**5**] and R. N. Pedersen [**6**]. These papers reduce the proofs to their very essence.

C. Olech [**2**] showed that the condition $f(x) > 0$ on $(0,a)$ is not necessary and that the cases of equality can be obtained.

THEOREM 2. *If f is absolutely continuous and $f(0) = f(a) = 0$, then (1.1) holds. There is equality if and only if $f(x) = cx$ for $0 \leq x \leq a/2$ and $f(x) = c(a-x)$ for $a/2 \leq x \leq a$, where c is a constant.*

The form of this result suggest that there is an antecedent theorem on half the interval.

THEOREM 2′. *Let f be absolutely continuous on $[0,a]$ with $f(a) = 0$. Then*

$$\int_0^a |f(x)f'(x)|dx \leq \frac{a}{2}\int_0^a f'(x)^2\, dx \tag{1.2}$$

with equality if and only if $f(x) = cx$ for c a constant.

This result implies Theorem 2. Let f be absolutely continuous on $[0,a]$ with $f(0) = f(a) = 0$. Then we apply (1.2) to $f_1(x) = f(x)$ on $[0, a/2]$ and $f_2(x) = f(a-x)$ on $[0, a/2]$.

We obtain

$$\int_0^{a/2} |f(x)f'(x)|dx \leq \frac{a}{4}\int_0^{a/2} f'(x)^2\, dx$$

and

$$\int_0^{a/2} |f(a-x)f'(a-x)|dx \leq \frac{a}{4}\int_0^{a/2} f'(x-a)^2 dx.$$

Setting $a - x = t$ in the second inequality and adding we obtain (1.1) and the cases of equality.

We shall give Mallows' proof of Theorem 2′. First it is sufficient to prove the theorem when $f \geq 0$ and $f' \geq 0$ on $[0, a]$. Indeed if $f_1(x) \equiv \int_0^x |f'(t)|dt$ then $|f(x)| \leq |f_1(x)|$ and $\int_0^a |f(x)f'(x)|dx \leq \int_0^a |f_1(x)f_1'(x)|dx$. We may assume $f(x) \geq 0$ and $f'(x) \geq 0$. By the Buniakowski-Schwarz inequality

$$(1.3)\qquad 2\int_0^a |f(x)f'(x)|dx = f(a)^2 = \left(\int_0^a f'(x)dx\right)^2 \leq a\int_0^a f'(x)^2 dx$$

with equality if and only if $f'(x)$ is a constant.

2. Opial's inequality can be generalized by using a different measure on the right.

THEOREM 1. *Let $p(x)$ be continuous and positive on $(0, a)$ such that $\int_0^a p(x)^{1-q'}dx < \infty$, where $q' > 1$.*

Let f be absolutely continuous with $f(0) = 0$. Then

$$(2.1)\qquad \int_0^a |f(x)f'(x)|dx \leq \frac{1}{2}\left(\int_0^a p(x)^{1-q'}dx\right)^{2/q'}\left(\int_0^a p(x)|f'(x)|^q\right)^{1/q}$$

where $\frac{1}{q} + \frac{1}{q'} = 1$ and equality holds if and only if $f'(x) = c\int_0^x p(t)^{1-q'}dt$ for some constant c.

PROOF: As above it is sufficient to prove the theorem when $f(x) \geq 0$ and $f'(x) \geq 0$ on $[0, a]$. Then as in (1.3) we write

$$2\int_0^a |f(x)f'(x)|dx = \left(\int_0^a \left(p(x)^{1/q}f'(x)\right)p(x)^{-1/q}dx\right)^2$$

and apply Hölder's inequality to bound this by

$$\left(\int_0^a p(x)|f'(x)|^q\,dx\right)^{2/q}\left(\int_0^a p(x)^{1-q'}dx\right)^{2/q'}.$$

The cases of equality are $|p(x)^{1/q}f'(x)|^q = c|p(x)^{-1/q}|^{q'}$ for some constant c.

P. R. Beesack [3] has obtained this result in the case $q = 2$ but with a proof that preceded the discovery of Mallow's proof of Opial's inequality. He also obtained a similar result which is a direct generalization of Theorem 1. G. S. Yang [7] simplified Beesack's proofs and at the same time gave a generalization of Theorem 1.

THEOREM 2. *Let p be a positive function on $[0, a]$ such that $\int_0^a dx/p(x) < \infty$, and let q be a positive, bounded and nonincreasing function on $[0, a]$. If f is absolutely continuous on $[0, a]$ with $f(0) = 0$, then*

$$\int_0^a q(x)|f(x)f'(x)|dx \leq \frac{1}{2}\int_0^a \frac{dx}{p(x)}\int_0^a p(x)q(x)f'(x)^2dx$$

with equality if and only if $q(x) =$ constant and $f(x) = c\int_0^x dt/p(t)$ for some constant c.

The introduction of the measure p, which may not be symmetric, means that a version of Theorem 1 of Section 1 does not follow from Theorem 2. Beesack found the right generalization.

THEOREM 3. *Let p be positive and continuous on $(0, a)$ with $\int_0^a dx/p(x) < \infty$. Let f be absolutely continuous on $[0, a]$ so that*

$$f(x) = \int_0^x f'(t)dt \quad \text{and} \quad f(x) = -\int_x^a f'(x)dt. \tag{2.2}$$

Suppose that $f(x)^2 = O\left(\int_0^x \frac{dt}{p(t)}\right)$ as $x \to 0^+$ *and* $f(x)^2 = O\left(\int_x^a \frac{dt}{p(t)}\right)$ as $x \to a^-$. *Then*

$$\int_0^a |f(x)f'(x)|dx \le \frac{\kappa}{2}\int_0^a p(x)f'(x)^2 dx,$$

where a' *and* κ *are determined by*

$$\int_0^{a'} \frac{dt}{p(t)} = \int_{a'}^a \frac{dt}{p(t)} = \kappa.$$

Equality holds in (2.2) if and only if

$$f(x) = A\int_0^x \frac{dt}{p(t)} \text{ on } [0,a'] \quad \text{and} \quad f(x) = B\int_x^a \frac{dt}{p(t)} \text{ on } [a',a],$$

where A *and* B *are constants.*

As above this theorem can be generalized with weaker hypothesis using Mallows' proof.

THEOREM 4. *Let* p *be positive and continuous on* $[0,a]$ *with* $\int_0^a dx/p(x) < \infty$ *and* $\int_0^a p(x)^{1-q'}dx < \infty$. *Let* f *be absolutely continuous on* $[0,a]$ *with* $f(0) = f(a) = 0$. *Then*

$$\int_0^a |f(x)f'(x)|dx \le \frac{\kappa}{2}\left(\int_0^a p(x)|f'(x)|^q\right)^{1/q},$$

where a' *and* κ *are defined by* $\left(\frac{1}{q}+\frac{1}{q'}=1\right)$

$$\int_0^{a'} p(x)^{1-q'}dx = \int_{a'}^a p(x)^{1-q'}dx = \kappa. \tag{2.3}$$

Equality holds if $f(x) = A\int_0^x p(t)^{1-q'}dt$ *on* $[0,a']$ *and* $f(x) = B\int_x^a p(t)^{1-q'}dt$.

PROOF: Define a' and κ by (2.3). As above we may assume that $f(x) \ge 0$ and $f'(x) \ge 0$ on $[0,a']$ and we get (2.1). On $[a',a]$ we may assume $f(x) \ge 0$ and

$f'(x) \le 0$. Then

$$\int_{a'}^{a} |f(x)f'(x)|dx = \frac{f(a')^2}{2} = \frac{1}{2}\left(\int_{a'}^{a} f'(x)dx\right)^2$$

and we again apply Hölder's inequality. The details are the same as for Theorem 1.

If $p(x) \equiv 1$ we get Opial's inequality.

J. M. Holt [8] gave a generalization of Theorem 2 which we do not reproduce here.

3. We may generalize Opial's inequality in a different way by considering powers of f and f'.

THEOREM 1. *If f is absolutely continuous on $[0, a]$ and $f(0) = 0$, then for $p \ge 0$, $q \ge 1$,*

$$\int_0^a |f(x)|^p |f'(x)|^q dx \le \frac{q}{p+q} a^p \int_0^a |f'(x)|^{p+q} dx. \tag{3.1}$$

Various authors have considered this inequality. L. K. Hua [9] showed the validity of (3.1) if $q = 1$ and p is a positive integer. G. S. Yang [7] gave a proof for $p, q \ge 1$ but P. R. Beesack observed that the proof is valid for $p \ge 0$, $q \ge 1$. For $q = 1$ and $p \ge 0$ this inequality was generalized by J. Calvert [10] which we will quote below. J. S. W. Wong [11] gave a short proof of that result.

Just as Opial's Theorem can be obtained from Theorem 2′ the above result also yields the next theorem.

THEOREM 2. *If f is absolutely continuous on $[0, a]$ and $f(0) = f(a) = 0$, then for $p \ge 0$ and $q \ge 1$,*

$$\int_0^a |f(x)|^p |f'(x)|^q dx \le \frac{q}{p+q} \left(\frac{a}{2}\right)^p \int_0^a |f'(x)|^{p+q} dx.$$

P. R. Beesack and K. M. Das [12] obtained more results in this genre, obtaining inequalities of the form $\int_0^a s(x)|f(x)|^p|f'(x)|^q dx \le k(a,p,q) \int_0^a r(x)|f'(x)|^{p+q} dx$. We quote one.

THEOREM 3. *Let p, q be real numbers such that $pq > 0$ and either $p+q > 1$ or $p+q < 0$. Suppose that r, s are nonnegative measurable functions on $(0, a)$ such that $\int_0^a r(x)^{-\frac{1}{p+q-1}} dx < \infty$ and that the constant κ is finite, where*

$$\kappa(a,p,q) = \left(\frac{q}{p+q}\right)^{\frac{q}{q+p}} \left(\int_0^a s(t)^{\frac{p+q}{p}} r(t)^{-\frac{q}{p}} \left(\int_0^t r(u)^{\frac{-1}{p+q-1}} du \right)^{p+q-1} dt \right)^{\frac{p}{p+q}}.$$

If f is absolutely continuous on $[0, a]$ with $f(0) = 0$ and f' of constant sign, then

$$\int_0^a s(x)|f(x)|^p |f'(x)|^q dx \le \kappa(a,p,q) \int_0^a r(x)|f'(x)|^{p+q} dx.$$

Equality holds if and only if either $q > 0$ and $f \equiv 0$ or

$$s(x) = k_1 r(x)^{\frac{q-1}{p+q-1}} \left(\int_0^x r(t)^{\frac{-1}{p+q-1}} dt \right)^{\frac{p(1-q)}{q}}$$

and $f(x) = k_2 \int_0^x r(t)^{\frac{-1}{p+q-1}} dt$ for some constants k_1, k_2 with $k_1 \ge 0$.

REMARK: As with Mallows' proof, the condition "f' as of one sign" may be omitted.

4. We have mentioned J. Calvert's paper [**10**] above and now we present two of his results.

THEOREM 1. *Let f be absolutely continuous on $[0, a]$ with $f(0) = 0$. Let g be a continuous complex valued function defined on the range of f and for $\int_0^a |f'(x)|dx$. Suppose that $|g(t)| \le g(|t|)$ for all t and that g is real for $t > 0$ and is nondecreasing there. Let r be a positive continuous function in $L^{1-q}(a,b)$ where $\frac{1}{p} + \frac{1}{q} = 1$ $(p > 1)$. If $F(x) = \int_0^x g(t)dt$ for real $x > 0$. Then*

$$\int_0^a |g(x)f'(x)|dx \le F\left(\left(\int_0^a r(x)^{1-q} dx \ dx \right)^{1/q} \left(\int_0^a r(x)|f'(x)|^p dx \right)^{1/p} \right),$$

with equality if $f(x) = c\int_0^a r(t)^{1-q}dt$.

The same result holds if $f(a) = 0$ and equality holds if $f(x) = c\int_x^a r(t)^{1-q}dt$.

THEOREM 2. *Let* f, g, r *be as in Theorem 1. If* $p < 1$ *and* $\frac{1}{p} + \frac{1}{q} = 1$, $f(0) = 0$, *and* $G(x) = \int_0^x dt/g(t)$, *then*

$$\int_0^a \left|\frac{f'(x)}{g(f)}\right| dx \geq G\left(\left(\int_0^a r(x)^{1-q}dx\right)^{1/q}\left(\int_0^a r(x)|f'(x)|^p dx\right)^{1/p}\right),$$

with equality if $f(x) = c\int_0^x r(t)^{1-q}dt$.

P. R. Beesack noticed that the conditions for equality in the above two theorems are sufficient but not necessary.

E. K. Godunova and V. I. Levin [**13**] assumed slightly more about g to get better results.

THEOREM 3. *Let* f *be absolutely continuous on* $[0, a]$, *with* $f(0) = f(a) = 0$. *Let* $r(x) > 0$ *and* $\int_0^a r(x)dx = 1$. *If* g *and* h *are convex increasing functions for* $x > 0$ *and* $g(0) = 0$, *then*

$$\int_0^a g'(|f(x)|)\,|f'(x)|dx \leq 2g\left(h^{-1}\left(\int_0^a r(x)h\left(\frac{|f'(x)|}{2r(x)}\right)dx\right)\right). \tag{4.1}$$

THEOREM 4. *Let* f *be absolutely continuous on* $[0, a]$ *with* $f(0) = 0$. *If* g *is a convex increasing function for* $x \geq 0$ *with* $g(0) = 0$, *then*

$$\int_0^a g'(|f(x)|)\,|f'(x)|dx \leq g\left(\int_0^a |f'(x)|dx\right).$$

A particular case of (4.1) is of interest. Let $g(t) = t^2/2$ and $h(t) = t^p$, $p \geq 1$, with $r(x) = s(x)^{\frac{1}{1-p}}\left(\int_0^a s(t)^{\frac{1}{1-p}}dt\right)^{-1}$ where $s(x) \geq 0$. Then (4.1) becomes

$$\int_0^a |f(x)f'(x)|dx \leq C\left(\int_0^a s(x)|f'(x)|^p\right)^{2/p} \tag{4.2}$$

with $C = \frac{1}{4}\left(\int\limits_0^a s(t)^{\frac{1}{1-p}}\right)^{\frac{2(p-1)}{p}}$.

P. Maroni [**14**] gave inequality (4.2) for $1 < p \leq 2$ with C replaced by $C \cdot 2^{\frac{2}{p-1}}$.

5. There is a connection between Opial's inequality and differential equations. D. W. Boyd and J. S. W. Wong [**15**] made this connection.

THEOREM 1. *Let ω and σ be nonnegative continuously differentiable functions on $[0, a]$ such that the boundary value problem*

$$(\sigma(t)u'(t)^p)' = \lambda\omega'(t)u(t)^p \tag{5.1}$$

$$u(0) = 0 \quad \sigma(a)u'(a)^p = \lambda\omega(a)u(a)^p, \quad u' > 0 \text{ on } [0, a] \tag{5.2}$$

has a solution for some $\lambda > 0$.

If f is absolutely continuous on $[0, a]$ and $f(0) = 0$, then for all $p > 0$

$$\int\limits_0^a |f'(x)f(x)^p|\omega(x)dx \leq \frac{1}{\lambda_0(p+1)}\int\limits_0^a |f'(x)|^{p+1}\sigma(x)dx$$

where λ_0 is the largest positive eigenvalue of the boundary value problem (5.1)–(5.2).

6. Note that with $\sigma = \omega \equiv 1$ and $p = 1$ this reduces to Opial's inequality. We have incorporated the review of this paper in our quotation of the theorem. See Math. Reviews **35**(1968), 563 by J. V. Ryff.

7. R. Redheffer [**16**] replaced the product $f(x)f'(x)$ by a weighted sum of the factors.

THEOREM 1. *If r, s, u and f are absolutely continuous on $[0, a]$ and*

$$\frac{r'(x)}{s(x)}f(x)^2 + \frac{r(x)}{s'(x)}f'(x)^2 \leq 0, \quad r(x)s'(x) \geq 0, \ s'(x) \neq 0, \ s(x) > 0$$

on $[0, a]$, then

$$\int\limits_0^a \frac{r(x)}{s'(x)}\left[f(x)^2u'(x)^2 + \frac{1}{2}\left(f(x)^2\right)'\left(u(x)^2\right)'\right]dx \geq B - A,$$

where

$$B = \overline{\lim_{x\to a^-}}\ u(x)^2f(x)^2\frac{r(x)}{s(x)} \text{ and } A = \underline{\lim_{x\to 0+}}\ u(x)^2f(x)^2\frac{r(x)}{s(x)}.$$

8. The replacement of $f'(x)$ by $f^{(n)}$ is a natural generalization of Opial's inequality.

THEOREM 1. *Let $f^{(n-1)}$ exist and be absolutely continuous on $[0,a]$ with $f^{(j)}(0) = 0$, $j = 0,\dots,n-1$, $(n \geq 1)$. There exists a constant C_n such that*

$$\int_0^a |f(x)f^{(n)}(x)|dx \leq C_n\, a^n \int_0^a |f^{(n)}(x)|^2 dx. \tag{8.1}$$

D. Willett [17] initiated the study of this inequality using it to prove the existence and uniqueness of solutions of linear differential equations of nth order. He proved that $C_n \leq \frac{1}{2}$. K. M. Das [18] improved this to show that $C_n < \frac{1}{2n!}\left(\frac{n}{2n-1}\right)^{1/2}$ unless $n = 1$ when this reduces to Theorem 2′ of Section 1.

D. W. Boyd [19] determined the best possible constants in (8.1). For $n = 2m+1$, the best possible constant is $\frac{\lambda_0}{2^n(n-1)!}$, where λ_0 is the largest positive eigenvalue of the $(m+1)\times(m+1)$ matrix

$$A = (a_{ij}) \text{ with } a_{ij} = \binom{2m}{2i}(2m-2j+2i+1)^{-1}, \quad (i,j = 0,\dots,m).$$

For $n = 2m$ the best possible constant is $(\lambda_0^n 2^n(n-1)!)^{-1}$ where λ_0 is the smallest positive solution of Det $B(\lambda) = 0$ where $B(\lambda) = (b_{ij}(\lambda))$ is an $m \times m$ matrix with

$$b_{ij}(\lambda) = \omega^{2ij}\lambda^{2i} \sum_{k=0}^{2m-2i-1} (-1)^k \frac{g_j^{(k)}(1)}{k}, \quad (i,j = 0,\dots,m-1),$$

where $g_j = \cosh(\omega^j \lambda x)$ and $\omega = \exp\left(\frac{\pi i}{m}\right)$.

These best possible constants are not easy to compute but Boyd gave good estimates.

THEOREM 2. *Let C_n denote the best possible constants in (8.1). Then $C_n = b_n/2n!$ where*

$$\frac{1}{2} \leq b_n \leq \left[\frac{n}{4n-2} + \binom{2n}{n}^{-1}\right]^{1/2}$$

so that $b_n \to 1/2$ as $n \to \infty$.

9. D. W. Boyd [20] made a connection between Opial's inequality and Wirtinger's inequality by proving a single inequality that contains them as a special case. See 28. of Chapter II.

10. We may also have discrete versions of Opial's inequality. J. S. W. Wong [11] proved the following version of inequality (3.1) for $q = 1$.

THEOREM 1. *Let u_i be a nondeceasing sequence of nonnegative real numbers with $u_0 = 0$. For $p \geq 1$*

$$\sum_{i=1}^{n}(u_i - u_{i-1})u_i^p \leq \frac{(n+1)^p}{p+1}\sum_{i=1}^{n}(u_i - u_{i-1})^{p+1}.$$

Cheng-Ming Lee [**21**] improved this by reintroducing q.

THEOREM 2. *Let u_i be a nondecreasing sequene of nonnegative real numbers. If $p, q > 0$, $p + q \geq 1$ or $p, q < 0$, we have*

$$\sum_{i=1}^{n}(u_i - u_{i-1})^q u_i^p \leq K_n \sum_{i=1}^{n}(u_i - u_{i-1})^{p+q} \qquad (u_0 = 0),$$

where

$$K_0 = \frac{q}{p+q}$$

and

$$K_n = \max\left(K_{n-1} + \frac{pn^{p-1}}{p+q}, \frac{q(n+1)^p}{p+q}\right) \quad (n = 1, 2, \dots).$$

If $p > 0$, $q < 0$, $p + q \leq 1$, $p + q \neq 0$ or $p < 0$, $q > 0$, $p + q \geq 1$, then

$$\sum_{i=1}^{n}(u_i - u_{i-1})^q u_i^p \geq C_n \sum_{i=1}^{n}(u_i - u_{i-1})^{p+q} \quad (\text{with } u_0 = 0),$$

where

$$C_0 = \frac{q}{p+q} \text{ and } C_n = \min\left(C_{n-1} + \frac{pn^{p-1}}{p+q}, \frac{q(n+1)^p}{p+q}\right) \quad (n = 1, 2, \dots).$$

If $p, q \geq 1$,

$$K_n = \frac{q(n+1)^p}{p+q};$$

if $p \leq 0$, $q < 0$, then

$$K_1 = 1, \text{ and } K_n = 1 + \frac{p}{p+q}\sum_{i=2}^{n} i^{p-1} \quad (n = 2, 3, \dots);$$

if $p \geq 0$, $p+q<0$, *then*

$$C_1 = 1 \text{ and } C_n = 1 + \frac{p}{p+q}\sum_{i=2}^{n} i^{p-1} \quad (n = 2, 3, \ldots).$$

Related discrete analogues are considered by P. R. Beesack also in [22].

11. The papers of D. O. Banks [23], O. Aramă [24] and A. Lupaş [25] are also connected to Opial's inequality.

12. R. Redheffer [26] has given a generalization of the following inequality of E. Hlawka:

If $u(0) = 0$, $a > 0$, and $u \not\equiv 0$, then

$$\int_0^a e^{-2x}(u')^2 dx > \int_0^a e^{-2x} uu' \, dx.$$

13. D. W. Boyd [27] gave a generalization of Opial's inequality which includes an integral operator.

14. The following result was given by P. R. Beesack [28]:

THEOREM. *Suppose that* r *and* s *are positive with* r *and* s' *continuous on* $(0, a)$, *and that equation*

$$(ry'^p)' = (p+1)^{-1}s'y^p$$

has a solution y *such that*

$$y'(x) > 0 \text{ and } y(x) = \int_0^x y' dt \text{ for } 0 < x < a, \text{ while}$$

$$r(x)\left[y'(x)/y(x)\right]^p \geq (p+1)^{-1}s(x) \text{ for } x \text{ near } a \quad (x > x_0), \text{ and}$$

$$\{r(x)[y'(x)/y(x)]^p - (p+1)^{-1}s(x)\}\left(\int_0^x r^{-1/p}\, dt\right)^p = O(1) \text{ as } x \to 0^+.$$

Let f *be any function such that* $f(x) = \int_0^x f'(t)dt$ *and* $\int_0^a r|f'|^{p+1}dx < \infty$. *Then*

$$\int_0^a s|f^p f'|dx \leq \int_0^a r|f'|^{p+1}dx \quad (p > 0). \tag{14.1}$$

Equality holds in (14.1) if and only if $f \equiv cy$, *where* $c = 0$ *unless both* $\int_0^a r y'^{p+1} dx < \infty$ and $\lim_{x\to a} y(x)\{r(x)y'^p(x) - (p+1)^{-1}s(x)y^p(x)\} = 0$. *The inequality (14.1) is sharp if either*

$$\int_0^x r y'^{p+1} dx = \infty \text{ and } \underline{\lim}_{x\to 0}\, r y y'^p < \infty,$$

or

$$\int_x^a r y'^{p+1} dx = \infty, \quad \underline{\lim}_{x\to a}\, y\{r y'^p - (p+1)^{-1} s y^p\} < \infty, \text{ and } r y'^p = O(1)$$

as $x \to a$.

15. A paper related to Opial's inequality is one by H. Herold [**29**].

16. In 1972 G. I. Rozanova published four related papers (see [**30**]–[**33**]).

In [**30**] and [**31**] a discrete analogue and a generalization for a class of functions with an integral representation are given respectively. The following two theorems are proved in [**32**] and [**33**] respectively:

THEOREM 1. *Let* $y(x)$ *be an absolutely continuous function on* $[0,a]$ *and let* $y(0) = 0$, $r(x) \geq 0$, $r'(x) > 0$, $r(0) = 0$. *Let, also,* $\phi(u)$ *and* $F(u)$ *be convex and increasing functions for* $u > 0$ *with* $F(0) = 0$. *Then*

$$\int_0^a r'\phi\left(\frac{|y'|}{r'}\right) F'\left(r\phi\left(\frac{|y|}{r}\right)\right) dx \leq F\left(\int_0^a r'\phi\left(\frac{|y'|}{r'}\right) dx\right), \tag{16.1}$$

with equality for $y(x) = cr(x)$, *where* c *is an arbitrary constant.*

THEOREM 2. *Let* $y(x)$ *be an absolutely continuous function on* $[0,a]$, $y(0) = 0$ *with* $r(x)$ *increasing, and* $r(0) = 0$. *Let, also,* $\phi(u)$ *and* $F(u)$ *be convex increasing functions for* $u > 0$, $F(0) = 0$, *and let* $Q(u)$ *be convex increasing and* $\Psi(u)$ *increasing with* $\Psi(0) = 0$. *If*

$$F'(z(x))\, z'(x) \Psi\left(\frac{1}{z'(x)}\right) \leq \left(\frac{F(z(a))}{z(a)}\right) \Psi'\left(\frac{x}{z(a)}\right), \tag{16.2}$$

holds, where $z(x) = \int_0^x r'\phi(|y'|/r')\, dt$, *then*

$$\int_0^a F'\left(r\phi\left(\frac{|y|}{r}\right)\right) G\left(r'\phi\left(\frac{|y'|}{r'}\right)\right) dx \leq H\left(\int_0^a r'\phi\left(\frac{|y'|}{r'}\right)\right) dx, \tag{16.3}$$

where $G(u) = uQ(\Psi(1/u))$, $H(u) = F(u)Q(\Psi(a/u))$, *with equality if and only if* $y(x) = Ax$, $r(x) = Bx$, $\Psi(u) = CF(\lambda u)$, *where* A, B, C *and* λ *are constants.*

For $y(x) = f(x)$, $f'(x) > 0$, $f(0) = 0$, $f(a) = b$, $\phi(u) = u$, $F(u) = \Phi(u) = u^2$, and $Q(u) = \sqrt{1+u}$, we get from (16.3) the following inequality of G. Pólya [**34**]

$$2\int_0^a f(x)\left(1 + f'(x)^2\right)^{1/2} \le b\left(a^2 + b^2\right)^{1/2} \tag{16.4}$$

with the weaker condition $f(x)/f'(x) \le x$ instead $f''(x) \ge 0$. Equality is valid in (16.4) iff $f(x) = (b/a)x$.

17. Multidimensional generalizations of Opial's inequality are given by I. D. Nečaev [**35**]. Here, we shall give two special cases as listed by the author.

Let V be an n-dimensional sphere of radius R_0, and let $u\colon V \to R$ be differentiable with $u(0) = 0$.

(1) If $pq > 0$, $p+q > 1$ and $p+q-n-\lambda > 0$, or $p < 0$, $q < 0$, $p+1-n-\lambda < 0$, then

$$\int_V r^{\frac{(p+q)(1-q)(p-\lambda)-p(n-q)}{q(p+q-1)}} |u|^p |\nabla u|^q dv \le K_1 \int_V r^{\lambda} |\nabla u|^{p+q} dv,$$

where

$$r = \left(\sum_1^n x_i^2\right)^{1/2},$$

$$K_1 = q(p+q)^{-1}\left(\frac{(p+q-1)}{p+q-n-\lambda)}\right)^p R_0^{p(p+q-n-\lambda)((q(p+q-1))^{-1}}.$$

(2) If $0 < p \le 1$, $0 \le s < p$, then

$$\left(\int_V r^{p^{-1}((n+\lambda)(1-p)(p-s)+p(p-n))} |\nabla u|^p |u|^{-s} dv\right)^{1/(p-s)}$$

$$\le K_2\left(\int_V r^{\lambda - p(n+\lambda-1)} |\nabla u|^p dv\right)^{1/p},$$

where

$$K_2 = \left(\frac{2\pi^{n/2}}{\Gamma(n/2)}\right)^{s/p(p-s)} \left(\frac{p}{(p-s)}\right)^{\frac{1}{p-s}} (n+\lambda)^{\frac{s(p-1)}{(p-s)}}.$$

18. G. G. Vrănceanu [**36**] proved the following result.

Let a function $f\colon [0,a] \to H$ (H is a Hilbert space with the inner product $\langle\ ,\ \rangle$) have a continuous derivative and $f(0) = 0$. Then

$$\int_0^a |\langle f(x), f'(x)\rangle|dx \le \frac{a}{2}\int_0^a \|f'(x)\|^2 dx.$$

19. The following result of D. T. Shum [**37**] is a generalization of a result from [**3**]:

Let f be an absolutely continuous function on $[0,a]$, and $f(0) = 0$. If $p > 0$, and $\int_0^a |f'|^{p+1}dx < \infty$, then

$$\int_0^a |f|^p|f'|dx + \frac{pa^p}{p+1}\int_0^a \frac{g_2(x)dx}{x^{p+1}} \le \frac{a^p}{p+1}\int_0^a |f'|^{p+1}dx, \tag{19.1}$$

where $g_2(x) = (p+1)\int_0^x |f|^p|f'|dx - |f(x)|^{p+1}$ (≥ 0) for $0 \le x \le a$. If either $p < -1$ and both $\int_0^a |f|^p|f'|dx < \infty$, and $\int_0^a |f'|^{p+1}dx < \infty$, or $-1 < p < 0$ and $\int_0^a |f|^p|f'|dx < \infty$, the reverse inequality holds.

For $p > 0$, equality holds in (19.1) if and only if $f(x) = cx$ for some constant c; for $p < -1$ equaliy never holds; for $-1 < p < 0$ equality holds if and only if $f(x) = cx$ for some constant $c \neq 0$.

20. In [**38**] D. T. Shum considered inequalities of the form

$$\int_a^b S|f|^p|f'|^q dx \le \frac{q}{p+q}\int_a^b R|f'|^{p+q}dx \tag{20.1}$$

valid for integrals $f(x) = \int_a^x f'dt$ having $f(a) = 0$. Here, $-\infty \le a < b \le +\infty$, $p \ge 0$, $q \ge 1$ (or $p < 0$, $q < 0$ if f' does not change sign), and $R(x) = r(x)\,(w(x))^{(p+q)(1-q)/q}$, $S(x) = s(x)\,(w(x))^{q-1}\left(\int_a^x w^{-1}dt\right)^{p(1-q)/q}$ where w is a positive, measurable function on (a,b) such that $\int_a^x w^{-1}dt < \infty$ for $x \in (a,b)$

while $r > 0$, $s \geq 0$ are the coefficients in a boundary value problem involving the differential equation

$$\left(r(y')^{p/q}\right)' = s'y^{p/q}, \tag{20.2}$$

and appropriate boundary conditions at a and b. It is assumed that this problem has a solution $y(x) = \int_a^x y'dt$ having $y' > 0$ on (a, b).

He also considered inequalities of the form

$$\left(\frac{p+q}{q}\right)^{\frac{k}{p+q}} C \left(\int_a^b W|f|^p|f'|^q dx\right)^{\frac{k}{p+q}} + \int_a^b S|f|^k dx \leq \int_a^b R|f'|^k dx \tag{20.3}$$

valid for integrals $f(x) = \int_a^x f'dt$. In (20.3),

$$W(x) = (w(x))^{q-1} \left(\int_a^x w^{-1} dt\right)^{\frac{p(1-q)}{q}},$$

$$R(x) = r(x)\,(w(x))^{\frac{k(q-1)}{q}},$$

and

$$S(x) = s(x) \left(\int_a^x w^{-1} dt\right)^{\frac{k(1-q)}{q}}$$

where w is as in (20.1), and $r > 0$, $s \geq 0$ are now the coefficients in a boundary value problem involving the differential equation

$$\left(r(y')^{\frac{k-q}{q}}\right)' + sy^{\frac{k-q}{q}} = 0 \tag{20.4}$$

and appropriate boundary conditions at a, b including $\lim_{x \to b-} r(y'/y)^{(k-q)/q} = C$ $(0 < C < \infty)$. In addition $p > 0$, $k > q \geq 1$, or f' is of one sign on (a, b), $p < 0$,

$k < q < 0$ or $p > 0$, $k < 0 < q \le 1$ provided also $\int_a^b w^{-1} dt < \infty$ and $|f(b)| < \infty$ in the latter case. As before, it is assumed that the boundary value problem has a solution $y(x) = \int_a^x y'\, dt$ with $y' > 0$ on (a,b) (and also $y(b) < \infty$ if $k/q < 0$).

21. In 1976, G. I. Rozanova gave two new papers with realted results ([**39**],[**40**]). For example, the following results are proved in [**40**]:

Let y and r be absolutely continuous functions which have n absolutely continuous derivatives ($n \ge 1$), with $y^{(i)}(0) = r^{(i)}(0) = 0$, $i = 0,1,\dots,m-1$, $1 \le m \le n$, $y^{(n)}(a) = r^{(n)}(a) = 0$, $y^{(n+1)}(x)$ and $r^{(j)}(x)$ ($j = 0,1,\dots,n+1$) do not change sign for $x \in [0,a]$.

Let $s(x) > 0$, and let $\phi(u)$ be a positive convex increasing function, $0 < q < 1$, $p > -q$. Then

$$\int_0^a s(x)\phi(|y/r|)\,|y^{(n)}|^p|y^{(n+1)}|^q\,dx$$

$$\le \frac{1}{(m-1)!}\left(\frac{q}{p+q}\right)^q \int_0^a g(x)\phi\left(|y^{(m)}/r^{(m)}|\right)|y^{(n)}|^{p+q}dx,$$

where

$$g(x) = |r^{(m)}(x)|\left(\int_x^a \left(\frac{(t-x)^{m-1}s(t)}{|r(t)|}\right)^{\frac{1}{1-q}} dt\right)^{1-q}.$$

If also Ψ is an increasing function for $u > 0$, with $\Psi(0) = 0$, then

$$\int_0^a |r(x)|\phi(|y/r|)\,\Psi'\left(|y^{(m)}|\right)|y^{(n+1)}|dx$$

$$\le \frac{1}{(m-1)!}\int_0^a |r^{(m)}(x)|(a-x)^{m-1}\phi\left(|y^{(m)}/r^{(m)}|\right)\Psi\left(|y^{(n)}|\right)dx.$$

22. Elementary proofs of some Opial-type integral inequalities are given by P. R. Beesack in [**41**]. Connections to Opial's inequality are given in a paper [**42**] of M. S. Hou.

23. The following theorems are proved by C. S. Lee [**43**].

THEOREM 1. *Let $r(x)$ be positive on $[0,a]$ with $\int_0^a r(x)^{-q}dr < \infty$. Let $s(x)$ be positive and nonincreasing on $[0,a]$. If $f(x)$ is absolutely continuous with $f(0)=0$ then for all $p \geq 0$, $q \geq 1$*

$$(p+q)\int_0^a r(x)^{q(q-1)}s(x)^q|f(x)|^p|f'(x)|^q dx$$

$$\leq q\left(\int_0^a r(x)^{-q}dx\right)^p\left(\int_0^a r(x)^{q(p+q-1)}s(x)^q|f'(x)|^{p+q}dx\right).$$

There is equality if and only if f' does not change sign, s is a positive constant, $q=1$ and $f(x) = A\int_0^x r(t)^{-q}dt$.

THEOREM 2. *Let $r(x)$ be positive on $[0,a]$ with $\int_0^a r(x)^{-q}dx < \infty$. Let $s(x) > 0$, be nonincreasing and both $\int_0^a r(x)^{q(q-1)}s(x)^q|f(x)|^p|f'(x)|^q dx < \infty$ and $\int_0^a r(x)^{q(p+q-1)}s(x)^q|f'(x)|^{p+q}dx < \infty$. If f is absolutely continuous and $f(0)=0$ then for all $p < -q$, $q \geq 1$*

$$\int_0^a q\, r(x)^{q(q+p-1)}s(x)^q|f'(x)|^{p+q}dx$$

$$\geq \frac{p+q}{\left(\int_0^a r(x)^{-q}\right)^p}\int_0^a r(x)^{q(q-1)}s(x)^q|f(x)|^p|f'(x)|^q dx$$

$$+p\int_0^a \frac{r(x)^{-q}h(x)}{\left(\int_0^x r^{-q}(t)dt\right)^{p+q}}dx,$$

where

$$h(x) = (p+q)\left(\int_0^x r(t)^{-q}dt\right)^{q-1}\int_0^x r(t)^{q(q-1)}s(t)^q|f(t)|^p|f'(t)|^q dt$$

$$- s(x)^q|f(x)|^{p+q} \ (\leq 0),$$

unless $f \equiv 0$.

24. G. V. Milovanović and I. Ž. Milovanović [44] used a method of P. R. Beesack ([28]) to prove some Opial-type inequalities. The main result is:

Let p be positive and continuous on $(0,a)$ with $\int_0^a p\,dt = P < +\infty$. Set $r(x) = 1/p(x)$, and let $a < \xi < b$. Then for all $F \in W_r^2[a,b]$ (the space of all functions u which are locally absolutely continuous on $(0,a)$ with $\int_0^a ru'^2 dx < +\infty$, where $x \mapsto r(x)$ is a given weight positive function) the inequality

(24.1)

$$\int_0^a s(x)|F'(x)(F(x)-F(\xi))|dx \le \frac{1}{2A^2}\max\left(\left(\int_0^\xi p\,dt\right)^2, \left(\int_\xi^a p\,dt\right)^2\right)\int_0^a rF'(x)^2 dx$$

holds, where the function s is defined by

$$s(x) = \int_x^\xi p(t)dt \qquad (0 \le x \le \xi),$$

$$s(x) = \int_\xi^x p(t)dt \qquad (\xi \le x \le a),$$

and the number A is a positive solution of the equation $\coth x = x$.

Conditions for equality are given in [44].

25. T. X. He and S. C. Wang [45] proved the following result:

Let f be absolutely continuous on $[0,a]$ with $f(0) = 0$. Then for every real $\lambda > 0$ and any numbers α, β with $0 \le \alpha < \beta \le a$ we have

$$\int_\alpha^\beta |f(x)|^\lambda |f'(x)|dx \le \frac{\beta^\lambda}{\lambda+1}\int_0^\beta |f'(x)|^{\lambda+1}dx - \frac{\alpha^\lambda}{\lambda+1}\int_0^\alpha |f'(x)|^{\lambda+1}dx.$$

This inequality is best possible and includes Opial's inequality and its generalizations due to S. L. Wang [46] and Z. J. Liang [47].

26. Y. Hong, H. Yang and D. Du [**48**] obtained an inequality for convex increasing functions. As a corollary they obtained a result from [**8**].

27. Results from [**12**] are extended to a Fourier type integral operator in a paper [**49**] of M. Dighe and V. M. Bhise.

28. G. S.Yang [**50**] gave some applications to integrodifferential inequalities of the Gronwall type.

29. R. P. Agarwal and E. Thandapani [**51**] gave the following modification of Das' result from [**18**]:

Let $p > 0$, $q > 0$ with $p + q > 1$. Let $f \in C^{n-1}[0, a]$ be such that $f^{(k)}(0) = 0$ for $k = 0, 1, \ldots, n-1$, where $n \geq 1$. Let $f^{(n-1)}$ be absolutely continuous and $\int_0^a |f^{(n)}|^{p+q} dx < +\infty$. Then

$$\int_0^a |f^{(k)}(x)|^p |f^{(n)}(x)|^q dx \leq M(k) a^{(n-k)p} \int_0^a |f^{(n)}(x)|^{p+q} dx, \tag{29.1}$$

where $M(k) = \lambda q^{\lambda q} \left(\frac{(n-k)(1-\lambda)}{n-k-\lambda} \right)^{p(1-\lambda)} ((n-k)!)^{-p}$ and $\lambda = 1/(p+q)$.

For $k = 0$ we have Das' result, but the reverse implication can be easily proved.

Agarwal and Thandapani used (29.1) in obtaining some results for Gronwall-type inequalities.

30. W. Z. Chen [**52**] gave an exposition of Opial type inequalities.

31. Inequalities of Opial-type in two variables are given by G. S. Yang [**53**].

THEOREM 1. *If $f(s,t)$, $f_1(s,t)$ $(= (\partial/\partial s) f(s,t))$ and $f_{12}(s,t)$ $(= (\partial^2/\partial s \partial t) f(s,t))$ are continuous functions on $[0,a] \times [0,b]$, and if $f(0,t) = f_1(s,0) = 0$ for $0 \leq s \leq a$, $0 \leq t \leq b$, then*

$$\int_0^a \left(\int_0^b |f(s,t)||f_{12}(s,t)| dt \right) ds \leq \frac{ab}{2} \int_0^a \left(\int_0^b |f_{12}(s,t)|^2 dt \right) ds.$$

THEOREM 2. *If $f(s,t)$, $f_1(s,t)$ and $f_{12}(s,t)$ are continuous functions on $[0,a] \times [0,b]$, and if $f(0,t) = f(a,t) = f_1(s,0) = f_1(s,b) = 0$, for $0 \leq s \leq a$, $0 \leq t \leq b$,*

then

$$\int_0^a \left(\int_0^b |f(x,t)||f_{12}(s,t)|dt \right) ds$$
$$\leq \frac{ab}{8} \int_0^a \left(\int_0^b |f_{12}(s,t)|^2 dt \right) ds.$$

Generalizations of these results are obtained by C. T. Lin and G. S. Yang [**54**].

THEOREM 3. *Let the conditions of Theorem 1 be fulfilled.*
Then

$$\int_0^a \left(\int_0^b |f(s,t)|^m |f_{12}(s,t)|^n dt \right) ds$$
$$\leq \frac{n}{m+n} a^m b^m \int_0^a \left(\int_0^b |f_{12}(s,t)|^{m+n} dt \right) ds,$$

where $m, n \geq 1$.

THEOREM 4. *Let the conditions of Theorem 2 be fulfilled.*
Then

$$\int_0^a \left(\int_0^b |f(s,t)|^m |f_{12}(s,t)|^n dt \right) ds$$
$$\leq \frac{n}{m+n} \left(\frac{a}{2}\right)^m \left(\frac{b}{2}\right)^m \int_0^a \left(\int_0^b |f_{12}(s,t)|^{m+n} dt \right) ds,$$

where $m, n \geq 1$.

Weighted versions of the above results are also given in [**54**].

32. Some discrete inequalities of Opial's type are obtained by G. V. Milovanović and I. Ž. Milovanović [**55**]. The following result is a special case of their general result:

Let $x_1, \ldots, x_n$ be real numbers, $x_0 = 0$, then

$$2 \sin^2 \left(\frac{\pi}{2(n+1)} \right) \sum_{k=1}^n x_k^2 \leq \sum_{k=1}^n x_k (x_k - x_{k-1}) \tag{32.1}$$
$$\leq 2 \cos^2 \left(\frac{\pi}{2(n+1)} \right) \sum_{k=1}^n x_k^2.$$

It is shown that this result is related to an extremal problem occurring in the investigation of approximative properties of positive polynomial operators.

33. Z. Qi [**56**] gave an inequality analogous to (3.1) for the integral $\int_0^a P(|f(x)|)\, Q(|f'(x)|)\, dx$, where, for $u \geq 0$, $P(u)$ is nondeaceasing and $Q(u)$ is nondecreasing and convex. The short proof uses Jensen's inequality.

34. A. B. Fagbohun and C. O. Imoru [**57**],[**58**] gave a generalization of Theorem 3 from 3. An application in integrodifferential inequalities is given in [**58**].

35. C. T. Lin and G. S. Yang [**59**] and B. G. Pachpatte [**60**] gave Opial type inequalities with two functions. For example, the following result is given in [**59**]:

THEOREM 1. *Let p be a bounded, positive and non- increasing function on $[0,a]$, and let f and g be absolutely continuous on $[0,a]$, with $f(0) = f(a) = 0$, $g(0) = g(a) = 0$. Then*

$$\int_0^a p(x)|f(x)|^\alpha |g(x)|^\alpha \left(|f(x)|^\beta |g'(x)|^\beta + |f'(x)|^\beta |g(x)|^\beta\right) dx$$

$$\leq \frac{\beta}{2(\alpha+\beta)} \left(\frac{a}{2}\right)^{2\alpha+\beta} \int_0^a p(x) \left(|f'(x)|^{2(\alpha+\beta)} + |g'(x)|^{2(\alpha+\beta)}\right) dx,$$

where $\alpha \geq 0$ and $\beta \geq 1$.

Applications to Grownwall-type inequaliites are also given.

An analogous result with three functions in the case $\beta = 1$ (as in [**60**]) is given in paper [**61**] of B. G. Pachpatte.

Related results for functions of two variables are given by B. G. Pachpatte [**62**] and C. T. Lin and G. S. Yang [**63**]. The main result from [**63**] is given by the following theorem.

THEOREM 2. *Let p be a positive function on $[0,a] \times [0,b]$, with $\int_0^a \left(\int_0^b p^{-1}(s,t)dt\right) ds < +\infty$, q be a bounded, positive and nonincreasing (with respect to each variable) function on $[0,a] \times [0,b]$, and let $f(0,t) = f(a,t) = f_1(s,0) = f_1(s,b) = g(0,t) = g_1(s,b) = 0$, for $(s,t) \in [0,a] \times [0,b]$.*

Then

$$\int_0^a \left(\int_0^b q(s,t)|f(s,t)g(s,t)|^m \right.$$
$$\left. \times \left(|f_{12}(s,t)g(s,t)|^n + |f(s,t)g_{12}(s,t)|^n\right) dt \right) ds$$
$$\leq \frac{n}{2(m+n)} \left(\frac{ab}{4}\right)^{2m+n-1} K \int_0^a \int_0^b p(s,t)q(s,t) \left(|f_{12}(s,t)|^{2(m+n)}\right.$$
$$\left. + |g_{12}(s,t)|^{2(m+n)}\right) dtds$$

where $m \geq 0$, $n \geq 1$, *and* $K = \max\{A_1, A_2, A_3, A_4\}$, *where*

$$A_1 = \int_0^X \left(\int_0^Y p^{-1}(s,t)dt \right) ds,$$

$$A_2 = \int_X^a \left(\int_0^Y p^{-1}(s,t)dt \right) ds,$$

$$A_3 = \int_0^X \left(\int_Y^b p^{-1}(s,t)dt \right) ds,$$

$$A_4 = \int_X^a \left(\int_Y^b p^{-1}(s,t)dt \right) ds,$$

$$X = \frac{a}{2} \quad \text{and} \quad Y = \frac{b}{2}.$$

For $f = g$, $2m + n = \alpha$ and $n = \beta$, we have a result of C. T. Lin from [**64**].

35. If $p \geq 1$, $q > 0$, $q_k \geq 0$ for $k = 0, 1, \ldots, n-1$ with $\sum_{k=0}^{n-1} q_k = 1$, and if f, $M(k)$ and λ are as in 29, then

$$\int_0^a \left(\prod_{k=0}^{n-1} |f^{(k)}(x)|^{q_k} \right)^p |f^{(n)}(x)|^q dx$$
$$\leq \sum_{k=0}^{n-1} M(k) q_k a^{(n-k)p} \int_0^a |f^{(n)}(x)|^{p+q} dx.$$

This result of G. S. Yang [**65**] is a simple consequence of a result from 29. and the arithmetic-geometric mean inequality.

A related result is given by B. G. Pachpatte [**66**].

Recently, L. Ju-Da [**67**] proved the following similar result:

Let p_i $(i = 0, 1, \ldots, n)$ be nonnegative real numbers and satisfy $p = \sum_{i=0}^{n} p_i > p_n > 0$, $p > 1$, $f(x)$ be of class C^n on $0 \le x \le a$ and satisfy $f(0) = f'(0) = \cdots = f^{(n-1)}(0) = 0$. Suppose that $h(x)$, $r(x)$ are positive functions on $0 \le x \le a$, such that

$$R_i(x) = \int_0^x (x-t)^{(n-i-1)\frac{p}{p-1}} \cdot r^{-\frac{1}{p-1}} dt < +\infty,$$

$$K(h,r) = \left(\int_0^a h^{\frac{p}{p-p_n}} r^{-\frac{p_n}{p-p_n}} \prod_{i=0}^{n-1} R_i^{\frac{p_i(p-1)}{p-p_n}} dx \right)^{\frac{p-p_n}{p}} < +\infty.$$

Then we have

$$\int_0^a h|f^{(n)}|^{p_n} \prod_{i=0}^{n-1} |f^{(i)}|^{p_i} dx \le \left(\frac{p_n}{p}\right)^{\frac{p_n}{p}} \frac{K(h,r)}{Q} \int_0^a r|f^{(n)}|^p dx,$$

where $Q = \prod_{i=0}^{n-1} ((n-i-1)!)^{p_i}$.

36. D. S. Mitrinović and J. E. Pečarić [**68**] noted that (32.1) is a consequence of Ozeki's discrete inequality given in Chapter II on Wirtinger's inequality.

37. We shall say that function $f : [0,a] \to R$ belongs to the class $U(v,K)$ if it can be represented in the following form.

$$f(x) = \int_0^x K(x,t)v(t)dt,$$

where v is a continuous function and K is an arbitrary nonnegative kernel. Let $f \in U(v,K)$ where $\left(\int_0^x K(x,t)^p dt\right)^{1/p} \le M$, and let ϕ: $[0,\infty) \to R$ be a differentiable function such that for $q > 1$ $(p^{-1} + q^{-1} = 1)$ the function $x \mapsto \phi(x^{1/q})$ is convex and $\phi(0) = 0$. Then

$$\int_0^a |f|^{1-q} \phi'(|f|)|v|^q \, dx \le \frac{q}{M^q} \phi \left(M \left(\int_0^a |v|^q \, dt \right)^{1/q} \right). \tag{37.1}$$

If the function $x \mapsto \phi(x^{1/q})$ is concave, then the reverse inequality in (37.1) is valid.

This is due to D. S. Mitrinović and J. E. Pečarić [**69**].

The following result is a special case of the above:

Let ϕ, q and p be defined as in the above result. If $f^{(n-1)} \in AC[0,a]$, with either $f^{(k)}(0) = 0$ for $k = 0, 1, \ldots, n-1$ or $f^{(k)}(a) = 0$ for such k, then

$$\int_0^a |f|^{1-q}\phi'(|f|)\,|f^{(n)}|^q\,dx \le \frac{q}{M^q}\phi\left(M\left(\int_0^a |f^{(n)}|^q\,dt\right)^{1/q}\right),$$

where $M = a^{n-1/q}/\left((n-1)!(np-p+1)^{1/p}\right)$.

38. J. E. Pečarić [**70**] proved the following result:

Let $f_i(t)$, $i = 1, \ldots, n$, be real-valued absolutely continuous funcitons on $[0,a]$ with $f_i(0) = 0$. Let $F(t_1, \ldots, t_n)$ be a nondecreasing function for $t_i \ge 0$, $i = 1, \ldots, n$, with $F(0, \ldots, 0) = 0$ such that all its the first partial derivatives F_i $(= \partial F/\partial t_i)$ are also nondecreasing functions. Then

$$\int_0^a \left(\sum_{i=1}^n F_i(|f_1(t)|, \ldots, |f_n(t)|)\,|f_i'(t)|\right) dt \le F\left(\int_0^a |f_1'(t)|dt, \ldots, \int_0^a |f_n'(t)|dt\right).$$

In the case $F(t_1, \ldots, t_n) = \sum_{i=1}^n F_i(t_i)$, we can obtain a result of B. G. Pachpatte [**71**].

39. C. H. Fitzgerald [**72**] proved the following result:

For each $n \ge 2$, there is a constant c_n such that for any real interval $[0,a]$ if $f \in C^n[0,a]$ and $f(0) = f'(0) = \cdots = f^{(n-1)}(0) = 0$ and $f(a) = f'(a) = \cdots = f^{(n-1)}(a) = 0$, then

$$\int_0^a |f(x)f'(x)|\,dx \le c_n a^{2n-1}\int_0^a f^{(n)}(x)^2\,dx.$$

Furthermore, the constant c_n is sharp for a polynomial spline of degree $2n-1$ which has a knot at the midpoint of $[0,a]$, is symmetric about the midpoint, and is in $C^{2n-2}[0,a]$.

40. Recent papers on Opial's inequality include those of C. T. Lin [**73**], [**74**], [**75**], L. M. Shieh [**76**] and Pachpatte [**77**]–[**82**]. Lin's papers give alternate proofs to the results of Pachpatte in [**77**] and [**78**], which are on discrete versions of Opial's inequality. Shieh [**76**] gives a weighted version.

41. W. S. Cheung [**83**] studies generalizations of Willett's version (8.1) with f on the left replaced by a product of derivatives, i.e.

$$\int_a^b w(x)\left(\prod_{k=0}^{n-1} |f^{(\beta)}(x)|^{n_k}\right)^p |f^{(n)}(x)|^q dx$$
$$\leq \sum_{k=0}^{n-1} M_k r_k (b-a)^{(n-k)p} \int_a^b w(x)|f^{(n)}(x)|^{p+q} dx$$

with $f^{(k)}(0) = 0$, $k = 0\ldots, n-1$, $w(x)$ nonincreasing and

$$M_k = \alpha q^{\alpha q}\left[\frac{(n-k)(1-\alpha)}{n-k-\alpha}\right]^{(1-\alpha)p}[(n-k)!]^{-p}, \quad \alpha = \frac{1}{(p+q)}.$$

References

1. OPIAL, Z., *Sur une inégalité*, Ann. Polon. Math. **8** (1960), 29–32.
2. OLECH, C., *A simple proof of a certain result of Z. Opial*, Ann. Polon. Math. **8** (1960), 61–63.
3. BEESACK, P. R. *On an integral inequality of Z. Opial*, Trans. Amer. Math. Soc. **104** (1962), 470–475.
4. LEVINSON, N., *On an inequality of Opial and Beesack*, Proc. Amer. Math. Soc. **15** (1964), 565–566.
5. MALLOWS, C. L., *An even simpler proof of Opial's inequality*, Proc. Amer. Math. Soc. **16** (1965), 173.
6. PEDERSEN, R. N., *On an inequality of Opial, Beesack and Levinson*, Proc. Amer. Math. Soc. **16** (1965), 174.
7. YANG, G.-S., *On a certain result of Z. Opial*, Proc. Japan Acad. **42** (1966), 78–83.
8. HOLT, J. M., *Integral inequalities related to non-oscillation theorems for differential equations*, SIAM J. **13** (1965), 767.
9. HUA, L.-K., *On an inequlaity of Opial*, Sci. Sinica **14** (1965), 789–790.

10. CALVERT, J., *Some generalizations of Opial's inequality*, Proc. Amer. Math. Soc. **18** (1967), 72–75.
11. WONG, J. S. W., *A discrete analogue of Opial's inequality*, Canad. Math. Bull. **10** (1967), 115–118.
12. BEESACK, P. R. and K. M. DAS, *Extensions of Opial's inequality*, Pacific J. Math. **26** (1968), 215-232.
13. GODUNOVA, E. K. and V. I. LEVIN, *On an inquality of Maroni* (Russian), Mat. Zametki **2** (1967), 221–224.
14. MARONI, P. M., *Sur l'inégalité d'Opial-Beesack*, C. R. Acad. Sci. Paris., A**264** (1967), 62–64.
15. BOYD, D. W. and J. S. W. WONG, *An extension of Opial's inequality*, J. Math. Anal. Appl. **19** (1967), 100-102.
16. REDHEFFER, R., *Inequalities with three functions*, J. Math. Anal. Appl. **16** (1966), 219–242.
17. WILLETT, D., *The existence-uniqueness theorem for an n-th order linear ordinary differential equation*, Amer. Math. Monthly **75** (1968), 174–178.
18. DAS, K. M., *An inequality similar to Opial's inequality*, Proc. Amer. Math. Soc. **22** (1969), 258–261.
19. BOYD, D. W., *Best constants in inequalities related to Opial's inequality*, J. Math. Anal. Appl. **25** (1969), 378–387.
20. __________, *Bcst constants in a class of integral inequalities*, Pacific J. Math. **30** (1969), 367–383.
21. LEE, Cheng-Ming, *On a discrete analogue of inequalities of Opial and Yang*, Canad. Math. Bull **11** (1968), 3–77.
22. BEESACK, P. R., *On certain discrete inequalities involving partial sums*, Canad. J. Math. **21** (1969), 222-234.
23. BANKS, D. O. *An integral inequality*, Proc. Amer. Math. Soc. **14** (1963), 823–829.
24. ARAMĂ, O., *Sur un problème d'interpolation relatif aux solutions des équations différentielles linéaires du quatrième orde*, Mathematica **33** 10 (1968), 5–15.
25. LUPAŞ, A., *An interpolation problem for the solutions of a class of linear differential equations*, Mathematica **35** 12 (1970), 87–104.
26. REDHEFFER, R., *Integral inequalities with boundary terms*, Inequalities II, Proc. II Symposium on Inequalities, Colorado (USA), August 14–22, 1967, ed. O. Shisha, New York and London, 1970, 261–291.
27. BOYD, D. W., *Inequalities for positive integral operators*, Pacif. J. Math. **38** (1971), 9–24.
28. BEESACK, P. R., *Integral inequalities involving a function and its derivative*, Amer. Math. Monthly **78** (1971), 705–741.

29. HEROLD, H., *Diskonjugierte Differentialgleichungen 2. Ordnung im Komplexen*, J. Reine Angew. Math. **257** (1972), 1–11.
30. ROZANOVA, G. I., *Diskretnyĭ analog neravenstva Bisaka i Dasa*, Uč. Zap. Mosk. Gos. Ped. In-ta im. Lenina **460** (1972), 52–57.
31. ________, *Neravenstva dlja istokoobrazno predstabimyh funkciĭ*, Uč. Zap. Mosk. Gos. Ped. In-ta im. Lenina **460** (1972), 66–72.
32. ________, *Integral'nye neravenstva s proizvodnysi i proizvol'nymi vypuklymi funkcijami*, Uč. Zap. Mosk. Gos. Ped. In-ta im. Lenina **460** (1972), 58–65.
33. ________, *Ob odnom integral'nom neravenstve, svjazannom s neravenstvom Polia*, Izvestija Vysš. Učebn. Zaved., Mat. **125** 10 (1975), 75–80.
34. PÓLYA, G., *Problem 4264*, Amer. Math. Monthly **54** (1947), 479.
35. NEČAEV, I. D., *Integral inequalities with gradients and derivatives* (Russian), Doklady Akad. Nauk SSSR **211** (1973), 1063–1066.
36. VRĂNCEANU, G. G., *On an inequality of Opial*, Bull. Math. Soc. Sci. Math. R. S. Roumanie **65** 17 (1973), 315–316.
37. SHUM, D. G., *A general and sharpened form of Opial's inequality*, Canad. Math. Bull. **17** 3 (1974), 385- -389.
38. ________, *On a class of new inequalities*, Trans. Amer. Math. Soc. **204** (1975), 299- -341.
39. ROZANOVA, G. I., *Sharp integral inequalities of order* $\alpha > 0$ (Russian), Mathematical physics, No. 3 (1976) (Russian), 97–103, Moskov. Gos. Ped. Inst. im. Lenina, Moscow, 1976.
40. ________, *Inequalities that contain derivatives of different orders* (Russian), Mathematical physics, No. 3 (Russian) (1976), 104–108, Moskov. Gos. Ped. Inst. im. Lenina, Moscow, 1976.
41. BEESACK, P. R., *Elementary proofs of some Opial-type integral inequalities*, J. Anal. Math. **36** (1979), 1–14.
42. HOU, M. S., *An inequality of Z. Opial* (Chinese), Kexue Tongbao **24** 6 (1979), 247–248.
43. LEE, C. S., *On some generalization of inequalities of Opial, Yang and Shum*, Canad. Math. Bull **23** 1 (1980), 71–80.
44. MILOVANOVIĆ, G. V. and I. Ž. MILOVANOVIĆ, *The best constant in some integral inequalities of Opial type*, Univ. Beograad. Publ. Elektrotehn. Fak. Ser. Mat. Fiz. No. 678–715 (1980), 48–53.
45. HE, T. X. and S. C. WANG, *A note on the Opial-Hua inequality* (Chinese), J. Math. Res. Exposition 1 (1982), 61–62.
46. WANG, S. L., Kexue Tongbao **8** (1980), 383–384.
47. LIANG, Z. J., J. Huazhong, Normal Acad. **2** (1980), 33–37.
48. HONG, Y., H. YANG and D. DU, *An inequality for convex functions*, Kexue Tongbao, Foreign Lang. Ed. **27** (1982), 1266–1270.

49. DIGHE, M. and V. M. BHISE, *On the generalisation of Beesack's extension of Opial inequality*, J. Indian Acad. Math. **3** (1981), 22–26.
50. YANG, G. S., *A note on an integrodifferential inequality*, Tamkang J. Math. **12** (1981), 257–264.
51. AGARWAL, R. P. and E. THANDAPANI, *On some new integrodifferential inequalities*, Anal. sti. Univ. "Al. I. Cuze" din Iaşi **28**, S. Ia, 1982, f.1, 123–126.
52. CHEN. W. Z. (with G. J. FENG; X. H. WANG), *Twenty years of Opial inequalities* (Chinese), J. Math. Res. Exposition **2** 4 (1982), 151–166.
53. YANG, G. S., *Inequality of Opial-type in two variables*, Tamkang J. Math. **13** (1982), 255–259.
54. LIN, C. T. and G. S. YANG, *A generalized Opial's inequality in two variables*, Tamkang J. Math. **15** (1984), 115–122.
55. MILOVANOVIĆ, G. V. and I. Ž. MILOVANOVIĆ, *Some discrete inequalities of Opial's type*, Acta Scient. Math. (Szeged) **47** (1984), 413–417.
56. QI, Z., *Further generalizations of Opial's inequality*, Acta Math. Sinica (N.S.) **1** 3 (1985), 196–200.
57. FAGBOHUN, A. B. and C. O. IMORU, *On an extension of Opial's inequality*, Simon Stevin, A Quarterly J. Pure Appl. Math. **59** (1985), Nr 1, 11–17.
58. __________, *A new class of integrodifferential inequalities*, Simon Stevin, A Quarterly J. Pure Appl. Math. **60** (1986), Nr 4, 301–311.
59. LIN, C. T. and G. S. YANG, *On Some Integrodifferential Inequalities*, Tamkang J. Math. **16** 4 (1985), 123–129.
60. PACHPATTE, B. G., *On Opial-type integral inequalities*, J. Math. Anal. Appl. **120** (1986), 547–556.
61. __________, *On certain integral inequalities related to Opial's inequality*, Period. Math. Hungar. **17** 2 (1986), 119–125.
62. __________, *On Opial-type inequalities in two independent variables*, Royal Soc. Edinburgh Proc. A **100** (1985), 263–270.
63. LIN, C. T. and G. S. YANG, *On some new Opial-type inequalities in two variables*, Tamkang J. Math. **17** 2 (1986), 31–36.
64. LIN, C. T., *A note on an integrodifferential inequality*, Tamkang J. Math. **23** (1985), 349–354.
65. YANG, G. S., *A note on an inequality similar to Opial inequality*, Tamkang. J. Math. **18** 4 (1987), 101–104.
66. PACHPATTE, B. G., *On Wirtinger-Opial type integral inequalities*, Tamkang J. Math. **17** 4 (1986), 1–6.
67. JU-DA, L., *Opial-type integral inequalities involving several higher order derivatives*, J. Math. Anal. Appl., (to appear).

68. MITRINOVIĆ, D. S. and J. E. PEČARIĆ, *On two lemmas of N. Ozeki*, J. Col. Arts Sci. Chiba Univ. B-21 (1988), 19–21.

69. ________, *Generalizations of two inequalities of Godunova and Levin*, Bull. Polish. Acad. Sci. Math. **36** (1988), 645–648.

70. PEČARIĆ, J. E., *Remark on the paper "Some inequalities similar to Opial's inequality" of B. G. Pachpatte*, (manuscript).

71. PACHPATTE, B. G. *Some inequalities similar to Opial's inequality*, (manus

72. FITZGERALD, C. H., *Opial-type inequalities that involve higher order derivatives*, in "General Inequalities," 4, ed. W. Walter, Basel, 1984, pp. 25–36.

73. LIN, C. T., *A note on an integrodifferential inequality*, Tamkang J. No. **23** (1985), 349–354.

74. ________, *Some generalizations of Opial's inequality*, Tamkang J. No. **24** (1986), 451–455.

75. ________, *A further generalization of Opial's integral inequality*, Tamkang J. No. **27** (1989), 491–493.

76. SHIEH, L. M., *On some generalized Opial inequalities*, Tamkang J. No. **23** (1985) 395–398.

77. PACHPATTE, B. G. *A note on discrete inequalities in several variables*, Tamkang J. Math. **19** (1988), 1–6.

78. ________, *A note on Opial and Wirtinger type discrete inequalities*, J. Math. Anal. Appl. **127** (1987), 470–474.

79. ________, *On a generalized Opial type inequality in two independent variables*, Anal. St. Univ. ACI Cuza Iaşi Matematika **35** (1989) 231--235.

80. ________, *On some new generalizations of Opial's inequality*, Demonstr. Math. **19** (1986), 281–291.

81. ________, *On two inequalities similar to Opial's inequality in two independent variables*, Period. Math. Hung. **18** (1987), 137–141.

82. ________, *On multidimensional Opial-type inequalities*, J. Math. Anal. Appl. **126** (1987), 85–89.

83. CHEUNG, W. S., *Some new Opial-type inequalities*, Mathematika **37** (1990), 136–142.

CHAPTER IV

HARDY'S, CARLEMAN'S AND RELATED INEQUALITIES

1. Some preliminaries.

THEOREM 1. *If* $p > 1$, $a_n \geq 0$, *and* $A_n = a_1 + a_2 + \cdots + a_n$, *then*

$$\sum_1^\infty \left(\frac{A_n}{n}\right)^p < \left(\frac{p}{p-1}\right)^p \sum_1^\infty a_n^p, \tag{1.1}$$

unless all the a are zero. The constant $\left(\frac{p}{p-1}\right)^p$ *is the best possible.*

The corresponding theorem for integrals is

THEOREM 2. *If* $p > 1$, $f(x) \geq 0$, *and* $F(x) = \int_0^x f(t)dt$, *then*

$$\int_0^\infty \left(\frac{F}{x}\right)^p dx < \left(\frac{p}{p-1}\right)^p \int_0^\infty f^p \, dx, \tag{1.2}$$

unless $f \equiv 0$. *The constant is the best possible.*

These theorems were first proved by G. H. Hardy [**1**], except that Hardy was unable to fix the constant in Theorem 1, and they are well-known as Hardy's inequalities. This imperfection was removed by E. Landau [**2**]. A great many alternative proofs of these theorems have been given by various writers, for example by T. A. A. Broadbent [**3**], E. B. Elliott [**4**], K. Grandjot [**5**], G. H. Hardy [**6**], Th. Kaluza and G. Szegö [**7**], and K. Knopp [**8**].

The following theorem is the well-known Carleman inequality [**9**]:

THEOREM 3.

$$\sum_1^\infty (a_1 a_2 \ldots a_n)^{1/n} < e \sum_1^\infty a_n, \tag{1.3}$$

unless (a_n) *is null. The constant e is the best possible.*

The less precise convergence theorem (without the constant e) was found independently by other writers, and there are a number of proofs of one form or the other of the theorem (see G. Pólya [**10**] , G. Valiron [**11**, p. 186, where there is a proof due to J. E. Littlewood], Kaluza and Szegö [**7**], Knopp [**8**] and A. M. Ostrowski [**12**].

The corresponding integral theorem is obtained in Knopp [**8**]:

THEOREM 4. *If f is not null, then*

$$\int_0^\infty \exp\left(\frac{1}{x}\int_0^x \log f(t)dt\right) dx < e\int_0^\infty f(x)dx. \tag{1.4}$$

The previous results are given in the well-known book of G. H. Hardy, J. E. Littlewood and G. Pólya [**13**] as Theorems 326, 327, 334 and 335 respectively.

2. Now, we shall list other related results from [**13**]:

(i) (Th. 328) If $p > 1$ and $F(x) = \int_x^\infty f(t)dt$, then

$$\int_0^\infty F^p\, dx < p^p \int_0^\infty (xf)^p\, dx, \tag{2.1}$$

unless $f \equiv 0$. The constant is the best possible.

(ii) (Th. 329) If $p > 1$, $r > 0$, and

$$f_r(x) = \frac{1}{\Gamma(r)}\int_0^x (x-t)^{r-1} f(t)dt, \tag{2.2}$$

then

$$\int_0^\infty \left(\frac{f_r}{x^r}\right)^p dx < \left(\frac{\Gamma\left(1-\frac{1}{p}\right)}{\Gamma\left(r+1-\frac{1}{p}\right)}\right)^p \int_0^\infty f^p\, dx, \tag{2.3}$$

unless $f \equiv 0$. If

$$f_r(x) = \frac{1}{\Gamma(r)}\int_x^\infty (t-x)^{r-1} f(t)dt, \tag{2.4}$$

then

$$\int_0^\infty f_r^p\, dx < \left(\frac{\Gamma\left(\frac{1}{p}\right)}{\Gamma\left(r+\frac{1}{p}\right)}\right)^p \int_0^\infty (x^r f)^p dx, \tag{2.5}$$

unless $f \equiv 0$. In each case the constant is the best possible.

REMARK: The function $f_r(x)$ of (2.2) is the Riemann-Liouville integral of $f(x)$ of order r, with "origin" 0. The function (2.4) is the Weyl integral of order r. Part of this result is proved by K. Knopp [**14**].

(iii) (Th. 330, Hardy [**15**]) If $p > 1$, $r \neq 1$, and $F(x)$ is defined by

$$F(x) = \int_0^x f(t)dt \quad (r > 1), \qquad F(x) = \int_x^\infty f(t)dt \quad (r < 1),$$

then

$$\int_0^\infty x^{-r} F^p \, dx < \left(\frac{p}{|r-1|}\right)^p \int_0^\infty x^{-r}(xf)^p dx, \tag{2.6}$$

unless $f \equiv 0$.

The constant is the best possible. It is also easy to verify that, when $p = 1$, the two sides of (2.6) are equal.

(iv) (Th. 331, E. T. Copson [**16**], Hardy [**18**]) If $p > 1$, then

$$\sum_1^\infty (a_n + a_{n+1} + \dots)^p < p^p \sum_1^\infty (na_n)^p, \tag{2.7}$$

unless (a_n) is null. The constant p^p is the best possible.

(v) (Th. 332, Hardy [**6**], Copson [**16**]) If $p > 1$, $a_n \geq 0$, $\lambda_n > 0$, and

$$\Lambda_n = \lambda_1 + \lambda_2 + \dots + \lambda_n, \quad A_n = \lambda_1 a_1 + \lambda_2 a_2 + \dots + \lambda_n a_n,$$

then

$$\sum_1^\infty \lambda \left(\frac{A_n}{\Lambda_n}\right)^p < \left(\frac{p}{p-1}\right)^p \sum_1^\infty \lambda_n a_n^p, \tag{2.8}$$

unless (a_n) is null.

REMARK: This is a weighted version of (1.1). Note that a weighted version of (2.7) is given in Copson [**16**]: If $p > 1$, $a_n \geq 0$, $\lambda_n > 0$, Λ_n and A_n are as above, then

$$\sum_1^\infty \lambda_n \left(\frac{\lambda_n a_n}{\Lambda_n} + \frac{\lambda_{n+1} a_{n+1}}{\Lambda_{n+1}} + \dots\right)^p \leq p^p \sum_1^\infty \lambda_n a_n^p. \tag{2.9}$$

(vi) (Th. 337) If $0 < p < 1$, $f(x) \geq 0$, $\int_0^\infty f^p dx < \infty$, $F(x) = \int_x^\infty f(t)dt$, then

$$\int_0^\infty \left(\frac{F}{x}\right)^p dx > \left(\frac{p}{1-p}\right)^p \int_0^\infty f^p \, dx, \tag{2.10}$$

unless $f \equiv 0$. The constant is the best possible.

(vii) (Th. 338) If $0 < p < 1$ and $\sum_1^\infty a_n^p < \infty$, then

$$\sum_1^{\infty}{}' \left(\frac{1}{n}(a_n + a_{n+1} + \dots)\right)^p > \left(\frac{p}{1-p}\right)^p \sum_1^\infty a_n^p,$$

unless (a_n) is null. The dash over the summation on the left- hand side implies that the term for which $n = 1$ is to be multiplied by $1 + \frac{1}{1-p}$. The constant is the best possible.

This result is due to E. B. Elliott (see [**13**, p. 252]).

(viii) (Th. 344) If $0 < p < 1$, then the reverse of (2.7) is valid. (See Copson [**17**].

(ix) (Th. 347, Hardy [**15**]) If r and F satisfy the conditions of (iii), but $0 < p < 1$, then the reverse inequality in (2.6) is valid.

(x) (Th. 349, Hardy [**6**]) If λ_n and Λ_n satisfy the conditions of (v), then

$$\sum_1^\infty \lambda_n \left(a_1^{\lambda_1} a_2^{\lambda_2} \dots a_n^{\lambda_n}\right)^{1/\Lambda_n} < e \sum_1^\infty \lambda_n a_n. \tag{2.11}$$

(xi) (Ths. 350, 351 and 352, Hardy [**19**]) If $p > 1$, $K(x) > 0$, and

$$\int K(x) x^{s-1} dx = \phi(s),$$

then

$$\int dx \left(\int K(xy) f(y) dy\right)^p < \phi^p\left(\frac{1}{p}\right) \int x^{p-2} f^p \, dx,$$

$$\int x^{p-2} \left(\int K(xy) f(y) dy\right)^p < \phi^p\left(\frac{1}{p'}\right) \int f^p \, dx.$$

In particular, when $K(x) = e^{-x}$, $F(x) = \int K(xy) f(y) dy$ is the Laplace transform of $f(x)$,

$$\int F^p \, dx < \Gamma^p\left(\frac{1}{p}\right) \int x^{p-2} f^p \, dx, \quad \int x^{p-2} F^p \, dx < \Gamma\left(\frac{1}{p'}\right) \int f^p \, dx.$$

If also $K(x)$ is a decreasing function of x, and $A(x) = \sum a_n K(nx)$, $A_n = \int a(x)K(nx)dx$, then

$$\int A^p(x)dx < \phi^p\left(\frac{1}{p}\right)\sum n^{p-2}a_n^p,$$

$$\sum A_n^p < \phi^p\left(\frac{1}{p}\right)\int x^{p-2}a^p(x)dx,$$

$$\int x^{p-2}A^p(x)dx < \phi^p\left(\frac{1}{p'}\right)\sum a_n^p,$$

$$\sum n^{p-2}A_n^p < \phi^p\left(\frac{1}{p'}\right)\int a^p(x)dx.$$

For $1 < p \le 2$, we have

$$\int F^{p'}\, dx \le \frac{2\pi}{p'}\left(\int^p \, dx\right)^{p'/p}.$$

(xii) (Th. 271, Hardy and Littlewood [**20**]) If $p > 1$ and $G(f)$ is the geometric mean of f over $(0, x)$, then

$$\int_0^\infty x^{p-1}G^p(f)dx < \frac{1}{p-1}\left(\Gamma\left(\frac{p}{p-1}\right)\right)^{1-p}\left(\int_0^\infty f\, dx\right)^p, \tag{2.12}$$

unless $f \equiv C\exp(-Bx^{p-1})$.

(xiii) (Th. 270, G. A. Bliss [**21**]) Let $n > m > 1$, and let f be a real nonnegative function on $[0, +\infty)$ such that the integral $\int_0^{+\infty} f(x)^m dx$ exists in the Lebesgue sense. Then the integral $F(x) = \int_0^x f(x)dx$ is finite for every x and

$$\int_0^{+\infty} x^{r-n}F(x)^n dx \le \frac{1}{n-r-1}\left(\frac{r\Gamma\left(\frac{n}{r}\right)}{\Gamma\left(\frac{1}{r}\right)\Gamma\left(\frac{n-1}{r}\right)}\right)^r\left(\int_0^{+\infty} f(x)^m dx\right)^{n/m}, \tag{2.13}$$

where $r = n/m - 1$.

REMARK: Inequality (2.11) was also proved by G. Sunouchi and N. Takagi [**22**]. They also prove the following inequality

$$\sum_1^\infty \frac{n}{\lambda_n}\left(e^{\lambda_n/\Lambda_n} - 1\right)\left(a_1^{\lambda_1}a_2^{\lambda_2}\dots a_n^{\lambda_n}\right)^{1/\Lambda_n} < e\sum_1^\infty \lambda_n a_n,$$

where e is the best possible constant, and where the conditions of (v) are satisfied.

3. The inequalities (1.1) and (1.3) are the special cases $\phi = x^t$ $(0 < t < 1)$, $\phi = \log x$, of (Knopp [**23**], [**13**, Th. 365]):

$$\sum \phi^{-1}\left(\frac{\phi(a_1) + \cdots + \phi(a_n)}{n}\right) < K(\phi) \sum a_n. \tag{3.1}$$

This inequality was also considered by H. P. Mulholland [**24**]:

Let $\phi(x)$ be a non-negative monotonic function whose lower bound is 0. Then a necessary and sufficient condition for $\phi(x)$ to satisfy (3.1) is that there should be some $k > 1$ which makes $\phi^{1/k}(x)$ quasi-convex.

(We say that a non-negative function $\phi(x)$, defined in $(-\infty, \infty)$, is quasi-convex if there is corresponding to it some convex function $\Psi(x)$ and some number $A > 1$ such that: $\Psi(x) \le \phi(x) \le A\phi(x)$.)

REMARK: Weighted versions of the above results are also considered in [**23**] and [**24**]. Some related results are given in Mulholland's papers [**25**], [**26**] and [**27**]. For example, in [**25**] is obtained a similar generalization of a part of Th. 346 from [**13**] (i.e. from Hardy and Littlewood [**28**], [**31**]).

4. The following result of V. Levin [**29**] is a generalization of results from Knopp [**14**] and J. G. van der Corput [**30**]:

$$\sum_{n=0}^{\infty} u^n \left(a_0^{\binom{n}{0}q^n} a_1^{\binom{n}{1}q^{n-1}} \cdots a_n^{\binom{n}{n}}\right)^{\frac{p}{(q+1)^n}} \le \frac{q\lambda + 1}{(1-\lambda)^{1-p}} \left(\sum_{n=0}^{\infty} a_n\right)^p,$$

where $a_n \ge 0$ (not all $a_n = 0$), $0 < q < 1$, $q > 0$, $0 < p \le 1$, and λ is defined by

$$0 < \lambda < 1, \quad u = \frac{(q+1)\lambda^{1-\frac{p}{q+1}}}{(q\lambda + 1)}.$$

5. In the following results under the (weighted) geometric mean of the nonnegative function $f(x)$ we define the function

$$G(f(x); p) = \exp\left(\frac{1}{p(x)} \int_0^x p'(t) \log f(t) dt\right),$$

where the weight-function $p(x)$ satisfies the following conditions: $p'(t) \ge 0$, $p(0) = 0$, so that $p(x) = \int_0^x p'(t)dt$. Furthermore we suppose that p' is the integral of p'', $p(x) \to \infty$ for $x \to \infty$ and $p(x)\log p'(x) \to 0$ for $x \to 0$. If $g(x) \geq 0$ is the integral of $g'(x)$ and $p(x)\log g(x) \to 0$ for $x \to 0$, V. Levin [**32**] proved that

$$\int_0^\infty g(x)G\left(f(x);p\right) dx < \int_0^\infty F(x)f(x)dx, \tag{5.1}$$

provided that the right-hand side is finite, where $F(x)$ is given by

$$F(x) = g(x)\exp\left(1 - \frac{p(x)p''(x)}{p'^2(x)} + \frac{p(x)g'(x)}{p'(x)g(x)}\right).$$

The inequality (5.1) is best possible.

Several special cases are given in [**32**]. For example, if σ, λ are real numbers with $\lambda > 0$, then

$$\int_0^\infty x^\sigma G\left(f(x);x^\lambda\right) dx < e^{(\sigma+1)/\lambda}\int_0^\infty x^\sigma f(x)dx, \tag{5.2}$$

where $e^{(\sigma+1)/\lambda}$ is the best possible constant.

In the case $g(x) \equiv 1$ we can consider the inequality (5.1) as the integral analogue of van der Corput's [**33**] generalization of Carleman's inequality.

Levin also gave another similar result, with the following special cases:

If $\lambda > 0$ and $e^{\lambda x}f(x) \in L(0,\infty)$,

$$\int_0^\infty xG\left(f(x)\right) dx \leq \frac{1}{\lambda}\int_0^\infty e^{\lambda x}f(x)dx, \tag{5.3}$$

with equality for $f = Ce^{-2\lambda x}$.

If $x^{-1}f(x) \in L(0,\infty)$ and $\lambda > 0$,

$$\int_0^\infty e^{-\lambda x}G\left(f(x)\right) dx \leq \frac{1}{2\lambda e}\int_0^\infty x^{-1}f(x)dx, \tag{5.4}$$

with equality for $f = Cxe^{-2\lambda x}$.

If $\lambda > 0$ and $e^{x^\lambda}f(x) \in L(0,\infty)$,

$$\int_0^\infty x^\lambda G\left(f(x)\right) dx \leq \int_0^\infty e^{x^\lambda}f(x)dx, \tag{5.5}$$

with equality for $f = C\exp(-x^{\lambda}(1+\lambda)/\lambda)$.

6. Y. C. Chow [**34**] has generalized Hardy's inequality (1.1) for several sequences.

7. T. M. Flett [**35**] proved the following generalizations of the Hardy inequality.

Let $f(x) \geq 0$ for $x \geq 0$ and set $F(x) = \int_0^x f\, dx$ $(\lambda > -1)$, $F(x) = \int_x^\infty f\, dt$ $(\lambda < -1)$.

If $q \geq p \geq 1$ and $\lambda \neq -1$, then

$$\left(\int_0^\infty x^{-1-q\lambda}\left(\frac{F(x)}{x}\right)^q dx\right)^{1/q} \leq B(p,q,\lambda)\left(\int_0^\infty x^{-1-p\lambda} f^p(x)\, dx\right)^{1/p}. \tag{7.1}$$

The case $q > p > 1$, $\lambda = -p^{-1}$ is due to Hardy and Littlewood [**20**], who conjectured the best (least) value of the constant $B(p,q,\lambda)$ in this case; this conjecture was proved by Bliss [**21**] (see (2.13)). The case $\lambda > -1$ is included in the following more general result involving the Riemann-Liouville integral of order r (see 2.2)).

If either $q \geq p \geq 1$ and $r > p^{-1} - q^{-1}$, or $q > p > 1$ and $r = p^{-1} - q^{-1}$; and if $\lambda > -1$, then

$$\left(\int_0^\infty x^{-1-q\lambda}\left(\frac{f_r(x)}{x^r}\right)^q dx\right)^{1/q}$$

$$\leq B_0(p,q,\lambda,r)\left(\int_0^\infty x^{-1-p\lambda} f^p(x) dx\right)^{1/p}. \tag{7.2}$$

In [**35**] no attempt is made to determine the best constant B or B_0 in (7.1) or (7.2). However, in (7.2), upper bounds for B_0 are obtained in the case $q \geq p \geq 1$, $r > p^{-1} - q^{-1}$. (Only the existence of B_0 is proved for the other case.)

Flett also proved some substitutes for (7.1) or (7.2) when one or more of p, q, λ, r takes a limiting value.

8. For some related results see papers [**36**] of N. Levinson and [**37**] of R. P. Boas. Boas gave generalizations of some results of A. A. Konyuškov [**38**].

N. Levinson [**36**] proved the following generalization of Hardy's inequality:

Let $p > 1$, $f(x) \geq 0$, $r(x) > 0$ for $x > 0$, and let r be an absolutely continuous function. If for some $\lambda > 0$ and for almost all x

$$\frac{p-1}{p} + \frac{xr'(x)}{r(x)} \geq \frac{1}{\lambda},$$

and if

$$H(x) = \frac{1}{xr(x)} \int_0^x r(t)f(t)dt,$$

then

$$\int_0^{+\infty} H(x)^p dx \leq \lambda^p \int_0^{+\infty} f(x)^p dx.$$

For $r(x) \equiv 1$ and $\lambda = p/(p-1)$ the inequality (1.2) is obtained.
Some similar results are also given in [**36**].

9. The following result was attributed by V. I. Levin and S. B. Stečkin [**39**, Th. D83] to V. V. Stepanov:

If $p > 1$, $f \geq 0$, $f \not\equiv 0$, $f \in L^p(-\infty, +\infty)$, $k > 0$, $F(x) = e^{kx} \int_x^\infty e^{-ku} f(u)du$, then

$$\int_{-\infty}^{+\infty} F^p(x)dx < k^{-p} \int_{-\infty}^{+\infty} f^p\, dx. \tag{9.1}$$

An extension of results from 1(iii). and 1(ix). is given by P. R. Beesack [**40**]: If $p < 0$ and $f > 0$ on R^+, and if $F(x) = \int_x^\infty f\, dt$ for $k > 1$, while $F(x) = \int_0^x f\, dt$ for $k < 1$, then

$$\int_0^\infty x^{-k} F^p\, dx < \left|\frac{p}{k-1}\right|^p \int_0^\infty x^{p-k} f^p\, dx. \tag{9.2}$$

In fact, Beesack considered inequalities of the form

$$\int_a^b s|u|^p dx \leq \int_a^b r|u'|^p dx \qquad (p < 0 \text{ or } p > 1) \tag{9.3}$$

or the same inequality with $\leq$ replaced by $\geq$ if $0 < p < 1$.
Such generalizations of Hardy's inequalities are given in Chapter II on Wirtinger type inequalities.

10. If we restrict the series to a finite number of terms in Carleman's inequality, we obtain an inequality with a smaller best possible bound λ_n:

$$\sum_{\nu=1}^n (a_1 \dots a_\nu)^{1/\nu} \leq \lambda_n \sum_{\nu=1}^n a_\nu \tag{10.1}$$

for all $a_1 > 0, \ldots, a_n > 0$. N. G. De Bruijn [**41**] established the asymptotic behavior of λ_n if $n \to \infty$.

He showed that

$$\lambda_n = e - \frac{2\pi^2 e}{(\log n)^2} + O\left(\frac{1}{(\log n)^3}\right). \tag{10.2}$$

H. S. Wilf [**42**] gives a similar result for Hardy's inequality in the case $p = 2$, i.e. for

$$\sum_{\nu=1}^{n} \left(\frac{a_1 + \cdots + a_\nu}{n}\right)^2 \leq \lambda_n \sum_{\nu=1}^{n} a_\nu^2. \tag{10.3}$$

He shows that

$$\lambda_n = 4 - \frac{16\pi^2}{(\log n)^2} + O\left(\frac{\log\log n}{(\log n)^3}\right). \tag{10.4}$$

See also H. S. Wilf [**153**].

11. Hardy's inequality was also considered in F. A. Sysoeva [**43**], and J. Kadlec and A. Kufner [**44**].

12. The following result was proved by E. K. Godunova [**45**]:

Let $F\colon (0,\infty) \to R$ be continuous, strictly increasing with $\lim\limits_{x\to 0+} F(x) = 0$ or $-\infty$, and F^{-1} convex; suppose further that a and b are two positive sequences and that for each $n \geq 1$, $w^{(n)}$ is a positive n-tuple with

$$\sum_{k=1}^{n} w_k^{(n)} = 1 \quad \text{and} \quad \sum_{n=k}^{\infty} w_k^{(n)} b_n \leq C, \quad k \geq 1.$$

Then

$$\sum_{n=1}^{\infty} b_n F_n(a; w^{(n)}) \leq C \sum_{n=1}^{\infty} a_n, \tag{12.1}$$

where $F_n(a,p)$ denotes the quasi-arithmetic mean of a with weight p, with respect to F. If further $C = \lim\limits_{n\to\infty} \frac{1}{n} \sum\limits_{k=1}^{n} b_k$, then the constant in (12.1) is best possible.

The following results are special cases of this theorem:

If a is a positive sequence and if $p, q > 1$, then:

(12.2)
$$\sum_{n=1}^{\infty} q^{-n}(q^n-1)^{1-p}(q-1)^p \left(\sum_{k=1}^{n} q^{k-1} a_k\right)^p \leq \sum_{k=1}^{\infty} a_k^p;$$

(12.3)
$$\sum_{n=1}^{\infty} (q^n-1)q^{-n} \left(\prod_{k=1}^{n} a_k^{q^{k-1}}\right)^{(q-1)/n} < \sum_{n=1}^{\infty} a_n;$$

(12.4)
$$\sum_{n=1}^{\infty} \frac{n}{(n+1)} \tan^p\left(\frac{\pi}{2n}\right) \left(\sum_{k=1}^{n} a_k \sin\left(\frac{k\pi}{n}\right)\right)^p < 2\pi \sum_{n=1}^{\infty} a_n^p;$$

(12.5)
$$\sum_{n=1}^{\infty} \frac{n}{(n+1)} \left(\prod_{k=1}^{n} a_k^{\sin(k\pi/n)}\right)^{\tan(\pi/2n)} < 2\pi \sum_{n=1}^{\infty} a_n;$$

(12.6)
$$\sum_{n=1}^{\infty} \frac{1}{(n+1)} \left(\frac{1}{n} \sum_{k=1}^{n} a_k\right)^p < \sum_{n=1}^{\infty} n^{-1} a_n^p;$$

(12.7)
$$\sum_{n=1}^{\infty} \frac{1}{(n+1)} \left(n! \prod_{k=1}^{n} a_k\right)^{1/n} < \sum_{n=1}^{\infty} a_n.$$

Since $e^{-1} < (n!)^{1/n}/(n+1)$, (12.7) implies Carleman's inequality. The inequality (12.7) is due to B. Akerberg [**46**].

Note that (12.1) is also a generalization of some results from [**29**]–[**31**]. Various integral inequalities are also obtained in [**45**].

Many discrete and integral inequalities of Carleman type are obtained in Godunova's paper [**47**]. For example, the following theorem is valid:

THEOREM. *If*

$$\int_0^{\infty} \Psi(x) \exp\left(\frac{1}{x}\int_0^x \log f(t)dt\right) dx \leq \int_0^{\infty} s(x)f(x)dx \tag{12.8}$$

and if $\lambda(x)$ *is a real function such that:* 1) $\lambda(0) = 0$, 2) $\lambda(\infty) = \infty$ *and* 3)

$\lambda'(x) > 0$ for $x > 0$, then

$$\int_0^\infty \Psi_1(y) \exp\left(\frac{1}{\lambda(y)} \int_0^y \lambda'(t) \log g(t) dt\right) dy \le \int_0^\infty s_1(y) g(y) dy, \tag{12.9}$$

where $s_1(y) = s(\lambda(y))\,\lambda'(y)$, $\Psi_1(y) = \lambda'(y)\Psi(\lambda(y))$.

If the inequality (12.8) is sharp, the same is valid for (12.9).

13. D. W. Boyd [48] used Hardy's inequality to prove the following result:

Let f be a nonnegative measurable function, and let $a \ge 0$, $\lambda > 0$, $q \ge 0$, $p \ge 0$ be real numbers with $p + q > 1$ and $q\lambda < 1$. Then

$$\int_a^\infty x^{(p+q)\lambda-2} f^p(x) \left(\int_x^\infty f^q(t) dt\right) dx \le (1 - q\lambda)^{-1} \int_a^\infty x^{(p+q)\lambda-1} f^{p+q}(x) dx \tag{13.1}$$

and the constant $(1 - q\lambda)^{-1}$ is best possible.

14. Let $K(t) \ge 0$ on $V_t = \{t = (t_1, \dots, t_n) \,|\, 0 < t_i < \infty,\ i = 1, \dots, n\}$, with $\int_{V_t} K(t) dV_t = 1$, and let V_x and V_y be similarly defined. Further, let $\phi(u)$ be a nonnegative convex function for $u \ge 0$, $f(y) \ge 0$, $y \in V_y$, $f \not\equiv 0$, and let $\phi(f(x))/(x_1 \dots x_n)$ be integrable on V_x. Then

$$\int_{V_x} (x_1 \dots x_n)^{-1} \phi\left((x_1 \dots x_n)^{-1} \int_{V_y} K\left(\frac{y_1}{x_1}, \dots, \frac{y_n}{x_n}\right) f(y_1, \dots, y_n) dV_y\right) dV_x \le \int_{V_x} (x_1 \dots x_n)^{-1} \phi(f(x))\, dV_x.$$

Hardy's inequality is a simple consequence of this result of E. K. Godunova [49].

15. R. P. Boas [50] considered the two cases $r = p$ and $r = 0$ of (1.6) (with $f(t)$ replaced by $t^{-1} f(t) \ge 0$ if $r = 0$), namely

$$\int_0^\infty x^{-p} F_1^p dx \le \left(\frac{p}{p-1}\right)^p \int_0^\infty f^p dx, \qquad \int_0^\infty F_2^p dx \le p^p \int_0^\infty f^p dx,$$

for $p > 1$, and noted that these inequalities suggested there may be others of the form

$$(15.1)\qquad \int_0^\infty x^{-p} F_1^p dx \le c_1 \int_0^\infty F_2^p dx \le c_1 c_2 \int_0^\infty x^{-p} F_1^p dx,$$

where $F_1(x) = \int_0^x f\, dt$, $F_2(x) = \int_x^\infty t^{-1} f\, dt$, and $c_i = c_i(p)$. In C. Bennett and K. Rudnick [51] inequalities (15.1) are proved with

$$c_1 = \left[\frac{p}{p-1}\right]^p - 1, \qquad c_2 = p^p - 1.$$

The inequalities (15.1) are the special case $s = 1$, $c = 0$ of the following more general inequalities [50], valid for $p > 1$, $s > 0$, $-1 < c < sp - 1$:

(15.2)
$$\int_0^\infty x^{c-sp} \left(\int_0^x u^s f(u) du\right)^p dx \le \left[\left(\frac{sp}{sp-c-1}\right)^p - 1\right] \int_0^\infty x^c \left(\int_x^\infty f\, du\right)^p dx,$$
$$\int_0^\infty x^c \left(\int_x^\infty f\, du\right)^p dx \le \left[\left(\frac{sp}{1+c}\right)^p - 1\right] \int_0^\infty x^{c-sp} \left(\int_0^x u^s f(u) du\right)^p dx.$$

If $0 < p < 1$, the inequalities (15.2) hold with $\le$ replaced by $\ge$. (To obtain (15.1), replace $f(y)$ by $u^{-1} f(u)$ in both of (15.2), with $s = 1$, $c = 0$.) For the case $p < 0$, Boas [50] also proved that if $s > 0$, $sp - 1 < c < -1$, then

(15.3)
$$\int_0^\infty x^{c-sp} \left(\int_0^x u^s f(u)\, du\right)^p dx \le \left(\frac{sp}{sp-c-1}\right)^p \int_0^\infty x^c \left(\int_x^\infty f\, du\right)^p dx,$$
$$\int_0^\infty x^c \left(\int_x^\infty f(u)\, du\right)^p dx \le \left(\frac{sp}{1+c}\right)^p \int_0^\infty x^{c-sp} \left(\int_0^x u^s f(u)\, du\right)^p dx.$$

In [50], it is noted that the constants in (15.2), (15.3) are not best possible. The proofs of (15.2), (15.3) are obtained from some general inequalities concerning integral operators and their convolution. These results are proved by using the

LEMMA (Boas [50]). *If φ is convex and continuous, f is nonnegative and measurable, λ is nondecreasing and bounded with $L = \lambda(\infty) - \lambda(0)$, then*

$$\int_0^\infty x^{-1}\varphi\left\{L^{-1}\int_0^\infty f(xu)d\lambda(u)\right\}dx \le \int_0^\infty x^{-1}\varphi(f(x))\,dx. \tag{15.4}$$

If φ is concave, then (15.4) holds with $\le$ replaced by $\ge$.

PROOF: By Jensen's inequality, if φ is convex,

$$\varphi\left\{L^{-1}\int_0^\infty f(xu)d\lambda(u)\right\} \le L^{-1}\int_0^\infty (f(xu))\,d\lambda(u),$$

so

$$\begin{aligned}\int_0^\infty x^{-1}\varphi\left\{L^{-1}\int_0^\infty f(xu)\;d\lambda(u)\right\}dx &\le L^{-1}\int_0^\infty\int_0^\infty x^{-1}\varphi(f(xu))\;d\lambda(u)dx\\ &= L^{-1}\int_0^\infty\int_0^\infty x^{-1}\varphi(f(xu))\;dx\;d\lambda(u)\\ &= L^{-1}\int_0^\infty\int_0^\infty t^{-1}\varphi(f(t))\;dt\;d\lambda(u)\\ &= \int_0^\infty t^{-1}\varphi(f(t))\;dt.\end{aligned}$$

REMARK: The inequality (15.4) includes both inequalities (2.6) as well as the corresponding inequalities for $p < 0$ and for $0 < p < 1$ as special cases. To see this, take $\varphi(x) \equiv x^p$, set $\alpha > 0$, $\beta > 0$, and

$$\lambda_1 = \begin{cases} \alpha^{-1}u^\alpha, & 0 \le u \le 1,\\ \alpha^{-1}, & u \ge 1,\end{cases} \qquad \text{so } L_1 = \alpha^{-1},$$

$$\lambda_2 = \begin{cases} 0, & 0 \le u \le 1,\\ \beta^{-1}\left[1 - u^{-\beta}\right], & u \ge 1,\end{cases} \qquad \text{so } L_2 = \beta^{-1}.$$

The inequality (15.4) reduces to

$$\int_0^\infty x^{-1-\alpha p}\left(\int_0^x t^{\alpha-1}f(t)\;dt\right)^p\;dx \le \alpha^{-p}\int_0^\infty x^{-1}f^p(x)\;dx,$$

or to

$$\int_0^\infty x^{-1+\beta p}\left(\int_1^\infty t^{-\beta-1}f(t)\,dt\right)^p dx \le \beta^{-p}\int_0^\infty x^{-1}f^p(x)\,dx$$

if $p > 1$ or $p < 0$. If $0 < p < 1$, these inequalities are reversed. In the first of these, let $f(t) = t^{1-\alpha}g(t)$ and $\alpha = (k-1)/p$, and in the second, let $f(t) = t^{1+\beta}g(t)$ and $\beta = (1-k)/p$. This gives

$$\int_0^\infty x^{-k}\left(\int_0^x g\,dt\right)^p dx \le \left(\frac{p}{k-1}\right)^p \int_0^\infty x^{p-k}\,g^p(x)\,dx \tag{15.5}$$

if $k > 1$, $p > 1$ or $k < 1$, $p < 0$, *and*

$$\int_0^\infty x^{-k}\left(\int_x^\infty g\,dt\right)^p dx \le \left(\frac{p}{1-k}\right)^p \int_0^\infty x^{p-k}g^p(x)\,dx \tag{15.6}$$

if $k < 1$, $p > 1$ *or* $k > 1$, $p < 0$. (The cases $p > 1$ give inequality (1) as stated.) Similarly, with $0 < p < 1$, we obtain

$$\int_0^\infty x^{-k}\left(\int_0^x g\,dt\right)^p dx \ge \left(\frac{p}{k-1}\right)^p \int_0^\infty x^{p-k}g^p(x)\,dx, \tag{15.7}$$

$$(0 <, p < 1,\ k > 1),$$

$$\int_0^\infty x^{-k}\left(\int_0^x g\,dt\right)^p dx \ge \left(\frac{p}{1-k}\right)^p \int_0^\infty x^{p-k}g^p(x)\,dx, \tag{15.8}$$

$$(0 < p < 1,\ k < 1).$$

16. G. Talenti [**52**], [**53**], G. Tomaselli [**54**], B. Muckenhoupt [**55**] have investigated the problem of finding necessary and sufficient conditions of measures μ and ν such that for a fixed $p \ge 1$ we have

$$\left(\int_0^\infty \left|\int_0^x f\,dt\right|^p d\mu\right)^{1/p} \le C\left(\int_0^\infty |f|^p\,d\nu\right)^{1/p} \tag{16.1}$$

for some finite constant C independent of f. In particular, Muckenhoupt showed that for Borel measures μ, ν such a condition was that

$$B = \sup_{r>0}\,[\mu(r,\infty)]^{1/p}\left(\int_0^r \left(\frac{d\tilde\nu}{dx}\right)^{-p'/p} dx\right)^{1/p'} < \infty, \tag{16.2}$$

where $\tilde{\nu}$ denotes the absolutely continuous part of ν. Moreover, if C is the least constant for which (16.1) holds, then $B \le C \le p^{1/p}(p')^{1/p'} B$ for $1 < p < \infty$ and $B = C$ for $p = 1$. Here $p' = p/(p-1)$ is the conjugate exponent of p. For absolutely continuous measures $d\mu(x) = u^p(x)dx$ and $d\nu(x) = v^p(x)dx$ the condition (16.2) reduces to

$$B = \sup_{r>0} \left(\int_r^\infty u^p\,dx \right)^{1/p} \left(\int_0^r v^{-p'}\,dx \right)^{1/p'} < \infty,$$

and this is the condition obtained earlier by Talenti and Tomaselli. For this case an analogous result in which $\int_0^x f\,dt$ is replaced by $\int_x^\infty f\,dt$ was also given by Muckenhoupt [**55**].

17. M. Izumi and S. I. Izumi [**56**] proved several results similar to Hardy's inequality. For example the following results are valid:

(i) Let $p > 1$ and $s < -1$ and let f be nonnegative and integrable on $(0, \pi)$. If $x^s f(x)^p$ is integrable, then

$$\int_0^\pi x^s G(x)^p\,dx \le \left(\frac{p}{-s-1}\right)^p \int_0^\pi x^s \left| f\left(\frac{x}{2}\right) - f(x)\right|^p dx, \tag{17.1}$$

where $G(x) = \int_{x/2}^{x} t^{-1} f(t)dt$.

(ii) Let $p > 1$ and $s \ne 1$ and let f be nonnegative, nonconstant and integrable on $(0, \infty)$. If $x^s f(x)^p$ is integrable on $(0, \infty)$, then

$$\int_0^\infty x^s G(x)^p\,dx \le |p(s+1)|^p \int_0^\infty x^s \left| f\left(\frac{x}{2}\right) - f(x)\right|^p dx, \tag{17.2}$$

where G is defined as in (i).

18. In [**57**, pp. 273–275], G. H. Hardy proved that the arithmetic mean of (a_n) in (1.1) can be replaced by a more general mean which contains the Euler mean, Cesàro mean or Hölder mean as particular cases. Another more general case has been studied in G. M. Petersen [**58**] and G. S. Davies and G. M. Petersen [**59**] where the following has been proved:

THEOREM 1. *Let $C = (c_{m,k})$ be a positive trianglular matrix (i.e. $c_{m,k} = 0$ for $k > m$, and $c_{m,k} > 0$ for $k \le m$ $(m = 1, 2, \dots)$), satisfying*

$$0 < \frac{c_{n,k}}{c_{m,k}} \le A_1 \qquad \text{for all } k \le m \le n.$$

Suppose there exists a sequence $f(k)$, $f(k) \nearrow \infty$, such that

$$0 < \frac{f(n)c_{n,k}}{f(m)c_{m,k}} \le A_2, \qquad \text{for all } k \le m \le n$$

and

$$\sum_{n=k}^{\infty} [f(n)]^{-p} \le A_3 [f(k)]^{1-p}, \qquad p > 1,$$

where A_1, A_2 and A_3 are constants.

If $a_n \ge 0$, and if $p > 1$, then

$$\sum_{m=1}^{\infty} \left(\sum_{k=1}^{m} c_{m,k} a_k \right)^p \le A \sum_{m=1}^{\infty} (a_m f(m) c_{m,m})^p . \tag{18.1}$$

The case $c_{m,k} = 1/m$, $k \le m$, $c_{m,k} = 0$, $k > m$, and $f(k) = k$ satisfies the condition of the theorem and (18.1) reduces to (1.1).

M. Izumi, S. I. Izumi and G. M. Petersen [**60**] investigated various generalizations of Hardy's inequality and of the above Theorem 1.

L. Leindler [**61**] proved among others the following similar result:

THEOREM 2. *Let $a_n \ge 0$ and $\lambda_n > 0$ $(n = 1, 2, \dots)$ be given. Then, using the notation*

$$A_{m,n} = \sum_{i=m}^{n} a_i \quad \text{and} \quad \Delta_{m,n} = \sum_{i=m}^{n} \lambda_i \quad (1 \le m \le n \le \infty),$$

we have

$$\left.\begin{aligned} \sum_{n=1}^{\infty} \lambda_n A_{1,n}^p \le p^p \sum_{n=1}^{\infty} \lambda_n^{1-p} \Delta_{n,\infty}^p a_n^p \\ \sum_{n=1}^{\infty} \lambda_n A_{n,\infty}^p \le p^p \sum_{n=1}^{\infty} \lambda_n^{1-p} \Delta_{1,n}^p a_n^p \end{aligned}\right\} \quad \text{for} \quad p \ge 1,$$

and

$$\left.\begin{aligned}\sum_{n=1}^{\infty} \lambda_n^{1-p} \Delta_{n,\infty}^p a_n^p \le 8 \sum_{n=1}^{\infty} \lambda_n A_{1,n}^p \\ \sum_{n=1}^{\infty} \lambda_n^{1-p} \Delta_{1,n}^p a_n^p \le 9 \sum_{n=1}^{\infty} \lambda_n A_{n,\infty}^p\end{aligned}\right\} \quad \text{for} \quad 0 < p \le 1.$$

Further genralizations are given by J. Németh [**62**],[**63**] and L. Leindler [**64**].

19. Let $p > 1$, $r \neq 1$, and $f(x)$ be nonnegative and Lebesgue integrable on $[0, b]$ or on $[a, \infty]$ according as $r > 1$ or $r < 1$, where $a > 0$, $b > 0$. If $F(x)$ is defined as in (iii) of 2, then if $\int_0^b x^{-r}(xf)^p\, dx < \infty$ in (18.1) and $\int_a^\infty x^{-r}(xf)^p\, dx < \infty$ in (18.2),

(19.1)
$$\int_0^b x^{-r} F^p\, dx + \frac{p}{r-1} b^{1-r} F^p(b) \le \left(\frac{p}{r-1}\right)^p \int_0^b x^{-r}(xf)^p\, dx,$$

for $r > 1$, and

(19.2)
$$\int_a^\infty x^{-r} F^p\, dx + \frac{p}{1-r} F^p(a) \le \left(\frac{p}{1-r}\right)^p \int_a^\infty x^{-r}(xf)^p\, dx,$$

for $r < 1$,

with equality in (18.1) or (18.2) only for $f \equiv 0$, where the constant $(p/(r-1))^p$ or $(p/(1-r))^p$ is the best possible when the left side of (18.1) or (18.2) is unchanged respectively. This result is due to D. T. Shum [**65**].

20. Let $\{\lambda_n\}$ be an arbitrary fixed sequence of real numbers. The aim of the paper [**66**] of B. Florkiewicz and A. Rybarski is to estimate the optimal constant ν in the inequality

(20.1)
$$\sum_{n=1}^{\infty} \left(\frac{\lambda_1 a_1 + \cdots + \lambda_n a_n}{\lambda_1 + \cdots + \lambda_n}\right)^2 \le \nu \sum_{n=1}^{\infty} a_n^2$$

which should hold for every sequence $\{a_n\}$ of real numbers. They proved the following results:

(i) If a sequence $\{\lambda_n\}$ is nondecreasing, then $\nu \le 4$.

(ii) Let $\sigma_n = \lambda_1 + \cdots + \lambda_n$. If $\sigma_n \to +\infty$ for $n \to \infty$ and $\sum_{n=1}^{\infty} \sigma_n^{-1} = +\infty$, then $\nu \ge 4$.

The same authors proved in [**67**] the following result:

THEOREM 1. *If a sequence* $\{a_k\}_{k=0}^{\infty}$ *of real numbers satisfies the condition*

$$0 < \frac{1}{2}a_0^2 + \sum_{k=1}^{\infty} a_k^2 < \infty,$$

then the inequality

$$a_0^2 + \sum_{k=1}^{\infty}\left(k+\frac{1}{2}\right)^{-2}\left(\frac{1}{2}a_0 + a_1 + \cdots + a_k\right)^2 < 4\left(\frac{1}{2}a_0^2 + \sum_{k=1}^{\infty} a_k^2\right)$$

is valid. The coefficient 4 cannot be diminished.

21. C. Bennett [**68**, Th. 6.3] proved the following results: If $\alpha > 0$, $1 \le q < \infty$, and f is nonnegative on $[0,1]$, then

$$\int_0^1 \left[(1-\log t)^{\alpha}\int_0^t f\,ds\right]^q d\mu(t) \le \alpha^{-q}\int_0^1 \left[t(1-\log t)^{1+\alpha} f(t)\right]^q d\mu(t),$$

$$\int_0^1 \left[(1-\log t)^{\alpha}\int_t^1 f\,ds\right]^q d\mu(t) \le \alpha^{-q}\int_0^1 \left[t(1-\log t)^{1-\alpha} f(t)\right]^q d\mu(t),$$

where $d\mu = dt/[t(1-\log t)]$. See also Bennett and K. Rudnick [**51**] for some related inequalities.

22. Some inequalities related to Hardy's inequality are obtained by L. Gluk [**69**], T. M. Flett [**70**] and K. F. Andersen [**71**].

23. Inequalities of Hardy's type involving two arbitrary nonnegative functions were proved by E. T. Copson [**72**], and we state these resuts as follows. In all cases we suppose that φ and f *are nonnegative measurable function on* R^+, *and that the integrals*

$$\Phi(x) = \int_0^x \varphi\,dt,\ F(x) = \int_0^x f\varphi\,dt\ \ (if\ c > 1),$$

$$or\ F(x) = \int_x^{\infty} f\varphi\,dt\ \ (if\ c < 1)$$

all exist for $x > 0$.

(i) *If* $0 < b \le \infty$, $p \ge 1$, $c > 1$ *and* $\int_0^b F^p \Phi^{-c} \varphi \, dx$ *converges at the lower limits of integration, then*

$$\int_0^b F^p \Phi^{-c} \varphi \, dx \le \left(\frac{p}{c-1}\right)^p \int_0^b f^p \Phi^{p-c} \varphi \, dx.$$

(ii) *If* $a > 0$, $0 < p \le 1$, $c > 1$ *and* $\lim_{x\to\infty} \Phi(x) = +\infty$, *while* $\int_a^\infty F^p \Phi^{-c} \varphi \, dx$ *exists, then*

$$\int_a^\infty F^p \Phi^{-c} \varphi \, dx \ge \left(\frac{p}{c-1}\right)^p \int_a^\infty f^p \Phi^{p-c} \varphi \, dx.$$

(iii) *If* $a > 0$, $p \ge 1$, $c < 1$ *and* $\int_a^\infty F^p \Phi^{-c} \varphi \, dx$ *exists, then*

$$\int_a^\infty F^p \Phi^{-c} \varphi \, dx \le \left(\frac{p}{1-c}\right)^p \int_a^\infty f^p \Phi^{p-c} \varphi \, dx.$$

(iv) *If* $0 < b \le \infty$, $0 < p \le 1$, $c < 1$, *and* $\int_0^b F^p \Phi^{-c} \varphi \, dx$ *converges, then*

$$\int_0^b F^p \Phi^{-c} \varphi \, dx \ge \left(\frac{p}{1-c}\right)^p \int_0^b f^p \Phi^{p-c} \varphi \, dx.$$

We note that if we take $\varphi(t) \equiv 1$ in these inequalities then (i), (iv) with $b = \infty$ reduce to cases of (2.6), as do (ii), (iii) on letting $a \to 0^+$.

In [72] Copson also considered the case $c = 1$ and obtained the following results: *let* φ, f *satisfy the preceding hypotheses, and* Φ *be defined as before. Set*

$$F(x) = \int_0^x f\varphi \, dt \ (p \ge 1), \qquad F(x) = \int_x^\infty f\varphi \, dt \ (0 < p \le 1),$$

and suppose these integrals exist for $x > 0$.

(v) *If $p \geq 1$, $b > 0$ and $\int_0^b F^p \Phi^{-1} \varphi \, dx$ converges, then*

$$\int_0^b F^p \Phi^{-1} \varphi \, dx \leq p^p \int_0^b f^p \Phi^{p-1} \left(\log \frac{\Phi(b)}{\Phi(x)} \right)^p \varphi \, dx.$$

(vi) *If $0 < p \leq 1$, $a > 0$ and $\int_a^\infty F^p \Phi^{-1} \varphi \, dx$ converges, then*

$$\int_a^\infty F^p \Phi^{-1} \varphi \, dx \geq p^p \int_a^\infty f^p \Phi^{p-1} \left(\log \frac{\Phi(x)}{\Phi(a)} \right)^p \varphi \, dx.$$

24. In [**73**], H. P. Heinig proved the following two inequalities. *Let $p > 0$ and $p + s > \lambda$. If $\int_0^\infty t^{\lambda - s} |f(t)| dt < \infty$, then*

$$\int_0^\infty x^\lambda \exp \left[p x^{-p} \int_0^x t^{p-1} \log |x^{-s} f(t)| dt \right] dx \tag{24.1}$$
$$\leq \frac{p e^{1/p}}{p + s - \lambda} \int_0^\infty t^{\lambda - s} |f(t)| dt.$$

Similarly, *if $p > 0$ and $2p - 1 > \lambda - sp$, while $\int_0^\infty t^{\lambda - sp} |f(t)|^p dt < \infty$, then*

$$\int_0^\infty x^\lambda \exp \left[p^2 x^{-p} \int_0^x t^{p-1} \log |x^{-s} f(t)| \, dt \right] dx \tag{24.2}$$
$$\leq \frac{ep}{(2+s)p - \lambda - 1} \int_0^\infty t^{\lambda - sp} |f|^p \, dt.$$

We illustrate by proving (24.2). We have $e = \exp \left[-p^2 \int_0^1 y^{p-1} \log y \, dy \right]$, so the change of variable $t = xy$ in (24.2) reduces it to

$$\int_0^\infty x^\lambda \exp \left[p^2 \int_0^1 y^{p-1} \log |x^{-s} y \, f(xy)| \, dy \right] dx \leq A \int_0^\infty t^{\lambda - sp} |f|^p \, dt,$$

where $A = p/[(2+s)p-\lambda-1]$. By Jensen's inequality the left side of this inequality does not exceed

$$\int_0^\infty x^\lambda \left\{ p\int_0^1 y^{p-1}|x^{-s}yf(xy)|^p dy \right\} dx$$

$$= p\int_0^1 y^{2p-1}\left\{\int_0^\infty x^{\lambda-sp}|f(xy)|^p\, dx\right\} dy$$

$$= p\int_0^1 y^{(s+2)p-\lambda-2}\left\{\int_0^\infty t^{\lambda-sp}|f(t)|^p\, dx\right\} dy,$$

on using Fubini's theorem. The results (24.2) now follows.

A similar application of Jensen's inequality was given in [**73**, Th. 7] to prove: *if $g \in L(0,\infty)$, then*

$$G(s) = \int_0^\infty e^{-st}\log|g(t)|dt$$

converges for $s > 0$, and

$$\int_0^\infty s^{-2}\exp(sG(s))\,ds \le \exp\left(-\int_0^\infty e^{-t}\log t\,dt\right)\int_0^\infty |g(x)|dx. \tag{24.3}$$

By setting $f = \log|g|$ in this result, it follows that: *if $\int_0^\infty e^f\,dx < \infty$, and $F(s) = \int_0^\infty e^{-st}f(s)ds$ is the Laplace transform of f, then $F(s)$ exists for $s > 0$, and*

$$\int_0^\infty \exp\left(s^{-1}F(s^{-1})\right)ds \le \exp\left(-\int_0^\infty e^{-t}\log t\,dt\right)\int_0^\infty e^f\,dx. \tag{24.4}$$

25. *Let f be continuous and nondecreasing on $[0,\infty)$ with $g(0) = 0$, $g(x) > 0$ for $x > 0$ and $g(\infty) = \infty$. Let $p \ge 1$, $k \ne 1$, and let f be nonnegative and Lebesgue-Stieltjes integrable with respect to g on $[0,b]$ or on $[a,\infty)$ according as $k > 1$ or $k < 1$, where $a > 0$ $b > 0$. Suppose*

$$F(x) = \int_0^x f\,dg\ (k>1),\ \textit{or}\ F(x) = \int_x^\infty f\,dg\ (k<1).$$

Then

$$\int_0^b g^{-k}F^p\,dg + \frac{p}{k-1}g(b)^{1-k}F^p(b) \le \left(\frac{p}{k-1}\right)^p \int_0^b g^{p-k}f^p\,dg, \quad (k>1),$$

$$\int_a^\infty g^{-k}F^p\,dg + \frac{p}{1-k}g(a)^{1-k}F^p(a) \le \left(\frac{p}{1-k}\right)^p \int_a^\infty g^{p-k}f^p\,dg, \quad (k<1),$$

with both inequalities reversed when $0 < p \le 1$. Equality holds in either case when $p = 1$ or $f \equiv 0$, and the constants on the right sides are best possible when the left sides are unchanged.

This result is due to C. O. Imoru [**74**]. Shum's results from 19. follow by taking $g(x) \equiv x$ in the above results.

26. An inequality for fractional integral of a function of two variables is given by H. Musielak [**75**].

27. The following inequality

$$\int_{-1}^{1}(1-x^2)f^2(x)dx + 2\int_{-1}^{1}F(x)F(-x)dx \ge 0 \tag{27.1}$$

is valid for all $f(x) \in L^2(-1,1)$, where $F(x) = -\int_x^1 f(t)dt$.

This inequality is due to D. C. Peaslee and W. A. Coppel [**76**].

They used (27.1) in a proof of the following inequality of E. Bombieri [**77**]:

$$\int_{-1}^{1}(1-x^2)f^2(x)dx + 2\iint_{x+y\ge 0}(x+y)f(x)f(y)dx\,dy \ge 0, \tag{27.2}$$

valid for all $f(x) \in L^2(-1,1)$.

Further generalizations are given by D. T. Shum [**78**].

28. Results from 16. were further extended by J. S. Bradley [**79**] to the case of mixed norms (two parameters p, q) and to values of $p, q < 1$ by H. Heinig [**80**], and Beesack and Heinig [**81**]. We present here a statement of the results in [**79**],[**81**].

In all cases u, v, f denote nonnegative, extended real- valued measurable functions on $R^+ = (0,\infty)$, and f is called admissible if the integral $\int_0^x f\,dt$ (or

$\int\limits_x^\infty f\,dt$ *in parts* (b) *below) exists (finite) for* $0 < x < +\infty$. *In all cases* p', p *denote conjugate exponents. In case* $k = +\infty$ *we interpret any integral norm* $\int\limits_A |g|^k\,dx^{1/k} = \operatorname{essup}_{x\in A} |g(x)|$, *and adopt the usual convention* $0 \cdot (+\infty) = 0$.

THEOREM 1. (Bradley [**79**]) (a) *Suppose that* $1 \le p, q \le +\infty$, and $\int\limits_0^r v^{-p'}dt$ *exists (finite) for* $0 < r < +\infty$. *If there exists a constant* $C > 0$ *such that*

$$\left\{\int\limits_0^\infty \left[u(x)\int\limits_0^x f\,dt\right]^q dx\right\}^{1/q} \le C\left\{\int\limits_0^\infty [v(x)f(x)]^p\,dx\right\}^{1/p} \tag{28.1}$$

holds for all admissible f, *then*

$$\sup_{r>0} K(r) \equiv \sup_{r>0}\left(\int\limits_r^\infty u^q\,dx\right)^{1/q}\left(\int\limits_0^r v^{-p'}\,dt\right)^{1/p'} = B < \infty. \tag{28.2}$$

Moreover, in case $p \le q$ *above, then* (28.2) *is also sufficient for* (28.1) *to hold for all admissible* f.

(b) *If* $1 \le p, q \le +\infty$ *and* $0 \le \int\limits_0^r v^{-p'}dt < \infty$ *for* $0 < r < \infty$, *then the existence of a constant* $C > 0$ *such that*

$$\left\{\int\limits_0^\infty \left[u(x)\int\limits_x^\infty f\,dt\right]^q dx\right\}^{1/q} \le C\left\{\int\limits_0^\infty [v(x)f(x)]^p\,dx\right\}^{1/p} \tag{28.3}$$

holds for all admissible f *implies that*

$$\sup_{r>0} J(r) \equiv \sup_{r>0}\left(\int\limits_0^r u^q\,dx\right)^{1/q}\left(\int\limits_r^\infty v^{-p'}\,dt\right)^{1/p'} = B < \infty \tag{28.4}$$

when $p \le q$, (28.4) *is also sufficient for* (28.3), *to hold for all admissible* f.

Finally, in all cases, if C *denotes the best* (= least) *constant in* (28.1) *or* (28.3) *then* $B \le C \le Bp^{1/q}(p')^{1/p'}$ *for* $1 < p \le q < \infty$, *while* $C = B$ *if* $p = 1$ *or* $q = +\infty$.

THEOREM 2. (Beesack and Heinig [**81**]) (a) *Suppose* $p, q < 0$ *and* $0 < \int\limits_0^r v^{-p'}\,dt < \infty$ *holds for all* $r > 0$. *If there exists* $C > 0$ *such that*

$$\left\{\int\limits_0^\infty \left[u(x)\int\limits_0^x f\,dt\right]^q\right\}^{1/q} \ge C\left\{\int\limits_0^\infty [v(x)f(x)]^p\,dx\right\}^{1/p} \tag{28.5}$$

holds for all admissible f, then

$$\inf_{r>0} K_1(r) \equiv \inf_{r>0} \left(\int_0^r u^q\, dx\right)^{1/q} \left(\int_0^r v^{-p'}\, dt\right)^{1/p'} = B > 0. \tag{28.6}$$

Moreover, in case $q \le p < 0$, and $0 < \int_0^r v^{-p'}\, dt < \infty$ for all $r > 0$, and $K_1(r)$ not only satisfies (28.6) but is nondecreasing on $\mathbf{R}^+$, then (28.5) holds for some $C > 0$ and all admissible f.

(b) *If $p, q < 0$, $0 < \int_r^\infty v^{-p'}\, dt < \infty$ for all $r > 0$ and*

$$\left\{\int_0^\infty \left[u(x) \int_x^\infty f\, dt\right]^q\right\}^{1/q} \ge C \left\{\int_0^\infty [v(x) f(x)]^p\, dx\right\}^{1/p} \tag{28.7}$$

holds for all admissible f (some $C > 0$), then

$$\inf_{r>0} J_1(r) \equiv \inf_{r>0} \left(\int_r^\infty u^q\, dx\right)^{1/q} \left(\int_r^\infty v^{-p'}\, dt\right)^{1/p'} = B > 0. \tag{28.8}$$

Moreover, in case $q \le p < 0$, and $0 < \int_r^\infty v^{-p'}\, dt < \infty$ for all $r > 0$, and $J_1(r)$ is nonincreasing on $\mathbf{R}^+$ and satisfies (28.8), then (28.7) holds for some $C > 0$ and all admissible f.

As in Theorem 1, the best (= largest) constant C in (28.5) or (28.7) satisfies $B \ge C \ge |p|^{1/q}(p')^{1/p'} B$ if $q \le p < 0$.

THEOREM 3. (Beesack and Heinig [**81**]) (a) *If $0 < p, q < 1$ and $0 < \int_r^\infty u^q\, dx$, $\int_r^\infty v^{-p'}\, dt < \infty$ for all $r > 0$, and if (28.5) holds for all admissible f, then (28.7) holds, and $B \ge C$. Moreover if $0 < q \le p < 1$, $0 < \int_r^\infty u^q\, dx$, $\int_r^\infty v^{-p'}\, dt < \infty$ for all $r > 0$, and if $J_1(r)$ is nonincreasing and satisfies (28.8), then (28.5) holds for all admissible f with $C \ge p^{1/q}|p'|^{1/p'} B$.*

(b) *Dually, if $0 < p, q < 1$, $0 < \int_0^r u^q\, dx$, $\int_0^r v^{-p'}\, dt < \infty$ for all $r > 0$, and if (28.7) holds for some $C > 0$ and all admissible f, then (28.6) holds with $B \ge C$.*

Moreover if $0 < q \le p < 1$, $0 < \int_0^r u^q\,dx$, $\int_r^\infty v^{-p'}\,dt < \infty$ *for all* $r > 0$, *and if* $K_1(r)$ *is nondecreasing and satisfies* (28.6), *then* (28.7) *holds for all admissible* f, *with* $C \ge p^{1/q}|p'|^{1/p'}B$.

Theorems 1,2,3 with $p = q$ and $(u, v) = \left(x^{-k/p},\ x^{1-(k/p)}\right)$ yield the three cases $p > 1$, $p < 0$, $0 < p < 1$ of Hardy's inequalities, respectively.

Another example, given by Bradley [**79**] for $1 < p \le q < \infty$ and by Heinig [**80**] for $q \le p < 0$ is

$$\left\{\int_0^\infty x^{\alpha q-1}\left(\int_0^x f\,dt\right)^q dx\right\}^{1/q} \le \frac{|p|}{|\alpha q||\alpha|^{q/p'}}\left\{\int_0^\infty x^{(\alpha+1)p-1} f^p(x)\,dx\right\}^{1/p}$$

in case $\alpha < 0$; and if $\alpha > 0$,

$$\left\{\int_0^\infty x^{\alpha q-1}\left(\int_x^\infty f\,dt\right)^q dx\right\}^{1/q} \le \frac{|p|}{|\alpha q||\alpha|^{q/p'}}\left\{\int_0^\infty x^{(\alpha+1)p-1} f^p(x)\,dx\right\}^{1/p}.$$

29. A. Kufner and H. Triebel [**82**, §2] also dealt with weighted norm inequalities of the form

$$\int_a^b \sigma_0(s)|u(x)|^p\,dx < \left(\frac{p}{p-1}\right)^p \int_a^b \sigma_1(x)|u'(x)|^p\,dx$$

from a quite different point of view, for the case $1 < p < \infty$. In [**82**], the approach was to express σ_0 and σ_1 in terms of a parameter function $\lambda = \lambda(x)$ in such a way that the inequality was valid for all $u \in C^1(a, b)$ having $u(a+) = 0$ (or $u(b-) = 0$). This allows examples to be easily constructed, and many such examples are given in [**82**].

30. L. Y. Chan [**83**] gave some special inequalities with mixed weights involving Stieltjes integrals. We shall state these results briefly. *Let* g *be continuous and nondecreasing on* $[0, \infty)$ *with* $g(0) = 0$, $g(+\infty) = +\infty$, $g(x) \ne 0$ *if* $x \ne 0$, $g(1) = 1$, $g(x) \ne 1$ *if* $x \ne 1$; *let* f *be nonnegative and measurable on* $[0, \infty)$ *and let* $F_i(x) = \int_{E_i} f\,dg$ *for* $i = 1, 2, 3, 4$ *where* $E_1 = [x, \infty)$ $(1 \le x < \infty)$, $E_2 = [1, x]$ $(1 < x < \infty)$, $E_3 = [0, x]$ $(0 < x \le 1)$, $E_4 = [x, 1]$ $(0 \le x < 1)$. *For real* $p, q \ne 0$, $r \ne 1$, *set* $A = (|q/(r-1)|)^{(p-1)q/p}(|p/(r-1)|)$, $\delta = [(q - r + 1)p/q] = 1$. *Then*

$$\int_1^\infty g^{-1}(\log g)^{-r} F_i^q\,dg \le A\left\{\int_1^\infty g^{p-1}(\log g)^\delta f^p\,dg\right\}^{q/p}$$

for $1 \le p \le q < \infty$, *or* $-\infty < q \le p < 0$ *where* $i = 1$ *if* $(r-1)/q < 0$ *and* $i = 2$ *if* $(r-1)/q > 0$. *Also, for the same* p, q, r,

$$\int_0^1 g^{-1}(-\log g)^{-r} F_i^q \, dg \le A \left\{ \int_0^1 g^{p-1}(-\log g)^{\delta} f^p \, dg \right\}^{q/p},$$

where $i = 3$ *if* $(r-1)/q < 0$ *and* $i = 4$ *if* $(r-1)/q > 0$. *In case* $0 < q \le p \le 1$ *these results hold with the inequality signs reversed.*

In [**83**] modifications of these results are given involving boundary terms as well as more complicated inequalities for the case $r = 1$.

In his previous paper [**84**] Chan considered the case $r = 1$ of (2.6). Other inequalities given there by Chan are

$$\int_1^\infty x^{-1} \left(\int_x^\infty f \, dt \right) dx \le p^p \int_1^\infty x^{p-1} (f(x) \log x)^p \, dx,$$

$$\int_1^\infty x^{-1} \left(\int_1^x f \, dt / \log x \right)^p dx \le \left(\frac{p}{p-1} \right)^p \int_1^\infty x^{p-1} f^p(x) \, dx,$$

$$\int_0^1 x^{-1} \left(\int_x^1 f \, dt / |\log x| \right)^p dx \le \left(\frac{p}{p-1} \right)^p \int_0^1 x^{p-1} f^p(x) dx,$$

all for $p > 1$. Some more general results involving boundary terms were also given in [**84**].

31. The unnatural hypotheses concerning the existence of the integrals $\int F^p \phi^{-c} \, dx$ in (i) and (iii) from 23. can be replaced by the assumption that the larger integral $\int f^p \phi^{p-c} \, dx$ converges. See also Beesack [**85**].

We shall give such a theorem for the case $p < 0$.

Let $p < 0$, *and let* ϕ, f *be positive measurable functions on* $\mathbf{R}^+$ *such that the integrals*

$$\Phi(x) = \int_0^x \varphi \, dt, \ F(x) = \int_0^x f\varphi \, dt \ (\text{if } c < 1), \ \text{or } F(x) = \int_x^\infty f\varphi \, dt \ (\text{if } c > 1)$$

exist for $x > 0$. *If* $0 < b < \infty$, *then for* $c > 1$ *we have*

$$\int_0^b F^p \Phi^{-c} \varphi \, dx + \frac{(-p)}{1-c} F^p(b) \Phi^{1-c}(b) \le \left(\frac{-p}{1-c} \right)^p \int_0^b f^p \Phi^{p-c} \varphi \, dx,$$

while for $c > 1$ we have for $0 < a < \infty$

$$\int_a^\infty F^p \Phi^{-c} \varphi \, dx + \frac{(-p)}{c-1} F^p(a) \Phi^{1-c}(a) \le \left(\frac{-p}{c-1}\right)^p \int_a^\infty f^p \Phi^{p-c} \varphi \, dx.$$

32. Shum's results from 19. are extended in Mitrinović, Beesack and Vasić [143]:

THEOREM 1. *Suppose $p \ne 0$ or 1 and $k \ne 1$. Let f be nonnegative (positive if $p > 0$) and Lebesgue integrable on $[0, b]$ if $(k-1)p > 0$, or on $[a, \infty)$ if $(k-1)p < 0$, where $a > 0$, $b > 0$, and let F be defined by*

$$F(x) = \int_0^x f \, dt \qquad \text{if } (k-1)p > 0; \tag{32.1}$$

$$\text{or } F(x) = \int_x^\infty f \, dt \qquad \text{if } (k-1)p < 0.$$

Then: (a) *if $p < 0$ or $p > 1$ and $\int_0^b t^{p-k} f^p \, dt < \infty$, we have for $(k-1)p > 0$,*

$$\int_0^b t^{-k} F^p \, dt \le \left|\frac{p}{k-1}\right|^p \int_0^b t^{p-k} f^p \, dt - \left|\frac{p}{k-1}\right| b^{1-k} F^p(b), \tag{32.2}$$

while if $p < 0$ or $p > 1$ and $\int_a^\infty t^{p-k} f^p \, dt < \infty$ we have for $(k-1)p < 0$,

$$\int_a^\infty t^{-k} F^p \, dt \le \left|\frac{p}{k-1}\right|^p \int_a^\infty t^{p-k} f^p \, dt - \left|\frac{p}{k-1}\right| a^{1-k} F^p(a). \tag{32.3}$$

(b) *if $0 < p < 1$ and $\int_0^b t^{-k} F^p \, dt < \infty$ we have for $(k-1)p > 0$,*

$$\int_0^b t^{-k} F^p \, dt \ge \left|\frac{p}{k-1}\right|^p \int_0^b t^{p-k} f^p \, dt - \left|\frac{p}{k-1}\right| b^{1-k} F^p(b), \tag{32.4}$$

while if $0 < p < 1$, $\int_a^\infty t^{-k} F^p \, dt < \infty$ *and* $(k-1)p < 0$ *we have*

$$\int_a^\infty t^{-k} F^p \, dt \geq \left| \frac{p}{k-1} \right|^p \int_a^\infty t^{p-k} f^p \, dt - \left| \frac{p}{k-1} \right| a^{1-k} F^p(a). \tag{32.5}$$

Moreover in all cases inequality can hold only if $f \equiv 0$ *(or never in case* $p < 0$*). The constant* $|p/(k-1)|^p$ *is best possible in each case.*

The following result was also proved in [**143**]:
If $k > 0$, *and* $p > 1$ *or* $p < 0$, *then*

$$\int_{-\infty}^\infty e^{kpx} \left(\int_x^\infty g \, dt \right)^p dx \leq k^{-p} \int_{-\infty}^\infty e^{kpx} g^p(x) \, dx; \tag{32.6}$$

if $k < 0$ *and* $p > 1$ *or* $p < 0$, *then*

$$\int_{-\infty}^\infty e^{kpx} \left(\int_{-\infty}^x g \, dt \right)^p dx \leq (-k)^{-p} \int_{-\infty}^\infty e^{kpx} g^p(x) \, dx. \tag{32.7}$$

If $0 < p < 1$, *then the inequalities* (32.6), (32.7) *hold with* $\leq$ *replaced by* $\geq$ *if* $k > 0$ *or* $k < 0$ *respectively. (If* $p = 1$, *both* (32.6) *and* (32.7) *hold, with equality.)*

Equality holds in (32.6) *or* (32.7) *only for* $g \equiv 0$ *(or never if* $p < 0$*) and the constants* k^{-p} *or* $(-k)^{-p}$ *are best possible.*

The inequality (32.6) is a generalization of (9.1).

33. Let the function space L_p and the norm $\|\cdot\|_p$ $(1 \leq p < \infty)$ be as usual. Let $K(t,s)$ denote a nonnegative, measurable function of $t, s > 0$ and let K be the operator defined by

$$(Kf)(t) = \int_0^\infty K(t,x) f(x) dx \qquad (t > 0,\ f \in L_p). \tag{33.1}$$

The following inequality is well-known as the Schur-Hardy inequality:
Let $K(t,s)$ be homogenous of degree -1, such that

$$A(p) = \int_0^\infty K(t,x) x^{-1/p} \, dx < \infty,$$

and let $f \in L_p$ $(1 \le p < \infty)$. Then $Kf \in L_p$ and

$$\|Kf\|_p \le A(p)\|f\|_p. \tag{33.2}$$

A generalization of this result is given by F. Fehér [**86**]. R. N. Mohapatra [**87**] gave some interesting special cases of results from [**86**]. For example, he proved the following result for an operator on the vector space of all measurable, Lebesgue integrable functions on $(0, \infty)$ defined by

$$T_r(f)(t) = t^{-r} \int_0^t f(x)\, dx.$$

Let $1/p \le r \le 1$, and let $f \in L_p$ $(1 \le p < \infty)$. Then

$$\|T_r f\|_p \le (1 + p(1-r))^{-1/p} \left(\frac{p}{(pr-1)}\right)^2 \|f\|_{p/(1+p(1-r))}.$$

For some related results see Butzer and Fehér [**88**], Fehér [**89**] and Love [**90**].

34. Various results connected to Hardy's inequality were given by E. R. Love [**91**], K. B. Stolarsky [**92**], V. M. Kokiašvili [**93**], F. A. Sysoeva [**94**], A. Kufner and H. Triebel [**95**] and Y. Ding [**96**]. For example, a paper [**96**] generalizes Hardy's inequality to the case of higher dimensions.

Let $p > 1$, $r > 2/p$, $F(x) - F(0) = F(\lambda, \theta) - F(0) = \int_0^\lambda f(t, \theta) dt$, where $f(x) = f(\lambda, \theta)$, $|x| = \lambda$, and $x \in R^2$. Then

$$\int_{R_2} |x|^{-pr} |F(x) - F(0)|^p\, dx \le \left(\frac{p}{pr-2}\right)^p \int_{R_2} |x|^{-p(r-1)} |f(x)|^p\, dx.$$

35. Let $f : I \to R^+$ be a convex function, $x_i \in I$ $(i = 1, 2, \dots)$, $c = (c_1, c_2, \dots)$ a positive sequence, and let for every $n \ge 1$, $q^{(n)} = \left(q_1^{(n)}, \dots, q_n^{(n)}\right)$ be a positive n-tuple such that $\sum_{k=1}^n q_k^{(n)} = 1$ $(n \ge 1)$. If d_k are positie numbers such that

$$\sum_{n=k}^{+\infty} c_n q_k^{(n)} \le d_k \qquad (k \ge 1), \tag{35.1}$$

then

$$\sum_{n=1}^{+\infty} c_n f\left(\sum_{k=1}^{n} q_k^{(n)} x_k\right) \leq \sum_{n=1}^{+\infty} d_n f(x_n). \tag{35.2}$$

If f is a concave function and if in (35.1) the reverse inequality holds, then the reverse inequality in (35.2) holds also.

This result is proved by P. M. Vasić and J. E. Pečarić [**97**]. The following result is also proved in [**97**]:

Let $p_i \geq 1$ $(i = 1, 2, \dots)$, $x_i > 0$ $(i = 1, 2, \dots)$ and let f be a positive function such that $x^{-1} f(x)$ decreases for $x > 0$. Suppose that $c = (c_1, c_2, \dots)$ is a positive sequence, such that

$$\sum_{n=k}^{+\infty} c_n \leq d_k \qquad (k \geq 1). \tag{35.3}$$

Then

$$\sum_{n=1}^{+\infty} c_n f\left(\sum_{k=1}^{n} p_k x_k\right) \leq \sum_{n=1}^{+\infty} p_n d_n f(x_n). \tag{35.4}$$

If $x^{-1} f(x)$ increases and if (35.3) is reversed, then (35.4) is reversed also.

36. Some inequalities for sequence spaces are established by P. Chandra and R. N. Mohapatra [**98**]. A proof uses results from [**59**].

37. Theorems obtained by R. N. Mohapatra and D. C. Russell [**99**] are essentially the integral analogues of the series inequalities of J. Németh [**62**], but they include, for example, all the inequalities given by E. T. Copson [**72**] and (for $p > 0$) those given by P. R. Beesack [**85**].

Here, we shall give only the following convolution inequality:

Let $s(t) \geq 0$ $(t > 0)$, $s(t) = 0$ $(t < 0)$, and for some constant $K \geq 1$, $s(x) \leq K s(y)$ for $x > y > 0$; suppose also that $S(x) = \int_0^x s(t)\, dt > 0$ for $x > 0$. If $f(t) \geq 0$ on R_+ and $1 < p < \infty$, then

$$\int_0^\infty \left(\frac{1}{S(x)} \int_0^x s(x-t) f(t)\, dt\right)^p dx \leq \left(\frac{p^2 K}{p-1}\right)^p \int_0^\infty (f(t))^p\, dt.$$

38. Further generalizations of a result from 25. are obtained by C. O. Imoru [**100**]–[**102**].

39. Hardy type integral inequalities are also considered by B. G. Pachpatte [**103**].

40. Various generalizations of the previous results are recently obtained by E. R. Love [**104**]–[**110**].

The paper [**105**] contains inequalities for weighted norms of the form $\|y_m\|_p = \left(\sum_{m=1}^{\infty} \lambda_m |y_m|^p\right)^{1/p}$, where $\lambda_m > 0$ and y_m are complex numbers. Theorem 1 is: If $p > 1$, $b(t)$ is nonnegative and decreasing in $(0,1]$, $\Lambda_m = \sum_{n=1}^{m} \lambda_n$, and the matrix (a_{mn}) is lower triangular with $|a_{mn}| \leq (\lambda_n/\Lambda_m)\, b\,(\Lambda_n/\Lambda_m)$ for $0 < n \leq m$, then

$$(40.1) \qquad \left\| \sum_{n=1}^{m} a_{mn} x_n \right\|_p \leq \left(\int_0^1 b(t) t^{-1/p}\, dt \right) \|x_m\|_p.$$

Theorems 2 and 3 from [**105**] are similar theorems for upper triangular and nontriangular matrices (a_{mn}). Theorems 1 and 2 include discrete Copson's generalizations of Hardy's inequalities.

(Note that (40.1) is related to (33.2).)

Generalizations of results from [**72**] and [**44**] and some analogous results are obtained in [**106**].

Hardy's inequality in Orlicz-type sequence spaces for operators related to generalized Hausdorff matrices is obtained in [**107**]. It is a generalization of results from A. Jakimovski, B. E. Rhoades and J. Tzimbalario [**111**] and D. Borwein [**112**].

J. A. Cochran and C. S.Lee [**113**] proved the following results:

(i) If λ and p are real numbers with $p > 0$, and $f(t)$ is measurable and nonnegative on $(0,\infty)$, then

$$(40.2) \qquad \int_0^{\infty} x^{\lambda} \exp\left(\frac{p}{x^p} \int_0^{x} t^{p-1} \log f(t)\, dt \right) dx$$

$$(40.2) \qquad \leq e^{(\lambda+1)/p} \int_0^{\infty} t^{\lambda} f(t)\, dt.$$

(ii) If λ and p are constants with $\lambda \geq 0$ and $p \geq 1$, and $0 < x_n \leq 1$, then

$$(40.3) \qquad \sum_{m=1}^{\infty} m^{\lambda} \left(\prod_{n=1}^{n} x_n^{np-1} \right)^{p/m^p} \leq e^{(\lambda+1)/p} \sum_{m=1}^{\infty} m^{\lambda} x_m.$$

Note that (40.2) is, in fact, Levin's inequality (4.2).

Generalizations of these results are given in **[108]**– **[110]**.

41. The above integral inequalities of Hardy type or related integral-differential inequalities (see (9.2)) have been applied in various branches of mathematics: theory of ordinary differential equations (oscillations of solutions of differential equations: R. A. Moore and Z. Nehari **[114]**, W. J. Kim **[115]**, D. B. Hinton and R. T. Lewis **[116]**; problems of approximation: A. Krzywicki and A. Rybarski **[117]**, A Rybarski **[118]**; the theory of differential operators in L^2: W. N. Everitt, M. Giertz and J. Weidmann **[119]**, W. N. Everitt **[120]**, R. J. Amos and W. N. Everitt **[121]**, J. S. Bradley, D. B. Hinton and R. M. Kauffman **[122]**); spectral theory of differential operators: C. R. Putnam **[123]**, W. N. Everitt **[124]**, D. B. Hinton and R. T. Lewis **[116]**; Sturm-Liouville problem: P. R. Beesack **[125]**,**[126]**, R. Redheffer **[127]**, D. C. Benson **[128]**; functional analysis (imbedding theorems for weighted Sobolev spaces): V. R. Portnov **[129]**, A. Kufner **[120]**, and particularly A. Kufner **[131]** and V. G. Maz'ya **[132]**; the theory of intermediate spaces: R. A. Hunt **[133]**, C. Bennett **[68]**; interpolation theorems for operators: R. A. Hunt and G. Weiss **[134]**, R. A. Hunt **[135]**; complex function theory (Z. Nehari **[136]**), theory of the Laplace transform (H. P. Heinig **[73]**), theory of Fourier series (V. M. Kokiašvili **[93]**, L. Y. Chan **[83]**), and differential geometry (H. Flanders **[137]**).

The Hardy's inequality (1.1) and its generalizations have been applied in differential geometry to obtain isoperimetric inequalities (see, for example, M. Janet **[138]**, B. A. Troesch **[139]**, D. C. Benson **[128]**, P. R. Beesack **[140]**, J. M. Feinberg **[141]**). They were also used to derive the Hardy-Littlewood maximal inequality and is generalizations (H. P. Heinig **[73]**, P. L. Butzer and F. Fehér **[88]**). The latter inequalities have numerous applications in different mathematical domains. Let us mention the following: complex function theory, singular integrals, harmonic functions, conjugate functions, ergodic theory, differentiation theorems for groups (cf. **[88]**).

The above "applications comment" of Hardy's inequality is given by B. Florkiewicz in his dissertation **[142]**.

42. Several recent papers on Hardy's inequality are **[144]** of B. G. Pachpatte and **[145]** of T. Ostrogorski, **[146]** of G. Sinnamon, **[147]** of G. Bennett, **[148]** of B. G. Pachpatte and E. R. Love, and **[149]** of R. N. Mohapatra and D. C. Russell, **[150]** of R. N. Mohapatra and K. Vajravelu, **[151]** of F. J. Martin-Reyes and E. Sawyer, and **[152]** of A. P. Pittenger.

43. H. S. Wilf **[153]** is a collection of theorems on finite sections of Hilbert's, Hardy's and Carleman's inequalities. In each case, if the sequences $\{a_n\}$ and

$\{b_n\}$ satisfy $a_n = b_n = 0$ for $n \geq N$, then the best constant λ_N is smaller than the best possible constant A for all sequences and λ_N converges to A. This book discusses the computation of λ_N and its rate of convergence to A. In most cases λ_N is an eigenvalue of a Hermitian matrix.

44. Weighted Hardy inequalities continue to be the object of research. Ě. N. Batuev and V. D. Stepanov [**154**] and G. Sinnamon [**155**] gives versions in dimension one and R^n respectively. Stepanov [**156**] gives a weighted inequality with a higher order derivative in R, while A. Wannebo [**157**] gives an n-dimensional version.

45. R. Brown and D. Hinton [**158**] give Hardy inequalities of the form

$$\left(\int_I |x^p f(x)|^p dx\right)^{1/p} \leq K \left(\int_I |x^\alpha f'(x)|^r dx\right)^{1/r}$$

for various cases:

i) $I = R^+$, $p \geq r$, $\alpha + r^{-1} \neq 1$, $\beta + p^{-1} = \alpha + r^{-1} - 1$;
ii) $I = [b, \infty)$, $\alpha + r^{-1} \neq 1$, $p \geq r$, $\beta + p^{-1} \leq \alpha + r^{-1} - 1$;
iii) $I = [b, \infty)$, $\{\alpha + r^{-1} = 1$ or $p < r\}$, $\beta + p^{-1} < \alpha + r^{-1} - 1$;
iv) $I = (0, b]$, $\alpha + r^{-1} \neq 1$, $p \geq r$, $\beta + p^{-1} \geq \alpha + n^{-1} - 1$; and
v) $I = (0, b]$, $\{\alpha + r^{-1} = 1$ or $p < r\}$, $\beta + p^{-1} > \alpha + r^{-1} - 1$.

46. As this book goes to press, the book [**159**] by B. Opic and A. Kufner has appeared. It is devoted to finding conditions under which there is a constant C such that

$$\left(\int_I |f(x)|^q w(x) dx\right)^{1/q} \leq C \left(\int_I |f'(x)|^p v(x) dx\right)^{1/p}$$

for all u absolutely continuous with a zero at one end of I. They consider various generalizations where f' on the right is replaced by f and f on the left by some operator acting on f, for examples, with $w(x) = 1/x$ and $v(x) \equiv 1$, we get (1.2). Furthermore, they consider appropriate multidimensional versions of this question. They give 84 references, [**160**] for example.

47. Lizorkin [**161**] gives the Hardy inequality

$$\int_0^\infty \left|\frac{1}{t}\int_{at}^{bt} g(h)dh\right|^\theta \frac{dt}{t^\lambda} \leq K \int_0^\infty |g(t)|^\theta \frac{dt}{t^\lambda},$$

for $1 \leq \theta \leq \infty$, $0 < a < b < \infty$, and $K = \left(\int_a^b s'^{\frac{1-1}{\theta}} ds \right)^{\theta}$.

48. P. D. Johnson and R. N. Mohapatra in a series of papers **[162]**–**[166]** have considered generalizations of the discrete version (1.1) of Hardy's inequality where $\|Ax\|_p \leq C(P, A)\|x\|_p$, $(Ax)_k = \sum_{j=1}^{\infty} a_{kj}x_j$, and A satisfies some additional properties.

49. B. G. Pachpatte **[167]** gave a generalization of (2.8). But his result is a simple consequence of both Jensen's inequality for convex functions and of (2.8) as was proved by D. S. Mitrinović and J. E. Pečarić **[168]**.

50. Weighted versions of various inequalities are given by H. P. Heinig **[169]**. In particular, he gives weighted versions of Theorem 3 of 1., (2.5), and the result of 4.. A weighted characterization of (2.5) was earlier discussed in **[151]** and independently by Stepanov in **[170]**. Best constants may also be found in **[10]** and **[19]**. Other references include Ding **[171]** and Lee **[172]**.

51. An additional reference is the paper of Carlson **[173]** and the alternate proofs of his result given in Ahiezer **[174]**.

References

1. HARDY, G. H., *Note on a theorem of Hilbert*, Math. Zeitschr. **6** (1920), 314–317.
2. LANDAU, E., *A note on a theorem concerning series of positive terms*, J. Lond. Math. Soc. **1** (1926), 38–39.
3. BROADBENT, T. A. A., *A proof of Hardy's convergence theorem*, J. Lond. Math. Soc. **3** (1928), 93–96.
4. ELLIOTT, E. B., *A simple exposition of some recently proved facts as a convergency*, J. Lond. Math. Soc. **1** (1926), 93–96.
5. GRANDJOT, K., *On some identities relating to Hardy's convergence theorem*, J. Lond. Math. Soc. **3** (1928), 114–117.
6. HARDY, G. H., *Notes on some points in the integral calculus (LX)*, Messenger of Math. **54** (1925), 150–156.
7. KALUZA, Th. and G. SZEGÖ, *Über Reihen mit lauter positiven Gliedern*, J. Lond. Math. Soc. **2** (1927), 266–272.

8. KNOPP, K., *Über Reihen mit positiven Gliedern*, J. Lond. Math. Soc. **3** (1928), 205–211.
9. CARLEMAN, T., *Sur les fonctions quasi-analytiques*, in "Conférences faites au cinquième congrès des mathématiciens scandinaves," Helsingfores, 1923, pp. 181–196.
10. PÓLYA, G., *Proof of an inequality*, Proc. London Math. Soc. **24** 2 (1926), Records of Proc. 1vii.
11. VALIRON, G., "Lectures on the General Theory of Integral Functions," Toulouse, 1923.
12. OSTROWSKI, A. M., *Über quasi-analytische Funktionen und Bestimmtheit asymptotischer Entwicklungen*, Acta Math. **53** (1929), 181–266.
13. HARDY, G. H., J. E. LITTLEWOOD and G. PÓLYA, "Inequalities," Cambridge, 1934.
14. KNOPP, K., *Über Reihen mit positiven Gliedern* (2te Mitteilung), J. Lond. Math. Soc. **5** (1930), 13–21.
15. HARDY, G. H., *Notes on some points in the integral calculus (LXIV)*, Messenger of Math. **57** (1928), 12–16.
16. COPSON, E. T., *Note on series of positive terms*, J. Lond. Math. Soc. **2** (1927), 9–12.
17. ________, *Note on series of positive terms*, J. Lond. Math. Soc. **3** (1928), 49–51.
18. HARDY, G. H., *Remarks on three recent notes in the Journal*, J. Lond. Math. Soc. **3** (1928), 166- -169.
19. ________, *The constants of certain inequalities*, J. Lond. Math. Soc. **8** (1933), 114–119.
20. HARDY, G. H. and J. E. LITTLEWOOD, *Notes on the theory of seris (XII): On certain inequalities connected with the calculus of variations*, J. Lond. Math. Soc. **5** (1930), 283–290.
21. BLISS, G. A., *An integral inequality*, J. Lond. Math. Soc. **5** (1930), 40–46.
22. SUNOUCHI, G. and N. TAKAGI, *A generalization of the Carleman inequality theorem*, Proceedings Phys.-Math. Soc. Japan **16** (1934), 164–166.
23. KNOPP, K.,*Neuere Sätze: Reihen mit positiven Gliedern*, Math. Zeitschr. **30** (1929), 387–413.
24. MULHOLLAND, H. P., *On the generalisation of Hardy's inequality*, J. Lond. Math. Soc. **7** (1934), 208- -214.
25. ________, *Concerning the generalization of the Young-Hausdorff theorem and the Hardy-Littlewood theorems on Fourier constants*, Proc. Lond. Math. Soc. **35** (1934), 257–293.

26. ________, *The generlization of certain inequality theorems involving powers*, Proc. Lond. Math. Soc. **33** 2 (1932), 481–516.
27. ________, *Some theorems on Dirichlet series with positive coefficients and related integrals*, Proc. Lond. Math. Soc. **29** (1927), 141–158.
28. HARDY, G. H. and J. E. LITTLEWOOD, *Elementary theorems concerning power series with positive coefficients and moment constants of positive functions*, J. f. Math. **157** (1927), 141–158.
29. LEVIN, V., *Two remarks on van der Corput's generalisations of Knopp's inequality*, Proc. Akad. Wet. Amsterdam **40** (1937), 429–431.
30. VAN DER CORPUT, J. G., *Generalizations of an inequality of Knopp*, Proc. Akad. Wet. Amsterdam **39** (1936), 1053–1055.
31. HARDY, G. H. and J. E. LITTLEWOOD, *Some new properties of Fourier constants*, Math. Annalen **97** (1927), 159–209 (199).
32. LEVIN, V., *O neravenstvah. III: neravenstva, vypolnjamye geometričeskim srednim neotricatel'noĭ funkcii*, Matem. Sbornik **4** 46 (1938), N.2, 325–331.
33. VAN DER CORPUT, J. G., *Generalisations of Carleman's inequality*, Proc. Acad. Wet. Amsterdam **39** (1936), 906–911.
34. CHOW, Y. C., *A note on Hardy's inequality and similar inequalities*, J. London Math. Soc. **14** (1939), 88- -93.
35. FLETT, T. M., *A note on some inequalities*, Proc. Glasgow Math. Assoc. **4** (1959), 7–15.
36. LEVINSON, N.,, *Generalizations of an inequality of Hardy*, Duke Math. J. **31** (1964), 389–394.
37. BOAS, R. P., *Inequalities for monotonic series*, J. Math. Anal. Appl. **1** (1960), 121–126.
38. KONYUŠKOV, A. A., *Beste approximationer durch trigonometrische Polynome und Fourierkoeffizienten*, Mat. Sb. (N.S.) **44** 86 (1958), 53–84.
39. LEVIN, V. I. and S. B. STEČKIN, *Inequalities*, Amer. Math. Soc. Translations, Series 2 **14** (1960), 1–29.
40. BEESACK, P. R., *Hardy's inequality and its extensions*, Pacif. J. Math. **11** (1961), 39–61.
41. DE BRUIJN, N. G., *Carleman's inequality for finite series*, Proc. Akad. Wet. Amsterdam Ser. A. **4** 4 (1963), 505–514.
42. WILF, H. S., *Finite sections of the classical inequalities*, Nederl. Akad. Wet. Amsterdam. Proc. Ser. A. **65** (1962), 340–343, (= Indagationes Mathematicae **24**).
43. SYSOEVA, F. A., *Verallgemeinerung einer Ungleichung von Hardy*, Izvestija Vysš. Učebn. Zaved., Mat. 1965 **49** 6 (1965), 140–141.

44. KADLEC, J. and A. KUFNER, *Characterization of functions with zero traces by integrals with weight functions, I*, Čas. Pešt. Mat **91** (1966), 463–470; II ibid, **92** (1967), 16–28.
45. GODUNOVA, E. K., *Inequalities with convex functions*, Izv. Vysš. Učebn. Zaved. Math. 4 (1965), 45- -63.
46. AKERBERG, B., *A variant of the proofs of some inequalities*, Proc. Cambridge Philos. Soc. **57** 1 (1961), 184–186.
47. GODUNOVA, E. K., *Nekotorye neravenstva tipa neravenstv Karlemans*, Ucen. Zap. Gos. Ped. Inst. im Lenina, Moskva 277 (1970), 176–191.
48. BOYD, D. W., *On the exponent of an osculatory packing*, Can. J. Math. **23** (1971), 355–363.
49. GODUNOVA, E. K., *Integral'nye neravenstva s vypuklymi funkcijami*, Mat. šestoǐ mež. fiz.-mat. nauč. Konf. Dal'nego Vostoka, Diff. int. ur. **3** (1967), 65–69.
50. BOAS, R. P., *Some integral inequalities related to Hardy's inequality*, Journal d'Analyse Math. **23** (1970), 53–63.
51. BENNETT, C and K. RUDNICK, *On Lorentz-Zygmund spaces*, (98 page preprint).
52. TALENTI, G., *Una diseguaglianza integrale*, Boll. Un. mat. Ital **21** 3 (1966), 25–34.
53. __________, *Sopra una diseguaglianza integrale*, Ann. Scuola Norm. Sup. Pisa **31** 3 (1967), 167–188.
54. TOMASELLI, G., *A class of inequalities*, Bol. Un. mat. Ital. IV Ser. **2** (1969), 622–631.
55. MUCKENHOUPT, B., *Hardy's inequality with weights*, Studia Math. **44** (1972), 31–38.
56. IZUMI, M. and S. I. IZUMI, *On some inequalities for Fourier series*, Journal d'Analyse Math. **21** (1968), 277–291.
57. HARDY, G. H., "Divergent Series," Cambridge, 1948.
58. PETERSEN, G. M., *An inequality of Hardy's*, Quart. J. Math. Oxford Ser. **13** 2 (1962), 237- -240.
59. DAVIES, G. S. and G. M. PETERSEN, *On an inequality of Hardy's (II)*, Quart. J. Math. Oxford Ser. **15** 2 (1964), 35–40.
60. IZUMI, M., S. I. IZUMI and G M. PETERSEN, *On Hardy's inequality and its generalization*, Tôhoku Math. J. **21** (1969), 601–613.
61. LEINDLER, L., *Generalization of inequalities of Hardy and Littlewood*, Acta Sci. Math. (Szeged) **31** (1970), 279–285.
62. NÉMETH, J., *Generalizations of the Hardy-Littlewood inequality*, Acta Sci. Math. (Szeged) **32** (1971), 295–299.
63. __________, *Generalizations of the Hardy-Littlewood inequality, II*, Acta Sci. Math. (Szeged) **35**, 127–134.

64. LEINDLER, L., *Inequalities of Hardy-Littlewood type*, Analysis Math. **2** (1976), 117–123.
65. SHUM, D. T., *On integral inequalities related to Hardy's*, Can. Math. Bull. **14** (1971), 225–230.
66. FLORKIEWICZ, B. and A. RYBARSKI, *A certain modification of Hardy's inequality*, Colloquium Math. **26**, (1972), 345–348.
67. __________, *On an integral inequality connected with Hardy's inequality (III)*, Colloquium Math. **27** (1973), 293–296.
68. BENNETT, C., *Intermediate spaces and the class* $L\log^+ L$, Arkiv for Math. **11** (1973), 215–228.
69. GLUK, L., *On an extension of Hardy's inequality and some weighted norm inequalities for the indefinite integral of conjugagte functions*, Notices Amer. Math. Soc. **1** 1 (1974), A–163.
70. FLETT, T. M., *Some elementary inequalities for integrals with applications to Fourier Transforms*, Proc. London Math. Soc. **29** 3 (1974), 538–556.
71. ANDERSEN, K. F., *On Hardy's inequality and Laplace transforms in weighted rearrangement invariant space*, Proc. Amer. Math. Soc. **39** (1973), 295–299.
72. COPSON, E. T., *Some integral inequalities*, Proc. Roy. Soc. Edinburgh A 75 **13** (1975-76), 157–164.
73. HEINIG, H., *Some extensions of Hardy's inequality*, SIAM J. Math. Anal. **6** (1975), 698–713.
74. IMORU, C. O., *On some integral inequalities related to Hardy's*, Canad. Math. Bull. **20** (1977), 307- -312.
75. MUSIELAK, H., *An inequality for double fractional integrals*, Funct. Approx., Comment. Math. **2** (1976), 183–195.
76. PEASLEE, D. C. and W. A. COPPEL, *On an inequality of Bombieri*, J. Australian Math. Soc. **9** (1969), 339–404.
77. BOMBIERI, E., *Sulla seconds variazione della fuzione di Koebe*, Boll. Un. Mat. Ital. **22** (1967), 25–32.
78. SHUM, D. T., *A Generalization of Bombieri's Inequality*, manuscript.
79. BRADLEY, J. S., *Hardy inequalities with mixed norms*, Canad. Math. Bull. **21** (1978), 405–408.
80. HEINIG, H., *Variations of Hardy's inequality*, Real Analysis Exchange **5** (1979-80), 61–81.
81. BEESACK, P. R. and H. HEINIG, *Hardy's inequalities with indices less than* 1, Proc. Amer. Math. Soc. **83** (1981), 532–536.
82. KUFNER, A. and H. TRIEBEL, *Generalizations of Hardy's inequality*, Conf. del Sem. di Mat. dell' Università di Bari **156** (1978), 1–21.

83. CHAN, L. Y., *Some further extensions of Hardy's inequality*, Canad. Math. Bull. **24** (1981), 393- -400.

84. _________, *Some extensions of Hardy's inequalities*, Canad. Math. Bull. **22** (1979), 165–169.

85. BEESACK, P. R., *On some integral inequalities of E. T. Copson*, in "General Inequalities 2," ed. E. F. Beckenbach, Basel, 1980, pp. 151–159 .

86. FEHÉR, F., *A generalized Schur-Hardy inequality on normed Köthe spaces*, in "General Inequalties 2," ed. E. F. Beckenbach, Basel, 1980, pp. 277–286.

87. MOHAPATRA, R. N., *Remarks on a generalization of the Schur-Hardy inequality*, in "General Inequalities 2," ed. E. F. Beckenbach, Basel, 1980, pp. 459–460.

88. BUTZER, P. L. and F. FEHÉR, *Generlized Hardy and Hardy-Littlewood inequalities in rearrangement-invariant spaces*, Comment. Math. Prace Mat. Tomus Specialis in Honorem Ladislai Orlicz I (1978), 41–64.

89. FEHÉR, F., *A note on a paper of E. R. Love*, Bull. Austral. Math. Soc. **19** (1978), 67–75.

90. LOVE, E. R. *Generlizations of a classical inequality* , Appl. Anal. **8** (1978), 47–59.

91. _________, *Links between some generalizations of Hardy's integral inequality*, in "General Inequalities 4," ed. W. Walter, Basel, 1984, pp. 47- -57.

92. STOLARSKY, K. B., *Another Proof of Carleman's Inequality*, manuscript.

93. KOKIAŠVILI, V. N., *O. neravenstvah Hardi v vesovyh prostranstvah*, Soob. Akad. Nauk Gruz. SSR, Matematika **96** 2 (1979), 37–40.

94. SYSOEVA, F. A., *Odno neravenstvo tipa Hardi- Litlvuda*, Sb. Tr. Mosk. Gos. Ped. Inst. im. Lenia **6** (1978), 136–141.

95. KUFNER, A. and A. TRIEBEL, *Ob odnom obobščenii neravenstva Hardi*, Akad. Nauk SSSR, Sib. Otd. Inst. Mat., *Teorija kubaturnyh formul i priloženija funkcional'nogo analiza k zadačam matematičeskoĭ fiziki*, Tr. Sem. S. L. Soboleva **1** (1978), 61–68.

96. DING, Y., *On a generalization of Hardy's inequality* (Chinese), J. Math. (Wuhan) **1** 1 (1981), 31–39.

97. VASIĆ, P. M. and J. E. PEČARIĆ, *Notes on some inequalities for convex functions*, Mat. Vesnik **6** 34 (1982), 185–193.

98. CHANDRA, P. and R. N. MOHAPATRA, *Inequalities which yield inclusions among sequence spaces containing l_p*, in "General Inequalities 3," ed. E. F. Beckenbach, Basel, 1983, 455–470.

99. MOHAPATRA, R. N. and D. C. RUSSELL, *Integral inequalities related to Hardy's inequality*, Aequationes Math. **28**, (1985), 199–207.

100. IMORU, C. O., *On some extensions of Hardy's inequality*, Internat. J. Math. Sci. **8** 1 (1985), 165–171; **9** 1 (1986), 207.

101. __________, *An extension of a type of generalized Hardy's integral inequality*, Simon Stevin, A Quart. J. Pure Appl. Math. **59** (1985), 109–112.

102. __________, *Generalizations of Hardy's integral inequality*, J. Math. Anal. Appl. **122** (1987), 7–15.

103. PACHPATTE, B. G., *On Hardy type integral inequalities*, Tamkang J. Math. **18** 2 (1987), 27–41.

104. LOVE, E. R. *Inequalities related to Carleman's inequality*, unpublished lecture to Inequalities Conference at Birmingham, July 1987.

105. __________, *Generlizations of Hardy's and Copson's inequalities*, J. London Math. Soc. **30** 2 (1984), 431–440.

106. __________, *Generalizations of Hardy's integral inequality*, Proc. Roy. Soc. Edinburgh **100A** (1985), 237–262.

107. __________, *Hardy's inequality in Orlicz-type sequence spaces for operators related to generalized Hausdorff matrices*, Math. Z. **193** (1986), 481–490.

108. __________, *Inequalities related to those of Hardy and of Cochran and Lee*, Math. Proc. Comb. Phil. Soc. **99** (1986), 395–408.

109. __________, *Some inequalities for geometric means*, in "General Inequalities 5," ed. W. Walter, Basel/Boston, 1987, pp. 87–93.

110. __________, *Inequalities related to Knopp's inequality*, J. Math. Anal. Appl. **137** (1987), 173–180.

111. JAKIMOVSKI, A., B. E. RHOADES and J. TZIMBALARIO, *Hausdorff matrices as bounded operators over l^p*, Math. Z. **138** (1974), 173–181.

112. BORWEIN, D., *Generlized Hausdorff matrices as bounded operators on l^p*, Math. Z. **183** (1983), 483–487.

113. COCHRAN, J. A. and C.-S. LEE,, *Inequalities related to Hardy's and Heinigs's*, Math. Proc. Cambridge Philos. Soc. **96** (1984), 1–7.

114. MOORE, R. A. and Z. NEHARI, *Nonoscillation theorems for a class of nonlienar differential equations*, Trans. Amer. Math. Soc. **93** (1959), 30–52.

115. KIM, W. J., *On the zeros of solutions of $y^{(n)} + py = 0$*, J. Math. Anal. Appl. **25** (1969), 189–208.

116. HINTON, D. B. and R. T. LEWIS, *Singular differential operators with spectra discrete and bounded below*, Proc. Roy. Soc. Edinburgh **84A** (1979), 117–134.

117. KRZYWICKI, A. and A. RYBARSKI, *On a linearization of an equation of an elastic rod (II)*, Zastosow. Mat. **7** (1964), 383–390.

118. RYBARSKI, A., *Angenäherte Schwingungsfrequenzformeln für konservative Systeme (2)*, Zastosow. Mat. **7** (1964), 255–269.
119. EVERITT, W. N., M. GIERTZ and J. WEIDMANN, *Some remarks on a separation and limit-point criterion of second order, ordinary differential expressions*, Math. Ann. **200** (1973), 335–346.
120. EVERITT, W. N., *An integral inequality with an application to ordinary differential operators*, Proc. Roy. Soc. Edinburgh **80A** (1978), 35–44.
121. AMOS, R. J. and W. N. EVERITT, *On integral inqualities and compact embeddings associated with ordinary differntial expressions*, Arch. Rational Mech. Anal. **71** (1979), 15–50.
122. BRADLEY, J. S. , D. B. HINTON and R. M. KAUFFMAN, *On the minimization of singular quadratic functionals*, Proc. Roy. Soc. Edinburgh **87A** (1981), 193–208.
123. PUTNAM, C. R., *An application of spectral theory to a singular calculus of variations problem*, Amer. J. Math. **70** (1948), 780–803.
124. EVERITT, W. N., *Integral inequalities and spectral theory, Spectral theory and differential equations*, in "Lecture Notes in Math." 448, Berlin, 1976, pp. 73–105.
125. BEESACK, P. R., *Integral inequalities of the Wirtinger type*, Duke Math. J. **25** (1958), 477–798.
126. __________, *Minimum properties of eigenvalues – elementary proofs*, in "General Inequalities 2," E. F. Beckenbach, ed., Basel, 1980, pp. 109--120.
127. REDHEFFER, R., *Inequalities with three functions*, J. Math. Anal. Appl. **16** (1966), 219–242.
128. BENSON, D. C., *Inequalities involving integrals of functions and their derivatives*, J. Math. Anal. Appl. **17** (1967), 292-308.
129. PORTNOV, V. R., *Two embedding theorems for spaes $L^1_{p,b}(\Omega \times R)$ and their applications* (Russian), Dokl. Akad. Nauk SSSR **155** (1964), 761–764.
130. KUFNER, A., *Imbedding theorems for general Sobolev weight spaces*, Ann. Scuola Norm. Sup. Pisa **23** (1969), 373–386.
131. __________, "Weighted Sobolev Spaces," Leipzig, 1980.
132. MAZ'YA, V. G., "Spaces of S. L. Sobolev" (Russian), Leningard, 1985.
133. HUNT, R. A., *On $L(p,q)$ spaces*, L'Enseignment Math. **12** (1966), 249–276.
134. HUNT, R. A. and G. WEISS, *The Marcinkiewicz interpolation theorem*, Proc. Amer. Math. Soc. **15** (1964), 996–998.
135. HUNT, R. A., *An extension of the Marcinkiewicz interpolation theorem to Lorentz spaces*, Bull. Amer. Math. Soc. **70** (1964), 803–807.

136. NEHARI, Z., *The Schwarzian derivative and schlicht functions*, Bull. Amer. Math. Soc. **55** (1949), 545- -551.
137. FLANDERS, H., *Remarks on Almgren's interior regularity theorems*, Illinois J. Math. **13** (1969), 707- -716.
138. JANET, M., *Sur la méthode de Legendre-Jacobi-Clebsch et quelquesunes de ses applications*, Bull. Sci. Math. **53** (1929), 144–160.
139. TROESCH, B. A., *An isoperimetric sloshing problem*, Comm. Pure Appl. Math. **18** (1965), 319–338.
140. BEESACK, P. R., *Integral inequalities involving a function and its derivative*, Amer. Math. Monthly **78** (1971), 705–741.
141. FEINBERG, J. M., *Some Wirtinger-like inequalities*, SIAM J. Math. Anal. **10** (1979), 1258–1271.
142. FLORKIEWICZ, B., *Integral inequalities of Hardy type*, Dissertation.
143. BEESACK, P. R., D. S. MITRINOVIĆ and P. M. VASIĆ, *Integral inequalities*, (The manuscript not achieved and not published – we used this manuscript in writing sections 15,16,21,23–25,28–32).
144. PACHPATTE, B. G., *On some extensions of Levinson's generalizations of Hardy's inequality*, Soochow J. Math. **13** 2 (1987), 203–210.
145. OSTROGORSKI, T., *Analogues of Hardy's inequality in R^n*, Studia Math. **88** 3 (1988), 209–219.
146. SINNAMON, G., *A note on the Stieltjes transformation*, Proc. Roy. Soc. Edinburgh Sect. A **110** 1- -2 (1988), 73–78.
147. BENNETT, G., *Some elementary inequalities II*, Quart J. Math. Oxford Ser. (2) **39** 156 (1988), 385–400.
148. PACHPATTE, B. G. and E. R. LOVE, *On some new inequalities related to Hardy's integral inequality*, J. Math. Anal. Appl. **149** (1990), 17–25.
149. MOHAPATRA, R. N. and D. C. RUSSELL, *Integral inequalities related to Hardy's inequality*, Aeq. Math. **28** (1985), 199–207.
150. MOHAPATRA, R. N. and K. VAJRAVELU, *Integral inequalities resembling Copson's inequality*, J. Aust. Math. Soc. Ser. A. **48** (1990), 124–132.
151. MARTÍN-REYES, F. J. and E. SAWYER, *Weighted inequalities for Riemann-Liouville fractional integrals of order one and greater*, Proc. Amer. Math. Soc. **106** (1989), 727- -733.
152. PITTENGER, A. P., *Note on a square function inequality*, Ann. Prob. **7** (1979), 907–908.
153. WILF, H. S., "Finite sections of some classical inequalities," Berlin, Heidelberg, New York, 1970.
154. BATUEV, Ě. N. and V. D. STEPANOV, *Weighted inequalities of Hardy type* (Russian), Sibirsk. Mat. Ž. **30** (1989), 13–22.
155. SINNAMON, G., *A weighted gradient inequality*, Proc. Roy. Soc. Edinburgh Sect. A. **111** (1989), 329–335.

156. STEPANOV, V. D., *A weighted inequality of Hardy type for higher-order derivatives* (Russian) Trudy Mat. Inst. Steklov. **187** (1989), 178–190.
157. WANNEBO, A., *Hardy inequalities*, Proc. Amer. Math. Soc. **109** (1990), 85–95.
158. BROWN, R. and D. HINTON, *Integral inequalities with power weights for functions of one variable*, preprint.
159. OPIC, B. and A. KUFNER, "Hardy-type inequalities," New York, 1990.
160. KUFNER, A., "Zur Hardyschen Ungleichung," Sem. anal. 1982/83, 1–18, Akad. Wiss. DDR, Berlin 1983.
161. LIZORKIN, P. I., *Operatoren, die mit der gebrochenen differentiation zusammenhängen und klassen differenzierbarer functionen*, Trudy Mat. Inst. AN SSSR **117** (1972), 212–243.
162. JOHNSON, P. D. and R. N. MOHAPATRA, *A Hardy-Davies-Petersen inequality for a class of matrices*, Canad. J. Math. **30** (1978),, 458–465.
163. ________, *Inequalities involving infinite matrices with nonnegative entries*, in "General Inequalities 2," ed. E. F. Beckenbach, Basel, 1980, pp. 55–80.
164. ________, *Inequalities involving lower-triangular matrices*, Proc. London Math. Soc. **41** (1980), 83–132.
165. ________, *Best possible results in a class of inequalities*, Pacific J. Math. **103** (1982), 432–436.
166. ________, *On inequalities related to sequences spaces as* $[p, q]$, in "General Inequalities 4," ed. W. Walter,, Basel, 1989, pp. 191- -201.
167. PACHPATTE, B. G., *A note on Copson's inequality involving series of positive terms*, Tamkang J. Math. **21** (1990), 13–19.
168. MITRINOVIĆ, D. S. and J. E. PEČARIĆ, *A note on Copson's inequality of B. G. Pachpatte*, (to appear).
169. HEINIG, H. P., *Weighted inequalities in Fourier analysis*, Proc. Summer School Roudnise, Czechoslovakie, May 1990 (Teubner).
170. STEPANOV, V. D., *Weighted inequalities of Hardy type for derivatives of higher order and their applications* (Russian), Dokl. AN SSSR (1988), No. **5**, 1059–1062.
171. DING, Y., *Weighted Hardy inequalty* (Chinese), J. Hangzhou Univ. Nat. Sci. Ed. **11** (1984), 168–174.
172. LEE, Kin-Chun and Gou-sheng YANG, *On generalization of Hardy's inequality*, Tamkang J. Math. **17** (1986), 109–119.
173. CARLSON, L., *A proof of an inequality of Carleman*, Proc. Amer. Math. Soc. **5** (1954), 932–933.
174. AHIEZER, H. I., *O vzvešennom približenii neperyonyh funkciǐ mnogočlenami na vseǐ čislovoǐ osi*, Uspehi Mat. Nauk t. **11**, byp 4 (70), 1956.

CHAPTER V

HILBERT'S AND RELATED INEQUALITIES

1. The inequality

$$(1.1)\qquad \sum_{m=1}^{\infty}\sum_{n=1}^{\infty}\frac{a_m b_n}{m+n}\le\pi\left(\sum_{m=1}^{\infty}a_m^2\right)^{1/2}\left(\sum_{n=1}^{\infty}b_n^2\right)^{1/2}$$

is known as Hilbert's double series theorem, or Hilbert's inequality. It was first proved by Hilbert in his lectures on integral equations, and a proof was published by H. Weyl [**1**] in his dissertation in 1908. The exact value π of the constant (which is best possible) was given by I. Schur [**2**]. He also gave an integral analogue of (1.1).

The following generalizations of (1.1) are given by G. H. Hardy and M. Riesz (see Hardy [**3**]):

THEOREM 1. *If $p>1$, $p'=p/(p-1)$, then*

$$(1.2)\qquad \sum_{m=1}^{\infty}\sum_{n=1}^{\infty}\frac{a_m b_n}{m+n}<\left(\frac{\pi}{\sin\left(\frac{\pi}{p}\right)}\right)\left(\sum_{m=1}^{\infty}a_m^p\right)^{1/p}\left(\sum_{n=1}^{\infty}b_n^{p'}\right)^{1/p'}$$

unless (a_n) or (b_n) is identically zero.

THEOREM 2. *If $p>1$, $p'=p/(p-1)$, then*

$$\int_0^{\infty}\int_0^{\infty}\frac{f(x)g(y)}{x+y}\,dx\,dy$$

$$(1.3)\qquad <\left(\frac{\pi}{\sin\left(\frac{\pi}{p}\right)}\right)\left(\int_0^{\infty}f^p(x)\,dx\right)^{1/p}\left(\int_0^{\infty}g^{p'}(x)\,dx\right)^{1/q}$$

unless $f\equiv 0$ or $g\equiv 0$.

These theorems, as well as the following are known as Hilbert's theorems (see [**4**, p. 234], Theorems 315, 316 and 323).

THEOREM 3. *If the conditions of Theorem 1 are satisfied, then*

$$\sum_{m=0}^{\infty}\sum_{n=0}^{\infty}\frac{a_m b_n}{m+n+1} < \left(\frac{\pi}{\sin\left(\frac{\pi}{p}\right)}\right)\left(\sum_{m=0}^{\infty} a_m^p\right)^{1/p}\left(\sum_{n=0}^{\infty} b_n^{p'}\right)^{1/p'}. \tag{1.4}$$

The constants $\frac{\pi}{\sin(\pi/p)}$ in all the above theorems are the best possible.

2. Other proofs, of the whole or of parts of Theorems 1–3, and generalizations in different directions have been given by L. Fejér and F. Riesz [**5**], E. C. Francis and J. E. Littlewood [**6**], G. H. Hardy [**7**], G. H. Hardy, J. E. Littlewood and G. Pólya [**8**], H. P. Mulholland [**9**],[**10**], P. M. Owen [**11**], G. Pólya and G. Szegö [**12**, I, 117, 290], I. Schur [**2**], and F. Wiener [**13**] (see [**4**, pp. 226–227]).

Note that (1.4) is sharper than (1.2), and several proofs of Hilbert's theorem, for example the proof of Mulholland [**9**], the proof of Schur (see for example [**4**, pp. 212–213]), the proof of Fejér and F. Riesz (see, for example, [**4**, p. 325]), and the proof of Pólya and Szegö [**12**, I, 290]), also give the result in this form. The last three are limited to the case $p = 2$ (see [**4**, pp. 234–235]).

Proofs of Hardy, Littlewood and Pólya [**8**] are based on general classes of bilinear forms (see Theorems 318, 319, 320, 322 from [**4**]). We shall give here only the following special case ([**4**, Th. 321]):

If $\lambda_m > 0$, $\nu_n > 0$, $\Lambda_m = \lambda_1 + \lambda_2 + \cdots + \lambda_m$, $M_n = \nu_1 + \nu_2 + \cdots + \nu_n$, $p > 1$, then

$$\sum_{m=1}^{\infty}\sum_{n=1}^{\infty}\frac{\lambda_m^{1/p'}\nu_n^{1/p}}{\Lambda_m + M_n}a_m b_n < \left(\frac{\pi}{\sin\left(\frac{\pi}{p}\right)}\right)\left(\sum_{m=1}^{\infty} a_m^p\right)^{1/p}\left(\sum_{n=1}^{\infty} b_n^{p'}\right)^{1/p'} \tag{2.1}$$

unless (a_n) or (b_n) is identically zero.

The cases $p = 2$ of these results (Theorems 318, 319, 320 and 322) were proved by Schur [**2**].

3. Suppose that $a_n \geq 0$, and that $A(x) = \sum\limits_{n=0}^{\infty} a_n x^n$, $A^*(x) = \sum\limits_{n=0}^{\infty} \frac{a_n x^n}{n!}$. Then

$$\pi \sum_{m=0}^{\infty} \sum_{n=0}^{\infty} \frac{(m+n)!}{m!n!} \frac{a_m a_n}{2^{m+n+1}} \geq \sum_{m=0}^{\infty} \sum_{n=0}^{\infty} \frac{a_m a_n}{m+n+1}, \tag{3.1}$$

$$\int_0^1 A^2(x)\, dx \leq \pi \int_0^1 \left(e^{-x} A^*(x)\right)^2 dx. \tag{3.2}$$

Inequality (3.1) is stronger than Hilbert's inequality (1.1) for $a_n = b_n$.

These inequalities are due to D. V. Widder [**15**]. G. H. Hardy [**16**] proved (3.2), which is equivalent to (3.1), by using Hilbert's inequality itself. He also proved the following generalization of (3.2):

If $p > 1$, then

$$\int_0^1 z^{p-2} (A(z))^p\, dz \leq \left(\Gamma\left(\frac{1}{p}\right)\right)^p \int_0^{\infty} z^{p-2} \left(e^{-z} A^*(z)\right)^p dz. \tag{3.3}$$

The following Hardy inequality was proved in both papers ([**15**],[**16**]):

$$\sum_{m=1}^{\infty} \sum_{n=1}^{\infty} \frac{\log(m/n)}{m-n} a_m a_n \leq \pi \sum_{m=1}^{\infty} \sum_{n=1}^{\infty} \frac{a_m a_n}{m+n}, \tag{3.4}$$

where the coefficient on the left-hand side is to be interpreted as $1/n$ when $m = n$.

Hardy [**16**] also gave the following inequality:

$$\sum_{m=1}^{\infty} \sum_{n=1}^{\infty} \frac{|\log(m/n)|}{\max(m,n)} a_m a_n \leq 2 \sum_{m=1}^{\infty} \sum_{n=1}^{\infty} \frac{a_m a_n}{\max(m,n)}. \tag{3.5}$$

The last two inequalities are consequence of the following Hardy's inequality for quadratic forms ([**16**], [**4**, p. 257]):

Suppose that $K_0(x) \geq 0$, that $K_1(x,y) = \int\limits_0^{\infty} K_0(xt) K_0(yt)\, dt$ (so that K_1 is symmetric and homogeneous of degree -1), and that

$$K_2(x,y) = \int_0^{\infty} K_1(x,t) K_1(y,t)\, dt.$$

Then

$$\sum_{m=1}^{\infty}\sum_{n=1}^{\infty} K_2(m,n)a_m a_n \le J \sum_{m=1}^{\infty}\sum_{n=1}^{\infty} K_1(m,n)a_m a_n, \tag{3.6}$$

where $J = \int_0^\infty \frac{K_1(w,1)}{\sqrt{w}}\, dw$.

The following results are connected to (3.4) and (3.5):

(i) ([4, Th. 341]) If a_m, b_n, $f(x)$, $g(y)$ are nonnegative and $p > 1$, then

$$\sum_1^\infty\sum_1^\infty \frac{a_m b_n}{\max(m,n)} < pp'\left(\sum_1^\infty a_m^p\right)^{1/p}\left(\sum_1^\infty b_n^{p'}\right)^{1/p'}, \tag{3.7}$$

$$\int_0^\infty\int_0^\infty \frac{f(x)g(y)}{\max(x,y)}\, dxdy < pp'\left(\int_0^\infty f^p(x)\, dx\right)^{1/p}\left(\int_0^\infty g^{p'}(x)\, dx\right)^{1/p'}, \tag{3.8}$$

unless (a_m) or (b_n) or $f(x)$ or $g(y)$ is identically zero.

The constants are the best possible.

(ii) ([4, Th. 342]) If $p > 1$, then

$$\sum_1^\infty\sum_1^\infty \frac{\log(m/n)}{m-n} a_m b_n < \pi^2 \operatorname{cosec}^2\left(\frac{\pi}{p}\right)\left(\sum_1^\infty a_m^p\right)^{1/p}\left(\sum_1^\infty b_n^{p'}\right)^{1/p'}, \tag{3.9}$$

$$\int_0^\infty\int_0^\infty \frac{\log(x/y)}{x-y} f(x)g(y)\, dxdy < \pi^2 \operatorname{cosec}^2\left(\frac{\pi}{p}\right)\left(\int_0^\infty f^p\, dx\right)^{1/p}\left(\int_0^\infty g^{p'}\, dy\right)^{1/p'}. \tag{3.10}$$

4. The paper [**10**] of H. P. Mulholland is concerned with a generalization of (1.2) in which a_m^p and b_n^p on the right are replaced by $\phi(a_m)$ and $\psi(b_n)$ respectively, where $\phi(x)$ and $\psi(x)$ are functions which increase strictly from 0 to ∞ with x and whose inverses are related by

$$\phi^{-1}(x)\psi^{-1}(x) = x, \tag{4.1}$$

and when ϕ is a convex function which satisfies

$$\phi(xy) \geq \phi(x)\phi(y). \tag{4.2}$$

Namely, if the integral

$$P = \int_0^\infty \frac{\phi^{-1}(x)}{x(1+x)}\,dx = \int_0^\infty \frac{dx}{\psi^{-1}(x)(1+x)} \tag{4.3}$$

is convergent, then

$$\sum_1^\infty \sum_1^\infty \frac{a_m b_n}{m+n} \leq P \frac{\phi^{-1}\left(\sum_1^\infty \phi(a_m)\right)}{\psi^{-1}\left(\dfrac{1}{\sum_1^\infty \psi(b_n)}\right)}. \tag{4.4}$$

H. P. Mulholland [**17**] also proved some related results.

Let $A(x) = \sum_2^\infty a_n n^{-1-x}$, and $A_n = \int_0^\infty a(x) n^{-x}\,dx$.

If $p > 1$, $q > 1$, $\nu = (1/p) + (1/q) - 1 \geq 0$, then

$$\int_0^\infty A(x)B(x)x^{-\nu}\,dx \leq K\left(\sum_2^\infty \frac{a_n^p}{n}\right)^{1/p}\left(\sum_2^\infty \frac{b_n^q}{n}\right)^{1/q}, \tag{4.5}$$

$$\sum_2^\infty A_n B_n n^{-1}(\log n)^{-\nu} \leq K\left(\int_0^\infty a^p(x)\,dx\right)^{1/p}\left(\int_0^\infty b^q(x)\,dx\right)^{1/q}. \tag{4.6}$$

When $\nu = 0$ and so $q = p'$, the best possible value of K is $\pi/\sin(\pi/p)$.

The special case of (4.5) is

$$\sum_2^\infty \sum_2^\infty \frac{a_m b_n}{\left(mn\,(\log(mn))^{1-\nu}\right)} \leq K\left(\sum_2^\infty \frac{a_n^p}{n}\right)^{1/p}\left(\sum_2^\infty \frac{b_n^q}{n}\right)^{1/q}, \tag{4.7}$$

and when $\nu = 0$, $K = \pi/\sin(\pi/p)$, and this is the smallest K for which the inequality holds.

H. P. Mulholland [18] gave some similar results for power series (in form related to (4.4)). These results are generalizations of results of G. H. Hardy and J. E. Littlewood [19].

5. Some related results are also given in Chapter X: Rearrangements of [4].

6. Related problems were considered by G. Szegö [20] and H. Frazer [21].

7. A. E. Ingham [22] proved the following result:

If $a_n \geq 0$ $(n = 0, 1, 2, \dots)$, $0 < \sum_0^\infty a_n^2 < \infty$ and $\lambda > 0$, then

$$\sum_{m=0}^\infty \sum_{n=0}^\infty \frac{a_m a_n}{m+n+\lambda} \leq M(\lambda) \sum_{n=0}^\infty a_n^2,$$

where

$$M(\lambda) = \frac{\pi}{\sin(\lambda\pi)}, \quad \left(0 < \lambda \leq \frac{1}{2}\right), \quad M(\lambda) = \pi, \quad \left(\lambda > \frac{1}{2}\right).$$

8. V. Levin [23] proved that in (1.4) for $p = q = 2$ the factor $(m+n+1)^{-1}$ can be replaced by the larger one

$(m+n)\log(m+n) + (m+n+2)\log(m+n+2) - 2(m+n+1)\log(m+n+1).$

9. By the Hilbert-Riesz inequality is meant the following inequality: let $p > 1$, $q > 1$, $p^{-1} + q^{-1} > 1$, $\lambda = 2 - p^{-1} - q^{-1} = p'^{-1} + q'^{-1}$ $(0 < \lambda < 1)$, $f(x) \geq 0$, $g(y) \geq 0$, and all integrals are finite, then

$$\int_{-\infty}^{\infty}\int_{-\infty}^{\infty} \frac{f(x)g(y)}{|x+y|^\lambda}\,dx\,dy \tag{9.1}$$

$$\leq K(p,q)\left(\int_{-\infty}^{\infty} f^p(x)\,dx\right)^{1/p}\left(\int_{-\infty}^{\infty} g^q(y)\,dy\right)^{1/q},$$

where $K(p, q)$ depends only on p and q.

The same result with integrals from 0 to ∞ is given in [4, Th. 340], as well as its discrete analogue (Th. 339). In [23], V. Levin showed that in the case with integrals from 0 to ∞ we have that

$$K(p,q) = \left(\pi \operatorname{cosec} \frac{\pi}{\lambda p'}\right)^\lambda = \left(\pi \operatorname{cosec} \frac{\pi}{\lambda q'}\right)^\lambda. \tag{9.2}$$

This constant is not the best possible. But some results with the best possible constants are given by V. Levin in [24] and [25]. In [25] he proved

THEOREM 1. *Let* $a^p = \sup_{-\infty<x<\infty} |x| f^{*p}(x)$, $b^q = \sup_{-\infty<x<\infty} |y| g^{*p}(y)$, *where* f^* *and* g^* *are rearrangements of* f *and* g *in symmetrical decreasing order,* $\delta = (q'-p)/(q'+p')$ $(0<\delta<1)$, $\sigma = (p'-q)/(p'+q')$ $(0<\sigma<1)$, *where* p, q, λ *are as for (9.1). Then*

(9.3)
$$\int_{-\infty}^{\infty}\int_{-\infty}^{\infty} \frac{f(x)g(x)}{|x+y|^{\lambda}}\, dx dy$$
$$< L(p,q) a^{\delta} b^{\sigma} \left(\int_{-\infty}^{\infty} f^p(x)\, dx\right)^{(1-\delta)/p} \left(\int_{-\infty}^{\infty} g^q(y)\, dy\right)^{(1-\sigma)/q},$$

where
$$L(p,q) = \frac{4}{\pi}\sin\frac{\lambda\pi}{2}\sin\frac{\pi}{2p}\sin\frac{\pi}{2q}\Gamma(1-\lambda)\Gamma\left(\frac{1}{p'}\right)\Gamma\left(\frac{1}{q'}\right)$$

is the best possible constant.

Note that the inequality (9.3) is slightly stronger than (9.1).

THEOREM 2. *Let* $p>1$, $q>1$, $p^{-1}+q^{-1}>1$, $\lambda = 2-p^{-1}-q^{-1}$; *let* r *and* s *be two arbitrary real numbers subjected to the only condition* $r+s=\nu\geq 0$ *and let* $f(x)\geq 0$, $g(y)\geq 0$, *with*

$$\int_{-\infty}^{\infty} \left(f(x)\left(\log' f(x)\right)^r\right)^p\, dx = A_r^p < \infty,$$
$$\int_{-\infty}^{\infty} \left(g(y)\left(\log' g(y)\right)^s\right)^q\, dy = B_s^q < \infty,$$

where $\log' t = \max(1, \log t)$ $(t>0)$.
Then

(9.4)
$$\int_{-\infty}^{\infty}\int_{-\infty}^{\infty} \left(\log'\left(\frac{1}{|x+y|}\right)\right)^{\nu} |x+y|^{-\lambda} f(x) g(y)\, dx dy$$
$$\leq K\left(\log' A_r^{-r}\right)^{|r|}\left(\log' B_s^{-s}\right)^{|s|} A_r B_s,$$

where K *depends only on* p, q, r *and* s.

10. Let a, b, c be real numbers and p, q, r be extended real numbers with $1 \le p, q, r < \infty$, and let p', q', r' be the usual conjugates of p, q, r. Let f, g, h be nonnegative measurable functions on $[0, \infty)$, and write

$$F^p = \int_0^\infty f^p(x)\,dx, \quad G^q = \int_0^\infty g^q(x)\,dx, \quad H^{r'} = \int_0^\infty h^{r'}(x)\,dx,$$

$$I = \int_0^\infty \int_0^\infty \frac{f(x)g(y)h(x+y)\,dxdy}{x^a y^b (x+y)^c},$$

$$u(x) = x^{-c} \int_0^x \frac{f(y)g(x-y)\,dy}{y^a (x-y)^b}, \quad U^r = \int_0^\infty u^r(x)\,dx,$$

with the usual conventions if p, q, r, or r' is infinite. Thus, for example, $F = \|f\|_p$ in all cases, so that if $p = \infty$, $F = \text{ess sup}\,|f(x)|$. Let p, q, r, a, b, c satisfy

$$\lambda = 2 - p^{-1} - q^{-1} - (r')^{-1} = a + b + c, \quad 0 \le \lambda \le 1.$$

Then there exists a constant $K = K(p, q, r, a, b)$ such that

$$U \le KFG \tag{10.1}$$

if and only if one of the following sets of conditions is satisfies:

(10.2) $\min(p, q, r') > 1, \quad \max(ap', bq') < 1, \quad \max(a, b) \le \lambda;$
(10.3) $\min(p, q, r') = 1, \quad \max(ap', bq') < 1, \quad \lambda > 0, \quad \max(a, b) < \lambda;$
(10.4) $\min(p, q, r') = 1, \quad \max(a, b) \le \lambda = 0.$

Observe that an application of Fubini's theorem shows that $I = \int_0^\infty u(t)h(t)\,dt$. Hence Hölder's inequality (and its converse) shows that (10.1) is equivalent to

$$I \le KFGH. \tag{10.1'}$$

This result was proved in 1938 by H. R. Pitt **[26]**, and is the integral analogue of a corresponding theorem for infinite series involving Cauchy products due to Hardy and Littlewood **[27]**. (The latter paper involves parameters α, β, γ related to a, b, c by $\alpha = a + p^{-1}$, $\beta = b + q^{-1}$, $\gamma = c + r^{-1}$.)

In **[26]** Pitt also states the following result as a corollary of the above (by expressing the new I as a sum of six integrals over $(-\infty, 0]$ or $[0, \infty)$):

Let p, q, r, a, b, c, λ be as above, and let f, g, h be nonnegative measurable functions on $(-\infty, \infty)$. Let

$$I = \int_{-\infty}^{\infty}\int_{-\infty}^{\infty} \frac{f(x)g(y)h(x+y)\,dxdy}{|x|^a|y|^b|x+y|^c},$$

$$u(x) = |x|^{-c}\int_{-\infty}^{\infty} \frac{f(y)g(x-y)\,dy}{|y|^a|x-y|^b},$$

and let $F = \|f\|_p$, $G = \|g\|_q$, $H = \|h\|_{r'}$, $U = \|u\|_r$. Then there exists $K = K(p,q,r,a,b)$ such that the inequalities (10.1) and (10.1′) are true (for all f, g, h) if any one of the following sets of conditions is satisfied:

(10.5) $\min(p,q,r') > 1$, $\max(ap',bq',cr) < 1$, $\max(a,b,c) \le \lambda$;
(10.6) $\min(p,q,r') = 1$, $\max(ap',bq',cr) < 1$, $\lambda > 0$, $\max(a,b,c) < \lambda$;
(10.7) $\min(p,q,r') = 1$, $a = b = c = \lambda = 0$.

11. Let $A^*(x)$ be defined as in 3. and let $B^*(x)$ be defined similarly. The following generalization of (3.1) and (3.2) is given by Y. C. Chow [**28**]:

(11.1)
$$\sum_{m=0}^{\infty}\sum_{n=0}^{\infty}\frac{a_m b_n}{m+n+1} \le \left(\frac{\pi}{\sin\left(\frac{\pi}{p}\right)}\right)\left(\int_0^{\infty} a^p(x)\,dx\right)^{1/p}\left(\int_0^{\infty} b^{p'}(x)\,dx\right)^{1/p'},$$

where $a(x) = e^{-x}A^*(x)$ and $b(x) = e^{-x}B^*(x)$.

Chow also proved

(11.2)
$$\sum_0^{\infty}\sum_0^{\infty}\cdots\sum_0^{\infty}\frac{a_{n_1}b_{n_2}\cdots c_{n_m}}{n_1+n_2+\cdots+n_m+1}\,\frac{(n_1+n_2+\cdots+n_m)!}{n_1!n_2!\ldots n_m!} \le \left(\sum a_{n_1}^{p_1}\right)^{1/p_1}\left(\sum b_{n_2}^{p_2}\right)^{1/p_2}\cdots\left(\sum c_{n_m}^{p_m}\right)^{1/p_m},$$

where $p_1 > 1$, $p_2 > 1, \ldots p_1^{-1} + p_2^{-1} + \cdots + p_m^{-1} = 1$, and the sequences are nonnegative.

12. H. Frazer [29] proved

$$\sum_{m=0}^{N}\sum_{n=0}^{N}\frac{a_m a_n}{m+n+1}\le\left(N\sin\left(\frac{\pi}{N}\right)\right)\sum_{n=0}^{N}a_n^2. \tag{12.1}$$

Note that $N\sin(\pi/N)<\pi$.

E. H. Corsey, H. Frazer and W. W. Sawyer [30] proved that in certain cases even this constant can be lessened.

13. O. Taussky [31] showed that

$$\max_{x_i}\sum_{1}^{n}\sum_{1}^{n}\frac{x_i x_j}{i+j}\Big/\sum_{1}^{n}x_i^2=\pi\{1+O(\log^{-1}n)\}.$$

14. S. B. Stečkin [32] proved that the constant in (9.2) is the best possible constant for a discrete analogue of (9.1), i.e.

$$\sum_{m=1}^{\infty}\sum_{n=1}^{\infty}\frac{a_m a_n}{(m+n)^{\lambda}}\le K(p,q)\left(\sum_{m=1}^{\infty}a_m^p\right)^{1/p}\left(\sum_{n=1}^{\infty}b_n^q\right)^{1/q}, \tag{14.1}$$

where $p>1$, $q>1$, $p^{-1}+q^{-1}\ge 1$, $\lambda=p'^{-1}+q'^{-1}$.

In fact, he proved the following more general results:

Let $w_{mn}\ge 0$ and for every $p>1$, a_m, b_n the following inequality

$$\sum_{1}^{\infty}\sum_{1}^{\infty}w_{mn}a_m b_n\le K(p)\left(\sum_{1}^{\infty}a_m^p\right)^{1/p}\left(\sum_{1}^{\infty}b_n^{p'}\right)^{1/p'} \tag{14.2}$$

be valid. Further, let $p>1$, $q>1$ and

$$p^{-1}+q^{-1}>1,\ \lambda=p'^{-1}+q'^{-1}\ (<1).$$

Then

$$\sum_{1}^{\infty}\sum_{1}^{\infty}w_{mn}^{\lambda}a_m b_n\le K^{\lambda}(\lambda q')\left(\sum_{1}^{\infty}a_m^p\right)^{1/p}\left(\sum_{1}^{\infty}b_n^q\right)^{1/q}. \tag{14.3}$$

Levin's result was also considered in F. F. Bonsall [33]. He used the following elementary application of Hölder's inequality.

Let p_i, p_i', q_i, λ be real numbers such that

$$(14.4)\qquad \begin{cases} p_i > 1,\ p_i^{-1} + (p_i')^{-1} = 1,\ (1 \le i \le n),\ \sum_1^n p_i^{-1} \ge 1, \\ \lambda = (n-1)^{-1} \sum_1^n (p_i')^{-1},\ q_i^{-1} = \lambda - (p_i')^{-1} \ge 0\ (1 \le i \le n). \end{cases}$$

Then

$$(14.5)\qquad 0 < \lambda \le 1,\ q_i^{-1} + (1-\lambda) = p_i^{-1},\ \sum_1^n q_i^{-1} + (1-\lambda) = 1,\ q_i > 1.$$

Hence, if $f_1, \ldots, f_n$, $u_1, \ldots, u_n$, w are nonnegative measurable functions and k is a positive constant such that

$$(14.6)\qquad u_1^{1/q_1} \ldots u_n^{1/q_n} w^{1-\lambda} \equiv k,$$

then

$$\int k f_1 \ldots f_n \, d\mu = \int (u_1 f_1^{p_1})^{1/q_1} \ldots (u_n f_n^{p_n})^{1/q_n} (w\ f_1^{p_1} \ldots f_n^{p_n})^{1-\lambda}\ d\mu$$

and so by Hölder's inequality

$$(14.7)\qquad \int k f_1 \ldots f_n \, d\mu \le \left(\int w f_1^{p_1} \ldots f_n^{p_n} d\mu \right)^{1-\lambda} \prod_{i=1}^n \left(\int u_i f_i^{p_i}\ d\mu \right)^{1/q_i}$$

In particular, that $n = 2$, $p_1 = p > 1$, $p_2 = q > 1$, $\lambda = (p') + (q') \le 1$. Then $q_1 = q'$ and $q_2 = p'$, so that

$$(14.7')\qquad \int k f q \, d\mu \le \left(\int w f^p g^q\ d\mu \right)^{1-\lambda} \left(\int u f^p\ d\mu \right)^{1/q'} \left(\int v g^q\ d\mu \right)^{1/p'}$$

holds for arbitrary u, v, w satisfying

$$(14.6')\qquad u^{1/q'} v^{1/p'} w^{1-\lambda} \equiv k.$$

Note, we can obtain Levin's inequality as follows:

In (14.6'), (14.7') take $k = 1$, $w \equiv 1$, $u \equiv \left(yx^{-1}\right)^{-t/p'} / (x+y)$ and $v \equiv \left(xy^{-1}\right)^{-t/p'} / (x+y)$ to obtain

$$I = \int_0^\infty \int_0^\infty \frac{f(x) g(y)}{(x+y)^\lambda}\, dx dy \le I_1^{1-\lambda} I_2^{1/q'} I_3^{1/p'},$$

where

$$I_1 = \int_0^\infty \int_0^\infty f^p(x) g^q(y)\, dx dy = \|f\|_p^p \|g\|_q^q,$$

$$I_2 = \int_0^\infty \int_0^\infty f^p(x) \left(\frac{y}{x}\right)^{-t/p'} \frac{dx dy}{x+y}$$

$$= \int_0^\infty f^p(x)\, dx \int_0^\infty \frac{u^{-t/p'}}{1+u}\, du \qquad (y = xu)$$

$$= \pi \csc\left(\frac{\pi t}{p'}\right) \|f\|_p^p, \qquad (\textit{if } 0 < t < p')$$

$$I_3 = \pi \csc\left(\frac{\pi t}{q'}\right) \|g\|_q^q, \qquad (\textit{if } 0 < t < q').$$

Hence

$$I \leq \left\{\pi \csc\left(\frac{\pi t}{p'}\right)\right\}^{1/q'} \left\{\pi \operatorname{cosec}\left(\frac{\pi t}{q'}\right)\right\}^{1/p'} \|f\|_p \|g\|_q.$$

By elementary calculus, the minimum value of this upper bound occurs for $t = \lambda^{-1}$, and this gives the constant factor preceding $\|f\|_p \|g\|_q$ the value $\left(\pi \csc \frac{\pi}{\lambda p'}\right)^{1/q'} \left(\pi \csc \frac{\pi}{\lambda q'}\right)^{1/p'} = \left(\pi \csc \frac{\pi}{\lambda \varphi'}\right)^{\lambda}$, since $\frac{1}{\lambda p'} + \frac{1}{\lambda q'} = 1$, so that $\pi - \frac{\pi}{\lambda p'} = \frac{\pi}{\lambda q'}$.

We can obtain a discrete analogue from the integral one by taking $f(x) = a_m$ for $m - 1 \leq x < m$ and $g(y) = b_n$ for $n - 1 \leq y < n$ $(m, n = 1, 2, \ldots)$, or by taking $d\mu$ to be a sum of point masses.

Taking $n = 3$ in (14.4) and proceeding in the same way as in the proof of Levin's inequality, Bonsall [**33**, Th. 8] proved that

$$\int_0^\infty \int_0^\infty \int_0^\infty \frac{f(x) g(y) h(z)}{(x+y+z)^{2\lambda}}\, dx dy dz \leq k \|f\|_{p_1} \cdot \|g\|_{p_2} \cdot \|h\|_{p_3} \tag{14.8}$$

with

$$k = \left\{\Gamma\left(\frac{1}{\lambda p_1'}\right) \Gamma\left(\frac{1}{\lambda p_2'}\right) \Gamma\left(\frac{1}{\lambda p_3'}\right)\right\}^{\lambda}.$$

Levin's inequality is the special case $K(x, y) \equiv (x+y)^{-1}$ of the following theorem.

THEOREM 1. (Bonsall [**33**, Th. 2]). *Let $p > 1$, $q > 1$, $\lambda = (p')^{-1} + (q')^{-1}$, let f, g be nonnegative and measurable on $[0, \infty)$, and let $K(x, y)$ be nonnegative and homogeneous of degree -1, with*

$$\int_0^\infty u^{-1/\lambda p'} K(1, u)\, du = k.$$

Then

$$\int_0^\infty \int_0^\infty f(x) g(y) K^\lambda(x, y)\, dx dy \le k^\lambda \|f\|_p \cdot \|g\|_q. \tag{14.9}$$

PROOF: In (14.6′), (14.7′) we take $k = 1$, $w \equiv 1$, $u \equiv \left(yx^{-1}\right)^{-t/p'} K(x, y)$ and $v \equiv \left(xy^{-1}\right)^{-t/q'} K(x, y)$, giving as in the proof of Levin's inequality

$$I = \int_0^\infty \int_0^\infty f(x) g(y) K^\lambda(x, y)\, dx dy \le I_1^{1-\lambda} I_2^{1/q'} I_3^{1/p'},$$

where

$$\begin{aligned} I_1 &= \|f\|_p^p \|g\|_q^q, \\ I_2 &= \int_0^\infty \int_0^\infty f^p(x) \left(\frac{y}{x}\right)^{-t/p'} K(x, y)\, dx dy \\ &= \|f\|_p^p \int_0^\infty u^{-t/p'} K(1, u)\, du, \\ I_3 &= \|g\|_q^q \int_0^\infty v^{-t/q'} K(v, 1)\, dv. \end{aligned}$$

Because K is homogeneous of degree -1, the change of variable $v = u^{-1}$ gives

$$I_3 = \|g\|_q^q \int_0^\infty u^{\frac{t}{q'} - 1} K(1, u) du.$$

The choice $t = \lambda^{-1}$ now gives (14.9) since $(\lambda p')^{-1} + (\lambda q')^{-1} = 1$.

Theorem 1 is the generalization to the nonconjugate case of a theorem due to I. Schur [**2**] in the case $p = q = 2$, and in the general conjugate case to Hardy, Littlewood, and Pólya [**4**, Th. 319].

REMARK 1: In the conjugate case ($q = p'$, $\lambda = 1$), it follows from Hölder's inequality and its converse that the inequality

$$\int_0^\infty \int_0^\infty f(x)g(y)K(x,y)\,dxdy \le k\|f\|_p \cdot \|g\|_q \tag{14.9'}$$

is equivalent to either of the inequalities

$$\left\{\int_0^\infty \left(\int_0^\infty K(x,y)f(x)\,dx\right)^p dy\right\}^{1/p} \le k\|f\|_p, \tag{14.10}$$

$$\left\{\int_0^\infty \left(\int_0^\infty K(x,y)g(y)dy\right)^q dx\right\}^{1/q} \le k\|g\|_q. \tag{14.11}$$

In this case, the constant k is best possible. Additional applications of (14.9′), (14.10), (14.11) are given in Hardy, Littlewood and Pólya [**4**, §9.9].

REMARK 2: Other inequalities giving upper bounds for the integrals

$$\int_0^\infty \int_0^\infty f(x)g(y)(x+y)^\lambda\,dxdy, \qquad \int_0^\infty \int_0^\infty f(x)g(y)|x-y|^{-\lambda}\,dxdy$$

are also given by Bonsall [**33**, Ths. 3, 4, 5].

REMARK 3: Bonsall also gave the following geralization of the Levin's result from 8. With the same conditions for p, q and λ,

$$\sum_{m=1}^\infty \sum_{n=1}^\infty \frac{a_m b_n}{(m+n-1)^\lambda}\left(1 + \frac{2}{4!}\,\frac{\lambda(\lambda+1)}{(m+n-1)^2} + \frac{2}{6!}\,\frac{\lambda(\lambda+1)(\lambda+2)(\lambda+3)}{(m+n-1)^4} + \dots\right)$$
$$\le \left(\pi \operatorname{cosec} \frac{\pi}{\lambda p'}\right)^\lambda \left(\sum_{m=1}^\infty a_m^p\right)^{1/p} \left(\sum_{m=1}^\infty b_n^q\right)^{1/q}.$$

15. In 1941, A. Beurling gave a proof (unpublished) of the following result:

If $a_n \geq 0$, $b_n \geq 0$ *for* $n = 1, 2, \ldots$, *and* $\sum_{m=1}^{\infty} m a_m^2 = A^2 < \infty$, $\sum_{n=1}^{\infty} n b_n^2 = B^2 < \infty$, *then*

$$\sum_{m=1}^{\infty} \sum_{n=1}^{\infty} a_m b_n / \log(m+n) \leq KAB \tag{15.1}$$

with $0 < K < 4e$.

Beurling also suggested the following generalization which was proved by A. E. Livingston [**34**]:

Let $h \colon R \to [0, +\infty)$ be nonnegative,, nondecreasing and continuous from the right. Let, furthermore $f, g \colon R \to [0, +\infty)$ be nonnegative, and let $1 < p < +\infty$, and $(1/p) + (1/q) = 1$. If

$$\int_0^\infty f(x)^p dh(x) = F^p < +\infty, \qquad \int_0^\infty g(y)^q dh(y) = G^q < +\infty,$$

then

$$\int_0^\infty \int_0^\infty \frac{f(x) g(y)}{h(x+y)} dh(x) dh(y) < KFG, \tag{15.2}$$

with $0 < K(h) < p + q$.

In particular, if $h(x+y) \geq h(x) + h(y)$, then $K(h) \leq \pi / (\sin(\pi/p))$, and if h is continuous, $h(0) = 0$, $\lim_{x \to +\infty} h(x) = +\infty$, and $h(x+y) \leq h(x) + h(y)$, then $K(h) \geq \pi / (\sin(\pi/p))$.

16. F. C. Hsiang [**35**] proved that, if $a_0, a_1, \ldots, a_N$ and $b_0, b_1, \ldots, b_N$ are sequences of nonnegative real numbers,

$$\sum_{m=0}^{N} \sum_{n=0}^{N} \frac{a_m b_n}{2m + 2n + 1} \leq (N+1) \sin \frac{\pi}{2(N+1)} \left(\sum_{m=1}^{N} a_m^2 \right)^{1/2} \left(\sum_{n=1}^{N} b_n^2 \right)^{1/2}.$$

Related to this inequality, see paper [**36**] of Q. A. M. M. Yahya.

17. In connection to Hilbert's inequality are also the papers [**37**] of W. W. Sawyer, and [**38**] of V. P. Il'in.

18. N. G. De Bruijn and H. S. Wilf [**39**] showed that the best possible constant C_N in

$$\sum_{n=1}^{N} \sum_{m=1}^{N} \frac{a_m a_n}{m+n} \leq C_N \sum_{n=1}^{N} a_n^2$$

is given by

$$C_N = \pi - \frac{1}{2}\pi^5 (\log N)^{-2} + O\left(\log\log N(\log N)^{-3}\right) \quad (N \to +\infty).$$

This is an extension of the result given in **[30]**, **[31]** and **[40]** (R. A. Fairthorne and J. C. P. Miller).

19. E. K. Godunova **[41]** proved the following generalization of Theorem 1 from 14.:

Let $p > 1$, $q > 1$, $p^{-1} + q^{-1} = 2 - \lambda > 1$, $p^{-1} + p'^{-1} = 1$, $q^{-1} + q'^{-1} = 1$; $u(t_1, t_2, \ldots, t_n) \in L_{1/\lambda}(V_t)$, $f(x_1, \ldots, x_n) \in L_p(V_x)$, $g(y_1, \ldots, y_n) \in L_q(V_y)$, where $V_x = \{(x_1, \ldots, x_n) \,|\, 0 < x_i < +\infty, \ i = 1, \ldots, n\}$; V_t and V_y are defined similarly. Then

$$\begin{aligned} \int_{V_x}\int_{V_y} u\left(\frac{y_1}{x_1}, \ldots, \frac{y_n}{x_n}\right)\left(\frac{y_1 \cdots y_n}{x_1 \cdots x_n}\right)^{\frac{1}{p-1}} (x_1, \ldots, x_n)^{-\lambda} \\ \times f(x_1, \ldots, x_n) g(y_1, \ldots, y_n) dV_x dV_y \\ < \|u\|_{1/\lambda} \|f\|_p \|g\|_q. \end{aligned}$$

20. Further related papers are R. Redheffer **[42]**, V. P. Il'in **[43]**, H. Davenport and H. Halberstam **[44]**, K. R. Matthews **[45]**, F. A. Sysoeva **[46]**,**[47]**,**[48]**, P. L. Walker **[49]**, A. Erdélyi **[50]**, F. Fehér **[51]** and A. Kufner **[52]**. For example, Walker **[49]** proved that the constant in (9.2) is asymptotically best possible.

21. The following result was proved by P. M. Vasić and J. E. Pečarić **[53]**:

Let $f\colon [k_1, k_2] \times [\ell_1, \ell_2] \to R_+$ be a real function, and let $a = (a_1, a_2, \ldots)$, $b = (b_1, b_2, \ldots)$ and $c = (c_1, c_2, \ldots)$ be three positive sequences. Suppose that for every $n \geq 1$, $q^{(n)} = \left(q_1^{(n)}, \ldots, q_n^{(n)}\right)$ is a positive n-tuple such that $\sum_{k=1}^{n} q_k^{(n)} = 1$.

(a) If $H(s,t) = f\left(K^{-1}(s), L^{-1}(t)\right)$ is a convex function and if d_k are positive numbers such that

$$\sum_{n=k}^{\infty} c_n q_k^{(n)} \leq d_k \qquad (k \geq 1) \tag{21.1}$$

then

$$\sum_{n=1}^{\infty} c_n (f\left(K_n\left(a, q^{(n)}\right), L_n\left(b; q^{(n)}\right)\right) \leq \sum_{n=1}^{\infty} d_n f(a_n, b_n), \tag{21.2}$$

where $K_n(a,w)$ and $L_n(b,w)$ are quasiarithmetic mean of a with weights w with respect to the functions K and L respectively.

(b) If $H(s,t)$ is concave and if in (21.1) the reverse inequality holds, then (21.2) is reversed.

22. Let S be a complex inner product space with inner product $\langle\cdot,\cdot\rangle$ and let λ be a real, nonnegative constant. If $x_n, y_n \in S$ for $m,n \in Z$ with $X^2 = \sum_{m=-\infty}^{\infty} \|x_m\|^2 < \infty$, $Y^2 = \sum_{n=-\infty}^{\infty} \|y_n\| < \infty$, then

$$\left|\sum_{m,n=-\infty}^{\infty} \frac{\langle x_m, y_n\rangle}{m-n+\lambda}\right| \le \frac{\pi XY}{|\sin \pi\lambda|}, \tag{22.1}$$

and equality holds if and only if either $X = 0$ or for some complex constant δ, $y_n = \delta \sum_{m=-\infty}^{\infty} x_m/(m-n+\lambda)$ for all $n \in Z$, where $X < \infty$.

This is the main result of the paper [**54**] of R. M. Redheffer and P. Volkmann, and is an extension of Schur's inquality [**2**], which is just (22.1) with $S = C$ and $\langle x_m, y_n\rangle = x_n\bar{y}_n$.

23. The Hilbert-Riesz inequality (9.1) can be given in equivalent form (see [4, Th. 382])

$$\int_{-\infty}^{\infty}\int_{-\infty}^{\infty} \frac{f(x)g(y)}{|x-y|^{\lambda}}\, dxdy \le K(p,q)\|f\|_p\|g\|_q. \tag{23.1}$$

(A discrete analogue is given as Theorem 381 in [**4**].)

In 1983 E. H. Lieb [**55**] obtained the best value for $K(p,q)$ in the inequality (23.1) when $p = q = 2/(2-\lambda)$. In this case

$$K(p,q) = \pi^{\lambda-1/2}\frac{\Gamma\left(\frac{1-\lambda}{2}\right)}{\Gamma\left(1-\frac{\lambda}{2}\right)}.$$

An application of this result is given by M. W. Wong in [**56**].

24. D. S. Mitrinović and J. E. Pečarić [**57**] gave several remarks about results from [**15**],[**16**] and [**28**].

1^0 First they proved that (11.1) can be proved by using Hilbert's inequality

(1.3). Indeed, we have

$$A(x) = \int_0^\infty e^{-t} A^*(xt)\, dt = \frac{1}{x} \int_0^\infty e^{-u/x} A^*(u)\, du,$$

$$B(x) = \frac{1}{x} \int_0^\infty e^{-u/x} B^*(u)\, du,$$

and so

$$\int_0^1 A(x)B(x)\, dx = \int_0^1 \frac{dx}{x^2} \int_0^\infty e^{-u/x} A^*(u)\, du \int_0^\infty e^{-u/x} B^*(u)\, du.$$

Putting $1/x = t + 1$, we have

$$\int_0^1 A(x)B(x)\, dx = \int_0^\infty dt \int_0^\infty e^{-tu} a(u)\, du \int_0^\infty e^{-tv} b(v)\, dv$$

$$= \int_0^\infty \int_0^\infty \frac{a(u)b(v)}{u+v}\, du dv$$

$$\leq \left(\frac{\pi}{\sin\left(\frac{\pi}{p}\right)}\right) \left(\int_0^\infty a^p(x)\, dx\right)^{1/p} \left(\int_0^\infty b^{p'}(x)\, dx\right)^{1/p'}$$

2^0 The following result is a simple consequence of (3.3):

THEOREM 1. *Let a and b be defined as in 11.*
Then

$$\sum\sum \frac{a_m b_n}{m+n+1} \leq \left(\frac{\pi}{\sin\left(\frac{\pi}{p}\right)}\right) \left(\int_0^\infty z^{p-2} a^p(z)\, dx\right)^{1/p} \times \left(\int_0^\infty z^{p'-2} b^{p'}(z)\, dz\right)^{1/p'}. \tag{24.1}$$

PROOF: We have

$$\int_0^1 A(z)B(z)\,dz = \left(\int_0^1 z^{-2}\, zA(z)\right)(zB(z))\,dz$$
$$\leq \left(\int_0^1 z^{-2}(zA(z))^p\,dz\right)^{1/p}\left(\int_0^1 z^{-2}(zB(z))^{p'}\,dz\right)^{1/p'}$$
$$= \left(\int_0^1 z^{p-2}A^p(z)\,dz\right)^{1/p}\left(\int_0^1 z^{p'-2}B^{p'}(z)\,dz\right)^{1/p'}$$
$$\leq \Gamma\left(\frac{1}{p}\right)\Gamma\left(\frac{1}{p'}\right)\left(\int_0^\infty z^{p-2}a^p(z)\,dz\right)^{1/p}$$
$$\times\left(\int_0^\infty z^{p'-2}b^{p'}(z)\,dz\right)^{1/p'},$$

which is equivalent to (24.1) (note that $\Gamma(1/p)\Gamma(1/p') = \pi/\sin(\pi/p)$).

3^0 The following resut is a generalization of Theorem 4 from [**16**]: (i.e a generalization of (3.6)):

THEOREM 2. *Suppose that* $K_0(x) \geq 0$, *that*

$$K_1(x,y) = \int_0^\infty K_0(xt)K_0(yt)\,dt$$

(so that K_1 *is symmetric and homogeneous of degree* -1*), and that*

$$K_2(x,y) = \int_0^\infty K_1(x,t)K_1(y,t)\,dt.$$

Then, for positive sequences a *and* b,

(24.2)
$$\left(\sum_1^\infty\sum_1^\infty K_2(m,n)a_mb_n\right)^2$$
$$\leq J^2\left(\sum_1^\infty\sum_1^\infty K_1(m,n)a_ma_n\right)\left(\sum_1^\infty\sum_1^\infty K_1(m,n)b_mb_n\right),$$

where

$$J = \int_0^\infty \frac{K_1(w,1)}{\sqrt{w}}\,dw.$$

PROOF: We have, in fact, putting $t = mw$ and using the symmetry and homgeneity of K_1,

$$\begin{aligned}\sum_1^\infty\sum_1^\infty K_2(m,n)a_m b_n &= \sum_1^\infty\sum_1^\infty a_m b_n \int_0^\infty K_1(m,t)K_1(n,t)dt\\ &= \sum_1^\infty\sum_1^\infty a_m b_n \int_0^\infty K_1(w,1)K_1(mw,n)\,dw\\ &= \int_0^\infty K_1(w,1)\Omega(w)\,dw,\end{aligned}$$

where $\Omega(w) = \sum_1^\infty\sum_1^\infty K_1(mw,n)a_m b_n$.

Further, we have

$$\Omega(w) = \sum_1^\infty\sum_1^\infty a_m b_n \int_0^\infty K_0(mwt)K_0(nt)\,dt = \int_0^\infty \phi_a(wt)\phi_b(t)\,dt,$$

where $\phi_a(t) = \sum_1^\infty a_n K_0(nt)$, $\phi_b(t) = \sum_1^\infty b_n K_0(nt)$, and this does not exceed

$$\begin{aligned}&\left(\int_0^\infty \phi_a(wt)^2\,dt \int_0^\infty \phi_b(t)^2\,dt\right)^{1/2}\\ &= \frac{1}{\sqrt{w}}\left(\int_0^\infty \phi_a(t)^2\,dt \int_0^\infty \phi_b(t)^2\,dt\right)^{1/2}\\ &= \frac{1}{\sqrt{w}}\left(\sum_1^\infty\sum_1^\infty K_1(m,n)a_m a_n\right)^{1/2}\left(\sum_1^\infty\sum_1^\infty K_1(m,n)b_m b_n\right)^{1/2}.\end{aligned}$$

By combining these results we obtain (24.2).

As in [**16**] we can give the following particular cases:

$$\sum_1^\infty\sum_1^\infty \frac{\log\left(\frac{m}{n}\right)}{m-n} a_m b_n \le \pi \left(\sum_1^\infty\sum_1^\infty \frac{a_m a_n}{m+n}\right)^{1/2} \left(\sum_1^\infty\sum_1^\infty \frac{b_m b_n}{m+n}\right)^{1/2},$$

where the coefficient on the left-hand side is to be interpreted as $1/n$ when $m = n$;

$$\sum_1^\infty\sum_1^\infty \frac{|\log\left(\frac{m}{n}\right)|}{\max(m,n)} a_m a_n \le 2 \left(\sum_1^\infty\sum_1^\infty \frac{a_m a_n}{\max(m,n)}\right)^{1/2} \left(\sum_1^\infty\sum_1^\infty \frac{b_m b_n}{\max(m,n)}\right)^{1/2}.$$

25. Extensions of results with nonconjugate parameters are given by D. S. Mitrinović and J. E. Pečarić [**58**].

If

$$p > 1, \quad q > 1, \quad \frac{1}{p} + \frac{1}{q} \ge 1, \tag{25.1}$$

and

$$\lambda > 0, \quad \frac{1}{p} + \frac{1}{q} > \lambda \ge \frac{1}{p'} + \frac{1}{q'}, \tag{25.2}$$

are valid, then

$$\sum_{m=1}^\infty\sum_{n=1}^\infty \frac{x_m y_n}{(m+n)^\lambda} \ge K(p,q,\lambda) \left(\sum_{m=1}^\infty x_m^p\right)^{1/p} \left(\sum_{n=1}^\infty y_n^q\right)^{1/q}, \tag{25.3}$$

where

$$K(p,q,\lambda) = \left(\pi \operatorname{cosec}\left(\frac{\pi}{2\lambda}\left(\frac{1}{p} - \frac{1}{q} + \lambda\right)\right)\right)^\lambda. \tag{25.4}$$

This follows from (1.2) and from the following general results:

THEOREM 1. *Let $a_{mn} \ge 0$ and for every $p > 1$ and nonnegative x_m, y_n the following inequality is valid*

$$\sum_1^\infty\sum_1^\infty a_{mn} x_m y_n \le K(p) \left(\sum_1^\infty x_m^p\right)^{1/p} \left(\sum_1^\infty y_n^{p'}\right)^{1/p'}. \tag{25.5}$$

If (25.1) and (25.2) are valid, then

(25.6)

$$\sum_1^\infty\sum_1^\infty a_{mn}^\lambda x_m y_n \le K^\lambda \left(\left(\frac{1}{2\lambda}\left(\frac{1}{p} - \frac{1}{q} + \lambda\right)\right)^{-1}\right) \left(\sum_1^\infty x_m^p\right)^{1/p} \left(\sum_1^\infty y_n^q\right)^{1/q}.$$

First, we shall prove the following

LEMMA. *Let*

$$a_{mn} \geq 0,\ b_{mn} \geq 0,\ u_1 > 0,\ v_1 > 0,\ u_2 > 0,\ v_2 > 0,$$

and let for every nonnegative x_m, y_n the following inequalities be valid:

$$\sum_1^\infty \sum_1^\infty a_{mn} x_m y_n \leq K_1 \left(\sum_1^\infty x_m^{1/u_1}\right)^{u_1} \left(\sum_1^\infty y_n^{1/v_1}\right)^{v_1}, \tag{25.7}$$

$$\sum_1^\infty \sum_1^\infty b_{mn} x_m y_n \leq K_2 \left(\sum_1^\infty x_m^{1/u_2}\right)^{u_2} \left(\sum_1^\infty y_n^{1/v_2}\right)^{v_2}, \tag{25.8}$$

Further, let

$$k_1 > 0,\ k_2 > 0,\ k_1 + k_2 \geq 1,\ u = k_1 u_1 + k_2 u_2,\ v = k_1 v_1 + k_2 v_2.$$

Then

$$\sum_1^\infty \sum_1^\infty a_{mn}^{k_1} b_{mn}^{k_2} x_m y_n \leq K_1^{k_1} K_2^{k_2} \left(\sum_1^\infty x_m^{1/u}\right)^{u} \left(\sum_1^\infty y_n^{1/v}\right)^{v}. \tag{25.9}$$

PROOF: Using a generalization of Hölder's inequality of J. L. W. V. Jensen (see for example [**AI**, pp. 52 and 78]) we get

$$\begin{aligned}\sum_1^\infty \sum_1^\infty a_{mn}^{k_1} b_{mn}^{k_2} x_m y_n &= \sum_1^\infty \sum_1^\infty \left(a_{mn} x_m^{u_1/u} y_n^{v_1/v}\right)^{k_1} \left(b_{mn} x_m^{u_2/u} y_n^{v_2/v}\right)^{k_2} \\ &\leq \left(\sum_1^\infty \sum_1^\infty a_{mn} x_m^{u_1/u} y_n^{v_1/v}\right)^{k_1} \left(\sum_1^\infty \sum_1^\infty b_{mn} x_m^{u_2/u} y_n^{v_2/v}\right)^{k_2}.\end{aligned} \tag{25.10}$$

As consequences of (25.7) and (25.8) we have

$$\begin{aligned}&\sum_1^\infty \sum_1^\infty a_{mn} x_m^{u_1/u} y_n^{v_1/v} \leq K_1 \left(\sum_1^\infty x_m^{1/u}\right)^{u_1} \left(\sum_1^\infty y_n^{1/v}\right)^{v_1}, \\ &\sum_1^\infty \sum_1^\infty b_{mn} x_m^{u_2/u} y_n^{v_2/v} \leq K_2 \left(\sum_1^\infty x_m^{1/u}\right)^{u_2} \left(\sum_1^\infty y_n^{1/v}\right)^{v_2}.\end{aligned} \tag{25.11}$$

So, (25.10) and (25.11) give

$$\begin{aligned}\sum_1^\infty\sum_1^\infty a_{mn}^{k_1}b_{mn}^{k_2}x_my_n &\le \left(K_1\left(\sum_1^\infty x_m^{1/u}\right)^{u_1}\left(\sum_1^\infty y_n^{1/v}\right)^{v_1}\right)^{k_1}\\ &\quad\times\left(K_2\left(\sum_1^\infty x_m^{1/u}\right)^{u_2}\left(\sum_1^\infty y_n^{1/v}\right)^{v_2}\right)^{k_2}\\ &= K_1^{k_1}K_2^{k_2}\left(\sum_1^\infty x_m^{1/u}\right)^{k_1u_1+k_2u_2}\left(\sum_1^\infty y_n^{1/v}\right)^{k_1v_1+k_2v_2}\\ &= K_1^{k_1}K_2^{k_2}\left(\sum_1^\infty x_m^{1/u}\right)^{u}\left(\sum_1^\infty y_n^{1/v}\right)^{v}.\end{aligned}$$

PROOF OF THEOREM 1: Theorem 1 follows from the Lemma if we set

$$\begin{gathered}b_{mn}=1,\ u_1=1/p_1,\ v_1=1/p_1',\ u_2=v_2=1,\ k_1=\lambda>0,\\ u=1/p,\ v=1/q.\end{gathered}$$

Then we have
$$\frac{1}{p}=\frac{\lambda}{p_1}+k_2,\qquad \frac{1}{q}=\frac{\lambda}{p_1'}+k_2,$$

wherefrom we get

$$k_2=\frac{1}{2}\left(\frac{1}{p}+\frac{1}{q}-\lambda\right)\quad\text{and}\quad \frac{1}{p_1}=\frac{1}{2\lambda}\left(\frac{1}{p}-\frac{1}{q}+\lambda\right).$$

The conditions of the Lemma are fulfilled since we have

$$k_2>0,\ k_1+k_2=\frac{1}{2}\left(\frac{1}{p}+\frac{1}{q}+\lambda\right)\ge\frac{1}{2}\left(\frac{1}{p}+\frac{1}{q}+\frac{1}{p'}+\frac{1}{q'}\right)=1.$$

We also have $0<1/p_1<1$, i.e. $0<\frac{1}{p}-\frac{1}{q}+\lambda<2\lambda$. Indeed, the first inequality is valid since $\lambda\ge 2-\frac{1}{p}-\frac{1}{q}>\frac{1}{q}-\frac{1}{p}$, i.e. $1>1/q$. Similarly we can obtain the second inequality.

In the case when

(25.12) $$0<\lambda=2-p^{-1}-q^{-1}=p'^{-1}+q'^{-1}<1$$

is valid, we have a result from 14. But, in this case we can also obtain an integral analogue, i.e. the following theorem is valid:

THEOREM 2. *Let $k(x,y) \geq 0$ and for every $p > 1$ and nonnegative $f(x)$, $g(y)$ the following inequality is valid*

$$\int_0^\infty \int_0^\infty k(x,y)f(x)g(y)\,dx\,dy \leq K(p)\left(\int_0^\infty f^p(x)\,dx\right)^{1/p}\left(\int_0^\infty g^{p'}(y)\,dy\right)^{1/p'}.$$

If (25.11) and (25.12) are valid, then

$$\int_0^\infty \int_0^\infty k^\lambda(x,y)f(x)g(y)\,dx\,dy$$
$$\leq K^\lambda(\lambda q')\left(\int_0^\infty f^p(x)\,dx\right)^{1/p}\left(\int_0^\infty g^q(y)\,dy\right)^{1/q}.$$

Since (1.3) is valid, Theorem 2 gives the integral inequality of V. Levin from [**24**] (see 9.).

Using Theorem 3 from 1. we get that the factor $(m+n)^{-\lambda}$ in (1.4) can be replaced by $(m+n-1)^{-\lambda}$. Similarly we can also given generalizations of Theorems 341 and 342 from [**1**] (see (i) and (ii) from 2.).

Of course, we can similarly obtain multidimentional generalizations of Theorems 1 and 2.

THEOREM 3. *Let $a_{mn\ldots s} \geq 0$ and for every $p_1 > 1, \ldots, p_r > 1$, such that $\sum_{i=1}^{r} p_i^{-1} = 1$ and nonnegative $x_m, y_n, \ldots, z_s$ the following inequality is valid*

$$\sum a_{mn\ldots s}x_m y_n \cdots z_s$$
$$\leq K(p_1,\ldots,p_r)\left(\sum_m x_m^{p_1}\right)^{1/p_1}\left(\sum_n y_n^{p_2}\right)^{1/p_2}\cdots\left(\sum_s z_s^{p_r}\right)^{1/p_r}.$$

If $q_i > 1$ $(i = 1,\ldots,r)$, $\lambda > 0$ and

$$(n-1)\lambda \geq \sum_{i=1}^{r}\frac{1}{q_i'}, \qquad \sum_{i=1}^{r}\frac{1}{q_i} \geq \lambda,$$

then

$$\sum a^\lambda_{mn\ldots s}x_m y_n \cdots z_s$$
$$\leq K(\underline{p}_1,\ldots,\underline{p}_r)\left(\sum_m x_m^{q_1}\right)^{1/q_1}\left(\sum_n y_n^{q_2}\right)^{1/q_2}\cdots\left(\sum_s z_s^{q_r}\right)^{1/q_r},$$

where $\frac{1}{p_i} = \frac{1}{n\lambda}\left(\frac{n}{q_i} - \sum_{j=1}^{r}(1/q_j) + \lambda\right)$, $i = 1, \ldots, r$.

THEOREM 4. *Let $k(x, y, \ldots, z) \geq 0$ and for every $p_1 > 1, \ldots, p_r > 1$, such that $\sum_{i=1}^{r}(1/p_i) = 1$, and nonnegative $f(x)$, $g(y)$, ..., $h(z)$ the following inequality is valid*

$$int \int \cdots \int k(x, y, \ldots, z)\, dx dy \ldots dz$$
$$\leq K(p_1, \ldots, p_r)\left(\int f^{p_1}(x)\, dx\right)^{1/p_1}\left(\int g^{p_2}(y)\, dy\right)^{1/p_2}$$
$$\ldots\left(\int h^{p_r}(z)\, dz\right)^{1/p_r}.$$

If $q_i > 1$, $i = 1, \ldots, r$, $0 < \lambda < 1$, with

$$\sum_{i=1}^{r}\left(\frac{1}{q_i}\right) \geq 1, \qquad (n-1)\lambda = \sum_{i=1}^{r}\left(\frac{1}{q_i'}\right),$$

then

$$\int\int \cdots \int k^{\lambda}(x, y, \ldots, z)\, dx dy \ldots dz$$
$$\leq K(\bar{p}_1, \ldots, \bar{p}_r)^{\lambda}\left(\int f^{q_1}(x)\, dx\right)^{1/q_1}\left(\int g^{q_2}(y)\, dy\right)^{1/q_2}$$
$$\ldots\left(\int h^{q_r}(z)\, dz\right)^{1/q_r},$$

where $\bar{p}_i = \lambda(\lambda - 1 + 1/q_i)^{-1}$.

REFERENCES

1. WEYL, H., *Singuläre Integralgleichungen mit besonderer Berücksichtingung des Fourierschen Integraltheorems*, Inaugural-Dissertation, Göttingen, 1908.
2. SCHUR, I., *Bemerkungen zur Theorie der beschränkten Bilinearformen mit unendlich vielen Veränderlichen*, Journal f. Math. **140** (1911), 1–28.

3. HARDY, G. H., *Note on a theorem of Hilbert concerning series of positive terms*, Proc. London Math. Soc. **23** 2 (1925), Records of Proc. XLV–XLVI.
4. HARDY, G. H., J. E. LITTLEWOOD and G. PÓLYA, "Inequalities," Cambridge, 1934.
5. FEJÉR, L. and F. RIESZ, *Über einige funktionentheoretische Ungleichungen*, Math. Zeitschr. **11** (1921), 305–314.
6. FRANCIS, E. C. and J. E. LITTLEWOOD, "Examples in infinite series with solutions," Cambridge, 1928.
7. HARDY, G. H., *Note on a theorem of Hilbert*, Math. Zeitschr. **6** (1920), 314–317.
8. HARDY, G. H., J. E. LITTLEWOOD and G. PÓLYA, *The maximum of certain bilinear form*, Proc. London Math. Soc. **25** 2 (1926), 265–282.
9. MULHOLLAND, H. P., *Note on Hilbert's double series theorem*, J. London Math. Soc. **3** (1928), 197–199.
10. ―――――, *A further generalisation of Hilbert's double series theorem*, J. London Math. Soc. **6** (1931), 100–106.
11. OWEN, P. M., *A generalisation of Hilbert's double series theorem*, J. London Math. Soc. **5** (1930), 270- -272.
12. PÓLYA, G. and SZEGÖ, "Aufgaben und Lehrsätze aus der Analysis, I, II" Berlin, 1925.
13. WIENER, F., *Elementarer Beweis eines Reihensatzes von Herrn Hilbert*, Math. Annalen **68** (1910), 361–366.
14. RIESZ, F., *Über die Randwerte einer analytischen Funktion*, Math. Zeitschr. **18** (1923), 87–95.
15. WIDDER, D. V., *An inequality related to one of Hilbert's*, J. London Math. Soc. **4** (1929), 194–198.
16. HARDY, G. H., *Remarks in addition to Dr. Widder's note on inequalities*, J. London Math. Soc. **4** (1929), 199-202.
17. MULHOLLAND, H. P., *Some theorems on Dirichlet series with positive coefficients and related integrals*, Proc. London Math. Soc. **29** 2 (1929), 281–292.
18. ―――――, *The generalizations of certain inequality theorems involving powers*, Proc. London Math. Soc. **33** 2 (1932), 481–516.
19. HARDY, G. H. and J. E. LITTLEWOOD, *Elementary theorems concerning power series with positive coefficients and moment constants of positive functions*, Journal f. Math. **157** (1927), 141–158.
20. SZEGÖ, G., *On some Hermitian forms associated with two given curves of the complex plane*, Trans. Amer. Math. Soc. **40** (1936), 450–461.
21. FRAZER, H., *On the moduli of regular functions*, Proc. London Math. Soc., **36** (1934), 532–546.

22. INGHAM, A. E., *A note on Hilber's inequality*, J. London Math. Soc. **11** (1936), 237–240.
23. LEVIN, V., *Two remarks on Hilbert's double series theorem*, J. Indian Math. Soc. **11** (1937), 111–115.
24. _________, *On the two-parameter extension and analogue of Hilbert's inequality*, J. London Math. Soc. **11** (1936), 119–124.
25. _________, *Notes on inequalities IV. On the Hilbert-Riesz inequality*, Bull. Acad. Sci. URSS, Sér. Math. Nr **5/6** (1938), 525–540 (Russian).
26. PITT, H. R., *On an inequality of Hardy and Littlewood*, J. London Math. Soc. **13** (1938), 95–101.
27. HARDY, G. H. and J. E. LITTLEWOOD, *An inequality*, Math. Zeit. **40** (1935), 1–40.
28. CHOW, Y. C., *On inequaities of Hilbert and Widder*, J. London Math. Soc. **14** (1939), 151–154.
29. FRAZER, H., *Note on Hilbert's inequality*, J. London Math. Soc. **14** (1946), 7–9.
30. CORSEY, E. H., H. FRAZER and W. W. SAWYER, *Empirical data on Hilbert's inequality*, Nature, London **161** (1948), 361.
31. TAUSSKY, O., *A remark concerning the characteristic roots of the finite segments of the Hilbert matrix*, Quart. J. Math. (Oxford) **20** (1949), 80–83.
32. STEČKIN, S. B., *On positive bilinear forms*, Dokl. Akad. Nauk SSSR (Russian) **65** (1949), 17–20.
33. BONSALL, F. F., *Inequalities with non-conjugate parameters*, Quart. J. Math. Oxford **2** 2 (1951), 135–150.
34. LIVINGSTON, A. E., *A genralization of an inequality due to Beurling*, Pacific J. Math. **4** (1954), 251- -257.
35. HSIANG, F. C., *An inequality for finite sequences*, Math. Scandinav. **5** (1957), 12–14.
36. YAHYA, Q. A. M. M., *On the generalization of Hilbert's inequality*, Amer. Math. Monthly **72** (1965), 518–520.
37. SAWYER, W. W., *On determinants associated with Hilbert's inequality*, J. London Math. Soc. **32** (1957), 133- -138.
38. IL'IN, V. P., *Nekotorye integral'nye neravenstva i ih primenenija v teorii differenciruemyh funkciĭ mnogih peremennyh*, Mat. Sb. No. 3 **54** 96 (1961), 331–380.
39. DE BRUIJN, N. G. and H. S. WILF, *On Hilbert's inequality in n dimensions*, Bull. Amer. Math. Soc. **68** (1962), 70–73.

40. FAIRTHORNE, R. A. and J. C. P. MILLER, *Hilbert's double series theorem and principal latent roots of the resulting matrix*, Math. Comput. **3** 26 (1949), 399–400.
41. GODUNOVA, E. K., *Obobščenie dvuparemetričeskogo neravenstva Gil'belta*, Izv. Vysš. Uč. Zaved. Matematika No. 1 **56** (1967), 35–39.
42. REDHEFFER, R., *An inequality for the Hilbert transform*, Proc. Nat. Acad. Sci. **61** (1968), 810–811.
43. IL'IN, V. P., *Über eine Inegralgleichung*, Mat. Zametki **6** (1969), 139–148 (Russisch).
44. DAVENPORT, H. and H. HALBERSTAM, Matematica **13** (1966), 91–96.
45. MATTHEWS, K. R., *On an inequality of Davenport and Halberstam*, J. London Math. Soc. **4** 2 (1972), 638–642.
46. SYSOEVA, F. A., *Neravenstva Gil'beroskogo tipa dlja funkciĭ s ograničennym smešannymi normami*, Moskov. Gos. Ped. Inst. Učen. Zap. **460** (1972), 43–51.
47. __________, *An inequality of Hardy-Littlewood type for a product of functions* (Russian), in "Differential equations and inequalities," Moskov. Gos. Ped. Inst. im. Lenina, Moskow, 1973, pp. 43–49.
48. __________, *Odno neravenstvo tipa Hardi-Litlevuda*, Mosk. Gos. Ped. Institut im. Lenina **6** (1978), 136–141.
49. WALKER, P. L., *A note on an inequality with non-conjugate parameters*, Proc. Edinburgh Math. Soc. **18**, ser. II (1973), 293–294.
50. ERDÉLYI, A., *An extension of a Hardy-Littlewood-Pólya inequality*, Proc. Edinburgh Math. Soc. **21** (1978), 11–15.
51. FEHÉR, F., *A Generalized Schur-Hardy inequality on normed Köthe spaces*, in "General Inequalities," **2**, ed. E. F. Beckenbahc, Basel, 1979.
52. KUFNER, A., *Zur Hardyschen Ungleichung*, in "Sem. Analysis 1982/83," IMath, Berlin, 1983, pp. 1–18.
53. VASIĆ, P. M. and J. E. PEČARIĆ, *Notes on some inequalities for convex functions*, Mat. Vesnik **6**, 19 34 (1982), 195–193.
54. REDHEFFER, R. M. and P. VOLKMANN, *Über einer Ungleichung von Schur*, Mh. Math. **95** (1983), 137–148.
55. LIEB, E. H., *Sharp constants in the Hardy-Littlewood-Sobolev and related inequalities*, Ann. Math. **118** (1983), 349–374.
56. WONG, M. W., *A lower bound for the spectrum of a one-dimensional pseudo-differential operator*, Math. Proc. Camb. Phil. Soc. **103** (1988), 317–320.
57. MITRINOVIĆ, D. S. and J. E. PEČARIĆ, *On inequalities of Hilbert and Widder*, (to appear in Proc. Edinburgh Math. Soc.).

58. ________, *A note on an inequality with nonconjugate parameters*, (to appear in Anzeig. Öster. Akad. Wiss.).
59. BEESACK, P. R., D. S. MITRINOVIĆ, and P. M. VASIĆ, *Integral inqualities*, (the manuscript not achieved and not published – we used this manuscript in writing sections 10 and 14).

CHAPTER VI

INEQUALITIES OF LYAPUNOV AND OF DE LA VALLÉE POUSSIN

1. Let p be a real continuous function of the real variable x on $[a, b]$. If the differential equation

$$y'' + p(x)y = 0$$

has a nontrivial solution y that vanishes at two points of $[a, b]$, then, according to A. M. Lyapunov [**1**], p is subject to the inequality

$$(1.1) \qquad (b-a)\int_a^b |p(x)|dx > 4.$$

This inequality is sharp in the sense that the constant 4 cannot be replaced by a larger number.

REMARK 1: This inequality may be applied in two ways. If one considers the eigenvalue problem

$$(1.2) \qquad y'' + \lambda p(x)y = 0, \qquad y(a) = y(b) = 0,$$

then if λ is the smallest eigenvalue one has from (1.1)

$$(1.3) \qquad \lambda > 4\left((b-a)\int_a^b |p(x)|dx\right)^{-1}.$$

This furnishes a lower bound. If more is known about p, then this inequality can be improved. This topic is disussed below in Section 9.

For the second interpretation, we consider a generalization to a scalar nth order linear operator of the form

$$Lx \equiv x^{(n)} + p_1(t)x^{(n-1)} + \cdots + p_n(t)x \tag{1.4}$$

with summable coefficients. The operator L is said to be disconjugate on an interval (a,b) if no non-trivial solution of $Lx = 0$ has n zeros on the interval. Disconjugacy is connected with solvability of boundary value problems because L is linear. Consider the boundary conditions

$$x^{(i)}(t_j) = c_{ij}, \quad j = 0,1,\ldots,J_i, \quad i = 1,2,\ldots,I, \tag{1.5}$$

where t_j is in (a,b) and $J_1 + J_2 + \cdots + J_I = n$.

The problem of finding a solution of $Lx = f(t)$ satisfying the conditions (1.5) has a unique solution for each choice of the c_{ij}, if and only if the problem of finding a solution of $Lx = 0$ satisfying the conditions (1.5) with all $c_{ij} = 0$ has a unique solution. If the operator L is disconjugate on (a,b) then this is satisfied for all choices of the t_j in (a,b).

In the special case of Lyapunov's inequality, we may conclude that if $(b - a)\int_a^b p(x)dx \leq 4$, then the operator $Ly = y'' + p(x)y$ is disconjugate on (a,b), and we may solve all two point boundary value problems on (a,b). In particular, sufficiently short intervals are intervals of disconjugacy. One tries to quantify this statement by deriving conditions of the form

$$\sum_{k=1} A_k(b-a)^{k-\frac{1}{r}}\|p_k\|_r < 1. \tag{1.6}$$

This discussion will be followed up in succeeding sections.

REMARK 2: The inequality (1.1) may be strengthened by replacing $|p|$ by $p^+ = \max\{p(t),0\}$ or by writing

$$\int_a^b (t-a)(b-t)p^+ \, dt > b - a. \tag{1.1'}$$

See Hartman [2, pp. 345].

REMARK 3: The inequality (1.1) may be given in the following way:

If f'' is continuous on $[0,1]$ and if $f(0) = f(1) = 0$, then

$$\int_0^1 |f''(x)/f(x)| \, dx > 4. \tag{1.7}$$

This inequality is the best possible.

2. In [**3**] de la Vallée Poussin (see also [**44**]) proved that (see (1.4)) $Lx = 0$ is disconjugate on the interval $[a, b]$ if

$$\sum_{k=1}^{n} \frac{1}{k!} P_k (b-a)^k \leq 1, \tag{2.1}$$

where $P_k = \max_{a \leq t \leq b} |p_k(t)|$.

A. Yu. Levin [**4**] (and, independently, V. G. Maz'ya) found a less restrictive condition than (2.1) for the disconjugacy of Ly:

$$\sum_{k=1}^{n-1} \frac{1}{k!} P_k h^k + \frac{(n-1)^{n-1}}{n^n n!} P_n h^n \leq 1. \tag{2.2}$$

The following result is obtained by A. Yu. Levin [**5**], [**6**]: Ly is disconjugate if

$$\sum_{k=1}^{n} \left(2^k k \left[\frac{k-1}{2}\right]! \left[\frac{k}{2}\right]! \right)^{-1} P_k h^k \leq 1. \tag{2.3}$$

Note that $\left(2^k k \left[\frac{k-1}{2}\right]! \left[\frac{k}{2}\right]!\right)^{-1} \leq \frac{1}{2} \frac{1}{k!}$, so that the coefficients of each power of h in inequality (2.1) are smaller than the corresponding coefficients in inequality (2.2).

It is easy to convince oneself that conditions (2.2) and (2.3) are independent (for $n > 2$).

Another refinement of (2.1) (and of (2.2)) is obtained by G. A. Bessmertnyh and A. Yu. Levin [**7**]:

$$\sum_{k=1}^{n} B_{k,n} P_k h^k < 1, \tag{2.4}$$

where

$$B_{k,n} = \frac{1}{k!}\left(1 - \frac{k}{n}\right), \quad k = 1, 2, \ldots, n-1; \quad B_{n,n} = \frac{1}{n!} \frac{(n-1)^{n-1}}{n^n}.$$

3. Hartman [**8**] considered the case of $\| \;\|_1$.

Introduce the constants $C_n = \frac{1}{n}\left[\frac{n}{2}\right]!$. The following are disconjugacy criteria for (2.1) on I:

Introduce the constants $C_n = \frac{1}{n}\left[\frac{n}{2}\right]!$. The following are disconjugacy criteria for (2.1) on I:

(i) $p_0, \ldots, p_{n-1}$ satisfy

$$\exp\left(\frac{1}{2}\|p_0\|_1\right)\sum_{k=1}^{n-1}\left(\frac{C_k}{4}\right)\left(\frac{k}{k+1}\right)^{k-1} h^k\|p_k\|_1 \le 1, \tag{3.1}$$

in particular,

$$\frac{1}{2}\|p_0\|_1 + \sum_{k=1}^{n-1}\left(\frac{C_k}{4}\right)\left(\frac{k}{k+1}\right)^{k-1} h^k\|p_k\|_1 \le 1; \tag{3.2}$$

(ii) If, in addition, $p_1(t) \equiv \cdots \equiv p_m(t) \equiv 0$ for some m, $1 \le m < n-1$, then (3.1) can be relaxed to

$$\exp\left(\frac{1}{2}\|p_0\|_1\right)\sum_{k=m}^{n-1}\left(\frac{C_k}{2^{m+1}}\right)\left(\frac{k}{k+m}\right)^{k-m} h^k\|p_k\|_1 \le 1; \tag{3.3}$$

(iii) $p_0, \ldots, p_{n-1}$ satisfy

$$\exp\left(\|p_0\|_1\right)\left(\sum_{k=1}^{n-1}\left(\left(\frac{C_k}{2^{k+1}}\right)h^k\|p_k\|_1\right)^2 + \frac{1}{2}\sum_{j\ne k} C_jC_k j^j k^k (j_k)^{-j-k} h^{j+k}\|p_j\|_1\|p_k\|_1\right) \le 1; \tag{3.4}$$

(iv) The quantities Q_1, Q_2 satisfy

$$\max(Q_1, Q_2) \le 1, \tag{3.5}$$

where

$$Q_1 = \exp\left(\int_a^{\frac{a+b}{2}} p_0^+\,dt\right)\sum_{k=1}^{n-1} C_k 2^{-k} h^k \int_a^{\frac{a+b}{2}} |p_k|\,dt, \tag{3.6}$$

$$Q_2 = \exp\left(\int_{\frac{a+b}{2}}^{b} p_0^-\,dt\right)\sum_{k=1}^{n-1} C_k 2^{-k} h^k \int_{\frac{a+b}{2}}^{b} |p_k|\,dt, \tag{3.7}$$

and p_0^+ and p_0^- are defined by $p_0^+(x) = \max(0, p(x))$, $p_0^-(x) = \max(0, -p_0(x))$.

4. We give the following result of D. Willett [**9**]: If

(4.1)

$$\left(\left[\frac{n-1}{2}\right]!\left[\frac{n}{2}\right]\right)^{-1} \int_a^b |r_n t)| \frac{(t-a)^{n-1}(b-t)^{n-1}}{(b-a)^{n-1}}\, dt$$

$$+ 2^{n-1}\int_a^b |r_1(t)|dt + \sum_{k=2}^{n-1} \frac{(2^{n-1}-1)}{(k-1)!} \int_a^b |r_k(t)| \frac{(t-a)^{k-1}(b-t)^{k-1}}{(b-a)^{k-1}}\, dt$$

$$\leq 1,$$

then

$$y^{(n)} = r_1(t)y^{(n-1)} + \cdots + r_n(t)y \tag{4.2}$$

is disconjugate on $[a, b)$.

5. Let us consider the differential operator

$$L_{n,p} = \frac{d^n}{dx^n} + \sum_{k=1}^{n} a_k(x)\frac{d^{n-k}}{dx^{n-k}},$$

where $a_k \in L^p[a,\ a+h]$, $k = 1, 2, \ldots, n$, $h > 0$, $1 \leq p \leq +\infty$. If we consider the linear differential equation

$$L_{n,p}[y] = 0, \tag{5.1}$$

we denote by Y_n the set of the solutions of the equation (5.1), that is the set of functions $y\colon [a,\ a+h] \to R$ with the properties

(i) y has a derivative of order $(n-1)$ which is absolutely continuous on $[a,\ a+h]$,

(ii) y verifies (5.1) almost everywhere on $[a,\ a+h]$.

DEFINITION 1: The differential operator $L_{n,p}$ has the interpolation property $I_n[a,\ a+h)$ if for each

$$P = (x_1, x_2, \ldots, x_n) \in [a, a+h)^n, \quad x_i \neq x_j \text{ for } i \neq j,$$

and for each

$$V = (y_1, y_2, \ldots, y_n) \in R^n,$$

DEFINITION 2: A differential operator $L_{n,p}$ has the interpolation property $H_n[a,\ a+h)$ if for each

$$P = (x_1, x_1, \ldots, x_1, x_2) \in [a,\ a+h)^n, \quad x_1 \neq x_2,$$

and for each

$$V = (y_1, y_2, \ldots, y_n) \in R^n$$

there is a unique solution $\bar{y} = \bar{y}(P, V; L_{n,p}) \in Y_n$ such that

$$\bar{y}(x_1) = y_1, \quad \bar{y}'(x_1) = y_2, \quad \ldots, \quad \bar{y}^{(n-2)}(x_1) = y_{n-1},$$
$$\bar{y}^{(w)}(x_2) = y_n,$$

where w is one of the numbers $0, 1, \ldots, n-2$.

Results from this Chapter gives some upper bounds for h, that is for the length of the interval of interpolation and hence intervals of disconjugacy. These upper bounds are expressed in the form of the following nonlinear inequalities

$$\sum_{k=1}^{n} R_{n-k+1} h^{n-k+1} M_p(a_{n-k+1}, h) \leq 1, \qquad \text{if } p < +\infty$$

or

$$\sum_{k=1}^{n} T_{n-k+1} h^{n-k+1} \|a_{n-k+1}\|_\infty, \qquad \text{if } p = +\infty,$$

where $M_p(a_k, h) = h^{-1/p}\|a_k\|_p$ is a mean of order p of function a_k, R_j, T_j, $j = 1, \ldots, n$, are constants.

Opial's inequality [**10**] (see also Chapter III) gives a condition that the operator $L_{2,\infty}$ has the interpolation property $I_2[a,\ a+h)$:

$$\frac{4}{\pi^2} h\|a_1\|_\infty + \frac{h^2}{\pi^2}\|a_2\|_\infty \leq 1. \tag{5.2}$$

For some related results see also papers [**11**],[**12**] of O. Aramă and D. Ripianu, [**13**],[**14**] of D. Ripianu, and [**15**] of B. Szmanda. For example, for $n = 3$, $p = 2$, O. Aramă and D. Ripianu [**12**] showed that if

$$\frac{1}{\pi^2\sqrt{2}} h^3 M_2(a_3, h) + \frac{\sqrt{2}}{\pi} h^2 M_2(a_2, h) + \frac{1}{\sqrt{2}} h M_2(a_1, h) \leq 1, \tag{5.3}$$

then L_{32} has the property $I_3[a,\ a+h)$.

O. Aramă [16] proves the inequality

$$\frac{1}{\pi^2\sqrt{6}}h^4M_2(a_4,h)+\frac{1}{\pi^2\sqrt{2}}h^3M_2(a_3,h)+\frac{\sqrt{2}}{\pi}h^2M_2(a_2,h) + \frac{1}{\sqrt{2}}hM_2(a_1,h)\le 1 \tag{5.4}$$

in order that the differential operator $L_{4,2}$ to have the property $H_4[a,\ a+h)$.

D. Bobrowski [17] obtained the following inequality, as a sufficient condition for the property $H_4[a,\ a+h)$ of the operator $L_{4,\infty}$,

$$\frac{1}{2\pi^2}h^4\|a_4\|_\infty+\frac{1}{2\pi^2}h^3\|a_3\|_\infty+\frac{1}{\pi^2}h^2\|a_2\|_\infty+\frac{1}{4}h\|a_1\|_\infty\le 1. \tag{5.5}$$

Several results of this kind are obtained by A. Lupaş [18]:

1) If

$$\frac{4}{3\pi^2}h^2\|a_2\|_\infty+\frac{1}{6}h\|a_1\|_\infty\le 1, \tag{5.6}$$

then the operator $L_{2,\infty}$ has the interpolation property $I_2[a,\ a+h)$.

2) If

$$16\pi\sqrt{3}\,(\Gamma(1/4))^{-4}\,h^2M_2(a_2,h)+\left(\frac{2}{\pi}\right)hM_2(a_1,h)\le 1, \tag{5.7}$$

then the differential operator $L_{2,2}$ has the interpolation property $I_2[a,a+h)$.

3) If

$$\frac{4}{\pi^4}h^4\|a_4\|_\infty+\frac{1}{2\pi^2}h^3\|a_3\|_\infty+\frac{1}{\pi^2}h^2\|a_2\|_\infty+\frac{1}{4}h\|a_1\|_\infty\le 1, \tag{5.8}$$

then the operator $L_{3,\infty}$ has property $H_4[a,\ a+h)$.

In the same paper, Lupaş also proved an result related to (5.4), as well as the case $n=3$ and $2\le p<+\infty$.

A. Lupaş [19] also considered the cases $n=5$, $p=+\infty$ and $n=5$, $p=2$. Here we shall give only the first case: If

$$\frac{8h^5}{\pi^5}\|a_5\|_\infty+\frac{4h^5}{\pi^4}\|a_4\|_\infty+\frac{h^3}{2\pi^2}\|a_3\|_\infty + \frac{h^2}{\pi^2}\|a_2\|_\infty+\frac{h}{4}\|a_1\|_\infty\le 1, \tag{5.9}$$

Related papers are **[20]** of P. Bailey and P. Waltman, **[21]**,**[22]** of A. Lasota, **[23]** of A. Lasota and Z. Opial, **[24]** of F. Lettenmeyer, **[25]** of B. Szmanda, **[26]** of U. Richard, and **[27]**,**[28]** of A. Yu. Levin.

6. D. Willett **[9]** considered the linear differential equation

$$Ly = r_1(t)y^{(n-1)} + \cdots + r_n(t)y, \tag{6.1}$$

where

$$Ly \equiv y^{(n)} + p_1(t)y^{(n-1)} + \cdots + p_n(t)y \tag{6.2}$$

where all functions are assumed to be continuous on a bounded interval $[a, b)$. He assumes that $Ly = 0$ is disconjugate on $[a, b)$ and derives a disconjugacy criterion for (6.1) of the form

$$\sum_{k=1}^{n} \int_a^b |r_k(t)|\nu_{n-k+1}(t)\, dt \leq 1. \tag{6.3}$$

The function ν_k is determined in terms of fundamental principal systems of solutions of $Ly = 0$, which is a concept defined in Willett **[29]**, and further described in **[9]**.

Connected to [3] and **[10]** are papers of U. Richard **[30]**,**[31]** and of G. Casadei **[32]**.

7. Inequalities of the form

$$\|f\|_p \leq H(n,p,q)(b-a)^{n+\frac{1}{p}+\frac{1}{q}} \|f^{(n)}\|_q \tag{7.1}$$

are very important in application to disconjugacy criteria for linear differential equations. This inequality (see Chapter II. on Wirtinger's inequality) is considered in papers **[33]** of J. Brink and **[34]**,**[35]** of A. M. Fink. Here, we shall give disconjugacy tests from **[35]**.

Consider differential equations of the form

$$y^{(n)} = a_1(x)y^{(n-1)} + \cdots + a_n(x)y \tag{7.2}$$

with integrable coefficients on $[a, b]$. If a non-trivial solution of (7.2) has n zeros on $[a, b]$, then there are sequences a_i and b_i such that

$$a \leq a_0 \leq \cdots \leq a_{n-1} = b_{n-1} \leq b_{n-2} \leq \cdots \leq b_0 \leq b \tag{7.3}$$

and

$$y^{(i)}(a_i) = y^{(i)}(b_i) = 0, \quad i = 0, \ldots, n-1. \tag{7.4}$$

Write $c = a_{n-1}$ and note that $a < c < b$. We want to derive necessary conditions for such a solution to exist so that the denial of these conditions give disconjagacy of the equation. For this purpose we take $a = a_0$ and $b = b_0$. Furthermore, let $\|f\|_p^* = \max\left\{\left(\int\limits_a^{\frac{a+b}{2}} |f|^p\right)^{1/p}, \left(\int\limits_{\frac{a+b}{2}}^b |f|^p\right)^{1/p}\right\}$ with the usual convention when $p = \infty$.

THEOREM 1. *Equation (7.2) is disconjugate on $[a, b]$ if one of the following inequalities holds:*

(i) $\sum\limits_{k=1}^n \|a_k\|_{p_i}^* H(k, \infty, p_i) 2^{\frac{1}{p}-k}(b-a)^{k-\frac{1}{p_i}} \leq 1,$

(ii) $\sum\limits_{k=1}^n \|a_k\|_\infty^* H(k, p, p) 2^{-k}(b-a)^k \leq 1,$

(iii) $\|a_1\|_n^* 2^{\frac{1}{p}-\frac{1}{r}}(b-a)^{1+\frac{1}{r'}-\frac{1}{p}}$
$+ \sum\limits_{k=2}^n \|a_k\|_r^* H(k-1, r', p) 2^{k+\frac{1}{r'}-\frac{1}{p}}(b-a)^{k+\frac{1}{r'}-\frac{1}{p}} \leq 1,$

$$1 \leq r' \leq p,$$

(iv) $\exp(\|a_1\|_1^*) \sum\limits_{k=2}^n \|a_k\|_{p_i}^* H(k-1, 1, p_k) 2^{\frac{1}{p_i}-k}(b-a)^{k-\frac{1}{p_i}} \leq 1$

(v) $(\exp\|a_1\|_1^*) \sum\limits_{k=2}^n \|a_k\|_{p_i}^* H(k-1, \infty, p_i) 2^{\frac{1}{p_i}-k}(b-a)^{k-\frac{1}{p_i}} \leq 1,$

or

(v) $(\exp\|a_1\|_1^*) \sum\limits_{k=2}^n \|a_k\|_\infty^* H(k-1, p_i, 1) 2^{-k}(1+p_i')^{-\frac{1}{p_1'}}(b-a)^k \leq 1.$

PROOF: Let there be a non-trivial solution with n zeroes on $[a, b]$. To prove (i) consider $[a, c]$ where c is determined by (7.3) and (7.4) with norms on $[a, c]$,

$$\begin{aligned}\|y^{(n)}\|_1 &< \sum_{k=1}^n \|a_k y^{(n-k)}\|_1 \leq \sum_{k=1}^n \|a_k\|_{p_i} \|y^{(n-k)}\|_{p_k'} \\ &\leq \sum_{k=1}^n \|a_k\|_{p_i} H(k, p_i', 1) \|y^{(n)}\|_1 (c-a)^{k+\frac{1}{p_i'}-1}.\end{aligned}$$

Cancelling $\|y^{(n)}\|_1$ one obtains

$$1 < \sum_{k=1}^n \|a_k\|_{p_i} H(k, \infty, p_i)(c-a)^{k-\frac{1}{p_i}}, \tag{7.5}$$

with a similar inequality on $[c, b]$. Now either $c - a$ or $b - c \leq (b-a)/2$ and so (i) follows by contraposition.

To prove (ii) write

$$\|y^{(n)}\|_p \leq \sum_{k=1}^{n} \|a_k y^{(n-k)}\|_p \leq \sum_{k=1}^{n} \|a_k\|_\infty \|y^{(n-k)}\|_p,$$

and finish as in (i).

To prove (iii), note that

$$\begin{aligned} y^{(n-1)}(t) &= \int_c^t y^{(n)}(s)\, ds \\ &\leq \sum_{k=1}^{n} \|a_k\|_r \|y^{(n-k)}\|_{r'}. \end{aligned}$$

If $r' \leq p$,

$$\begin{aligned} \|y^{(n-1)}\|_p &< \sum_{k=1}^{n-1} (c-a)\|a_k\|_r \|y^{(n-k)}\|_{r'} \\ &\leq (c-a)^{1+\frac{1}{r'}-\frac{1}{p}} \|a_1\|_r \|y^{(n-1)}\|_p \\ &\quad + \sum_{k=2} \|a_k\|_r H(k-1, r'p) \|y^{(n-1)}\|_p (c-a)^{k+\frac{1}{r'}-\frac{1}{p}} \end{aligned}$$

and one completes as in (i).

To prove (iv), (v) and (vi) write (7.2) as

$$\left[y^{(n-2)} \exp\left(-\int_c^x a_1\right)\right]' = \sum_{k=2}^{n} a_k y^{(n-k)} \exp\left(-\int_c^x a_1\right)$$

and integrate from x and c to obtain

$$y^{(n-1)} \exp\left(-\int_c^x a_1\right) \leq \sum_{k=2}^{n} \int_x^c |a_k y^{(n-k)}| \exp\left(-\int_c^t a_1\right) dt$$

and thus

$$|y^{(n-1)}(x)| \leq \exp\left(\int_a^c |a_1|\right) \sum_{k=2}^{n} \int_x^c |a_k y^{(n-k)}|. \tag{7.6}$$

To prove (iv) take max over $[a, c]$ and use Hölder's inequality on the right. To prove (v) and (vi) integrate (7.6) from a to c and replace $\int_a^c |t-a||a_k y^{(n-k)}|$ by $(c-a)\|a_k\|_{p_i}\|y^{(n-k)}\|_{p_i'}$ for (v) and by $\|a_k\|_\infty \|(t-a)\|_{p_i'}\|y^{(n-k)}\|_p$ for (vi).

All of the above proofs derive an expression of the form

$$1 < \sum_{k=1}^{n} \|a_k\|_{p_k} B_k (c-a)^{s_k} \tag{7.7}$$

on $[a, c]$ and a similar expression on $[a, b]$. The numbers B_k and s_k are independent of a and c, and are the same for the expression on $[c, b]$. An attempt to combine these was made in Zaiceva [**36**], but the last inequality was wrong. Her method was successfully completed by Hartman [**8**] who was inspired by the paper of Nehari [**37**]. His ideas with slight modification work here also.

THEOREM 2. *Suppose one has inequalities of the form (7.7) on $[a, c]$ and $[c, b]$. Let $m = \min_k s_k > 0$, then*

$$1 < \sum_{k=1}^{n} \|a_k\|_{p_k} B_k 2^{-m-\frac{1}{p}} \left(\frac{s_k}{s_k+m}\right)^{s_k - m} (b-a)^{s_k},$$

where norms are over $[a, b]$.

PROOF: Denote norms on $[a, c]$ by $\| \quad \|^a$ and on $[c, b]$ by $\| \quad \|^b$. Multiplying the two inequalities one gets

$$1 < \sum_{k=1}^{n}\sum_{j=1}^{n} \|a_k\|_{p_k}^a \|a_j\|_{p_j}^b B_k B_j (c-a)^{s_k}(b-c)^{s_j}. \tag{7.8}$$

Note that

$$(c-a)^{s_k}(b-c)^{s_j} \le (b-a)^{s_k+s_j} \frac{s_k^{s_k} s_j^{s_j}}{(s_k+s_j)^{s_k+s_j}}$$

$$= (b-a)^{s_k+s_j} \left[\frac{s_k}{s_k+s_j}\right]^{s_k-m} \left[\frac{s_j}{s_k+s_j}\right]^{s_j-m} \left[\frac{s_k s_j}{(s_k+s_j)^2}\right]^m$$

$$\le (b-a)^{s_k+s_j} \left[\frac{s_k}{s_k+m}\right]^{s_k-m} \left[\frac{s_j}{s_j+m}\right]^{s_j-m} 4^{-m}.$$

Dominating the right hand side of (7.8) by this expression one gets

$$1 \le \left\{\sum_{k=1}^{n} \|a_k\|_{p_k}^a B_k (b-a)^{s_k} 2^{-m} \left(\frac{s_k}{s_k+m}\right)^{s_k-m}\right\}$$

$$\times \left\{\sum_{j=1}^{n} \|a_j\|_{p_j}^b B_j (b-a)^{s_j} 2^{-m} \left(\frac{s_j}{s_j+m}\right)^{s_j-m}\right\}.$$

Now using $\|a_k\|_{p_k}^a + \|a_k\|_{p_k}^b \le 2^{\frac{1}{p'}} \|a_k\|_{p_k}$ and the geometric mean arithmetic mean inequality, we get the theorem.

From the last theorem and the proofs of Theorem 1 one can read off more disconjugacy tests. Some of these constants in the case of p-norms are better than those of Willett **[9]** or Martelli **[37]**.

Generally, for k near n these are small, while Willett's are small for k small.

For other theorems that give disconjugacy see Willett **[38]** and Kim **[39]** and **[40]**. One of Kim's theorems from **[39]** we may quote here. The differential equation (7.2) is disconjugate on $(-c, c)$ if

$$\sum_{k=1}^{n-1} |a_k(x)| \frac{(c+|x|)}{k!} + \frac{|a_n(x)|}{n!} (c-|x|)(c+|x|)^{n-1} \le 1.$$

An earlier result for a special equation

$$y^{(n)}(x) - \lambda p(x) y(x) = 0, \qquad y(a) = y(b) = 0, \tag{7.9}$$

is given by Makai **[41]**, **[42]**. For example, in **[42]**; if λ_n is the nth eigenvalue of the problem (7.9) then

$$\lambda_n < \left(\frac{\pi n}{\int_a^b \sqrt{p(x)}\, dx} \right)^2 \qquad \text{if } r(x) = 5p'(x)^2 - 4p(x)p''(x) \ge 0,$$

and

$$\lambda_n > \left(\frac{\pi n}{\int_a^b \sqrt{p(x)}\, dx} \right)^2 \qquad \text{if } r(x) \le 0.$$

For $n = 1$ these give Lyapunov type inequalities.

Many other papers give disconjugacy criteria but do not lead easily to Lyapunov type inequalities.

8. Let us consider the case of $n = 2$ in more detail. Let $x(t)$ be a nontrivial solution of

$$x'' + g(t)x' + f(t)x = 0, \quad x(0) = x(h) = 0. \tag{8.1}$$

The theorem of C. De La Vallée Poussin [3] gives

$$1 < 2\|g\|_\infty h + \|f\|_\infty \frac{h^2}{2}. \tag{8.2}$$

P. H. Hartman and A. Winter [43] improved this to

$$1 < \|g\|_\infty h + \|f\|_\infty \frac{h^2}{6}. \tag{8.3}$$

The best possible inequality of this type was obtained by Z. Opial [10] who proved that

$$\pi^2 \le 4\|g\|_\infty h + \|f\|_\infty h^2, \tag{8.4}$$

where equality can only hold when $g \equiv 0$.

The proof is elementary but long, so we give the proof of a somewhat weaker inequality (also due to Opial)

$$\pi^2 \le \pi\|g\|_\infty h + \|f\|_\infty h^2. \tag{8.5}$$

Let x solve (8.1) and multiply by x and integrate from 0 to h. We get

$$\int_0^h x''(t)x(t)\,dt + \int_0^h g(t)x'(t)x(t)\,dt + \int_0^h f(t)x(t)\,dt = 0.$$

However an integration by parts of the first term gives

$$0 \le \|f\|_\infty \int_0^h x^2(t)\,dt + \|g\| \int_0^h |x(t)|\,|x'(t)|\,dt - \int_0^h x'(t)^2\,dt. \tag{8.6}$$

Equality is possible only if $g \equiv 0$. Now using a Wirtinger inequality (see Sansone [44] or (2.1) in Chapter II.)

$$\int_0^h x(t)^2 dt < \frac{h^2}{\pi^2} \int_0^h x'(t)^2\,dt \tag{8.7}$$

and the Buniakowski-Schwarz inequality we get

$$\begin{aligned} \int_0^h |x(t)|\,|x'(t)|\,dt &\le \left(\int_0^h x'(t)^2\,dt \int_0^h x(t)^2\,dt \right)^{1/2} \\ &\le \frac{h}{\pi} \int_0^h x'(t)^2\,dt. \end{aligned} \tag{8.8}$$

Combining this with (8.6) yields

$$-\int_0^h x'(t)^2\,dt + \|g\|\frac{h}{\pi}\int_0^h x'(t)^2\,dt + \|f\|\frac{h^2}{\pi^2}\int_0^h x'(t)^2\,dt > 0 \tag{8.9}$$

giving the result.

A. M. Fink and D. F. St. Mary [**45**] proved that

$$h\int_0^h f^+(t)\,dt - 4\exp\left(-\frac{1}{2}\int_0^h |g|\,dt\right) > 0 \tag{8.10}$$

and a fortiori

$$h\int_0^h f^+(t)\,dt + 2\int_0^h |g(t)|\,dt > 4. \tag{8.11}$$

This inequality is sharp since it reduces to Lyapunov's inequality (1.1) when $g \equiv 0$. This inequality was also obtained by A. Yu. Levin [**46**].

H. Hochstadt [**47**] obtained the following result: If for $n \geq 2$

$$y^{(n)} - p(t)y^{(n-1)} - q(t)y = 0$$

has n zeros on $[a, b]$, then

$$\left((b-a)^{n-1}\int_a^b |q(t)|\,dt\right)^{1/n} + \frac{1}{n}\int_a^b |p(t)|\,dt \geq 2. \tag{8.12}$$

9. We return to the interpretation of (1.1) as an eigenvalue inequality. Recall that it may be improved if more is known about p. For example, A. M. Fink [**48**] showed that if $p > h > 0$ on $[a, b]$, then

$$\|p\|_1 \geq (b-a)h^2 + 2(b-a)^2 \cot g\frac{(b-a)h}{2} > 4.$$

Some similar results may be found in a paper [**49**] of W. Leighton.

D. Banks in a series of papers has gotten similar improvements by assuming that p is monotone, or convex, or bounded both above and below. As an example, if p is monotone then 4 may be replaced by 7.88... and by 9.397... if p is convex ([**50**]). See D. C. Barnes [**51**] for more results of this kind and a bibliography of related results.

A very interesting result is obtained by A. Galbraith [**52**]. He has shown that if a and b are successive zeroes of the differential equation

$$y'' + py = 0 \tag{9.1}$$

with p a nonnegative linear function, then

$$(b-a)\int_a^b p(x)\,dx \geq \pi^2.$$

In [**53**] A. M. Fink proved a general result wherefrom be obtained

$$\frac{9}{8}\lambda_0^2 \leq (b-a)\int_a^b p(x)\,dx \leq \pi^2 \qquad (\text{for } p \geq 0)$$

and that these are the best possible bounds. The constant $\frac{9}{8}\lambda_0^2 = 9.478132\ldots$ and $\pi^2 = 9.869604\ldots$ so that we have a delicate test for the spacing of zeroes for linear p. Furthermore, when the restriction that $p \geq 0$ is removed, π^2 is still the upper bound for he shows that the function for fixed a and b is an even function of the slope of p, that is decreasing for positive slopes.

He also considered inequalities when p is concave. Á. Elbert, in a series of papers [**54**]–[**58**] considered Lyapunov inequalities of various kinds for (9.1), especially when p is concave. For example, he showed [**55**] that

$$\lambda_0(b-a)\int_a^b p(x)dx \geq 8.550\ldots$$

if $p \geq 0$, concave, and monotone. In [**54**] he showed that if $p(x) \geq 0$ and continuous with $p(x)^\nu$ concave then the inequality

$$(b-a)\int_a^b p(x)dx \geq \frac{(2\nu+1)^2 n^2\pi^2}{4\nu(\nu+1)} \tag{9.2}$$

implies that every solution of (9.1) has n zeros on $[a, b]$. The coefficient of $n^2\pi^2$ cannot be replaced by a smaller number.

B. Schwarz [59] and P. R. Beesack and B. Schwarz [60] studied eigenvalue inequalities for (9.1) and for the related equation

$$y'' + p^* y = 0 \tag{9.3}$$

where p^* is an equimeasurable rearrangement of p. Since $\int_a^b p = \int_a^b p^*$, these lead to Lyapunov type inequalities if something is explictly known about (9.3).

Other research on Lyapunov inequalities for (9.1) can be found in S. B. Eliason [61] and D. F. St. Mary and S. B. Eliason [62] where the interval in $\left[-\frac{T}{2}, \frac{T}{2}\right]$ and $p(t) \leq p(-t)$ on $\left[0, \frac{T}{2}\right]$.

10. As generalizations of (1.6) we consider several results.

V. Levin [63] gave the following generalizations of a problem of Golab (Jber. Deutsch. Math.-Vereinig. **43**, 1 (1933)):

Let $f(x)$ be π-periodic twice differentiable function such that $f(x)+f''(x) \geq 0$. If

$$J_\lambda^\lambda = \frac{1}{\pi}\int_0^\pi \left(1 + \frac{f''(x)}{f(x)}\right)^\lambda dx,$$

then $J_\lambda \leq 1$ for $0 \leq \lambda \leq 1/2$, and $J_\lambda \geq 1$ for $1 \leq \lambda < +\infty$.

11. Some results related to (8.2) are obtained by T. Satô [64], M. Tumura [65] and G. Sansone [44].

12. G. Borg [66] proved the following two results:

(i) If $k < \pi^2/h^2$, $y(0) = y(h) = 0$, $y(x) > 0$ in $(0, h)$, then

$$\int_0^h |y''/y + k| dx \geq \begin{cases} 2\sqrt{k} \cot g \left(h\sqrt{k}/2\right), & k \geq 0, \\ 2\sqrt{-k} \coth \left(h\sqrt{-k}/2\right), & k < 0. \end{cases} \tag{12.1}$$

(ii) If $k > 0$, $y(0) = y(h) = 0$, $y \neq 0$ in $(0, h)$, then

$$\int_0^h (y''/y + k)^2 dx \geq \frac{64}{3h^3} K^3 \left(K + (\lambda^2 + 1)(K - 2E)\right), \tag{12.2}$$

where λ^2, K and E are defined by $k = 4(\lambda^2 + 1)K^2/h^2$

$$K = \int_0^{\pi/2} (1 - \lambda^2 \sin^2 \phi)^{-1/2} d\phi, \ E = \int_0^{\pi/2} (1 - \lambda^2 \sin^2 \phi)^{1/2} d\phi.$$

The constant in the right hand side of (12.2) is the best possible.

For some applications in ordinary differential equations, see Borg [67].

Also connected to Borg's inequalities are the papers of H. O. Cordes [68] and H. J. Cohen [69].

For other applications, as well as generalizations of (1.6) to complex differential equations and univalent functions, see Z. Nehari [70] and P. R. Beesack [71].

13. Lyapunov inequalities for nonlinear equations can be found in Á. Elbert [72] and S. B. Eliason [73], [74]. Elbert considers

$$y''(t) + p(t)y\left(q(t)\right) = 0$$

while Eliason [73], [74] considers

$$(r(t)y')' + p(t)yf(y) = 0 \tag{13.1}$$

with $y(a) = y(b) = 0$, where for each $\mu > 0$, there is a $\nu > 0$ so that $y^{\nu+1}f(y)$ is increasing on $(0, \mu]$ and $\lim\limits_{y\to 0+} y^{\nu+1}f(y) = 0$. If (13.1) has a solution y with consectutive zeros at a and b and $M = |y|_\infty$, then

$$f(M)\int_a^b r^{-1}(x)dx \int_a^b p_+(x)dx > \begin{cases} 16\nu(\nu+1)^{-1} & , \nu \geq 1, \\ 4 & , 0 < \nu < 1. \end{cases}$$

Neither constant can be replaced by a larger one.

14. Disconjugacy and Lyapunov inequalities are related by the boundary conditions $y(a) = y(b) = 0$. If one considers the boundary conditions $y(a) = y'(b) = 0$, then one says that $y'' + py = 0$ is disfocal on $[a, b]$ if no non-trivial solution statisfies these boundary conditions at c and d with $a \leq c \leq d \leq b$. St. Mary [75] observed that if

$$(r(t)y')' + p(t)y = 0 \tag{14.1}$$

has a non-trivial solution with $y(a) = y'(b) = 0$, then

$$\int_a^b r^{-1}(x)dx \int_a^b p_+(x)dx > 1, \quad \text{where } p_+(x) = \max\left(0, p(x)\right).$$

For the equation (14.1) with $r \equiv 1$, he shows that

$$(b-a)\int_a^b p(x)dx \geq \frac{\pi^2}{4}$$

if $\int_t^b p \geq 0$ on $[a, b]$ and p is nonincreasing, while

$$(b-a)\int_a^b p(x)dx \leq \frac{\pi^2}{4}$$

if p is nondecreasing. He also gives interesting comparison theorems.

Kwong [**76**] gave the following result. If

$$y'' + p(t)y = 0, \qquad y(a) = y'(b) = 0$$

has a non-trivial solution, then

$$\int_a^b \int_a^t p_+(x)ds\,dt > 1. \tag{14.2}$$

B. J. Harris [**77**] has extended these results. A simple version of his result is an interated versions of (14.2), to wit, that if a non trivial solution exists then

$$\int_a^b \int_a^t p_+(x) \int_x^b \int_a^t p_+(r)dr\,ds\,dx\,dt \geq 1.$$

15. R. P. Agarwal and G. V. Milovanović [**78**] improving on an earlier result of A. Bogar and G. Gustafson [**79**] gave a condition

$$1 \leq \sum_{k=0}^{5} C_k \|p_k\|_1 (b-a)^{6-k}$$

for the disconjugacy of $y^{(6)} = \sum_{i=0}^{5} p_i(x)y^{(i)}$ on $[a, b]$; $C_0 = \frac{1}{32805}$, $C_1 = \frac{25+34\sqrt{10}}{911250}$, $C_2 = \frac{1}{360}$, $C_3 = \frac{1}{30}$, $C_4 = \frac{1}{5}$ and $C_5 = \frac{2}{3}$.

16. A simple proof of (1.1) can be found in Leighton [**80**], and some details on (1.1) can be found in Kreĭn [**81**].

17. Lyapunov's inequality is discussed in the notes [**82**] by Á. Elbert.

18. Beesack [**83**] has proved some higher order versions of (1.6). For example, if $f^{(n)}$ and $f^{(n)}/f$ are continuous on $[a,b]$, $f(a)=f(b)=0$, and f has n zeros on $[a,b]$ counting those at a and b, then

$$\int_a^b \left|\frac{f^{(n)}(x)}{f(x)}\right| dx \geq \left(\frac{n}{n-1}\right)^{n-1} \frac{n!}{(b-a)^{n-1}}. \tag{18.1}$$

In case $f^{(2m)}$ and $f^{(2m)}/f$ are continuous on $[a,b]$ and f has an m-fold zero at a and at b, then

$$\int_a^b \left|\frac{f^{(2m)}(x)}{f(x)}\right| dx > \frac{(2m-1)!2^{2m}}{(b-a)^{2m-1}}, \tag{18.2}$$

and

$$\int_a^b [(x-a)(b-x)]^m \left|\frac{f^{(2m)}(x)}{f(x)}\right| dx > (2m-1)!(b-a). \tag{18.3}$$

Both (18.1) and (18.2) give (1.1) as special cases and (18.3) gives (1.1)′. The inequality (18.2) implies that if $y^{(2m)}(x)+p(x)y(x)=0$ has a solution with and m-fold zero at a and at b, then

$$\int_a^b p(x)dx > \frac{(2m-1)!2^{2m}}{(b-a)^{2m-1}}.$$

19. S. S. Cheng [**84**] establishes an inequality analogous to Lyapunov's inequality (1.1) in the theory of the difference equation

$$\Delta^2 x(k-1)+p(k)x(k)=0$$

with

$$x(0)=x(N+1)=0$$

and $p(k)$ nonnegative on the set $\{2,3,\ldots,N\}$. More general limiting conditions are also considered in [**34**].

REFERENCES

1. LYAPUNOV, A. M., *Problème général de la stabilité du mouvement*, Ann de la Faculté de Toulouse (2) **9** (1907), 406.
2. HARTMAN, P. H., "Ordinary Differential Equations," New York-London-Sydney, 1964.
3. DE LA VALLÉE POUSSIN, Ch., *Sur l'équation différentielle linéaire du second ordre*, J. Math. Pures Appl. **8** 9 (1929), 126–144.
4. LEVIN, A. Yu., Nauč. Dokl. Vysš. Školy Fiz.-Mat. Nauki **5** (1958), 34.
5. __________, *Some estimates for a differentiable function*, Soviet Math. Dokl **2** (1961), 523–524.
6. __________, *An estimate for a function with monotonically distributed zeros of successive derivatives*, Mat. Sb. **64** (1964), 396–409, (Russian).
7. BESSMERTNYH, G. A. and A. Yu. LEVIN, *Some estimates of differentiable function of one variable*, Dokl. Akad. Nauk. SSSR **144** (1963), 471–474.
8. HARTMAN, P., *On disconjugacy criteria*, Proc. Amer. Math. Soc. **24** (1970), 374–381.
9. WILLETT, D., *Generalized de la Vallée Poussin disconjugacy tests for linear differential equations*, Canad. Math. Bull. **14** 3 (1971), 419–428.
10. OPIAL, Z., *Sur une inégalité de C. de la Vallée Poussin*, Ann. Polon. Math. **6** (1959), 87–91.
11. ARAMĂ, O. and D. RIPIANU, *Quelques recherches actuelles concernant l'équation de Ch. J. de la Vallée Poussin, relative au problème polylocal dans la théorei des équations différentielles*, Mathematica **8** 31 (1966), 1, 18–28.
12. __________, *Sur l'équation de Ch. J. de la Vallée Poussin relative au problème polylocal de la théorie des équation différentielles*, Colloque sur la théorie de l'approximation des fonctions, Cluj 15–20 IX (1967).
13. RIPIANU, D., *Asupra inegalitátii lui de la Vallée Poussin in ca ecuatiilor diferentiale de ordinul al doilea*, Stuii si cercet. matem. (Cluj), **XIV** (1963), 123–150, 399–403.
14. __________, *O problemá de minimum in teoria interpolárii*, Studii si cercet de matematică (Cluj), **XIV** (1963), 365–397.
15. SZMANDA, B., *Oszacowanie przedzialu istnienia i jednoznacznosci rozwiaza'nm pewnego zagadnienia brzegowego*, Fasciculi Mathematici Nr **5** 74 (1970), 23–31.
16. ARAMĂ, O., *Sur un problème d'interpolation relatif aux solutions des équations différentielles linéaires du quatrième ordre*, Mathematica (Cluj) **10** 33 (1968), 1, 5–15.

17. BOBROWSKI, D., *O odleglosci miedzy zerami pewnych równania rózn zkowego liniowego czwatego rzedu*, Zeszyty Naukowe, Politichniki Poznanskiej 2 (Matematyka) (1965), 55–60.
18. LUPAŞ, A., *An interpolation problem for the solutions of a class of linear differential equations*, Mathematica **12** 35 (1970), 87–104.
19. ________, *On the length of the interpolation interval for a class of linear differential operators*, Proc. Intern. Conf. Construc. Funct. Theory, Varna (1870), 209–215.
20. BAILEY, P. and P. WALTMAN, *On the distance between constructive zeros for second order differential equations*, J. Math. Anal. Appl. **1** (1966), 23–30.
21. LASOTA, A., *Sur les problèmes linéaires aux limites pour un système d'équations différentilles ordinaires*, Bull. Acad. Polon. Sci. Sér. Sci. Math., Ast. et Phys. **10** (1962), 565–570.
22. ________, *Sur la distance entre les zéros de l'équation differtielle linéare du troisième orde*, Ann. Polon. Math. **13** 2 (1963), 129–132.
23. LASOTA, A. and Z. OPIAL, *L'application du principe de Pontriaguin à l'evaluation de l'intervalle d'existence et d'unicité des solutions d'un problème aux limites*, Bull. Acad. Polon. Sci., Sér Sci. Math., Ast. et Phys. **11** 2 (1963), 41–46.
24. LETTENMEYER, F., *Über die von einem punktausgehenden Integralkunven einer Differentialgleichung 2. Ordnung*, Deut. Math. **7** (1944), 56–74.
25. SZMANDA, B., *O odległości miedzu zerami pewnych rozwiazan rṕwnania rózniczkowego liniowego n-tego rzedu.*, Zesz. Nauk. Pol. Pozn. Matem. **4** (1969), 65–70.
26. RICHARD, U., *Sullz diseguaglianza di De La Vallée Poussin per le equasioni differenzialli di ordine n*, Rend. Mat. **10** (1977), 583–596.
27. LEVIN, A. Yu, *A fredholm equation with a smooth kernel and boundary-value problems for a linear differential equation*, Dokl. Akad. Nauk SSSR **159** (1964), (Soviet Math. Dokl. **5** (1964), 1415–1419).
28. ________, *Non- oscillation of solutions of the equation* $x^{(n)}+p_1(t)x^{(n-1)}+\cdots+p_n(t)x=0$, Uspehi Mat. Nauk **24** (1969), 43–96, (Russian Math. Surveys **24** (1969), 43–100).
29. WILLETT, D., *Asymptotic behavior of disconjugate nth order differential equations*, Canad. J. Math. **23** (1971), 293–314.
30. RICHARD, U., *Sur un théorème d'unicité de M. C. De La Vallée Poussin pour l'équation différentielle du second ordre*, Congrès Int. Math. Moscou 1966, Résumés 6 (Equations Différentielles Ordinaires).
31. ________, *Metodi diversi per ottenere diseguaglianze alla de la Vallée Poussin nelle equazioni differenziali ordinarie del secondo e terzo ordine*, Rend. Sem. Mat. Univ. Pol. Torino **27** (1967/68), 35–68.

32. CASADEI, G., *Sul teorema di unicita di De La Vallée Poussin per equzioni differenziali del terzo ordine*, Rend. Sem. Mat. Univ. di Padova.
33. BRINK, J., *Inequalities involving* $\|f\|_p$ *and* $\|f^{(n)}\|_q$ *for* f *with n-zeros*, Pacific J. Math. **42** 2 (1972), 289–311.
34. FINK, A. M., *Conjugate inequalities for functions and their derivatives*, SIAM J. Math. Anal. **5** 3 (1974), 399–411.
35. __________, *Differential inequalities and disconjugacy*, J. Math. Anal. Appl. **49** (1975), 758–772.
36. ZAICEVA, G. S., *A multipoint boundary value problem*, Soviet. Math. Dokl. **8** (1967), 1183–1185.
37. MARTELLI, M., *Sul criterio di univitá di de la Vallée Poussin*, Atti Accad. Naz. Lincei Rend. Cl. Sci. Fis. Mat. Natur. **45** 8 (1968), 7–12.
38. WILLETT, D., *Oscillation on finite or infinite intervals of second order linear differential equations*, Canad. Math. Bull. **14** 4 (1971), 539–550.
39. KIM, W. J., *Disconjugacy and nonoscillation criteria for linear differential equations*, J. Diff. Eq. **8** (1970), 163–172.
40. __________, *On the zeros of solutions of* $y^{(n)} + py = 0$, J. Math. Anal. Appl. **25** (1969), 189–208.
41. MAKAI, E., *Über Eigenwertabschätzungen bei gewissen homogenen linearen Differentialgleichungen zweiter Ordnung*, Composito Math. **6** (1939), 368–374.
42. __________, *Über die Nullstellen von Funktionen, die Lösung Sturm-Liouville'scher Differentialgleichungen sind*, Comment. Math. Helv. **16** (1943–4), 153–199.
43. HARTMAN, P. H., and A. WINTNER, *On an oscillation criterion of de la Vallée Poussin*, Quart. Appl. Math. **13** (1955), 330–332.
44. SANSONE, G., "Equazioni differenziali nel campo reale," Vol. I, 2nd ed., Bologna, 1948.
45. FINK, A. M. and D. S. ST. MARY, *On an inequality of Nehari*, Notices Amer. Math. Soc. **16** (1969), 91.
46. LEVIN, A. Yu, *On linear second order differential equations*, Dokl. Akad. Nauk SSSR **153** (1963), 1257–1260, (Soviet. Math. **4** (1963), 1814–1817).
47. HOCHSTADT, H., *On an inequality of Lyapunov*, Proc. Amer. Math. Soc. **22** (1969), 282–284.
48. FINK, A. M., *The functional* $T\int_0^T R$ *and the zeroes of a second order linear differential equation*, J. Math. Pures et Appl. **45** (1966), 387–394.
49. LEIGHTON, W., *On the zeroes of solutions of a second-order linear differential equation*, J. Math. Pures et Appl. **44** (1965), 296–310.
50. BANKS, D., *Bounds for the eigenvalues of some vibrating systems*, Pacific J. Math. **10** (1960), 439–474.

51. BARNES, D. C., *Bounds for the eigenvalues of some vibrating systems*, Pacific J. Math. **29** (1969), 43- -61.
52. GALBRAITH, A., *On the zeros of solutions of ordinary differential equations of the second order*, Proc. Amer. Math. Soc. **17** (1966), 333–337.
53. FINK, A. M. *On the zeroes of $y''+py=0$ with linear, convex and concave p*, J. Math. Pures et Appl. **46** (1967), 1–10.
54. ELBERT, Á., *On the zeros of the solutions of the differential equation $y''+q(x)y=0$, where $[q(x)]^{\nu}$ is concave*, Studia Sci. Math. Hungarica **2** (1967), 293–298.
55. ________, *On a functional connected with the zeros of the solutions of the differential equation $y''+q(t)y=0$, where $q(t)$ is a nonnegative, monotonic, concave function*, Ann. Univ. Sci. Budapest. Rol. Eot. nom. sect. Math. **11** (1968), 129–136.
56. ________, *On the zeros of solutions of ordinary second order differential equations*, Publ. Math. (Debrecen) **15** (1968), 13–17.
57. ________, *On the solutions of the differential equation $y''+q(x)y=0$, where $[q(x)]^{\nu}$ is concave*, I. Acta Math. Acad. Sci. Hung. **20** 1–2 (1969), 1–11.
58. ________, *On the solutions of the differential equation $y''+q(x)y=0$, where $[q(x)]^{\nu}$ is concave, II*, Acta Math. Acad. Sci. Hung. **4** (1969), 257–266.
59. SCHWARZ, B. *On the extrema of the frequencies of nonhomogeneous strings with equimeasurable density*, J. Math. and Mech. **10** (1961), 401–422.
60. BEESACK, P. R. and B. SCHWARZ, *On the zeros of solutions of second-order linear differential equations*, Canad. J. of Math. **8** (1956), 504–515.
61. ELIASON, S. B., *The integral $T\int_{-T/2}^{T/2} p(t)dt$ and the boundary value problem $x''+p(t)x=0$, $x(-T/2)=x(T/2)=0$*, J. Diff. Eq. **4** (1968), 646--659.
62. ST. MARY, D. F. and S. B. ELIASON, *Upper bounds of $T\int_{-T/2}^{T/2} p(t)dt$ and the differential equation $x''+p(t)x=0$*, J. Diff. Eq. **6** (1969), 154–160.
63. LEVIN, V., *On some integral inequalities involving periodic functions*, J. London Math. Soc. **10** (1935), 45–48.
64. SATÔ, T., *Seizi Senkei Bibunhôteisiki no Zero-ten no Bunpu ni tuite*, Kansû Hôteisiki **22** (1940), 39–43.
65. TUMURA, M., *Kôkai Zyôbibunhôteisiki ni tuite*, Kansû Hôteisiki **30** (1941), 20–35.

66. BORG, G., *Über die Stabilität geisser Klassen von linearen Differentialgleichungen*, Ark. Mat. Ast. Fys. **31** A (1944),No. 1, 31 pp.
67. ________, *On a Liapounoff crieterion of stability*, Amer.J. Math. **71** (1949), 67–70.
68. CORDES, H. O., *An inequality of G. Borg*, Amer. Math. Monthly **63** (1956), 27–29.
69. COHEN, H. J., *Problem E 1205*, Amer. Math. Monthly **63** (1956), 582–583.
70. NEHARI, Z., *Univalent functions and linear differential equations*, in "Lectures on Functions of a Complex Variable," Ann Arbor, 1955, pp. 148–151.
71. BEESACK, P. R., *Nonoscillation and disconjugacy in the complex domain*, Trans. Amer. Math. Soc. **81** (1956), 211–242.
72. ELBERT, Á., *Oboščenie odnogo neravenstva Ljapunova dlja differencial'nyh uravneiĭ s zapazdyvajuščim argumentom*, Ann. Univ. Sci. Budapest. Rol. Eöt. nom. sect. Mat. **12** (1969), 107–111.
73. ELIASON, S. B., *A Lypanov inequality for a certain second order nonlinear differential equation*, J. Lond. Math. Soc. **2** 2 (1970), 461–466.
74. ________, *Lypanov inequalities and bounds on solutions of certain second order equations*, Canad. Math. Bull. **17** (1974), 499–504.
75. ST. MARY, D. F., *Some oscillation and comparison theorems for* $(r(t)y')' + p(t)y = 0$, J. Differential Eq. **5** (1969), 314–323.
76. KWONG, M. K., *On Lyapunov's inequality for disfocality*, J. Math. Anal. Appl. **83** (1981), 486–494.
77. HARRIS, B. J., *On an inequality of Lyapunov for disfocality*, J. Math. Anal. Appl. **146** (1990), 495- -500.
78. AGARWAL, R. P. and G. V. MILOVANOVIĆ, *On an inequality of Bogar and Gustafson*, J. Math. Anal. Appl. **146** (1990), 207–216.
79. BOGAR, G. A. and G. B. GUSTAFSON, *Effective estimates of invertibility intervals for linear multipoint boundary value problems*, J. Differential Eq. **29** (1978), 180–204.
80. LEIGHTON, W., *On Liapunov's inequality*, Proc. Amer. Math. Soc. **33** (1972), 627–628.
81. KREĬN, M. G., *Ob odnom predpoloženii A. M. Ljapunova*, Funk. Anal. Prilož. **7** (1973), 45–54.
82. ELBERT, A., *Nekotorye zadači kačestvennoĭ teorii differencial'nyh uravneniĭ vtorogo porjadka*, Univ. Družbi narodov im. P. Lumumby, Moskva, 1971.
83. BEESACK, P. R., *On the Green's function of an N-point boundary value problem*, Pacific J. Math. **12** (1962), 801–812.
84. CHENG, S. S., *A discrete analogue of the inequality of Lyapunov*, Hokkaido Math. J. **12** 1 (1983), 105–112.

CHAPTER VII

ZMOROVIČ'S AND RELATED INEQUALITIES

1. In [**1**] V. A. Zmorovič has proved the following theorem:

THEOREM 1. *If the function $f: [a-h\ a+h] \to \mathbf{R}$ is twice continuously-differentiable, then*

$$\int_{a-h}^{a+h} (f''(x))^2\,dx \geq \frac{3}{2h^3}\,[f(a+h)-2f(a)+f(a-h)]^2, \tag{1.1}$$

with equality if and only if f is given by

$$f(x) = \begin{cases} C_1\{(h-a+x)^3+6h^2(a-x)\}+C_2x+C_3 & (x\in[a-h,\ a]), \\ C_1(h+a-x)^3+C_2x+C_3 & (x\in[a,\ a+h]), \end{cases}$$

where $C_1,\ C_2,\ C_3$ are arbitrary real constants.

The mentioned Zmorovič's result is an improvement of the inequality

$$\int_{a-h}^{a+h} (f''(x))^2\ dx \geq \frac{1}{2h^3}\,[f(a+h)-2f(a)+f(a-h)]^2, \tag{1.2}$$

which was obtained by M. A. Lavrent'ev (see [**1**]) through geometric considerations under stronger conditions. Lavrent'ev proved the last inequality under the condition that f is a four times continuously-differentiable function on $[a-h,\ a+h]$.

Zmorovič [**1**] also proved a weighted version of Theorem 1.

A similar result may be found in Hardy, Littlewood and Pólya [**2**] (Theorem 264):

THEOREM 2. *If*

$$f(-1)=-1,\ f(1)=1,\ f'(-1)=f'(1)=0$$

and k is a positive integer, then

$$\int_{-1}^{1} (f''(x))^{2k}\ dx \geq 2\left(\frac{4k-1}{2k-1}\right)^{2k-1}, \tag{1.3}$$

with inequality unless

$$f(x) = \frac{4k-1}{2k}\, x - \frac{2k-1}{2k}\, x^{(4k-1)/(2k-1)}.$$

2. Generalizations of Zmorovič's results are obtained by R. Ž. Djordjević and G. V. Milovanović [**3**].

We give their text with small modifications.

First introduce the notation

$$\Delta = \frac{1}{h^2}\,[f(a+h) - 2f(a) + f(a-h)]$$

and note that

$$\int_0^h (h-t)\,(f''(a-t) + f''(a+t))\,dt = h^2\Delta. \tag{2.1}$$

THEOREM 1. *Let the function $f\colon [a-h,\ a+h] \to \mathbf{R}$ be twice continuously-differentiable and $g\colon [a-h,\ a+h] \to \mathbf{R}^+$ continuous.*

Then

$$\int_{a-h}^{a+h} g(x)|f''(x)|^r\, dx \geq \frac{h^{2r}}{\delta(r)^{r-1}}\, |\Delta|^r \qquad (r > 1), \tag{2.2}$$

where

$$\delta((r) = \int_0^h (h-t)^{\frac{r}{r-1}} \left(g(a-t)^{\frac{1}{1-r}} + g(a+t)^{\frac{1}{1-r}}\right) dt. \tag{2.3}$$

Equality in (2.2) holds if and only if the function f is given by

$$f(x) = \begin{cases} A_1 \int\limits_a^x (x-t) \left|\frac{h-a+t}{g(t)}\right|^{\frac{1}{r-1}} dt + A_2 x + A_3 & (x \in [a-h,\ a]), \\ A_1 \int\limits_a^x (x-t) \left|\frac{h+a-t}{g(t)}\right|^{\frac{1}{r-1}} dt + A_2 x + A_3 & (x \in [a,\ a+h]), \end{cases} \tag{2.4}$$

where $A_1,\ A_2,\ A_3$ are arbitrary real constants.

PROOF: Put $\gamma(t) = \left(g(a-t)^{\frac{1}{1-r}} + g(a+t)^{\frac{1}{1-r}}\right)^{\frac{1-r}{r}}$. Then (2.3) becomes

$$\delta(r) = \int_0^h \left(\frac{h-t}{\gamma(t)}\right)^{\frac{r}{r-1}} dt. \tag{2.5}$$

Since

$$\int_{a-h}^{a+h} g(x)|f''(x)|^r\, dx = \int_0^h (g(a-t)|f''(a-t)|^r + g(a+t)|f''(a+t)|^r)\, dt,$$

putting

$$p_1 = g(a-t),\ p_2 = g(a+t),\ z_1 = f''(a-t),\ z_2 = f''(a+t),$$

and using the inequality (see [4],[5])

$$|z_1 + \cdots + z_n|^r \le \left(\sum_{i=1}^n p_i^{\frac{1}{1-r}}\right)^{r-1} (p_1|z_1|^r + \cdots + p_n|z_n|^r) \tag{2.6}$$

$$(z_i \in \mathbf{C},\ p_i > 0\ (i = 1, \ldots, n),\ r > 1),$$

we have

$$\int_{a-h}^{a+h} g(x)|f''(x)|^r\, dx \ge \int_0^h \frac{|f''(a-t) + f''(a+t)|^r}{\left(g(a-t)^{\frac{1}{1-r}} + g(a+t)^{\frac{1}{1-r}}\right)^{r-1}}\, dt$$

$$= \int_0^h (\gamma(t)|f''(a-t) + f''(a+t)|)^r\, dt.$$

In view of (2.5)

(2.7)

$$\int_{a-h}^{a+h} g(x)|f''(x)|^r\, dx$$

$$\ge \frac{1}{\delta(t)^{r-1}} \left(\int_0^h (\gamma(t)|f''(a-t) + f''(a+t)|)^r\, dt\right) \left(\int_0^h \left(\frac{h-t}{\gamma(t)}\right)^{\frac{r}{r-1}}\, dt\right)^{r-1}.$$

Applying Hölder's inequality to the right side in (2.7), we obtain

$$\int_{a-h}^{a+h} g(x)|f''(x)|^r\, dx \ge \frac{1}{\delta(r)^{r-1}} \left(\int_0^h (h-t)|f''(a-t) + f''(a+t)|dt\right)^r$$

$$\ge \frac{1}{\delta(r)^{r-1}} \left|\int_0^h (h-t)(f''(a-t) + f''(a+t))\, dt\right|^r$$

$$= \frac{h^{2r}\Delta^r}{\delta(r)^{r-1}}.$$

Since equality holds in (2.6) if and only if

$$p_1|z_1|^{r-1} = \cdots = p_n|z_n|^{r-1} \quad \text{and} \quad z_k\bar{z}_j \geq 0 \ (k,j = 1,\ldots,n),$$

and in Hölder's inequality

$$\int_\alpha^\beta |\Phi(x)\Psi(x)|\,dx \leq \left(\int_\alpha^\beta |\Phi(x)|^p\,dx\right)^{1/p} \left(\int_\alpha^\beta |\Psi(x)|^q\,dx\right)^{1/q} \quad \left(\frac{1}{p}+\frac{1}{q}=1,\ p>1\right)$$

if and only if $|\Phi(x)|^p = C|\Psi(x)|^q$, where C is a real constant (see [**6**, p. 54]), we conclude that equality in (2.2) holds if and only if

$$\begin{aligned} &g(a-t)^{\frac{1}{r-1}} f''(a-t) = g(a+t)^{\frac{1}{r-1}} f''(a+t), \\ &(\gamma(t)\left(f''(a-t)+f''(a+t)\right))^r = C\left(\frac{h-t}{\gamma(t)}\right)^{\frac{r}{r-1}}. \end{aligned} \tag{2.8}$$

From (2.8), if we put $C = A_1^r$, it follows that

$$f''(a-t) = A_1\left(\frac{h-t}{g(a-t)}\right)^{\frac{1}{r-1}}, \ f''(a+t) = A_1\left(\frac{h-t}{g(a+t)}\right)^{\frac{1}{r-1}} \quad (0 \leq t \leq h).$$

By integration, we obtain (2.4).

This completes the proof.

COROLLARY 1. *If the function* $f\colon [a-h,\ a+h] \to \mathbf{R}$ *is twice continuously-differentiable, then*

$$\int_{a-h}^{a+h} |f''(x)|^r\,dx \geq \left(\frac{2r-1}{2r-2}\right)^{r-1} h|\Delta|^r \quad (r>1). \tag{2.9}$$

Equality in (2.9) *holds if and only if the function* f *is given by*

$$f(x) = \begin{cases} C_1\left\{(h-a+x)^{\frac{2r-1}{r-1}} + \frac{4r-2}{r-1}\,h^{\frac{r}{r-1}}(a-x)\right\} + C_2x + C_3 & (x \in [a-h,\ a]) \\ C_1(h+a-x)^{\frac{2r-1}{r-1}} + C_2x + C_3 & (x \in [a,\ a+h]), \end{cases}$$

where C_i $(i=1,2,3)$ *are arbitrary real constants.*

Putting $g(x) \equiv 1$, the statement of the Corollary 1 follows from Theorem 1.

REMARK: For $r = 2$, The Corollary 1 reduces to Theorem 1 of Section 1.

Observing that the left side of inequality (2.2) is of the form

$$\int_{a-h}^{a+h} g(x)\Phi\left(f''(x)\right)\,dx,$$

where $\Phi(t)=|t|^r$ $(r>1)$, we conclude that Φ is a convex function. We are led to consider now general versions of Theorem 1.

First, we give the following definition.

DEFINITION: A continuous function $\Phi\colon \mathbf{R}\to\mathbf{R}^+$ belongs to the class M if there is a convex function $F\colon \mathbf{R}\to\mathbf{R}^+$ and real numbers λ and m so that for each $x\in\mathbf{R}$ the inequalities

$$F(x)\le\Phi(x)^{1/m}\le\lambda F(x)\quad(\lambda\ge 1;\ m>1)$$

are valid.

A convex function is continuous and a Jensen convex function as defined in [5].

On of the possible generalizations of Theorem 1 is:

THEOREM 2. *Let the functions* $\Phi\colon\mathbf{R}\to\mathbf{R}^+$, $f\colon[a-h,\ a+h]\to\mathbf{R}$, $g\colon[a-h,\ a+h]\to\mathbf{R}^+$ *satisfy the conditions:*

1^0 $\Phi\in M$;

2^0 f *is twice continuously-differentiable;*

3^0 g *continuous.*

Then

$$\int_{a-h}^{a+h} g(x)\Phi\left(f''(x)\right)\,dx\ge\frac{h^{2m}}{\lambda^m\delta(m)^{m-1}}\,\Phi(\Delta),$$

where δ *is given by* (2.3).

REMARK: Zmorovič's weighted version of (1.1) can be obtained from Theorem 1 for $r=2$ and $g(a-t)\equiv g(a+t)$.

3. Further generalizations of Theorems 1 and 2 are obtained by R.Ž.Djordvević, G. V. Milovanović and J. E. Pečarić [7].

We use the standard operators D (differentiation operator), δ (central difference operator), μ (averaging operator), which are defined by

$$Df(x) = h\frac{df(x)}{dx}, \quad \delta f(x) = f\left(x+\frac{h}{2}\right) - f\left(x-\frac{h}{2}\right),$$
$$\mu f(x) = \frac{1}{2}\left(f\left(x+\frac{h}{2}\right) + f\left(x-\frac{h}{2}\right)\right),$$

(3.1) $$\|\Phi\|_{r,p} = \left(\frac{1}{2h}\int_{a-h}^{a+h} p(x)|\Phi(x)|^r dx\right)^{1/r}, \quad \|\Phi\|_r = \|\Phi\|_{r,1} \quad (r>1),$$

$$\sigma_n(r) = \left(\frac{1}{2h}\int_0^h (h-t)^{\frac{(n-1)r}{r-1}}\left(p(a-t)^{\frac{1}{1-r}} + p(a+t)^{\frac{1}{1-r}}\right) dt\right)^{\frac{r-1}{r}},$$

THEOREM 1. *Let $f \in C^n[a-h,\ a+h]$ and the function $p\colon [a-h,\ a+h] \to \mathbf{R}^+$ be continuous. Then*

(3.2) $$\|f^{(n)}\|_{r,p} \ge \frac{(n-1)!}{h\sigma_n(r)}|M_n f(a)| \quad (r>1),$$

where

(3.3) $$M_n = \frac{1}{2}\delta^2 - \sum_{i=1}^{\frac{n-2}{2}} \frac{D^{2i}}{(2i)!} \qquad (n \text{ is even}),$$
$$= \mu\delta - \sum_{i=1}^{\frac{n-1}{2}} \frac{D^{2i-1}}{(2i-1)!} \qquad (n \text{ is odd}).$$

Equality in (3.2) holds if and only if the function f is given by

(3.4)
$$f(x) = \begin{cases} A_n \int_a^x (x-t)^{n-1}\left[\frac{(h-a+t)^{n-1}}{p(t)}\right]^{\frac{1}{r-1}} dt + \sum_{k=0}^{n-1} A_k(x-a)^k & (x\in[a-h,\ a]), \\ A_n \int_a^x (x-t)^{n-1}\left[\frac{(h+a-t)^{n-1}}{p(t)}\right]^{\frac{1}{r-1}} dt + \sum_{k=0}^{n-1} A_k(x-a)^k & (x\in[a,\ a+h]), \end{cases}$$

where A_k $(k = 0,1,\ldots,n)$ are arbitrary real constants.

For $p(x) \equiv 1$ we obtain the following result:

THEOREM 2. *If $f \in C^n[a-h,\ a+h]$ and $r > 1$, then:*

$$\|f^{(n)}\|_r \geq \frac{(n-1)!}{h^n}\left(\frac{nr-1}{r-1}\right)^{\frac{r-1}{r}} |M_n f(a)|. \tag{3.5}$$

The last Theorem (for $n = 2$) represents the generalization of a result from [3].

From Theorem 1 immediately follows:

THEOREM 3. *Let functions f and p satisfy the conditions as in Theorem 1 and let $r > 1$.*

1^0 *If $n = 2k$ and $f^{(2i)}(a) = 0$ $(i = 1, \dots, k-1)$, then*

$$\|f^{(2k)}\|_{r,p} \geq \frac{(2k-1)!}{2h\sigma_{2k}(r)}|f(a+h) - 2f(a) + f(a-h)|;$$

2^0 *If $n = 2k+1$ and $f^{(2i-1)}(a) = 0$ $(i = 1, \dots, k)$, then*

$$\|f^{(2k+1)}\|_{r,p} \geq \frac{(2k)!}{2h\sigma_{2k+1}(r)}|f(a+h) - f(a-h)|.$$

The following result is also proved in [7]:

Let $n \in N$, $r > 1$ and $\gamma_n(r) = \left(\frac{1}{2h}\int\limits_{a-h}^{a+h} \frac{|t-a|^{\frac{(n-1)r}{r-1}}}{p(t)^{\frac{1}{r-1}}}\, dt\right)^{\frac{r-1}{r}}$.

THEOREM 4. *Let $f \in C^n[a-h,\ a+h]$, $p \in C[a,b]$, $p(x) > 0$ $(x \in [a,b])$. Then*

$$\|f^{(n)}\|_{r,p} \geq \frac{(n-1)!}{2h\gamma_n(r)}\left|\sum_{k=0}^{n-1} \frac{h^k}{k!}\left(f^{(k)}(a-h) - (-1)^k f^{(k)}(a+h)\right)\right|, \tag{3.6}$$

with equality if and only if

$$f^{(n)}(t) = A\left(\frac{(t-a)^{n-1}}{p(t)}\right)^{\frac{1}{r-1}} \quad (A \in \mathbf{R}), \tag{3.7}$$

when n is odd;

$$f^{(n)}(t) = A\left(\frac{|t-a|^{n-1}}{p(t)}\right)^{\frac{1}{r-1}} \operatorname{sgn}(t-a) \quad (A \in \mathbf{R}), \tag{3.8}$$

when n is even.

Theorem 5 follows from Theorem 4.

THEOREM 5. *Let $f \in C^n[a-h,\ a+h]$ and*

$$f^{(k)}(a-h) = (-1)^k f^{(k)}(a+h) \qquad (k = 1, \dots, n-1), \tag{3.9}$$

then

$$\|f^{(n)}\|_r \geq \frac{(n-1)!}{2h^n} \left(\frac{nr-1}{r-1}\right)^{\frac{r-1}{r}} |f(a+h) - f(a-h)|. \tag{3.10}$$

REMARK 1: For $n = 2$ and $r = 2k$, this resut reduces to the Theorem 264 from [**2**], i.e. to Theorem 2 from 1.

REMARK 2: If $f \in C^1[a,b]$, according to Theorem 5 (for $n = 1$), we have

$$\int_a^b |f'(x)|^r \, dx \geq \frac{1}{(b-a)^{r-1}} |f(b) - f(a)|^r \qquad (r > 1).$$

This result is a particular case of the inequality (4.12) from [**8**].

4. Important generalizations of Zmorovič's inequality are given by A. Lupaş [**9**], and we give his results without modifications. If $-\infty < a < b < +\infty$ and $f\colon [a,b] \to R$ let us denote by $[x_0, x_1, \dots, x_n; f]$ the divided difference at the system of knots

$$d\colon \quad a = x_0 < x_1 < \cdots < x_{n-1} < x_n = b.$$

For $f \in C^n[a,b]$, $g \in C[a,b]$, $g > 0$ and $p \in (1, +\infty)$, we use the notation

$$\|f^{(n)}\|_p = \left(\frac{1}{b-a} \int_a^b |f^{(n)}(t)|^p \, dt\right)^{1/p} \quad \text{and } \|f^{(n)}\|_{g,p} = \left(\int_a^b g(t) |f^{(n)}(t)|^p \, dt\right)^{1/p}.$$

The aim is to find the best upper bound $K_n^* = K_n^*(d, g, p)$ in the inequality

$$\|[a, x_1, \dots, x_{n-1}, b; f]\| \leq K_n^* \|f^{(n)}\|_{g,p}, \qquad n \geq 2.$$

We define $w(x) = w(x\colon d) = \prod_{j=0}^{n} (x - x_j)$ and

$$Q_n(t) = Q_n(t\colon d) = \frac{1}{(n-1)!} \sum_{i=0}^{n} \frac{(x_i - t)_+^{n-1}}{w'(x_i)}, \quad t \in [a,b], \tag{4.1}$$

where

$$(x-t)_+^{n-1} = \begin{cases} (x-t)^{n-1}, & a \leq t \leq x \leq b \\ 0, & a \leq x < t \leq b. \end{cases}$$

The case $g = 1$ and $p = 2$ is the easiest.

THEOREM 1. *If $f \in C^{(n)}[a,b]$ and $a < x_1 < \cdots < x_{n-1} < b$, then*

$$\|[a, x_1, \ldots, x_{n-1}, b; f]\| \leq K_n(d)\|f^{(n)}\|_2, \tag{4.2}$$

where

$$K_n(d) = \left(\frac{(-1)^n(b-a)}{(2n-1)!}\sum_{k=0}^{n-1}\sum_{i=k+1}^{n}\frac{(x_i - x_k)^{2n-1}}{w'(x_i)\cdot w'(x_k)}\right)^{1/2}. \tag{4.3}$$

If

$$f(t) = f_*(t) = C\sum_{i=1}^{n}\frac{(x_i - t)_+^{2n-1}}{w'(x_i)} + P_{n-1}(t), \tag{4.4}$$

P_{n-1} being an arbitrary polynomial of the degree $n-1$ and C a real constant, then equality holds in (4.2).

PROOF: If $f \in C^{(n)}[a,b]$, then

$$f(x) = \sum_{k=0}^{n-1}\frac{f^{(k)}(a)}{k!}(x-a)^k + \int_a^x f^{(n)}(t)\frac{(x-t)^{n-1}}{(n-1)!}\,dt.$$

In other words

$$f(x) = P_{n-1}(x) + \int_a^b f^{(n)}(t)\frac{(x-t)_+^{n-1}}{(n-1)!}\,dt,$$

where P_{n-1} is a polynomial of the degree $n-1$.

Since $[a, x_1, \ldots, x_{n-1}, b; P_{n-1}] = 0$ one obtains

$$[a, x_1, \ldots, x_{n-1}, b; f] = \int_a^b f^{(n)}(t)Q_n(t)\,dt, \tag{4.5}$$

where Q_n is define as in (4.1). If in (4.4) $C = \frac{(-1)^n}{(2n-1)!}$, then $f_*^{(n)} = Q_n$. Therefore (4.5) enables us to write

$$\int_a^b Q_n^2(t)\,dt = [a, x_1, \ldots, x_{n-1}, b; f_*]$$

$$= \frac{(-1)^n}{(2n-1)!}\sum_{k=0}^{n-1}\sum_{i=0}^{n}\frac{(x_i - x_k)_+^{2n-1}}{w'(x_i)\cdot w'(x_k)} = \frac{1}{b-a}\left[K_n(d)\right]^2.$$

According to (4.5)

$$|[a, x_1, \ldots, x_{n-1}, b; f]| \leq \left(\int_a^b Q_n^2(t)\, dt\right)^{1/2} \cdot \left(\int_a^b |f^{(n)}(t)|^2\, dt\right)^{1/2}$$
$$= K_n(d) \cdot \|f^{(n)}\|_2,$$

with equality if and only if $f^{(n)}(t) = C_1 \cdot Q_n(t)$ where C_1 is a real constant.

Taking into account that $K_2(d) = 1/\sqrt{3}$, the following proposition is true.

THEOREM 2. *If $f \in C^{(2)}[a, b]$ and $a < x < b$, then*

$$|[a, x, b; f]| \leq \frac{1}{\sqrt{3}} \|f''\|_2,$$

with equality if and only if f is given by

$$f(t) = \begin{cases} C\frac{(b-t)^3}{b-x} + C\frac{(t-x)^3(b-a)}{(b-x)(x-a)} + C_1 t + C_2, & t \in [a, x] \\ C\frac{(b-t)^3}{b-x} + C_1 t + C_2 & t \in (x, b] \end{cases}$$

where C, C_1, C_2 are arbitrary real constants.

This theorem is an extension of inequality (1.1). Let us assume that $f(a) = f(b) = 0$. From the above theorem one find the inequality

$$|f(x)| \leq \frac{(x-a)(b-x)}{\sqrt{3}} \|f''\|_2.$$

Taking into account that $(x-a)(b-x) \leq \frac{(b-a)^2}{4}$, $x \in [a, b]$, we obtain $|f(x)| \leq \frac{(b-a)^2}{4\sqrt{3}} \|f''\|_2$, $x \in [a, b]$. Therefore, with the notation $\|f\|_\infty = \max\limits_{x \in]a,b]} |f(x)|$ we have prove the following

THEOREM 3. *If $f \in C^{(2)}[a, b]$, $f(a) = f(b) = 0$, then*

$$\|f\|_\infty \leq \frac{(b-a)^2}{4\sqrt{3}} \|f''\|_2,$$

the equality case being valid for

$$f(t) = \begin{cases} c(t-a)^3 - \frac{3}{4}c(t-a)(b-a)^2, & t \in [a,\ (a+b)/2] \\ c(b-t)^3 - \frac{3}{4}c(b-t)(b-a)^2, & t \in ((a+b)/2,\ b], \end{cases}$$

where c is an arbitrary real constant.

This result is the equation $K(2,1,\infty,2)=\frac{1}{4\sqrt{3}}$ (see 18. of Chapter II).

Another particular case of the inequality (4.2) is the following:

if the points $x_k\in(d)$ are equidistant, that is $x_k=a+\frac{k}{n}(b-a)$, $k=0,1,\ldots,n$, then

$$[a,x_1,\ldots,x_{n-1},b;f]=\frac{n^n}{n!(b-a)^n}\Delta^n f(a),$$

where $\Delta^n f(a)=\sum\limits_{k=0}^{n}(-1)^{n-k}\binom{n}{k}f\left(a+\frac{k}{n}(b-a)\right)$. From Theorem 1 one finds the following proposition

THEOREM 4. *$f\in C^{(n)}[a,b]$, then*

$$|\Delta^n f(a)|\le(b-a)^n\cdot C_n^*\|f^{(n)}\|_2,$$

where

$$C_n^*=\frac{1}{n^{n-1}}\left(2\sum_{k=1}^{n}\frac{(-1)^{n-k}k^{2n-1}}{(n+k)!(n-k)!}\right)^{1/2}.$$

Moreover,

$$C_2^*=\frac{1}{\sqrt{12}},\quad C_3^*=\frac{1}{18}\sqrt{\frac{11}{15}},\quad C_4^*=\frac{1}{384}\sqrt{\frac{151}{35}},\quad C_5^*=\frac{1}{45,000}\sqrt{\frac{15,619}{35}}.$$

Let $p>1$ and $Q_n(t)=Q_n(t\colon d)$ be defined as in (1). Taking into account that $Q_n\ge 0$ (see [**10**]) denote

$$K_n^*=K_n^*(d,g,p)=\left(\int_a^b\frac{[Q_n(t)]^{p/(p-1)}}{[g(t)]^{1/(p-1)}}\,dt\right)^{(p-1)/p},\tag{4.6}$$

where $g\in C[a,b]$ is a positive function. By means of Hölder's inequality from (4.5) we obtain

$$\|[a,x_1,\ldots,x_{n-1},b]\|\le\int_a^b|f^{(n)}(t)|[g(t)]^{1/p}\cdot\frac{Q_n(t)}{[g(t)]^{1/p}}\,dt$$

$$\le\|f^{(n)}\|_{g,p}\left(\int_a^b\frac{[Q_n(t)]^q}{[g(t)]^{q/p}}\right)^{1/q},\quad\frac{1}{p}+\frac{1}{q}=1.$$

Therefore the following result is true:

THEOREM 5. *If $f \in C^{(n)}[a,b]$ and (d): $a < x_1 < \cdots < x_{k-1} < b$ is an arbitrary system of knots, then*

$$\|[a, x_1, \dots, x_{n-1}, b; f]\| \le K_n^* \|f^{(n)}\|_{g,p}, \tag{4.7}$$

where K_n^* is defined in (4.6). If

$$f^{(n)}(t) = C\left(\frac{Q_n(t:\, d)}{g(t)}\right)^{1/(p-1)},$$

then equality holds in (4.7).

The following special cases are interesting:

THEOREM 6. *If $f \in C^{(2)}[a,b]$ and $a < x < b$, $p > 1$, then*

$$\|[a, x, b; f]\| \le \left(\frac{p-1}{2p-1}\right)^{\frac{p-1}{p}} \cdot \|f''\|_p, \tag{4.8}$$

with equality for

$$f(t) = \begin{cases} C\left(\frac{(t-a)^{2p-1}}{x-a}\right)^{\frac{1}{p-1}} + C_1 t + C_2, & a \le t \le x \\ C\left(\frac{(b-t)^{2p-1}}{b-x}\right)^{\frac{1}{p-1}} + C_1 t + C_2 + C(b-a)\left(2t - a - b + \frac{t-x}{p-1}\right), & x < t \le b. \end{cases}$$

THEOREM 7. *If $f \in C^{(2)}[a,b]$, $f(a) = f(b) = 0$, then*

$$\|f\|_\infty \le \frac{(b-a)^2}{4}\left(\frac{p-1}{2p-1}\right)^{\frac{p-1}{p}} \cdot \|f''\|_p$$

with equality for

$$f(t) = C\begin{cases} \left(\frac{2}{b-a}\right)^{\frac{1}{p-1}} (t-a)^{\frac{2p-1}{p-1}} - (t-a)(b-a)\frac{2p-1}{2(p-1)}, & a \le t \le \frac{a+b}{2} \\ \left(\frac{2}{b-a}\right)^{\frac{1}{p-1}} (b-t)^{\frac{2p-1}{p-1}} - (b-t)(b-a)\frac{2p-1}{2(p-1)}, & \frac{a+b}{2} < t \le b. \end{cases}$$

This result is $K(n,1,\infty,p) = \frac{1}{4}\left(\frac{p-1}{2p-1}\right)^{\frac{p-1}{p}}$ (See 18. of Ch. II).

5. Generalizations of Lupaş' results are given by J. E. Pečarić in [**11**].

Let $u_0, u_1, \ldots, u_n$ be of class $C^n[a,b]$, and

$$u_0(t) = w_0(t)$$

$$u_1(t) = w_0(t)\int_a^t w_1(s_1)ds_1,$$

$$u_2(t) = w_0(t)\int_a^t w_1(s_1)\int_a^{s_1} w_2(s_2)ds_2ds_1,$$

$$\vdots$$

$$u_n(t) = w_0(t)\int_a^t w_1(s_1)\int_a^{s_1} w_2(s_2)\ldots\int_a^{s_{n-1}} w_n(s_n)ds_n\ldots ds_1,$$

where $w_i(t)$ are positive functions of class C^{n-1} on the closed interval $[a,b]$. Then $\{u_t\}_0^n$ is called Extended Complete Tchebycheff System or an ECT-system.

A function f defined on the open interval (a,b) is convex with respect to $\{u_t\}_0^n$ if

$$U\begin{pmatrix} u_0, u_1, \ldots, u_n, f \\ t_0, t_1, \ldots, t_n, t_{n+1}\end{pmatrix} = \begin{vmatrix} u_0(t_0) & u_0(t_1) & \ldots & u_0(t_{n+1}) \\ u_1(t_0) & u_1(t_1) & \ldots & u_1(t_{n+1}) \\ \vdots & \vdots & & \vdots \\ u_n(t_0) & u_n(t_1) & \ldots & u_n(t_{n+1}) \\ f(t_0) & f(t_1) & \ldots & f(t_{n+1}) \end{vmatrix} \geq 0$$

for all choice of $\{t_i\}_0^{n+1}$ satisfying $a < t_0 < t_1 < \cdots < t_{n+1} < b$. In symbols we write

$$f \in C(u_0, u_1, \ldots, u_n).$$

Note also that [**12**, pp. 382 and 395] the function

$$f(t) = f_n(t;x) = \begin{cases} w_0(t)\int_x^t w_1(s_1)\int_x^{s_1} w_2(s_2)\ldots\int_x^{s_{n-1}} w_n(s_n)ds_n\ldots ds_1, & x \leq t \leq b, \\ 0, \quad a \leq t \leq x, \end{cases}$$

is contained in $\bigcap_{k=0}^n C(u_0, u_1, \ldots, u_k)$. For $n = 0$, $f_n(t;x) = w_0(t)$ for $x \leq t \leq b$ and equals zero otherwise. (Notice that $f_n(t;a) = u_n(t)$.)

Let D_j, $j = 0, 1, \ldots, n$, denote the first-order differential operator

$$(D_jf)(t) = \frac{d}{dt}\left(\frac{f(t)}{w_j(t)}\right).$$

Suppose that the function f satisfies the condition that $D_n D_{n-1} \ldots D_0 f$ is continuous on $[a, b]$, then (see[**12**, pp. 388–389])

$$f(t) = \sum_{i=0}^{n} a_i u_i(t) + \int_a^b f_n(t; x)(d_n \ldots D_0 f)(x)dx, \tag{5.1}$$

where $a_0 = f(a)/w_0(a)$ and $a_i = ((D_{i-1} \ldots D_0 f)(a))\, w_i(a)$, $i = 1, \ldots, n$, and

$$\left\| f - \sum_{i=0}^{n} a_i u_i \right\| \le C_0 \|D_n \ldots D_0 f\|, \tag{5.2}$$

where $\|f\| = \sup_{a \le t \le b} |f(t)|$ and $C_0 = \int_a^b f_n(b; x)dx$.

We use the notation

$$\|D_n \cdots D_0 f\|_p = \left(\frac{1}{b-a} \int_a^b |(D_n \cdots D_0 f)(t)|^p \, dt \right)^{1/p}$$

$$\|D_n \cdots D_0 f\|_{g,p} = \left(\int_a^b g(t)|D_n \cdots D_0 f)(t)|^p \, dt \right)^{1/p},$$

where $g(t) > 0$ ($\forall t \in [a, b]$).

The following results are also proved in [**10**]:

THEOREM 1. *Let $f\colon [a, b] \to \mathbf{R}$ be a real function such that $D_n \cdots D_0 f$ is continuous on $[a, b]$ and $a < t_1 < \cdots < t_n < b$, then*

$$\left| U \begin{pmatrix} u_0, u_1, \ldots, u_n, f \\ a, t_1, \ldots, t_n, b \end{pmatrix} \right| \le K_{g,n} \|D_n \cdots D_0 f\|_{g,p}, \tag{5.3}$$

where

$$\begin{aligned} K_{g,n} &= \left(\int_a^b \left(Q_n(x)^q / g(x)^{q/p} \right) dx \right)^{1/q}, \quad p^{-1} + q^{-1} = 1, \; p \in (1, +\infty), \\ Q_n(x) &= U \begin{pmatrix} u_0, u_1, \ldots, u_n, f_n \\ a, t_1, \ldots, t_n, b \end{pmatrix}. \end{aligned} \tag{5.4}$$

If the function f satisfies $(D_n \cdots D_0 f)(x) = C\,(Q_n(x)/g(x))^{1/(p-1)}$, where C is an aribtrary real constant, then equality holds in (5.3).

THEOREM 2. *If f satisfies the condition of Theorem 1 and $a = t_0 < t_1 < \cdots < t_n < t_{n+1} = b$, then*

$$\left| U\begin{pmatrix} u_0, u_1, \ldots, u_0, f \\ a, t_0, \ldots, t_n, b \end{pmatrix} \right| \leq K_n \| D_n \cdots D_0 f \|_2, \tag{5.5}$$

where

$$K_n = \left[(b-a) \sum_{j=0}^{n} \sum_{k=j+1}^{n+1} (-1)^{n+1+j+k} F_k F_j g_k(t_j) \right]^{1/2}, \qquad \text{and}$$

$$g_k(x) = \begin{cases} w_0(x) \int\limits_x^{t_k} w_1(a_1) \int\limits_{a_1}^{t_k} w_2(a_2) \ldots \int\limits_{a_n}^{t_k} f_k(t_k, x)\, dx\, da_n \ldots da_1, & x \leq t_k, \\ 0, & x > t_k. \end{cases}$$

If

$$f(x) = f_*(x) = C \sum_{k=1}^{n+1} (-1)^k F_k g_k(x) + \sum_{i=0}^{n} a_i u_i(x),$$

where $C, a_0, \ldots, a_n$ are arbitrary real constants, then equality holds in (5.5).

THEOREM 3. *Let $f\colon [a,b] \to \mathbf{R}$ be a real function such that $D_n \cdots D_0 f$ is continuous on $[a,b]$ and*

$$(D_k \cdots D_0 f)(a) = \ldots (D_n \cdots D_0 f)(a) = 0. \tag{5.6}$$

If $a = t_0 < t_1 < \cdots < t_k < t_{k+1} = b$, then

$$\left| U\begin{pmatrix} u_0, u_1, \ldots, u_k, f \\ a, t_1, \ldots, t_k, b \end{pmatrix} \right| \leq K_{g,n,k} \| D_n \cdots D_0 f \|_{g,p},$$

where

(5.7)

$$K_{g,n,k} = \left(\int\limits_a^b \left(Q_{n,k}(x)^q / g(x)^{q/p} \right) dx \right)^{1/q}, \quad p^{-1} + q^{-1} = 1, \ p \in (1, +\infty),$$

$$Q_{n,k}(x) = U\begin{pmatrix} u_0, u_1, \ldots, u_k, f_n \\ a, t_1, \ldots, t_k, b \end{pmatrix}.$$

THEOREM 4. *Let $f\colon [a,b] \to \mathbf{R}$ be a real function such that $D_n \cdots D_0 f$ is continuous on $[a,b]$.*

(i) *If $a = t_0 < t_1 < \cdots < t_{n+1} = b$, then*

$$\left| U\begin{pmatrix} u_0, u_1, \ldots, u_n, f \\ a, t_1, \ldots, t_n, b \end{pmatrix} \right| \leq \|D_n \cdots D_0 f\| \int_a^b Q_n(x)dx,$$

where Q_n is given by (5.4) with equality if and only if

$$f(t) = \sum_{i=0}^{n} a_i u_i(t) + Cw_0(t) \int_a^t w_1(s_1) \ldots \int_a^{s_{n-1}} s_n w_n(s_n) ds_n \ldots ds_1.$$

(ii) *Let $a = t_0 < t_1 < \cdots < t_k < t_{k+1} = b$. If (5.6) holds, then*

$$\left| U\begin{pmatrix} u_0, u_1, \ldots, u_k, f \\ a, t_1, \ldots, t_k, b \end{pmatrix} \right| \leq \|D_n \cdots D_0 f\| \int_a^b Q_{n,k}(x)dx,$$

where $Q_{n,k}$ is given by (5.7), with equality if and only if

$$f(t) = \sum_{t=0}^{k} a_i u_i(t) + CW_0(t) \int_a^t w_1(s_1) \cdots \int_a^{s_{n-1}} s_n w_n(s_n) ds_n \ldots ds_1.$$

$C, a_0, \ldots, a_n$ are arbitrary real constants.

Now, we shall give an inequality similar to inequality (5.2):

THEOREM 5. *Let $f\colon [a,b] \to \mathbf{R}$ be a real function such that $D_n \cdots D_0 f$ is continuous on $[a,b]$. Then for $p \in (1, +\infty)$,*

$$\left\| f - \sum_{i=0}^{n} a_i u_i \right\| < C_n \|D_n \cdots D_0 f\|_p,$$

where

$$a_0 = f(a)/w_0(a),\ a_i = ((D_{i-1} \cdots D_0 f)(a))/w_i(a) \quad (i = 1, \ldots, n),$$

and

$$C_n = \left((b-a)^{q/p} \int_a^b f_n(b,x)^q dx \right)^{1/q}, \quad p^{-1} + q^{-1} = 1.$$

PROOF: From (5.1) we have

$$\begin{aligned}\left\|f-\sum_{i=1}^{n}a_iu_i\right\| &= \sup_{a\le t\le b}\left|f(t)-\sum_{i=0}^{n}a_iu_i(t)\right| \\ &\le \sup_{a\le t\le b}\int_a^b f_n(t;x)|(D_n\cdots D_0f)(x)|\,dx \\ &\le \sup_{a\le t\le b}\left((b-a)^{q/p}\int_a^b f_n(t;x)^q dx\right)^{1/q} \\ &\quad\times\left(\frac{1}{b-a}\int_a^b |(D_n\cdots D_0f)(x)|^p\,dx\right)^{1/p} \\ &= C_n\|D_n\cdots D_0f\|_p.\end{aligned}$$

The following result was proved by J. E. Pečarić, G. H. Tudor, B. Crstici and B. Savić [**13**].

THEOREM 6. *Let $f\in C(u_0,u_1,\dots,u_n)$. If $D^nf(x)$ exists and it is an increasing function on (c,b) with*

$$D^{i-1}f(c)=0\quad (i=0,1,\dots,n;\ n\ge 1;\ D^{-1}f\equiv f),$$

then

$$0\le f(t)\le D^nf(t)\int_c^t W_n(t;x)dx\qquad (\forall t\in(c,b)).$$

A simple consequence of Theorem 6 is

THEOREM 7. *Let $f\colon [a,b]\to \mathbf{R}$ be n times differentiable function on (a,b). If $f(c)=f'(c)=\cdots=f^{(n-1)}(c)=0$, $f^{(n)}(x)>0$ $(\forall x\in(c,b),\ c>a)$, and if $f^{(n)}(x)$ is an increasing function on (c,b), then*

$$0<f(x)<f^{(n)}(x)\frac{(x-c)^n}{n!}\qquad (\forall x\in(c,b)).$$

6. R. Farwig and D. Zwick [**14**] proved that if $f^{(n)}$ is convex, then

$$\frac{f^{(n)}(\bar{x})}{n!}\le [x_0,\dots,x_n]f\le \sum_{i=0}^{n}\frac{f^{(n)}(x_0)}{(n+1)!} \tag{6.1}$$

where $[x_0, \ldots, x_n]f$ is the n-th order divided difference of f at the points $x_0, \ldots, x_n$ and $\bar{x}$ is the center of mass of these points $\bar{x} = \frac{1}{n+1} \sum_{i=0}^{n} x_i$. For further generalizations see E. Neumann [19].

A multivariable version of this is also true, E. Neuman and J. Pečarić [15]. They proved that for $x_i \in \mathbf{R}^n$,

$$f(\bar{x}) \leq \frac{1}{\operatorname{vol}(\sigma)} \int_{\sigma} f(x)dx \leq \frac{1}{n+1} \sum_{i=0}^{n} f(x_i) \tag{6.2}$$

if f is convex, $\bar{x}$ is as above and σ is the convex hull of the points $\{x_0, \ldots, x_n\}$. Equality holds if and only if f is a polynomial of degree 1. The differential version of this is the following. Let D^α be a differential operator of order n ($|\alpha| = n$). Suppose $D^\alpha f$ is convex on σ, then

$$D^\alpha f(\bar{x}) \leq n! \left([x^0, \ldots, x^n]f \right)_\alpha \leq \frac{1}{n+1} \sum_{i=1}^{n} D^\alpha f(x_i). \tag{6.3}$$

This generalizes (6.1).

7. Other papers related to inequalities of this chapter are given by V. A. Zmorovich and N. I. Cherneǐ [16], A. Isihara [17], and A. Erdélyi [18].

REFERENCES

1. ZMOROVIČ, V. A., *On some inequalities* (Russian), Izv. Polytehn. Inst. Kiev **19** (1956), 92–107.
2. HARDY,G. H., J. E. LITTLEWOOD and G. PÓLYA, "Inequalities," Cambridge, 1934.
3. DJORDJEVIĆ, R. Ž. and G. V. MILOVANOVIĆ, *On some generalizations of Zmorovič's inequality* , Univ. Beograd. Publ. Elektrotehn. Fak. Ser. Mat. Fiz. No. **544–576** (1976), 25–30.
4. VASIĆ, P. M. and J. D. KEČKIĆ, *Some inequalities for complex numbers*, Math. Balkanica **1** (1971), 282–286.
5. BULLEN, P. S., *A note on a recent paper of P. M. Vasić and J. D. Kečkić*, Math. Balkanica **2** (1972), 1–2.
6. MITRINOVIĆ, D. S., (In cooperation with P. M. VASIĆ), " Analytic inequalities," Berlin–Heidelberg–New York, 1970.

7. DJORDJEVIĆ, R. Ž., G. V. MILOVANOVIĆ and J. E. PEČARIĆ, *Some estimates of L^r norm on the set of continuously-differentiable functions*, Univ. Beograd. Publ. Elektrotehn. Fak. Ser. Mat. Fiz. No. **634–677** (1979), 57- -61.
8. MILOVANOVIĆ, G. V., *O nekim funkcionalnim nejednakostima*, Univ. Beograd. Publ. Elektrotehn. Fak. Ser. Mat. Fiz. No. **599** (1977), 1–59.
9. LUPAŞ, A., *Inequalities for divided differences*, Univ. Beograd. Publ. Elektrotehn. Fak. Ser. Mat. Fiz. No. **678–715** (1980), 24–28.
10. IONESCU, D. V., "Cuadraturi numerice," Bucuresti, 1957.
11. PEČARIĆ, J. E., *Concerning Extended Complete Tchebychehh System*, J. Math. Anal. Appl. **102**, (1984), 385–392.
12. KARLIN, S. J. and W. J. STUDDEN, "Tchebycheff Systems: with Applications in Analysis and Statistics, " New York/London/Sydney, 1966.
13. PEČARIĆ, J. E., G. H. TUDOR, B. CRSTICI and B. SAVIĆ, *Note on Taylor's Formula and Some Applications*, J. Approx. Theory **51** (1987), 47–53.
14. FARWIG, R. and D. ZWICK, *Some divided difference inequalities for n-convex functions*, J. Math. Anal. Appl. **108** (1985), 430–437.
15. NEUMAN, E. and J. E. PEČARIĆ, *Inequalities involving multivariate convex functions*, J. Math. Anal. Appl. **137**, (1989), 541–599.
16. ZMOROVICH, V. A. and N. I. CHERNEĬ, *Some integral inequalities* (Russian), Dokl. Akad. Nauk. Ukrain. SSR Ser. A. (1983), No. 6, 13–16.
17. ISIHARA, S., *Information loss and entropy increase*, J. Math. Anal. Appl. **39** (1972), 314–317.
18. ERDÉLYI, A., *An extension of Hardy-Littlewood-Polya inequality*, Proc. Edinburgh Math. Soc., II. Ser. **21** (1978), 11–15.
19. NEUMANN, E., *Inequalities involving multivariate convex functions II*, Proc. Amer. Math. Soc. **109** (1990), 965–974.

CHAPTER VIII
CARLSON'S AND RELATED INEQUALITIES

1. Let $a_1, a_2, \ldots$ be a sequence of nonnegative real numbers not all equal to zero, and let $\sum_{k=1}^{+\infty} k^2 a_k^2 < +\infty$. Then

$$\left(\sum_{k=1}^{+\infty} a_k\right)^4 \leq \pi^2 \left(\sum_{k=1}^{+\infty} a_k^2\right) \left(\sum_{k=1}^{+\infty} k^2 a_k^2\right), \tag{1.1}$$

where π^2 is the best possible constant.

The above inequality has the following integral analogue: let f be a nonnegative real-valued function on $[0, +\infty)$, such that $f(x)^2$ and $x^2 f(x)^2$ are integrable functions on the same interval. Then

$$\left(\int_0^{+\infty} f(x)dx\right)^4 \leq \pi^2 \left(\int_0^{+\infty} f(x)^2 dx\right) \left(\int_0^{+\infty} x^2 f(x)^2 dx\right), \tag{1.2}$$

where π^2 is also the best possible constant.

Without any difficulty, from (1.1) we can pass to (1.2), and conversely.

Inequalities (1.1) and (1.2) are called Carlson's [**1**].

2. Two related results are given in the books Pólya and Szegö [**2**, II, pp. 114,218, and I, pp. 57, 214, respectively] and Hardy, Littlewood and Pólya [**3**, p. 166]:

1⁰ If $a \geq 0$, $b \geq 0$, $a \neq b$, and ϕ is nonnegative and decreasing, then

$$\left(\int_0^{+\infty} x^{a+b}\phi\, dx\right)^2 < \left\{1 - \left(\frac{a-b}{a+b+1}\right)^2\right\} \int_0^{+\infty} x^{2a}\phi\, dx \int_0^{+\infty} x^{2b}\phi\, dx, \tag{2.1}$$

unless $\phi = C$, where $C > 0$, in $(0, \lambda)$, and $\phi = 0$ in $(\lambda, +\infty)$ (λ may be 0).

(The case $a = 0$, $b = 2$ was mentioned by Gauss in connection with the Theory of Errors: see [**4**, IV, p. 12].)

2^0 If $a \geq 0$, $b \geq 0$, $a \neq 1$, and ϕ is nonnegative and increasing, then

$$\left(\int_0^1 x^{a+b}\phi\, dx\right)^2 > \left\{1 - \left(\frac{a-b}{a+b+1}\right)^2\right\} \int_0^1 x^{2a}\phi\, dx \int_0^1 x^{2b}\phi\, dx, \tag{2.2}$$

unless $\phi = C$.

A. M. Fink and Max Jodeit Jr. [**5**] showed that (2.2) is valid if $a, b > -1/2$ even if $a = 1$. Indeed (2.2) is equivalent to

$$\int_0^1\int_0^1 \phi(x)\phi(y)K(x,y)dxdy \geq 0, \tag{2.3}$$

where

$$K(x,y) \equiv (a+b+1)^2 x^{a+b}y^{a+b} - \frac{(2a+1)(2b+1)}{2}\left[x^{2a}y^{2b} + x^{2b}y^{2a}\right].$$

If $G(s,t) = \int_s^1\int_t^1 K(x,y)dxdy$ then $G(s,t) \geq 0$ on $[0,1]\times[0,1]$ and $G(s,t) = 0$ only if $(s,t) = (0,0)$ or $s = 1$ or $t = 1$. We prove this.

With $\alpha = a + \frac{1}{2}$, $\beta = b + \frac{1}{2}$, $t_1 = t^\alpha$, $t_2 = t^\beta$, $s_1 = s^\alpha$ and $s_2 = s^\beta$, $2G = (s_1 - s_2)^2 + (t_1 - t_2)^2 - (s_1t_1 - s_2t_1)^2$. But $(s_1t_2 - s_2t_1) = \frac{t_1+t_2}{2}(s_2 - s_1) - \frac{s_2-s_1}{2}(t_1 - t_2)$ so

$$\begin{aligned} 2G = (s_1 - s_2)^2\left[1 - \left[\frac{t_1+t_2}{2}\right]^2\right] + (t_1 - t_2)^2\left[1 - \left[\frac{s_1+s_2}{2}\right]^2\right] \\ + 2\left[\frac{t_1+t_2}{2}\right]\left[\frac{s_1+s_2}{2}\right](s_1 - s_2)(t_1 - t_2). \end{aligned}$$

Since $s_1 - s_2 = s^\alpha - s^\beta$ and $t_1 - t_2 = t^\alpha - t^\beta$ have the same sign each of the three terms is nonnegative. If neither s nor t are 1 and both are nonzero, one of the three terms is positive. This property of G is used in the following way. Since φ is increasing, $\varphi(x) = \varphi(0) + \int_0^x d\mu(t)$ for some nonnegative measure and we may write $\varphi(x) = \int_0^x d\sigma(t)$ where σ has a mass of $\varphi(0)$ at zero. The inequality (2.3) may be written as

$$\int_0^1\int_0^1 K(x,y)\int_0^x\int_0^y d\sigma(t)\sigma(s)\; dxdy \geq 0. \tag{2.4}$$

By Fubini's Theorem, (2.4) is equivalent to

$$\int_0^1\int_0^1 G(t,s)d\sigma(t)d\sigma(s) \geq 0. \tag{2.5}$$

This proves (2.2) and gives the cases of equality. For if equality holds in (2.5) the measure $d\sigma(t)d\sigma(s)$ must be zero except at (0,0), that is $\varphi(t) \equiv \varphi(0)$.

J. E. Pečarić [6] has observed that the above proof gives

$$\begin{aligned} &2(a+b+1)^2 \int_0^1\int_0^1 f(x,y)x^{a+b}y^{a+b}dxdy \\ &\geq (2a+1)(2b+1)\int_0^1 f(x,y)\left(x^{2a}y^{2b}+x^{2b}y^{2a}\right)dxdy. \end{aligned} \tag{2.6}$$

when $f(0,0)$, $f_1(x,y)$, $f_2(x,y)$ and $f_{12}(x,y) \geq 0$ on $[0,1]\times[0,1]$, and $a,,b \geq -1/2$.

To see this, we observe that (2.6) is equivalent to

$$\int_0^1\int_0^1 f(x,y)K(x,y)dxdy \geq 0, \tag{2.7}$$

where K is given by (2.3). If we write

$$f(x,,y) = \int_0^x\int_0^y d\lambda(s,t),$$

then as above (2.7) is equivalent to

$$\int_0^1\int_0^1 G(s,t)d\lambda(s,t) \geq 0, \tag{2.8}$$

and the result follows if $d\lambda \geq 0$. But

$$f(s,t) = f(0,0) + \int_0^s f_1(\xi,0)d\xi + \int_0^t f_2(0,\eta)d\eta + \int_0^s\int_0^t f_{12}(\xi,\eta)d\xi d\eta.$$

So that $d\lambda(s,t) \geq 0$.

The special case when $f(x,y) \equiv \varphi(x)\psi(y)$, with φ and ψ increasing and non-negative, yields

$$2(a+b+1)^2 \int_0^1 \varphi(x)x^{a+b}dx \int_0^1 \psi(y)y^{a+b}dy$$
$$> (2a+1)(2b+1)\left(\int_0^1 \varphi(x)x^{2a}dx \int_0^1 \psi(y)y^{2b}dy + \int_0^1 \varphi(x)x^{2b}dx \int_0^1 \psi(y)y^{2a}dy\right)$$

unless $\varphi(x) \equiv \varphi(0)$ and $\psi(y) \equiv \psi(0)$.

3. Two simple proofs of (1.1) were given by G. H. Hardy [**7**]. E. Landau proved a stronger inequality than (1.1):

$$\left(\sum_{k=1}^{+\infty} a_k\right)^4 < \pi^2 \left(\sum_{k=1}^{+\infty} a_k^2\right)\left(\sum_{k=1}^{+\infty}\left(k-\frac{1}{2}\right)^2 a_k^2\right).$$

4. The following generalization of (1.1) was given by R. M. Gabriel [**8**]:

$$\sum a_n < G(p,\lambda)\left(\sum n^{p-1-\lambda}a_n^p\right)^{\frac{1}{2p}}\left(\sum n^{p-1+\lambda}a_n^p\right)^{\frac{1}{2p}}, \tag{4.1}$$

if $\sum n^{p-1+\lambda}a_n^p < +\infty$, where $p > 1$, $\lambda > 0$, and

$$G(p,\lambda) = 2^{1/p}\left\{\frac{\Gamma^2\left(\frac{1}{2p-1}\right)}{2\lambda\Gamma\left(\frac{1}{p-1}\right)}\right\}^{\frac{p-1}{p}}.$$

The constant $G(p,\lambda)$ is the best possible.

Inequality (4.1) has an integral analogue. Let $f(x) \geq 0$, $0 < x < +\infty$. Then

$$\int_0^\infty f(x)dx \leq G(p,\lambda)\left\{\int_0^\infty x^{p-1-\lambda}f^p(x)dx\right\}^{\frac{1}{2p}}\left\{\int_0^\infty x^{p-1-\lambda}f^p(x)dx\right\}^{\frac{1}{2p}}. \tag{4.2}$$

5. A. Beurling [**9**] proved (1.2).

6. A note [**10**] of V. I. Levin contains a proof of the following inequality:

Let $a_n \geq 0$, $n = 1, 2, \ldots$; then, provided that the right-hand side is finite,

$$\left(\sum_{0}^{\infty} a_n\right)^{(k+1)(2k+1)} < \prod_{\nu=1}^{2k} \nu^{2\nu-2k-1} \cdot \prod_{u=0}^{2k} \left(\sum_{0}^{\infty} n^u a_n^{k+1}\right), \tag{6.1}$$

unless all $a_n = 0$. The constant is the best possible.

7. Further generalizations of Carlson's inequality were given by W. B. Caton [**11**] and B. Sz.-Nagy [**12**]. For example, B. Sz.-Nagy demonstrated the following inequalities

$$\left|\sum_{k=1}^{+\infty} a_k\right|^q + \left|\sum_{k=1}^{+\infty} (-1)^k a_k\right|^q < \frac{\pi q}{2} \left(\sum_{k=1}^{+\infty} k^{p_1} |a_k|^{p_1}\right)^{\frac{1}{p_1}} \left(\sum_{k=1}^{+\infty} |a_k|^{p_2}\right)^{\frac{1}{(p_2-1)p_1}}$$

and

$$\left|\sum_{k=1}^{+\infty} a_k\right|^q < \frac{\pi q}{2} \left(\sum_{k=1}^{+\infty} \left(k - \frac{1}{2}\right)^{p_1} |a_k|^{p_1}\right)^{\frac{1}{p_1}} \left(\sum_{k=1}^{+\infty} |a_k|^{p_2}\right)^{\frac{1}{(p_2-1)p_1}},$$

where $a_1, a_2, \ldots$ are real numbers and, $q = 1 + \frac{p_2}{p_1(p_2-1)}$, $1 \leq p_1 \leq 2$ and $1 < p_2 \leq 2$.

We assume that the series in question converge.

He also proved two related inequalities. In [**13**] B. Sz.Nagy showed that (1.2) can be derived from one result which is connected to Kolmogorov's inequality studied in Chapter I.

8. R. Bellman [**14**] showed that Gabriel's and Caton's generalizations of Carlson's inequality can be derived by the same elementary method (see, for example Beckenbach and Bellman [**15**, pp. 175–177]).

The best possible constant in such inequalities was proved by V. I. Levin [**16**], who showed that, if $p > 1$, $q > 1$, $\lambda > 0$, $\mu > 0$ and if f is a real valued nonnegative function on $[0, +\infty)$, such that $x^{p-1-\lambda} f(x)^p$ and $x^{q-1+\mu} f(x)^q$ are integrable functions on the considered interval, then

(8.1)

$$\int_0^{+\infty} f(x)dx \leq C \left(\int_0^{+\infty} x^{p-1-\lambda} f(x)^p dx\right)^{\frac{\mu}{p\mu+q\lambda}} \left(\int_0^{+\infty} x^{q-1+\mu} f(x)^q dx\right)^{\frac{\lambda}{p\mu+q\lambda}}$$

with

$$C = \left(\frac{1}{ps}\right)^s \left(\frac{1}{qt}\right)^t \left(\frac{\Gamma\left(\frac{s}{1-s-t}\right)\Gamma\left(\frac{t}{1-s-t}\right)}{(\lambda+\mu)\Gamma\left(\frac{s+t}{1-s-t}\right)}\right)^{1-s-t}, \tag{8.2}$$

$$s = \frac{\mu}{p\mu + q\lambda}, \qquad t = \frac{\lambda}{p\mu + q\lambda}.$$

The constant C is the best possible.

PROOF: Let $p' = \frac{p}{p-1}$, $q' = \frac{q}{q-1}$ $(p' > 1,\ q' > 1)$.

Consider the function

$$z = z(k) = k^{-\frac{p'q}{p'q\lambda + pq'\mu}}(1-k)^{\frac{pq'}{p'q\lambda + pq'\mu}} = \left\{\frac{(1-k)^{p-1}}{k^{q-1}}\right\}^{\frac{1}{\delta}}, \tag{8.3}$$

where $\delta = p\mu + q\lambda - \mu - \lambda > 0$. In the interval $0 < k < 1$ the function $z(k)$ is continuous and strictly monotonic. Therefore its continuous inverse function $k = k(z)$ exists on $0 < z < +\infty$.

Consider the identity

$$\int_0^\infty f(x)dx \equiv \int_0^\infty k(yx)x^{-\frac{p-1-\lambda}{p}}x^{\frac{p-1-\lambda}{p}}f(x)dx + \int_0^\infty [1 - k(yx)]\, x^{-\frac{q-1+\mu}{q}}x^{\frac{q-1+\mu}{q}}f(x)dx, \tag{8.4}$$

where $y > 0$ is arbitrary. Set

$$P = \left\{\int_0^\infty x^{p-1-\lambda}f^p(x)dx\right\}^{\frac{1}{p}}, \qquad Q = \left\{\int_0^\infty x^{q-1+\mu}f^q(x)dx\right\}^{\frac{1}{q}},$$

using Hölder's inequality we get

$$\int_0^\infty k(yx)x^{-\frac{p-1-\lambda}{p}}x^{\frac{p-1-\lambda}{p}}fx)\,dx \tag{8.5}$$

$$\leq \left\{\int_0^\infty k^{p'}(yx)x^{-\frac{p-1-\lambda}{p-1}}\,dx\right\}^{\frac{1}{p'}} P = y^{-\frac{\lambda}{p}} I_1 P,$$

$$\int_0^\infty [1-k(yx)]\,x^{-\frac{q-1+\mu}{q}}x^{\frac{q-1+\mu}{q}}f(x)\,dx \tag{8.6}$$

$$\leq \left\{\int_0^\infty [1-k(yx)]^{q'}\,x^{-\frac{q-1+\mu}{q-1}}\,dx\right\}^{\frac{1}{q'}} Q = y^{\frac{\mu}{q}} I_2 Q,$$

where

$$I_1 = \left\{\int_0^\infty k^{p'}(z)z^{-1+\frac{\lambda}{p-1}}\,dz\right\}^{\frac{1}{p'}}, \quad I_2 = \left\{\int_0^\infty [1-k(z)]^{q'}\,z^{-1-\frac{\mu}{q-1}}\,dx\right\}^{\frac{1}{q'}}.$$

Now, from (8.4), for $y = \left(\frac{qt}{ps}\,\frac{I_1 P}{I_2 Q}\right)^{\frac{pq}{p\mu+q\lambda}}$ we get

$$\int_0^\infty f(x)\,dx \leq y^{-\frac{\lambda}{p}} I_1 P + y^{\frac{\mu}{q}} I_2 Q = \left(\frac{1}{ps}\right)^{ps}\left(\frac{1}{qt}\right)^{qt} I_1^{ps} I_2^{qt} P^{ps} Q^{qt}. \tag{8.7}$$

If we change arguments in integrals I_1 and I_2, we get

$$I_1^{p'} = \frac{p'q}{p'p\lambda + pq'\mu}\int_0^\infty k^{\frac{s}{1-s-t}}(1-k)^{\frac{t}{1-s-t}}\,dk$$

$$+ \frac{pq'}{p'q\lambda + pq'\mu}\int_0^\infty k^{\frac{s}{1-s-t}+1}(1-k)^{\frac{t}{1-s-t}-1}\,dk$$

$$= \frac{ps}{\lambda+\mu}\,\frac{\Gamma\left(\frac{s}{1-s-t}\right)\Gamma\left(\frac{t}{1-s-t}\right)}{\Gamma\left(\frac{s+t}{1-s-t}\right)},$$

and analogously

$$I_2^{q'} = \frac{qt}{\lambda+\mu}\,\frac{\Gamma\left(\frac{s}{1-s-t}\right)\Gamma\left(\frac{t}{1-s-t}\right)}{\Gamma\left(\frac{s+t}{1-s-t}\right)}.$$

Now, from (8.7) we obtain (8.1) with (8.2).

We should also prove that the constant (8.2) is the best possible. Indeed equality in (8.1) with constant (8.2) is valid if

$$\begin{aligned} f(x) &= y^{\frac{\lambda}{p-1}} k^{\frac{1}{p-1}}(yx)x^{-\frac{p-1-\lambda}{p-1}} \\ &= y^{-\frac{\mu}{q-1}}\left[1-k(yx)\right]^{\frac{1}{q-1}} x^{-\frac{q-1+\mu}{q-1}}, \end{aligned}$$

where $y > 0$ is arbitrary.

In a limiting case, Levin also obtains the following inequality

$$\int_0^\infty f(x)\,dx < \left\{\int_0^\infty x^{-\lambda} f(x)\,dx\right\}^{\frac{\mu}{\lambda+\mu}} \left\{\int_0^\infty x^{\mu} f(x)\,dx\right\}^{\frac{\lambda}{\lambda+\mu}}.$$

9. The problem of determining the best possible constant K in a more general inequality

$$\text{(9.1)} \qquad \int_0^{+\infty} x^{\alpha}|f(x)|^{\beta} \le K\left(\int_0^{+\infty} x^{\alpha_1}|f(x)|^{\beta_1}\,dx\right)^{a_1}\left(\int_0^{+\infty} x^{\alpha_2}|f(x)|^{\beta_2}\,dx\right)^{a_2},$$

where the exponents are real numbers subject to certain conditions, was considered by B. Kjellberg in [**17**]–[**19**]. (In fact, by a simple transformation we can obtain (9.1) from the analogous inequality for $\alpha = \beta = 0$.)

10. For Carlson's inequality and its generalizations see also Levin and Stečkin [**20**].

11. If f is a nonnegative and nonincreasing real-valued function on $(0, +\infty)$, if $1 \le q_1 \le q_2 < +\infty$ and $1 \le p < +\infty$, then

$$\begin{aligned} &\left(\int_0^{+\infty}\left(t^{\frac{1}{p}} f(t)\right)^{q_2}\frac{dt}{t}\right)^{\frac{1}{q_2}} \\ &\qquad \le \left(\frac{q_1}{p}\right)^{\frac{1}{q_1}-\frac{1}{q_2}}\left(\int_0^{+\infty}\left(t^{\frac{1}{p}} f(t)\right)^{q_1}\frac{dt}{t}\right)^{\frac{1}{q_1}}. \end{aligned}$$

This inequality was attributed to A. P. Calderon by O'Neil [**21**], and was used later by Hunt and Weiss to give a short proof of the interpolation theorem of Marcinkiewicz on quasi-linear operators [**22**].

12. Inequality (6.1) was further generalized by V. I. Levin and E. K. Godunova [**23**]. In fact, they proved the following two results:

1^0 If $\lambda_i > 0$ $(i = 1, 2, \ldots, k)$ and $p = (1/k)\sum_{i=1}^{k} \lambda_i$, then

$$\left(\sum_{n=1}^{+\infty} a_n\right)^{(p+1)k} < A^{pk} k^k \left\{\prod_{\nu=1}^{k} \sum_{n=1}^{+\infty} a_n^{p+1} n^{\lambda_\nu}\right\},$$

where A is defined by

$$A = \min_{1 \le s \le k-1} \frac{1}{r_2 - r_1} \left[s^s (k-s)^{k-s}\right]^{\frac{1}{kp}} \cdot \frac{\Gamma\left(\frac{k-s}{kp}\right)\Gamma\left(\frac{s}{kp}\right)}{\Gamma\left(\frac{1}{p}\right)};$$

and $\sum_{i=1}^{n} \lambda_i = sr_1$, $\sum_{i=s+1}^{k} \lambda_i = (k-s)r_2$, $0 < s < k$, $r_2 > r_1$.

2^0 If $\lambda_\nu = \frac{k}{2} + h\left(\nu - \frac{k}{2}\right)$, then

$$\left(\sum_{n=1}^{\infty} a_n\right)^{\frac{(k+1)(k+2)}{2}} < h^{-\frac{k(k+1)}{2}} C(k) \prod_{\nu=0}^{k} \sum_{n=1}^{\infty} a_n^{\frac{k+2}{2}} n^{\lambda_\nu},$$

where

$$C(k) \cdot h^{-\frac{k(k+1)}{2}} = h^{-\frac{k(k+1)}{2}} (k+1)^{k+1} \prod_{\mu=0}^{k} C_k^\mu = h^{-\frac{k(k+1)}{2}} \prod_{\mu=1}^{k+1} \mu^{2\mu-k-1}$$

is the best possible constant.

13. G. M. Pigolkin [**24**] extends the Carlson inequality to higher dimensions. Namely, he proved the following two results:

1^0 Let f be a nonnegative real valued function on R^{n-1}, (μ_{ij}) a nonsingular $(n-1) \times (n-1)$-matrix with determinant G and positive row sums σ_j, $1 \le j \le n-1$, for its inverse. Then if $0 < \gamma < 1$

$$\left(\int f\right)^{\frac{\sigma+\gamma}{\sigma(1-\gamma)}} \le C\left(\int f^{\frac{\sigma+\gamma}{\sigma}}\right)$$
$$\times \prod_{i=1}^{n-1}\left(\int f^{\frac{\sigma+\gamma}{\sigma}}(x_1,\dots,x_{n-1})\prod_{j=1}^{n-1}|x_j|^{\mu_{ij}}\,dx_1,\dots,dx_{n-1}\right)^{\nu_i},$$

where $\sigma = \sum_{j=1}^{n-1}\sigma_j$, $\nu_1 = \left(\frac{\gamma}{1-\gamma}\right)\frac{\sigma_1}{\sigma}$, and

$$C = 2^{\frac{(n-1)\gamma}{\sigma(1-\gamma)}}\frac{\left(\frac{1}{1-\gamma}\right)^{\frac{1}{1-\gamma}}}{\prod_{i=1}^{n-1}\nu_i^{\nu_i}}\left\{\frac{\Gamma\left(\frac{1-\gamma}{\gamma}\sigma\right)}{|G|\Gamma\left(\frac{\sigma}{\gamma}\right)}\prod_{i=1}^{n-1}\Gamma(\sigma_i)\right\}^{\frac{\gamma}{\sigma(1-\gamma)}}.$$

Further, equality occurs if and only if

$$f(x_1,\dots,x_{n-1}) = f_0(x_1,\dots,x_{n-1})$$
$$= \alpha_0\left(\sum_{i=1}^{n-1}\left\{\alpha_i + \frac{1}{a_i}\prod_{j=1}^{n-1}|x_j|^{\mu_{ij}}\right\}\right)^{-\frac{\sigma}{\gamma}},$$

with $\alpha_0 \ge 0$, $\alpha_i > 0$, $1 \le i \le n-1$. Carlson's inequality is obtained if $n = 2$, $\gamma = \sigma = \frac{1}{2}$. This result can be obtained from the following result.

2^0 If Ω_n is any domain in R^n, and ω_i, $0 \le 1 \le n-1$, are such that $\omega = \sum_{i=1}^{n-1}\omega_i \ne n$, then

$$\left(\int_{\Omega_n}|x_n|^{\omega_0}\prod_{i=1}^{n-1}|x_i|^{\omega_i}\,dx_1,\dots,dx_n\right)^{\frac{\sigma+\gamma}{\sigma(1-\gamma)}}$$
$$\le 2^{\frac{\gamma}{\sigma(1-\gamma)}}\left(1+\frac{\gamma}{\sigma}\right)^{\frac{1}{1-\gamma}}|n+\omega|^{-\frac{\gamma}{\sigma(1-\gamma)}}$$
$$\times C\left(\int_{\Omega_n}|x_n|^{\bar\omega_c}\prod_{j=1}^{n-1}|x_j|^{\bar\omega_j}\,dx_1,\dots,dx_n\right)$$
$$\times \prod_{i=1}^{n-1}\left\{\int_{\Omega_n}|x_n|^{(\bar\omega_0-\bar\sigma_i)}\prod_{j=1}^{n-1}|x_j|^{(\omega_j+\mu_{ij})}dx_1,\dots,dx_n\right\}^{\nu_i}$$

where $\bar{\omega}_0 = \gamma_n/\sigma + \omega_0(1+\gamma/\sigma)$, $\bar{\omega}_i = \omega_i(1+\nu/\sigma)$, $1 \le i \le n-1$, $\bar{\sigma}_i = \sum_{j=1}^{n} \mu_{ij}$, and the matrix (μ_{ij}), C, γ, σ, ν_i are as above. Further for equality it is necessary and sufficient that for all points $(x_1, \ldots, x_n) \in \Omega_n$,

$$(x_1^2 + \cdots + x_n^2)^{\frac{n+\omega}{2}} \le f_0(x_1, \ldots, x_{n-1}) \left(1 + \sum_{i=1}^{n-1} x_i^2\right)^{\frac{n+\omega}{2}} \prod_{i=1}^{n-1} |x_i|^{-\omega_i}.$$

14. Let $f(x)$ be a nonnegative integrable function in $(0, +\infty)$. Then

$$\int_0^{+\infty} f(x)\,dx \le 2\left(\int_0^{+\infty} \sqrt{x}\,f(x)\,dx\right)^{1/2} \left(\int_0^{+\infty} f(x)^2\,dx\right)^{1/4}. \tag{14.1}$$

This result is due to B. Klefsjö [**25**] (we suppose that all integrals are finite).

15. A generalization of (2.1) is given by V. N. Volkov [**26**]. Volkov proved a general result (see 16.) wherefrom he obtained the following special cases:

1^0 If $a \ge 0$, $b \ge 0$, $p > 1$, $p^{-1} + q^{-1} = 1$, and g is nonnegative and decreasing function, then

$$\int_0^{+\infty} x^{a+b} g(x)\,dx \le C\left(\int_0^{+\infty} x^{ap} g(x)\,dx\right)^{1/p} \left(\int_0^{+\infty} x^{bq} g(x)\,dx\right)^{1/q}, \tag{15.1}$$

where

$$C = \frac{(ap+1)^{1/p}(bq+1)^{1/q}}{(1+a+b)}.$$

2^0 If a, b, p, q are defined as in 1^0 and if g is a nonnegative, nonincreasing convex function, then (15.1) holds with constant

$$C = \frac{((ap+1)(ap+2))^{1/p}\,((aq+1)(aq+2))^{1/q}}{(1+a+b)(2+a+b)}.$$

16. Results of Volkov and Klefsjö are generalized by D. S. Mitrinović and J. E. Pečarić [**27**].

The following result is a generalization of Volkov's result.

THEOREM 1. *Let f_i on R^+, $i = 1,2,3,4$, be nonnegative functions and let g be a real function which has the following representation*

$$g(x) = \int_0^{+\infty} K(x,t)dh(t), \tag{16.1}$$

where $K(x,,t) \geq 0$, when $x \in (0,+\infty)$, $t \in (0,+\infty)$ and h is a nondecreasing function. If p, q are two real numbers such that $p^{-1} + q^{-1} = 1$, $p > 1$, then

(16.2)

$$\int_0^{+\infty} f_1(x)f_2(x)g(x)\,dx \leq C\left(\int_0^{+\infty} f_3(x)g(x)\,dx\right)^{1/p}\left(\int_0^{+\infty} f_4(x)g(x)\,dx\right)^{1/q},$$

where

(16.3)

$$C = \sup_t \left\{\left(\int_0^{+\infty} K(x,t)f_1(x)f_2(x)\,dx\right)\left(\int_0^{+\infty} K(x,t)f_3(x)\,dx\right)^{-1/p} \times \left(\int_0^{+\infty} K(x,t)f_4(x)\,dx\right)^{-1/q}\right\}.$$

PROOF: By using (16.1) we have

$$\int_0^{+\infty} f_1(x)f_2(x)g(x)\,dx = \int_0^{+\infty}\int_0^{+\infty} f_1(x)f_2(x)K(x,t)dh(t)\,dx$$

$$= \int_0^{+\infty} dh(t) \left\{ \int_0^{+\infty} K(x,t) f_1(x) f_2(x)\, dx \right\}$$

$$\leq C \int_0^{+\infty} dh(t) \left(\int_0^{+\infty} K(x,t) f_3(x) dx \right)^{1/p} \left(\int_0^{+\infty} K(x,t) f_4(x)\, dx \right)^{1/q}$$

$$\leq C \int_0^{+\infty} dh(t) \left(\frac{\lambda^P}{p} \int_0^{+\infty} K(x,t) f_3(x)\, dx + \frac{\lambda^{-q}}{q} \int_0^{+\infty} K(x,t) f_4(x)\, dx \right)$$

$$= C \left(\frac{\lambda^p}{p} \int_0^{+\infty} f_3(x) g(x)\, dx + \frac{\lambda^{-q}}{q} \int_0^{+\infty} f_4(x) g(x) dx \right).$$

For

$$\lambda = \left(\int_0^{+\infty} f_3(x) g(x)\, dx \right)^{-\frac{1}{pq}} \left(\int_0^{+\infty} f_4(x) g(x)\, dx \right)^{\frac{1}{pq}}$$

we obtain (16.2).

REMARK: For $f_3(x) = f_1(x)^p$, $f_4(x) = f_2(x)^q$ we get a result of V. N. Volkov **[26]**.

If we set

(16.4)
$$K(x,t) = \begin{cases} 1, & 0 \leq x < t, \\ 0, & x > t, \end{cases}$$

then g is a nonincreasing function, and (16.2) holds with

(16.5)
$$C = \sup_t \left\{ \left(\int_0^t f_1(x) f_2(x)\, dx \right) \left(\int_0^t f_3(x)\, dx \right)^{-1/p} \left(\int_0^t f_4(x)\, dx \right)^{-1/q} \right\}.$$

By using this result we can obtain results from **[26]**, and also many other results. For example, the following inequality is valid:

(16.6)
$$\int_0^{+\infty} (x+1) e^x g(x)\, dx$$

$$\leq p^{1/p} q^{1/q} \left(\int_0^{+\infty} x^{p-1} g(x)\, dx \right)^{1/p} \left(\int_0^{+\infty} e^{qx} g(x) dx \right)^{1/q}.$$

Now, we shall prove:

THEOREM 2. *Let $f(x)$ be a nonnegative integrable function on $(0, +\infty)$, and let a, b, u, v be real numbers such that $a > 0$, $b + v > 0$, $u > 0$, $v > 0$, $u + v = 1$. Then,*

(16.7)
$$\left(\int_0^{+\infty} f(x)\,dx\right)^{a+b+v} \le C\left(\int_0^{+\infty} x^a f(x)\,dx\right)^{b+v}\left(\int_0^{+\infty} x^{-b/u} f(x)^{1/u}\,dx\right)^{ua},$$

where

$$C = (a+b+v)^{a+b+v} v^{va} a^{-a} (b+v)^{-(b+v+va)}. \tag{16.8}$$

PROOF: For $t > 0$ we have

$$\int_t^{+\infty} f(x)\,dx \le \frac{1}{t^a}\int_t^{+\infty} x^a f(x)\,dx \le \frac{1}{t^a}\int_0^{+\infty} x^a f(x)\,dx. \tag{16.9}$$

By the inequality of Buniakowski-Schwarz, we have for $t > 0$

(16.10)
$$\begin{aligned}\int_0^{t} f(x)\,dx &\le \left(\int_0^{t} x^{b/v}\,dx\right)^{v}\left(\int_0^{t} x^{-b/u} f(x)^{1/u}\,dx\right)^{u}\\ &\le t^{b+v}\left(\frac{v}{b+v}\right)^{v}\left(\int_0^{+\infty} x^{-b/u} f(x)^{1/u}\,dx\right)^{u}.\end{aligned}$$

Combining (16.9) and (16.10), we find

$$\int_0^{+\infty} f(x)\,dx \le t^{b+v}\left(\frac{v}{b+v}\right)^{v}\left(\int_0^{+\infty} x^{-b/u} f(x)^{1/u}\,dx\right)^{u} + t^{-a}\left(\int_0^{+\infty} x^a f(x)\,dx\right),$$

whence for

$$t = \left\{a v^{-v}(b+v)^{-1+v}\left(\int_0^{+\infty} x^a f(x)\,dx\right)\left(\int_0^{+\infty} x^{-b/u} f(x)^{1/u}\,dx\right)^{-u}\right\}^{1/(a+b+v)}$$

we obtain (16.7).

For $b = 0$ we get

$$\left(\int_0^{+\infty} f(x)\,dx\right)^{a+v} \le C \left(\int_0^{+\infty} x^a f(x)\,dx\right)^v \left(\int_0^{+\infty} f(x)^{1/u}\,dx\right)^{ua}, \tag{16.11}$$

where

$$C = \frac{(a+v)^{a+v}}{(a^a v^v)}. \tag{16.12}$$

Finally, for $a = u = v = 1/2$ we get the inequality of B. Klefsjö.

References

1. CARLSON, F., *Une inégalité*, Ark. Mat. Astr. Fys. **25B** 1 (1935), 1–5.
2. PÓLYA, G. and G. SZEGÖ, "Aufgaben und Lehrsätze aus der Analysis, I, II," Berlin, 1925.
3. HARDY, G. H., J. E. LITTLEWOOD and G. PÓLYA, "Inequalities," Cambridge, 1934.
4. GAUSS, C. F., "Werke, Göttingen, 1863–1929."
5. FINK, A. M. and Max JODEIT, Jr., *Jensen inequalities for functions with higher monotonicities*, Aeq. Math. J. **40** (1990), 26–43.
6. PEČARIĆ, J. E., Result not published.
7. HARDY, G. H., *Note on two inequalities*, J. London Math. Soc. **11** (1936), 167–170.
8. GABRIEL, R. M., *An extension of an inequality due to Carlson*, J. London Math. Soc. **12** (1937), 130–132.
9. BEURLING, A., *Sur les intégrales de Fourier absolument convergentes et leur application à une transformation fonctionelle*, in "Congrès des mathématiciens scandinaves 1938," Helsingfors, 1939.
10. LEVIN, V. I., *Notes on inequalities. I – Some inequalities between series* (Russian), Mat. Sbornik (N.S.) **3** (1938), 341–345.
11. CATON, W. B., *A class of inequalities*, Duke Math. J. **6** (1940), 442–461.
12. SZ.-NAGY, B., *On Carlson's and related inequalities* (Hungarian), Mat. Fiz. Lapok **48** (1941), 162- -175.
13. __________, *Über Integralungleichungen zwischen einer Funktion und ihrer Ableitung*, Acta Sci. Math. Szeged **10** (1941), 64–74.

14. BELLMAN, R., *An integral inequality*, Duke Math. J. **9** (1943), 547–550.
15. BECKENBACH, E. F. and R. BELLMAN, "Inequalities," 2nd ed., Berlin-Heidelberg-New York, 1965.
16. LEVIN, V. I., *Exact constant in the inequalities of Carlson's type* (Russian), Doklady Akad. Nauk SSSR (N.S.) **59** (1948), 635–638.
17. KJELLBERG, B., *Ein Momentenproblem*, Ark. Mat. Astr. Fys. **29 A** No. 2 (1942), 1–33.
18. _________, *On some inequalities*, C. R. 10e Congrès Math. Scandinaves, Copenhagen 1946, (1947), 333-340.
19. _________, *A note on an inequality*, Ark. Math. **3** (1956), 293–294.
20. LEVIN, V. I. and S. B. STEČKIN, *Inequalities*, Amer. Math. Soc. Transl. **14** 2 (1960), 1- -29. See, in particular, pp. 5–11.
21. O'NEIL, R., *Convolution opertors and $L(p,q)$ spaces*, Duke Math. J. **30** (1963), 136.
22. HUNT, R. A. and W. WEISS, *The Marcinkiewicz interpolation theorem*, Proc. Amer. Math. Soc. **15** (1964), 996–998.
23. LEVIN, V. I. and E. K. GODUNOVA, *A generalization of Carlson's inequality* (Russian), Mat. Sb. **67** 109 (1965), 643–646.
24. PIGOLKIN, G. M., *Mehrdimensionale Ungleichungen Carlsonschen Type und solche, die mit ihnen zusammenhängen* (Russisch), Sibir. Mat. Žurn. **11** (1970), 151–160.
25. KLEFSJÖ, B., *Problem 2808*, Elementa (Stockholm) **54** (1971), 128.
26. VOLKOV, V. N., *Utočnenie neravenstva Gel'dera*, Mosk. Gos. Ped. Inst. im V. I. Lenina. Učen. Zap. – Neravenstva **460** (1972), 110–115.
27. MITRINOVIĆ, D. S. and J. E. PEČARIĆ, *Two integral inequalities*, Southeast As. Bul. Math., to appear.

CHAPTER IX

INEQUALITIES INVOLVING KERNELS

1. Let (a_{ij}) be an $m \times n$ matrix of nonnegative terms. Then

$$mn \sum_{i=1}^{m} \sum_{j=1}^{n} a_{ij} \sum_{r=1}^{m} a_{rj} \sum_{s=1}^{n} a_{is} \geq \left(\sum_{i=1}^{m} \sum_{j=1}^{n} a_{ij} \right)^3 , \tag{1.1}$$

with equality if and only if all the row sums are equal, or all the column sums are equal, or both.

F. V. Atkinson, G. A. Watterson, and P. A. P. Moran [**1**] proved the following result. They used (1.1) to obtain

$$\sum_{i=1}^{m} \sum_{j=1}^{n} a_{ij} p_i q_j \sum_{r=1}^{m} a_{rj} p_r \sum_{s=1}^{n} a_{is} q_s \geq \left(\sum_{i=1}^{m} \sum_{j=1}^{n} a_{ij} p_i q_j \right)^3 , \tag{1.2}$$

where $p_i \geq 0$, $q_i \geq 0$, $\sum\limits_{i=1}^{m} p_i = \sum\limits_{i=1}^{n} q_i = 1$.

Putting in (1.2) $m = n$, $a_{ij} = a_{ji}$, and $p_i = q_i$ one obtains an inequality conjectured by S. P. H. Mandel and I. M. Hughes [**2**] on genetical grounds.

2. There is an obvious integral analogue of (1.1). If $K(x,y) \geq 0$ is integrable on the rectangle $0 \leq x \leq a$, $0 \leq y \leq b$, then

$$ab \int_0^a \int_0^b \int_0^a \int_0^b K(x,t) K(x,y) K(s,y) dx\, dy\, ds\, dt \geq \left(\int_0^a \int_0^b K(x,y) dx\, dy \right)^3 . \tag{2.1}$$

This version gives us the title of this chapter.

If $a = b$ and $K(x,y) = K(y,x)$, inequality (2.1) can be written in the form

$$(2.2) \qquad a^2 \int_0^a \int_0^a K_3(x,y)dx\,dy \geq \left(\int_0^a \int_0^a K(x,y)dx\,dy \right)^3 ,$$

where K_3 is the third-order interate of K in the sense of the theory of iterated kernels of integral equations.

3. Inequality (2.2) suggests a more general result

$$(3.1) \qquad a^{n-1} \int_0^a \int_0^a K_n(x,y)dx\,dy \geq \left(\int_0^a \int_0^a K(x,y)dx\,dy \right)^n .$$

This inequality was proved in [**1**] for all n of the form $2^r 3^s$, and was conjectured for other positive integral values of n.

J. F. C. Kingman [**3**] gave a more direct proof of (1.2) which is based on the convexity of $f(x) = x^k$ for $k \geq 1$.

A generalization to products of sums of $a_{ijk\ldots}$ taken over subsets of the index set $\{i, j, k, \ldots\}$ was also obtained in [**3**]. For example

$$\sum_{i=1}^{m} \sum_{j=1}^{n} \sum_{k=1}^{\ell} a_{ijk} p_i q_j r_k \sum_{s=1}^{n} \sum_{t=1}^{\ell} a_{ist} q_s r_t \sum_{s=1}^{m} \sum_{t=1}^{\ell} a_{sjt} p_s r_t \sum_{s=1}^{m} \sum_{t=1}^{n} a_{stk} p_s q_t$$
$$\geq \left(\sum_{i=1}^{m} \sum_{j=1}^{n} \sum_{k=1}^{\ell} a_{ijk} p_i q_j r_k \right)^4$$

for $p_i \geq 0$, $q_i \geq 0$, $r_i \geq 0$, and $\sum_{i=1}^{m} p_i = \sum_{i=1}^{n} q_i = \sum_{i=1}^{\ell} r_i = 1$.

Noting that the notion of a "partial average" is a particular example of a Radon-Nikodým derivative, J. F. C. Kingman [**4**] established an inequality involving such derivatives, which is a generalizaion of (1.2).

Using Kingman's method of proof, P. R. Beesack [**5**] proved the somewhat weaker inequality than (3.1), namely

$$(3.2) \qquad \int_0^a \int_0^a K_n(x,y)dx dy \geq a^{n+1} \exp \left(\frac{n \int_0^a \int_0^a \log K(x,y)dx\,dy}{a^2} \right) ,$$

for $n \geq 1$ and $K(x,y) \geq 0$. In case $K = \text{const} > 0$, both (3.1) and (3.2) are true with equality holding, but in general (3.1) is better than (3.2), P. R. Beesack in [5] showed that the conjecture (3.1) is true for all $n \geq 1$ and nonnegative symmetric kernels K. He used a result for symmetric matrices with nonnegative elements due to H.P. Mulholland and C. A. B. Smith [6]. More generally

$$\left(\int_0^a v(x)^2 dx\right)^{n-1} \int_0^a \int_0^a v(x)v(y)K_n(x,y)dx\,dy \geq \left(\int_0^a \int_0^a v(x)v(y)K(x,y)dx\,dy\right)^n$$

holds for arbitrary nonnegative continuous functions v, and $K(x,y) \geq 0$ and symmetric; setting $v(x) \equiv 1$, (3.1) is obtained.

In Beesack's paper [5] inequality (3.2) was formulated in a more general form.

4. P. R. Beesack and J. E. Pečarić [7] extended results from [5] to general mean values of kernels and also obtained some reverse or complementary inequalities involving such means. Here, we shall give their results.

Let μ be a nonnegative measure on a σ-algebra of subsets of A with $\mu(A) < \infty$. Write $A_n = A \times \cdots \times A$ (n times) and $d\mu_n$ for the product measure. $\mu \times \cdots \times \mu$ on subsets of A_n. Suppose $K(x,y)$ is a nonnegative μ_2 measurable function and ψ is a strictly monotone continuous function. A sequence of generalized kernels $\psi_n K(x,y)$ are defined as follows:

$$\text{(4.1)} \qquad \begin{cases} \psi_1 K(x,y) \equiv \psi\left(K(x,y)\right), \\ \psi_{n+1} K(x,y) \equiv \int_A \psi_n\left(K(x,z)\right)\psi_1\left(K(z,y)\right) d\mu(z). \end{cases}$$

If ψ is the identity function then $\psi_n K(x,y) = K_n(x,y)$ is the ordinary n-th iterated kernel in the theory of integral equations. It is easy to see that

$$\text{(4.2)} \qquad \int_{A_2} \psi_n K d\mu_2 = \int_{A_{n+1}} \left\{\prod_{j=1}^n \psi\left(K(x_{j-1}, x_j)\right)\right\} d\mu_{n+1}$$

if $\psi_1 K \in L_2(A_2)$ and thus also

$$\text{(4.3)} \qquad \psi_{n+1} K(x,y) = \int_A \psi_1 K(x,z) \psi_n K(z,y) d\mu(z).$$

We may also define the n-th order means of K with respect to ψ. For $n = 1, 2, \ldots$ they are

$$\alpha_{n,\psi}(K) = \psi^{-1}\left\{\int_{A_2} \psi_n K(x,y) d\mu_2(x,y)/\mu(A)^{n+1}\right\}. \tag{4.4}$$

In [7], there is a careful discussion on the validity of the results when some integrals are $+\infty$. We refer to that paper for this. Here we will assume all the integrals are finite.

If χ is another continuous and strictly monotonic function with $\chi_1 K \in L_2(A_2)$ then one can form $\alpha_{n,\chi}(K)$.

THEOREM 1. *With the above hypotheses and supposing $\chi(u) > 0$ for $u > 0$, let $\phi(u) = \chi\left(\psi^{-1}(u)\right)$ on R^+. Then for $n \geq 1$*

$$\alpha_{n,\psi}(K) \leq \alpha_{n,\chi}(K) \tag{4.5}$$

if either χ is increasing and ϕ is convex and submultiplicative or χ is decreasing and ϕ is concave and supermultiplicative.

PROOF: In the first alternative, we have by (4.2) and Jensen's inequality (with $\hat{K} = \psi(K)$)

$$\begin{aligned}
&\phi\left(\int_{A_2} \psi_n K \, d\mu/\mu(A)^{n+1}\right) \\
&= \phi\left(\int_{A_{n+1}} \prod_1^n \hat{K}(x_{j-1}, x_j) d\mu_{n+1}/\mu(A)^{n+1}\right) \\
&\leq \int_{A_{n+1}} \phi\left(\prod_1^n \hat{K}(x_{j-1}, x_j)\right) /\mu(A)^{n+1} d\mu_{n+1} \\
&\leq \int_{A_{n+1}} \prod_1^n \phi\left(\hat{K}(x_{j-1}, x_j)\right) /\mu(A)^{n+1} \, d\mu_{n+1} \\
&= \int_{A_{n+1}} \prod_1^n \chi_n K \, d\mu_2/\mu(A)^{n+1}.
\end{aligned}$$

The last inequality is the submultiplicative structure of ϕ. By applying χ^{-1} to both sides one completes the proof. For the second alternative, the inequalities are reversed.

The above results are extended in the following way. If r is a non-zero real number let

$$(4.7)\qquad \begin{cases} K_1^{[r]}(x,y) = (K(x,y))^r, \\ K_{n+1}^{[r]}(x,y) = \int\limits_A K_n^{[r]}(x,z)K_1^{[r]}(z,y)d\mu(z), \quad n = 1,\ldots \end{cases}$$

and

$$(4.8)\qquad \alpha_n^{[r]}(K) = \left\{ \int\limits_{A_2} K_n^{[r]} d\mu_2/\mu(A)^{n+1} \right\}^{1/r}.$$

For $r = 0$ let

$$(4.9)\qquad \alpha_n^{[0]}(K) = \exp\left\{ n \int\limits_{A_2} \log K d\mu_2/\mu(A)^2 \right\},$$

if either $\int\limits_A \log^+ K d\mu_2$ or $\int\limits_A \log^- K d\mu_2$ is finite, with

$$\alpha_n^{[0]}(K) = +\infty \quad \text{if} \int\limits_{A_2} \log^+ K\, d\mu_2 = +\infty,$$

and

$$\alpha_n^{[0]}(K) = 0 \qquad \text{if} \int\limits_{A_2} \log^- K\, d\mu_2 = -\infty.$$

Note that if $\psi(u) = u^r$, $r \neq 0$, then $K_n^{[r]} = \psi_n K$. There is no ψ such that $\alpha_n^{[0]} K = \alpha_{n,\psi}(K)$ but

$$\begin{aligned} \int\limits_{A_{n+1}} \log \prod_1^n K(x_{j-l}, x_j) d\mu_{n+1} &= \sum_1^n \int\limits_{A_{n+1}} \log K(x_{j-1}, x_j) d\mu_{n+1} \\ &= n\mu(A)^{n-1} \int\limits_{A_2} \log K\, d\mu_2 \end{aligned}$$

so that

$$\alpha_n^{[0]}(K) = \exp\left\{ \int\limits_{A_{n+1}} \log \prod_1^n K(x_{j-1}, x_j) d\mu_{n+1}/\mu(A)^{n+1} \right\}$$

with the above conventions.

THEOREM 2. *Let ψ be continuous and strictly monotone and let $\phi(u) = \log \psi^{-1}(u)$ for $u > 0$. If either $\psi(K) \in L_2(A_2)$ or $\psi \geq 0$, then*

$$\alpha_{n,\psi}(K) \leq \alpha_n^{[0]}(K) \tag{4.9}$$

if ϕ is convex, ψ^{-1} is submultiplicative, and $\int_{A_2} \log^+ K d\mu_2 < \infty$. If ϕ is concave, ψ^{-1} is supermultiplicative, and either $\psi \geq 0$ and $\int_{A_2} \psi(K) d\mu_2 < \infty$ or $\psi(K) \in L_2(A_2)$, then inequality in (4.9) is reversed.

The above two theorems have a corollary dealing with the means (4.8).

COROLLARY 1. *Let K be measurable and nonnegative, then for $-\infty < s \leq r < \infty$*

$$\alpha_n^{[s]}(K) \leq \alpha_n^{[r]}(K) \tag{4.10}$$

provided $\alpha_1^{[t]}(K)$ is finite for some $t > 0$.

PROOF: (Sketch only.) For $s < r$ with $r \neq 0$, take $\chi(u) = u^r$, $\psi(u) = u^s$ so that $\phi(u) = u^{r/s}$ is both sub- and supermultiplicative. If $s < r < 0$ we get the second alternative of Theorem 1 while if $s < 0 < r$ or $0 < s < r$ the first alternative is correct.

If $s < 0 = r$ we apply Theorem 2 with $\psi(u) = u^s$ so that $\phi(u) = \left(\frac{1}{s}\right) \log u$ is convex and ψ^{-1} is multiplicative. Here $\int_{A_2} K^t d\mu_2$ is finite for some $t > 0$, and $\int_{A_2} \log^{-1} K d\mu_2$ is finite.

The last case $s = 0 < r$ follows from Theorem 2 in the same way with $\phi(u) = \left(\frac{1}{r}\right) \log u$ and $\psi(u) = u^r$.

REMARK 1: The special case of (4.10) for $r = 1$ and $s = 0$ appears in Beesack [5] with the hypothesis that $K \in L_2(A_2)$.

A second special case is noted in the following Corollary.

COROLLARY 2. *If $\chi \geq 0$ and either*

a) *χ is increasing and $\phi = \chi(\psi^{-1})$ is convex and submultiplicative or*
b) *χ is decreasing and ϕ is concave and supermultiplicative, then*

$$\psi^{-1}\left(\psi_n\left(K(x_0, x_n)\right)/\mu(A)^{n-1}\right) \leq \chi^{-1}\left(\chi_n K(x_0, x_n)/\mu(A)^{n-1}\right). \tag{4.11}$$

In particular for $\psi(u) \equiv u$, and K_n the usual iterated kernel,

$$\chi_n K(x_0, x_n) \leq \mu(A)^{n-1} \chi\left[K_n(x_0, x_n)/\mu(A)^{n-1}\right] \tag{4.12}$$

if (b) holds and

$$\chi_n K(x_0, x_n) \geq \mu(A)^{n-1} \chi\left[K_n(x_0, x_n)/\mu(A)^{n-1}\right] \tag{4.13}$$

if (a) holds.

For Volterra type kernels with $A = [0, a]$ and $K(x, y) = 0$ for $x < y$ the above can be applied with $\psi(0) = 0$ and increasing on $[0, \infty)$ and continuous. One can easily show that the iterated kernels $\psi_n K$ are also of the Volterra type and that the representations

$$\begin{aligned} \psi_{n+1}K(x,y) &= \int_y^x \psi_n K(x,z)\psi_1 K(z,u)d\mu(z) \\ &= \int_y^x \psi_1 K(x,z)\psi_n K(z,u)d\mu(z) \end{aligned} \tag{4.14}$$

or for $(x_0 \leq x_n)$

$$\psi_n K(x_n, x_0) = \int_{x_0}^{x_n} \cdots \int_{x_0}^{x_2} \prod_1^n \psi\left(K(x_j, x_{j-1}\right) d\mu(x_1) \cdots d\mu(x_{n-1}) \tag{4.15}$$

hold. Note that $\mu(A)^{n+1}$ is replaced by

$$\int_0^a \int_{x_0}^a \int_{x_0}^{x_n} \cdots \int_{x_0}^{x_2} d\mu(x_1) \ldots d\mu(x_n) d\mu(x_0).$$

Introduce the following notation:

$$\begin{aligned} x_1(\alpha, \beta, n) &= \min\left(\alpha^n, \beta^n, \alpha^{n-1}, \alpha\beta^{\alpha-1}\right) \\ x_2(\alpha, \beta, n) &= \max\left(\alpha^n, \beta^n \beta\right), \end{aligned} \tag{4.16}$$

and let λ be determined as follows: set $\mu = [\phi(x_2) - \phi(x_1)]/(x_2 - x_1)$ and if $\mu \neq 0$ let $x = \bar{x}$ be the unique solution in $[x_1, x_2]$ of the equation $\mu\phi(x) = \phi'(x)\left[\phi(x_1) + \mu(x - x_1)\right]$; then $\lambda = \mu/\phi'(\bar{x})$.

The following two theorems are also proved in [**7**]:

THEOREM 3. *Let K be μ_2-measurable on A_2 with $0 < m \leq K(x, y) \leq M$ for all $(x, y) \in A_2$. Let ψ, χ: $R^+ \to R$ be continuous, strictly monotonic functions with inverse functions ψ^{-1}, χ^{-1}, and suppose that $\chi(R^+) = R^+$, and $\psi(R^+) = R^+$ if all $\psi(x) > 0$, or $\psi(R^+) = R$ if some $\psi(x) \leq 0$.*

Let $\alpha = \min(\psi(m), \psi(M))$ and $\beta = \max(\psi(m), \psi(M))$ and let $x_1 = x_1(\alpha, \beta, n)$, $x_2 = x_2(\alpha, \beta, n)$ be defined by (4.16). Set $\phi(u) = \chi\left(\psi^{-1}(u)\right)$ for $u \in \psi(R^+)$ and let $\mu = [\phi(x_2) - \phi(x_1]/(x_2 - x_1)$.

(a) *Suppose $\phi''(x) \geq 0$ with equality for at most isolated points of $[x_1, x_2]$ and let $\lambda > 1$. Then if ϕ and χ are both supermultiplicative, we have*

$$\alpha_{n,\chi}(K) \leq \chi^{-1}(\lambda)\alpha_{n,\psi}(K) \tag{4.17}$$

if χ is increasing, while the opposite inequality holds if χ is decreasing. Moreover, if ϕ is supermultiplicative but χ is submultiplicative, then

$$\alpha_{n,\chi}(K) \leq \left[\chi^{-1}(\lambda^{-1}]^{-1} \alpha_{n,\psi}(K) \tag{4.18}$$

if χ is increasing while the opposite holds if χ is decreasing.

(b) *Suppose $\phi''(x) \leq 0$ with equality for at most isolated points of $[x_1, x_2]$, and let $\lambda \in (0,1)$. Then if ϕ and χ are both supermultiplicative we have*

$$\alpha_{n,\chi}(K) \geq \chi^{-1}(\lambda)\alpha_{n,\psi}(K) \tag{4.19}$$

if χ is increasing, and the reverse inequality if χ is decreasing. Moreover, if ϕ is submultiplicative but χ is supermultiplicative, then

$$\alpha_{n,\chi}(K) \geq \left[\chi^{-1}(\lambda^{-1}]^{-1} \alpha_{n,\psi}(K) \tag{4.20}$$

if χ is increasing with the opposite inequality if χ is decreasing.

THEOREM 4. *Let $K, m, M, \alpha, \beta, x_1, x_2$ be as in* Theorem 3, *where ψ is continuous and strictly monotonic on R^+ with $\psi(R^+) = R^+$ if all $\psi(x) > 0$, or $\psi(R^+) = R$ if some $\psi(x) \leq 0$. Set $\phi(u) = \log \psi^{-1}(u)$ and $\mu = [\phi(x_2) - \phi(x_1)]/(x_2 - x_1)$.*

(a) *If ψ^{-1} is supermultiplicative, and ϕ' is strictly increasing on $I = [x_1, x_2]$, and $\lambda > 0$ is determined as follows: let $x = \bar{x}$ be the unique solution of the equation $\phi'(x) = \mu$ $(x_1 < \bar{x} < x_2)$; then $\lambda = \phi(x_1) - \phi(\bar{x}) + \mu(\bar{x} - x_1)$,* and

$$\alpha_n^{[0]}(K) \leq e^{\lambda}\alpha_{n,\psi}(K). \tag{4.21}$$

(b) *If ψ^{-1} is submultiplicative, ϕ' is strictly decreasing on* **R** *and $\lambda > 0$ is determined by $\lambda = \phi(\bar{x}) - \phi(x_1) - \mu(\bar{x} - x_1)$, where $x = \bar{x}$ is the unique solution of $\phi'(x) = \mu$ $(x_1 < \bar{x} < x_2)$, then*

$$\alpha_{n,\psi}(K) \leq e^{\lambda}\alpha_n^{[0]}(K). \tag{4.22}$$

Theorem 4 is complementary to Theorem 2 only in *the case that ψ^{-1} is multiplicative.* In this case (4.9) and (4.21) give

$$\alpha_{n,\psi}(K) \leq \alpha_n^{[0]}(K) \leq e^{\lambda}\alpha_{n,\psi}(K)$$

provided $0 < m \leq K(x,y) \leq M$, $\psi(\mathbf{R}^+) = \mathbf{R}$, and ϕ' is strictly increasing, while the opposite of (4.9) and (4.22) give

$$\alpha_{n\psi}(K) \geq \alpha_n^{[0]}(K) \geq e^{-\lambda}(K)\alpha_{n,\psi}(K) \tag{4.23}$$

provided $0 < m \le K(x,y) \le M$, $\psi(\mathbf{R}^+) = \mathbf{R}$, and ψ' is strictly decreasing.

As an application of Theorems 3 and 4 we again consider the r- means $\alpha_n^{[r]}(K)$ defined by (4.8) and (4.9). For $r, s \ne 0$ the functions χ, ψ, ϕ considered in Corollary 1 are all actually multiplicative.

Let $0 < m \le K(x,y) \le M$ for $(x,y) \in A_2$ and for $r, s \ne 0$, $n \ge 1$ let

$$\mu = (M^{nr} - m^{nr})(M^{ns} - m^{ns})^{-1},$$

$$B_{r,s}^{(n)} = \left(\frac{\mu s}{r}\right)^{1/s} \left\{ \frac{m^{nr}M^{ns} - m^{ns}M^{nr}}{[1-(r/s)](M^{rs} - m^{ns})} \right\}^{1/r-1/s}. \tag{4.24}$$

Then

$$\alpha_n^{[r]}(K) \le B_{r,s}^{(n)} \alpha_n^{[s]}(K) \quad \text{if} \quad s < r \ (s, r \ne 0). \tag{4.25}$$

This follows from Theorem 3.

We also have

$$\alpha_n^{[r]}(K) \le B_{r,n} \alpha_n^{[0]}(K) \quad (r > 0) \tag{4.26}$$

where $B_{r,n} = (r\mu_r e)^{-1/r} \exp(\mu_r m^{nr})$, with

$$\begin{gathered} \mu_r = n \log(M/m)/(M^{nr} - m^{nr}), \qquad \text{and} \\ a_n^{[0]}(K) \le B_{s,n}^{-1} \alpha_n^{[s]}(K) \quad (s < 0). \end{gathered} \tag{4.27}$$

REMARK 2: The special case $n = 1$ of Theorems 3 and 4 and of the inequalities (4.26) and (4.27) was proved in [**8**]. In the case $n = 1$ the analogue of Theorem 3 alone was used to give the analogues of (4.26) and (4.27), because no questions of sub- or supermultiplicativity of ϕ arose.

REMARK 3: Theorems 3 and 4 are simple consequences of conversions of Jensen's inequality obtained in [**8**] as integral analogues of results from [**9**]. (For some functional analogues see [**10**].)

REFERENCES

1. ATKINSON, F. V., G. A. WATTERSON and P. A. P. MORAN, *A matrix inequality* Quart. J. Math. Oxford Ser. (2) **11** (1960), 127–140.
2. MANDEL, S. P. H. and I. M. HUGHES, *Change in mean viability at a multiallelic locus in a population under random mating*, Nature (London) **182** (1958), 63–64.
3. KINGMAN, J. F. C., *On an inequality in partial averages*, Quart. J. Math. Oxford Ser. (2) **12** (1961), 78–80.

4. __________, *An inequality involving Radon-Nikodým derivatives*, Proc. Cambridge Phil. Soc. **63** (1967), 195–198.
5. BEESACK, P. R., *Inequalities involving iterated kernels and convolutions*, Univ. Beograd. Publ. Elektrotehn. Fak. Ser. Mat. Fix. No. **274–301**, (1969), 11–16.
6. MULHOLLAND, H. P. and C. A. B. SMITH, An inequality arising in genetical theory, Amer. Math. Monthly **66** (1959), 673–683.
7. BEESACK, P. R. and J. E. PEČARIĆ, *Inequalities involving nth-order means*, J. Math. Anal. Appl. **99**(1984), 127–149.
8. BEESACK, P. R., *On inequalities complementary to Jensen's*, Canad. J. Math. **35**(1983), 324–338.
9. MITRINOVIĆ, D. S. and P. M. VASIĆ, *The centroid method in inequalities*, Univ. Beograd. Publ. Eletrotehn. Fak. Ser. Mat. Fiz. No. **498–541** (1975), 3–16.
10. BEESACK, P. R. and J. E. PEČARIĆ, *On the Jensen's inequality for convex functions*, J. Math. Anal. Appl. **110**(1985), 536–552.

CHAPTER X

CONVOLUTION, REARRANGEMENT AND RELATED INEQUALITIES

1. If $\lambda > 0$, $\nu > 0$, $\lambda + \nu < 1$, and z is defined by $z_n = \sum_{i+j=n} x_i y_j$, $\mathfrak{G}_r(x) = (\sum x_i^r)^{1/r}$, then

$$\mathfrak{G}_{\frac{1}{1-\lambda-\nu}}(z) \leq \mathfrak{G}_{\frac{1}{1-\lambda}}(x)\mathfrak{G}_{\frac{1}{1-\nu}}(y), \tag{1.1}$$

with equality only if all the x, or all the y, or all the x but one and all the y but one, are zero.

This inequality is due to W. H. Young **[1]**–**[3]**, and it is very important in the theory of Fourier series. (Young does not consider the question of equality, Hardy, Littlewood and Pólya Inequalities, 1934.) The proof uses Hölder's inequality. Here, we shall first give generalizations of Young's inequality given in Hardy, Littlewood and Pólya's book as theorems 278–284:

1^0 (278.) *If $\lambda > 0$, $\mu > 0, \ldots, \nu > 0$ $\lambda + \mu + \cdots + \nu < 1$, and*

$$w_n = \sum_{i+j+\cdots+k=n} x_i y_j \ldots z_k,,$$

then

$$\mathfrak{G}_{\frac{1}{1-\lambda-\mu-\cdots-\nu}}(w) \leq \mathfrak{G}_{\frac{1}{1-\lambda}}(x)\mathfrak{G}_{\frac{1}{1-\mu}}(y)\ldots\mathfrak{G}_{\frac{1}{1-\nu}}(z),$$

unless all numbers of one set, or all but one of every set, are zero.

2^0 (279.) *If*

$$c_n = \sum_{i_1+i_2+\cdots+i_n} a_{i_1}a_{i_2}\ldots a_{i_k},$$

then

$$\sum c_n^2 \leq \left(\sum a_n^{\frac{2k}{2k-1}}\right)^{2k-1},$$

unless all a but one are zero.

Theorems 277-279, "being homogeneous in $\sum$", have integral analogues.

3^0 (280.) *If* $\lambda > 0$, $\mu > 0$, $\lambda + \mu < 1$, *and*

$$h(x) = \int_{-\infty}^{\infty} f(t)g(x-t)dt,$$

then

$$\left(\int_{-\infty}^{\infty} h^{\frac{1}{1-\lambda-\mu}}\, dx\right)^{1-\lambda-\mu} \le \left(\int_{-\infty}^{\infty} f^{\frac{1}{1-\lambda}}\, dx\right)^{1-\lambda} \left(\int_{-\infty}^{\infty} g^{\frac{1}{1-\mu}}\, dx\right)^{1-\mu}$$

unless f *or* g *is null.*

4^0 (281.) *If* $\lambda > 0$, $\mu > 0$, $\lambda + \mu < 1$, *and*

$$h(x) = \int_{0}^{x} f(t)g(x-t)dt,$$

then

$$\left(\int_{0}^{\infty} h^{\frac{1}{1-\lambda-\mu}}\, dx\right)^{1-\lambda-\mu} \le \left(\int_{0}^{\infty} f^{\frac{1}{1-\lambda}}\, dx\right)^{1-\lambda} \left(\int_{0}^{\infty} g^{\frac{1}{1-\mu}}\, dx\right)^{1-\mu},$$

unless f *or* g *is null.*

5^0 (282.) *If* k *is an integer and*

$$\phi(x) = \int_{-\infty}^{\infty} \cdots \int_{-\infty}^{\infty} f(x_1)f(x_2)\dots f(x_{k-1})$$
$$\times f(x - x_1 - \cdots - x_{k-1})dx_1 dx_2 \dots dx_{k-1},$$

then

$$\int_{-\infty}^{\infty} \phi^2(x)dx \le \left(\int_{-\infty}^{\infty} f^{\frac{2k}{2k-1}}(x)dx\right)^{2k-1}.$$

6^0 (283.) *If*

$$\phi(x) = \int f(x_1)\dots f(x_{k-1})f(x - x_1 - \cdots - x_{k-1})dx_1 \dots dx_{k-1},$$

the integration being defined by

$$x_i \geq 0, \qquad \sum x_i \leq x,$$

then

$$\int_0^\infty \phi^2(x)dx \leq \left(\int_0^\infty f^{\frac{2k}{2k-1}}(x)dx \right)^{2k-1}.$$

7⁰ (284.) *If $f(x)$ has the period 2π, and*

$$\phi(x) = \frac{1}{(2\pi)^{k-1}} \int_{-\pi}^{\pi} \cdots \int_{-\pi}^{\pi} f(x_1)f(x_2)\ldots f(x_{k-1})$$
$$\times f(x - x_1 - \cdots - x_{k-1})dx_1 dx_2 \ldots dx_{k-1},$$

then

$$\frac{1}{2\pi} \int_{-\pi}^{\pi} \{\phi(x)\}^2 dx \leq \left(\frac{1}{2\pi} \int_{-\pi}^{\pi} f^{\frac{2k}{2k-1}}(x)dx \right)^{2k-1}.$$

The inequality (1.1) is equivalent to one involving three functions, namely

$$(1.2) \qquad \left| \iint f(x)g(y-x)h(y)dx\, dy \right| \leq \|f\|_p \|g\|_q \|h\|_t,$$

valid for $f \in L^p$, $g \in L^q$, $h \in L^t$ with $p, q, t > 1$ and $p^{-1} + q^{-1} + t^{-1} = 2$ (so $r^{-1} + t^{-1} = 1$). The unit constant on the right side of (1.1), (1.2) is not best possible; it was only shown in 1975 by W. Beckner [5], that Gaussians give the best constant in (1.2) when $1 \leq p, q, t \leq 2$. In [6] the following very general and powerful result is proved (Theorem 1). Let n, k be integers with $1 \leq n \leq k$, let p_j, $1 \leq j \leq k$ be extended real numbers such that $1 \leq p_j \leq \infty$ and $\sum_{1}^{n} p_j^{-1} = n$, and let $a^{(j)}$, $1 \leq j \leq k$, be given vectors in $\mathbf{R}^n$. Then there exists a constant D, depending on $p_1, \ldots, p_n$, $a^{(1)}$, $a^{(2)}$, $\ldots$, $a^{(n)}$, such that for all $f_j \in L^{p_j}(\mathbf{R})$,

$$(1.3) \qquad \left| \int_{\mathbf{R}^n} \prod_{j=1}^{k} f_j\left(\langle a^{(j)}, x\rangle\right) dx \right| \leq D \prod_{j=1}^{k} \|f_j\|_{p_j},$$

where $\langle a^{(j)}, x\rangle = \sum_{i=1}^{n} a_i^{(j)} x_i$. The proof of the theorem depends on an earlier inequality of Brascamp, Lieb and Luttinger [7] (this result will be given later).

2. Many papers prove convolution inequalities, see for example papers of Subrahmanyam [**8**] (he finds the best possible constants in some convolution inequalities), Boas and Imoru [**9**], Beesack and Pečarić [**10**]. Here we shall give results of Beesack and Pečarić from [**10**].

Let f and g be nonnegative Lebesgue measurable functions on $(0,\infty)$ and for each $x \geq 0$ consider the measurable nonnegative convolution $f * g$ defined by the formula

$f*g(x) = \int_0^x f(t)g(x-t)dt = \int_0^x g(t)f(x-t)dt$. If $\psi\colon (0,\infty) \to (0,\infty)$ is continuous, onto, and strictly monotonic, we consider bounds for $\psi f_1 * \psi f_2 * \cdots * \psi f_n(x)$, where $\psi f_j \equiv \psi(f_j)$, as well as in inequalities involving generalized means of $\psi f_1 * \cdots * \psi f_n$. To this end introduce the definitions

(2.1)
$$\beta_{n\psi}(f_1,\ldots,f_n;x) = \psi^{-1}\{(n-1)!\psi f_1 * \cdots * \psi f_n(x)/x^{n-1}\},$$

(2.2)
$$\beta_n^{[0]}(f_1,\ldots,f_n;x) = \exp\left\{(n-1)\int_0^x (x-t)^{n-2}\sum_1^n \log f_j(t)dt/x^{n-1}\right\},$$

where, for simplicity, we always assume $\int_0^x \log^+ f_j(t)dt < \infty$ for $1 \leq j \leq n$ $(n \geq 2)$.

THEOREM 1. *Let f_j $(1 \leq j \leq n)$ be nonnegative measurable functions on $R^+ = (0,\infty)$, and let $\phi, \chi\colon R^+ \to R^+$ be continuous, onto, and strictly monotonic. Let $\phi(u) = \chi\left(\psi^{-1}(u)\right)$ for $u > 0$. Then, for $n \geq 2$ and $x > 0$ we have*

$$\beta_{n,\psi}(f_1,\ldots,f_n;x) \leq \beta_{n,\chi}(f_1,\ldots,f_n;x) \tag{2.3}$$

provided either (a) χ is increasing and ϕ is both convex and submultiplicative, or (b) χ is decreasing and ϕ is both concave and supermultiplicative.

Moreover, if $\int_0^x \log^+ f_j dt < \infty$ for $1 \leq j \leq n$, and we set $\phi_1(u) = \log\left(\psi^{-1}(u)\right)$, $u > 0$, then for $n \geq 2$, $x > 0$,

$$\beta_{n,\psi}(f_1,\ldots,f_n;x) \leq \beta_n^{[0]}(f_1,\ldots,f_n;x) \tag{2.4}$$

provided ϕ_1 is convex and ψ^{-1} is submultiplicative, while the opposite inequality holds if ϕ_1 is concave and ψ^{-1} supermultiplicative.

PROOF: The proof depends on the representations

$$\psi f_1 * \cdots * \psi f_n(x_n) = \int_0^{x_n} \cdots \int_0^{x_2} \prod_1^n \psi f_j(x_j - x_{j-1}) dx_1 \ldots dx_{n-1},$$

$$\frac{x_n^{n-1}}{(n-1)!} = \int_0^{x_n} \cdots \int_0^{x_2} dx_1 \ldots dx_{n-1},$$

$$\frac{1}{(n-2)!} \int_0^{x_n} (x_n - t)^{n-2} \sum_1^n \log f_j(t) dt$$

$$= \int_0^{x_n} \cdots \int_0^{x_2} \sum_1^n \log f_j(x_j - x_{j-1}) dx_1 \ldots dx_{n-1},$$

valid for $n \geq 2$, $x_n \geq 0$ (where $x_0 = 0$) under our hypotheses. Note that all integrals exist, finite or infinite, under these hypotheses. The last identity is clearly true for $n = 2$ and can be proved by induction. We omit further details of the proof, only noting that the main step is the application of Jensen's inequality with $D = \{(x_1, \ldots, x_{n-1}): \ 0 \leq x_1 \leq x_2 \leq \cdots \leq x_{n-1} \leq x_n\}$, $\nu =$ Lebesgue measure in R^{n-1}, and $q = 1/\nu(D) = (n-1)! x_n^{1-n}$.

COROLLARY. *For $r \neq 0$, $n \geq 2$, and nonnegative measurable f_j $(1 \leq j \leq n)$ on R^+ define the r- means*

$$\beta_n^{[r]}(f_1, \ldots, f_n; x) = \left\{(n-1)! f_1^r * \cdots * f_n^r(x)/x^{n-1}\right\}^{1/r}. \tag{2.5}$$

Then for $-\infty < s < r < \infty$ we have for $n \geq 2$, $x > 0$,

$$\beta_n^{[s]}(f_1, \ldots, f_n; x) \leq \beta_n^{[r]}(f_1, \ldots, f_n; x). \tag{2.6}$$

This follows by taking $\chi(u) = u^r$, $\psi(u) = u^s$ if $-\infty < s < r < \infty$ with $r, s \neq 0$ in Theorem 1, etc. In this case the proof (when $r = 0$ or $s = 0$) is simpler since we are assuming the finiteness of all the integrals $\int_0^x \log^+ f_j(t) dt$.

The special case $s = 0$, $r = 1$ of (2.6) was proved in [**11**, Theorem 2].

From the above results, we can obtain bounds for $\chi f_1 * \cdots * \chi f_n$ in terms of the ordinary convolution $f_1 * \cdots * f_n$ by taking $\psi(u) = u$ in (2.3). We find that

$$\chi f_1 * \ldots \chi f_n(x) \leq \frac{x^{n-1}}{(n-1)!} \chi\left((n-1)! \frac{f_1 * \cdots * f_n(x)}{x^{n-1}}\right) \tag{2.7}$$

if $\chi > 0$ is decreasing, concave and supermultiplicative on $\mathbf{R}^+$, while the opposite inequality holds if $\chi > 0$ is increasing, convex and submultiplicative.

In the same paper, Beesack and Pečarić give some converses of the above results.

3. Rearrangements of vectors and of functions are considered in various places, see [**4**, Ch. 10] for example. There are several monotonic rearrangements, and here we shall give some of these results.

DEFINITION 1: Let $a = (a_1, \dots, a_n)$ be a real vector. By $\underline{a} = (\underline{a}_1, \dots, \underline{a}_n)$ we denote nonincreasing rearrangement of vector a, i.e. $\underline{a}_1 \geq \underline{a}_2 \geq \cdots \geq \underline{a}_n$. Similarly, $\bar{a} = (\bar{a}_1, \dots, \bar{a}_n)$ is nondecreasing rearrangement of vector a, i.e. $\bar{a}_1 \leq \bar{a}_2 \leq \cdots \leq \bar{a}_n$.

The best known theorem on rearrangement of vectors is:

THEOREM 1. *Let a and b be two real vectors. Then*

$$\sum_{i=1}^{n} \bar{a}_i \underline{b}_i \leq \sum_{i=1}^{n} a_i b_i \leq \sum_{i=1}^{n} \bar{a}_i \bar{b}_i. \tag{3.1}$$

PROOF: ([**12**]) Without loss in generality we can suppose that $b = \bar{b}$ and $\bar{b}_1 > 0$ (by adding a constant). Now, using the Abel identity we get

$$\begin{aligned}\sum_{i=1}^{n} a_i \bar{b}_i &= \bar{b}_1 \sum_{i=1}^{n} a_i + \sum_{i=2}^{n} \left(\sum_{j=i}^{n} a_j\right) (\bar{b}_i - \bar{b}_{i-1}) \\ &\leq \bar{b}_1 \sum_{i=1}^{n} \bar{a}_i + \sum_{i=2}^{n} \left(\sum_{j=i}^{n} \bar{a}_j\right) (\bar{b}_i - \bar{b}_{i-1}) = \sum_{i=1}^{n} \bar{a}_i \bar{b}_i,\end{aligned}$$

and

$$\sum_{i=1}^{n} a_i b_i \geq \bar{b}_1 \sum_{i=1}^{n} \underline{a}_i + \sum_{i=2}^{n} \left(\sum_{j=i}^{n} \underline{a}_j\right) (\bar{b}_i - \bar{b}_{i-1}) = \sum_{i=1}^{n} \underline{a}_i \bar{b}_i.$$

REMARK: For a version of Theorem 1 where a and b are in R^n but the sums are from $i = 1$ to k, see 13.

The case of m vectors was considered in [**13**]:

THEOREM 2. *Let $a_j = (a_{ji}, \dots, a_{jn})$ $(j = 1, \dots, m)$ be nonnegative vectors. Then*

$$\sum_{i=1}^{n} \left(\prod_{j=1}^{m} a_{j1}\right) \leq \sum_{i=1}^{n} \left(\prod_{j=1}^{m} \bar{a}_{ji}\right). \tag{3.2}$$

PROOF: ([12]) For $m = 2$, (3.2) follows from (3.1). Suppose that (3.2) is valid for $m-1$. Then, if we suppose that $a_m = \bar{a}_m$, we have

$$\begin{aligned}\sum_{i=1}^{n}\left(\prod_{j=1}^{m-1} a_{ji}\right)\bar{a}_{mi} &= \bar{a}_{m1}\sum_{i=1}^{n}\left(\prod_{j=1}^{m-1} a_{ji}\right) + \sum_{i=2}^{n}\left(\sum_{k=i}^{n}\left(\prod_{j=1}^{m-1} a_{jk}\right)\right)(\bar{a}_{mi} - \bar{a}_{m,i-1}) \\ &\le \bar{a}_{m1}\sum_{i=1}^{n}\left(\prod_{j=1}^{m-1} \bar{a}_{ji}\right) + \sum_{i=2}^{n}\left(\sum_{k=i}^{n}\left(\prod_{j=1}^{m-1} \bar{a}'_{jk}\right)\right)(\bar{a}_{mi} - \bar{a}_{m,i-1}) \\ &\le \bar{a}_{m1}\sum_{i=1}^{n}\left(\prod_{j=1}^{m-1} \bar{a}_{ji}\right) + \sum_{i=2}^{n}\left(\sum_{k=i}^{n}\left(\prod_{j=1}^{m-2} \bar{a}_{jk}\right)\right)(\bar{a}_{mi} - \bar{a}_{m,i-1}) \\ &= \sum_{i=1}^{n}\left(\prod_{j=1}^{m} \bar{a}_{ji}\right),\end{aligned}$$

where $\bar{a}'_{jk}$ $(i \le k \le n)$ is the nondecreasing rearrangement of $(a_{ji}, \ldots, a_{jn})$.

We can formulate the following generalization of Theorems 1 and 2 using results from [14]–[19].

THEOREM 3. *(i) Let $F(x, y)$ be a positive set function on $I_x \times I_y$, i.e. a function such that*

$$F(x+h, y+k) - F(x+h, y) - F(x, y+k) + F(x, y) \ge 0$$

for $h, k \ge 0$ (or $h, k \le 0$) with $x+h, y+k \in I_x \times I_y$, and for arbitrary choices of $x, y \in I_x \times I_y$.

If $x = (x_1, \ldots, x_n) \in I_x^n$, $y = (y_1, \ldots, y_n) \in I_y^n$, then

$$\sum_{i=1}^{n} F(\underline{x}_i, \bar{y}_i) \le \sum_{i=1}^{n} F(x_i, y_i) \le \sum_{i=1}^{n} F(\bar{x}_i, \bar{y}_i). \tag{3.3}$$

If $F(x, y)$ is a negative set function (the reverse sign in the above definition inequality is valid), then the reverse inequality in (3.3) is valid.

(ii) Let $F(x_1, \ldots, x_m)$ be a positive set function for each pair of variables, i.e. F is I-superadditive function, on $I_{x_1} \times \cdots \times I_{x_m}$. If $x_j \in I_{x_j}^n$ $(j = 1, \ldots, m)$, then

$$\sum_{i=1}^{n} F(x_{1i}, \ldots, x_{mi}) \le \sum_{i=1}^{n} F(\bar{x}_{1i}, \ldots, \bar{x}_{mi}). \tag{3.4}$$

If F is a postive set function for every pair of variables x_i, x_j when $i, j > k$ or $i, j \leq k$ and also a negative set function when $i \leq k$ and $j > k$, then (3.4) becomes

$$\sum_{i=1}^{n} F(x_{1i}, \ldots, x_{ki}, x_{k+1,i}, \ldots, x_{mi}) \tag{3.5}$$
$$\leq \sum_{i=1}^{n} F(\underline{x}_{1i}, \ldots, \underline{x}_{ki}, \bar{x}_{k+1,i}, \ldots, \bar{x}_{mi}).$$

REMARK: Let $\phi(a, b)$ be a function of two variables. Let us define the matrix $\phi(x, y)$ where x, y are two vectors, by

$$\phi(x,y) = \begin{bmatrix} \phi(x_1, y_1) & \ldots & \phi(x_1, y_n) \\ \vdots & & \\ \phi(x_n, y_1) & \ldots & \phi(x_n,, y_n) \end{bmatrix}.$$

For a positive function ϕ we say that it is totally positive of order 2 (TP_2) if every subdeterminant of the matrix $\phi(x, y)$ is nonnegative. It is known that the function ϕ is TP_2 iff $F(a, b) = \log \phi(a, b)$ is a positive set function. In this case (3.4) becomes

$$\prod_{i=1}^{n} \phi(x_{1i}, \ldots, x_{mi}) \leq \prod_{i=1}^{n} \phi(\bar{x}_{1i}, \ldots, \bar{x}_{mi}); \tag{3.6}$$

where ϕ is positive function which is TP_2 as a function of every pair of variables.

EXAMPLES: 1^0 The function $\phi(x_1, \ldots, x_m) = \min_{1 \leq i \leq m} x_i$, $x_i > 0$, is TP_2 and a positive set function in every pair of variables. So the following inequalities are valid

$$\prod_{i=1}^{n} \min(\bar{x}_i, \underline{y}_i) \leq \prod_{i=1}^{n} \min(x_i, y_i) \leq \prod_{i=1}^{n} \min(\bar{x}_i, \bar{y}_i) \qquad ([\mathbf{19}]); \tag{3.7}$$

$$\prod_{i=1}^{n} \min_{1 \leq j \leq m} x_{ji} \leq \prod_{i=1}^{n} \lim_{1 \leq j \leq m} \bar{x}_{ji} \qquad ([\mathbf{20}]); \tag{3.8}$$

$$\sum_{i=1}^{n} \min_{1 \leq j \leq m} x_{ji} \leq \sum_{i=1}^{n} \min_{1 \leq j \leq m} \bar{x}_{ji} \qquad ([\mathbf{20}]). \tag{3.9}$$

2^0 If f is a convex function, then

$$\sum_{i=1}^{n} f(\underline{x}_i + \bar{y}_i) \leq \sum_{i=1}^{n} f(x_i + y_i) \leq \sum_{i=1}^{n} f(\bar{x}_i + \bar{y}_i) \qquad ([\mathbf{16}]); \tag{3.10}$$

$$\sum_{i=1}^{n} f(\bar{x}_i - \bar{y}_i) \leq \sum_{i=1}^{n} f(x_i - y_i) \leq \sum_{i=1}^{n} f(\bar{x}_i - \underline{y}_i) \qquad ([\mathbf{18}]). \tag{3.11}$$

If f is a concave function, the reverse inequalities are valid.

(Of course, we suppose that all expressions are well defined.)

3^0 If $f'(z) + zf''(z) \underset{(<)}{\geq} 0$, then

$$(3.12) \quad \sum_{i=1}^{n} f(\underline{x}_i \bar{y}_i) \underset{(>)}{\leq} \sum_{i=1}^{n} f(x_i y_i) \underset{(>)}{\leq} \sum_{i=1}^{n} f(\bar{x}_i \bar{y}_i) \qquad ([16]);$$

$$(3.13) \quad \sum_{i=1}^{n} f(\bar{y}_i/\bar{x}_i) \underset{(>)}{\leq} \sum_{i=1}^{n} f(y_i/x_i) \underset{(>)}{\leq} \sum_{i=1}^{n} f(\bar{y}_i/\underline{x}_i) \qquad ([16],[21],[22]).$$

4^0 If $f'(1+z) + zf''(1+z) \underset{(<)}{\geq} 0$, then

$$\sum_{i=1}^{n} f(1+\bar{y}_i/\bar{x}_i) \underset{(>)}{\leq} \sum_{i=1}^{n} f(1+y_i/x_i) \underset{(>)}{\leq} \sum_{i=1}^{n} f(1+\bar{y}_i/\underline{x}_i) \qquad ([16],[21]).$$

5^0

$$\sum_{k=1}^{n} (1+\bar{y}_k)^{\underline{x}_k} \leq \prod_{k=1}^{n} (1+y_k)^{x_k} \leq \prod_{k=1}^{n} (1+\bar{y}_k)^{\bar{x}_k} \qquad ([16]).$$

6^0

$$\sum_{k=1}^{n} \underline{x}_k^{\bar{y}_k} \leq \sum_{k=1}^{n} x_k^{y_k} \leq \sum_{k=1}^{n} \bar{x}_k^{\bar{y}_k}.$$

7^0 ([**18**]) The following example is a generalization of a result from [**23**].

Let F be a positive function, a and b two vectors from R^n, and y a vector from R^{2n} defined by $y_{2i-1} = a_i$, $y_{2i} = b_i$, $i = 1, \ldots, n$. If F is also a symmetric function, then

$$(3.14) \qquad \sum_{i=1}^{n} F(a_i, b_i) \leq \sum_{i=1}^{n} F(\underline{a}_i, \underline{b}_i) \leq \sum_{i=1}^{n} F(\underline{y}_{2i-1}, \underline{y}_{2i}).$$

If f is a positive and symmetric function, and ϕ is defined by

$$\phi(y) = \prod_{i=1}^{n} f(y_{2i-1}, y_{2i}),$$

then

$$\phi(y) \leq \phi(\underline{y}) \quad \text{for every } y \in R^{2n}$$

iff f is TP_2. In a special case

$$f(x,y)=(xy)^m+a_1(xy)^{m-1}+\cdots+a_m \quad (x,y\geq 0,\ a_i\geq 0,\ i=1,\ldots,m)$$

is TP_2, so the expression ([**23**])

$$\prod_{i=1}^{n}\left[(y_{2i-1}y_{2i})^m+a_1(y_{2i-1}y_{2i})^{m-1}+\cdots+a_m\right]$$

has maximum when y is arranged in decreasing order.

REMARK: A special case of (3.10), for $f(x)=\log x$, is given in [**24**]. The case of m sequences was first proved in [**13**]:

$$\prod_{i=1}^{n}\sum_{j=1}^{m}a_{ji}\geq\prod_{i=1}^{n}\sum_{j=1}^{m}\bar{a}_{ji} \tag{3.15}$$

where vectors a_j are defined as in Theorem 2.

REMARK: A. Kovačec [**25**],[**26**] gave new proofs of (3.1) and (3.2). He also considered generalizations of these inequalities. For example, in [**25**] he considered the inequality

$$\sum_{i=1}^{n}a_ib_{\pi(i)}\leq\sum_{i=1}^{n}a_ib_{\sigma(i)},$$

where $\pi(i)$ and $\sigma(i)$ are two permutations of the set $\{1,\ldots,n\}$.

Z. Daróczy [**27**] proved the following two results with several functions:

THEOREM 4. *The function* $F(x)=\sum\limits_{i=1}^{n}g_i(x_i)$ $(x_i\in(a,b),\ i=1,2,\ldots,n)$ *satisfies the inequality*

$$F(\bar{x})\leq F(x)\leq F(\underline{x}), \tag{3.16}$$

for all choices of $x_i\in(a,b)$ *iff the functions* $t\mapsto g_k(t)-g_{k+1}(t)$ $(k=1,2,\ldots,n-1)$ *are nondecreasing on the interval* (a,b).

THEOREM 5. *Let* $g_j(t)$, $j=2,3,\ldots,2n$, *be real-valued functions defined on interval* $(0,\infty)$. *For any two* n*-tuples* x *and* y $(x_i,y_i\in(0,+\infty),\ i=1,\ldots,n)$ *we have*

$$\sum_{i=1}^{n}\sum_{k=1}^{n}g_{i+k}(x_iy_k)\leq\sum_{i=1}^{n}\sum_{k=1}^{n}g_{i+k}(\underline{x}_i\underline{y}_i) \tag{3.17}$$

if the functions $g_j(t)-g_{j+1}(t)$ $(j=2,3,\dots,2n-1)$ *are nondecreasing on interval* $(0,+\infty)$.

Theorem 5 is, in fact, a generalization of an inequality of F. Wiener: If $c_2 \geq c_3 \geq \cdots \geq c_{2n} \geq 0$, and x and y are two nonnegative n-tuple Then

$$\sum_{i=1}^{n}\sum_{k=1}^{n} c_{i+k}x_iy_k \leq \sum_{i=1}^{n}\sum_{k=1}^{n} c_{i+k}\underline{x}_i\underline{y}_k. \tag{3.18}$$

The following result was proved in [**28**]:

Let a and b be two n-tuples such that $0 \leq a_1 \leq \cdots \leq a_n$, and $0 \leq b_1 \leq \cdots \leq b_n$. Then the inequality (3.1) can be written in the form

$$\sum_{k=1}^{n} a_kb_{n-k+1} \leq \sum_{k=1}^{n} a_kb_{\sigma k} = \sum_{k=1}^{n} a_kb_k, \tag{3.19}$$

where σ denotes any permutation of the natural numbers $1,2,\dots,n$.

Further, let be

$$m_\sigma = \min_{1\leq k\leq n} a_kb_{\sigma k}, \qquad M_\sigma = \max_{1\leq k\leq n} a_kb_{\sigma k},$$

then

$$\min_\sigma m_\sigma = a_1b_1, \qquad \max_\sigma M_\sigma = a_nb_n,$$

$$\max_\sigma m_\sigma = \min_{1\leq k\leq n} a_kb_{n-k+1}, \qquad \min_\sigma M_\sigma = \max_{1\leq k\leq n} a_kb_{n-k+1}.$$

4. Let $a=(a_{-n},\dots,a_0,\dots,a_n)$. Define the following rearrangements of a: The sequence a^+ is defined by: $a_0^+ \geq a_1^+ \geq a_{-1}^+ \geq a_{-2}^+ \geq \cdots$; the sequence ${}^+a$ is defined by: ${}^+a_0 \geq {}^+a_{-1} \geq {}^+a_1 \geq {}^+a_{-2} \geq {}^+a_2 \geq \cdots$, and the sequence a^* is defined by: $a_0^* \geq a_1^* \geq a_{-1}^* \geq a_{-2}^* \geq \cdots$. The set a^* is said to be symmetrically decreasing.

The following theorem of R. M. Gabriel [**29**] is a generalization of results from [4],[**30**] (see Hardy, Littlewood, and Pólya, Inequalities, pp. 265–272).

THEOREM 1. *If a, b, c are three sets, of which c is symmetrical, then*

$$\sum_{r+s+t=0} a_r b_s c_t \leq \sum_{r+s+t=0} a_r^{++} b_s c_t^{*} = \sum_{r+s+t=0} {}^{+}a_r b_s^{+} c_t^{*}.$$

Proof of this theorem , generalizations for several sequences, and a series of similar results are given in [**4**, pp. 265- -276].

Symmetrical decreasing rearrangements are also considered in [**23**], i.e., if $y = (y_1, \ldots, y_n)$, then y^- and ${}^-y$ are defined by $y_1^- \leq y_n^- \leq y_2^- \leq \cdots = y_{[(n+2)/2]}^-$ and ${}^-y_n \leq {}^-y_1 \leq {}^-y_{n-1} \leq \cdots \leq {}^-y_{[(n+1)/2]}$. Further, we have circular-symmetric rearrangement of sequence y if its circular rearrangement (one of its circular rearrangements) is symmetrical decreasing; i.e. if $y_1 \leq y_2 \leq y_n \leq y_3 \leq y_{n-1} \leq \cdots \leq y_{[(n+3)/2]}$, or $y_2 \leq y_1 \leq y_3 \leq y_n \leq y_{n-1} \leq \cdots \leq y_{[(n+4)/2]}$ holds.

A result from [**23**], given in Example 7^0, is a special case of the following result:

THEOREM 2. *Let $(u, v) \mapsto F(u, v)$ be symmetrical real-valued function defined for $a < u$, $v < b$, $-\infty \leq a < b \leq +\infty$, and suppose that the function G defined by*

$$G(u, v, w) = F(u, v) + F(u, w) - F(v, w) \qquad (a < u, v, w < b),$$

is nonincreasing with respect to u and w for $u \leq \min(v, w)$.

If x is an n-tuple with elements from (a, b), then

$$\sum_{i=1}^{n} F(x_i, x_{i+1}) \qquad (x_{n+1} = x_1) \tag{4.1}$$

is maximal if x is arranged in circular-symmetric sense.

Moreover, if G is decreasing in v and w for $u < \min(v, w)$, if

$$F(u, u) = G(u, v, u) = G(u, u, w) > G(u, v, w) \qquad (u < \min(v, w))$$

and if there exist no three elements of x with the same value, then (4.1) has its maximum only if x is arranged in circular- symmetric sense.

If $f(u, v) = f(|u - v|)$, where $x \mapsto f(x)$ is a concave and nonincreasing function for $x \geq 0$, we have a result of A. L. Lehman [**31**]. (Note that a result of S. Abramovich from Example 7^0 is a generalization of one result from [**32**].)

5. Similar results are valid for functions:

DEFINITION: If f is a measurable function on a measure space (X, μ), then for $y \geq 0$,

$$m(f, y) = \mu(E_y)$$

where

$$E_y = \{x\colon |f(x)| > y\}.$$

Furthermore, since $m(f,y)$ is a non-increasing function of y we let $f^*(x)$ be that function which is inverse to $m(f,y)$ and continuous to the left.

The following results are valid: (a) the functions f and f^* are "equimeasurable" (see [**4**, p. 277] and [**33**]), so if $f \geq 0$ they have equal integrals over $(0,1)$ and $\int_0^1 F(f^*)\,dx = \int_0^1 F(f)\,dx$, for any measurable F for which the integrals exist [**4**, p. 277].

(b) A monotonic rearrangement of a positive concave function is also a concave function.

An integral analogue of (3.4) is given by G. G. Lorentz [**14**].

THEOREM 1. *The necessary and sufficient conditions that a continuous function $\phi(x,u_1,\ldots,u_n)$ defined for $0 < x < 1$, $u_k \geq 0$, $k = 1,2,\ldots,n$, satisfies the inequality*

$$\int_0^1 \phi(x, f_1(x), \ldots, f_n(x))\,dx \leq \int_0^1 \phi(x, f_1^*(x), \ldots, f_n^*(x))\,dx \tag{5.1}$$

for all nonnegative measurable $f_i(x)$, are

$$\phi(u_i+h, u_j+h) - \phi(u_i+h, u_j) - \phi(u_i, u_j+h) + \phi(u_i,u_j) \geq 0, \tag{5.2}$$

(5.3)

$$\int_0^\delta \{\phi(x-t, u_i+h) + \phi(x+t, u_i) - \phi(x+t, u_i+h) - \phi(x-t, u_i)\}dt \geq 0$$

for every $0 < x < 1$, $u_k \geq 0$, $k = 1,\ldots,n$, $h > 0$, $0 < \delta < x$, $\delta < 1-x$, and $i \neq j$. If ϕ is continuous the second partial derivatives in all variables,the conditions (5.2) and (5.3) are equivalent to

$$\frac{\partial^2\phi}{\partial u_i \partial u_j} \geq 0, \tag{5.2a}$$

$$\frac{\partial^2\phi}{\partial x \partial u_i} \leq 0. \tag{5.3a}$$

We can also consider the symmetric decreasing rearrangement of a function f, i.e. $f^\dagger$. Suppose that f is defined for all real, or almost all real x, and that

$m(f,y)$ is finite for all positive y. We may define an even function $f^\dagger(x)$ by agreeing that

$$f^\dagger\left(\frac{1}{2}m(f,y)\right) = y,$$

and that $f^\dagger(-x) = f^\dagger(x)$; or, that is the same thing, that $f^\dagger$ is even and $f^\dagger(x) = f^*(2x)$ for positive x. Then $f^\dagger$ is the symmetric decreasing rearrangement of f.

The following result is valid ([**33**], [**4**, pp. 279- -287]):

THEOREM 2. *If f, g, h are nonnegative functions defined on R, then*

$$\int_{-\infty}^{+\infty}\int_{-\infty}^{+\infty} f(x)g(y)h(-x-y)\,dxdy \le \int_{-\infty}^{+\infty}\int_{-\infty}^{+\infty} f^\dagger(x)g^\dagger(y)h^\dagger(-x-y)\,dxdy. \tag{5.4}$$

We may plainly suppose that none of f, g, h is null. We may also replace $-x-y$ by $\pm x \pm y$ without changing the significance of the inequality.

The following generalization of this result is given in [**34**]:

THEOREM 3. *The following inequality is valid*

$$\int_{R^n} \prod_{i=1}^{n} |F_i(x_i)H_i(x_i - x_{i+1})|\,dx \le \int_{R^n} \prod_{i=1}^{n} F_i^\dagger(x_i)H_i^\dagger(x_i - x_{i-1})\,dx,$$

where all the functions are nonnegative and defined on R, and $x_{n+1} = x_1$.

In the same paper the authors suppose the validity of one more general result, which is proved in [**7**], namely the inequality

$$\int_{\mathbf{R}^n} \prod_{j=1}^{k} f_j\left(\langle a^{(j)}, x\rangle\right)\,dx \le \int_{\mathbf{R}^n} \prod_{j=1}^{k} f_j^\dagger\left(\langle a^{(j)}, x\rangle\right)\,dx$$

as well as a generalization of this inequality to functions of several variables.

6. The following result is proved in [**35**]:

THEOREM 1. *The folowing inequality is valid*

$$\sum_{k=1}^{n-1} |\Delta \underline{a}_k|^p \le \sum_{k=1}^{n-1} |\Delta a_k|^p, \qquad p \ge 1, \tag{6.1}$$

with equality iff the sequence a is monotonic.

The following integral analogue is also valid [**35**], [**36**]:

THEOREM 2. *Let f be a.e. differentiable function in $[0,b]$. Then*

$$\int_0^b |f^{*\prime}(x)|^p\,dx \le \int_0^b |f'(x)|^p\,dx, \qquad p>1, \tag{6.2}$$

with equality only if f is a monotonic function.

(6.2) is also valid for $0<p\le 1$, and for $p<0$ the reverse inequality is valid (see [**37**]).

Generalizations of (6.2) are obtained in [**38**], namely:

THEOREM 3. *Let f be differentiable almost everywhere in $[0,b]$ and let $G(y)$ be a nondecreasing function for $y>0$. Then*

$$\int_0^b G\left(|f^{*\prime}|\right)\,dx \le \int_0^b G\left(|f'(x)|\right)\,dx. \tag{6.3}$$

PROOF: As in [**37**], [**39**] let the multiplicity $n(y)$ of f at the level y be the number of roots $x_k=x_k(y)$, $k=1,\dots,n(y)$, of the equation $y=f(x)$, in $[0,b]$. The basic relation connecting the derivatives of f is obtained in [**37**]:

$$|f^{*\prime}(x)|^{-1}=\sum_{k=1}^{n}|f'(x_k)|^{-1}.$$

Using x^* as independent variable for the rearranged function f^*, we have as in [**39**]

$$|f^{*\prime}(x^*)|=|dy/dx^*|,$$

hence from the basic relation

$$dx^*=|f^{*\prime}(x^*)|^{-1}\,dy=\sum_{k=1}^{n}|f'(x_k)|^{-1}\,dy=\sum_{k=1}^{n}dx_k.$$

Therefore

$$\begin{aligned}G\left(|f^{*\prime}|\right)\,dx^* &= G\left(\left(\sum_{k=1}^{n}|f'(x_k)|^{-1}\right)^{-1}\right)\sum_{k=1}^{n}|f'(x_k)|^{-1}\,dy\\ &= H\left(\sum_{k=1}^{n}|f'(x_k)|^{-1}\right)\,dy,\end{aligned}$$

where $H(x) = xG(1/x)$. It is obvious that the function $H(x)/x$ is nonincreasing, so the following inequality is valid (see [**4**, p. 83])

$$H\left(\sum x\right) \le \sum H(x).$$

Thus we obtain

$$\begin{aligned} G\left(|f^{*\prime}(x^*)|\right) dx^* &= H\left(\sum_{k=1}^{n} |f'(x_k)|^{-1}\right) dy \\ &\le \sum_{k=1}^{n} H\left(|f'(x_k)|^{-1}\right) dy \\ &= \sum_{k=1}^{n} |f'(x_k)|^{-1} G\left(|f'(x_k)|\right) dy \\ &= \sum_{k=1}^{n} G\left(|f'(x_k)|\right) dx_k \end{aligned}$$

using in the last step the formula $dy = |f'(x_k)| dx_k$.

Integration over the domain $[0, b]$ now yields the starred result, as the integral elements based on the dx_k exactly cover the interval $[0, b]$ once when the summation over all integral elements based on dx^* is performed. This completes the proof of Theorem 3.

Recall from [**37**] that a function $f(x) = f(x_1, \dots, x_m)$ has a spherically symmetric equimeasurable decreasing rearrangement which is essentially a function f^* of volume or of radial distance only. Let

$$\mu(z) = \text{meas.}\{(x_1, \dots, x_m) \mid f(x_1, \dots, x_m) > z\},$$

and let

$$f^*(x) = \mu^{-1}(x).$$

The basic relation for $f \in PC^1$ is derived by integration over the level surface $f = f^*$ in the domain D involved. If dn denotes the inward normal differential, and ∇f the gradient, then $|\nabla f| dn = df$. Since

$$\mu(z) = \int_{f \ge z} dV = \int_{f \ge z} dn dS,$$

we find

$$d\mu = \sum \int dn dS = \sum \int \frac{df}{|\nabla f|} dS.$$

The summation runs over all components of the level surface $f = f^*$. However,

$$d\mu = -\frac{df^*}{|f^{*\prime}(x)|}$$

while $|df| = |df^*|$. Comparing, we obtain the m-dimensional basic relation (see [**37**],[**39**]):

$$\frac{1}{|f^{*\prime}(x)|} = \sum \int_{f=f^*} \frac{dS}{|\nabla f|}.$$

The m-dimensional analogue of the multiplicity function is now the level surface area

$$S = \sum \int_{f=f^*} dS.$$

THEOREM 4. *Let $G(x)$ be nondecreasing for $x > 0$. Then*

$$\int G\left(|f^{*\prime}|\right) dV \leq \int G\left(|\nabla f|\right)\, dV. \tag{6.4}$$

PROOF: The integral on the left has the differential

$$\begin{aligned}
G\left(|f^{*\prime}(x)|\right) dV &= G\left(\left(\sum \int \left(|\nabla f|\right)^{-1}\, dS\right)^{-1}\right) \sum \int \, dSdn \\
&= G\left(\left(\sum \int \left(|\nabla f|\right)^{-1}\, dS\right)^{-1}\right) \sum \int |\nabla F|^{-1}\, dSdf \\
&= H\left(\sum \int |\nabla f|^{-1}\, dS\right)\, df \\
&\leq \sum \int_{f=f^*} H\left(|\nabla f|^{-1}\right)\, dSdf \\
&= \sum \int_{f=f^*} \left(|\nabla f|\right)\, dSdf \\
&= \sum \int_{f=f^*} G\left(|\nabla f|\right)\, dSdn \\
&= \sum \int_{f=f^*} G\left(|\nabla F|\right)\, dV.
\end{aligned}$$

The result now follows, as in the one dimensional case, by integration over the domain D.

The following result, related to Theorem 2, was also proved in [**37**]:

THEOREM 5. *Let $n(y)$ be the multiplicity of f at the level y, and $f \in C^1[0,b]$. Then the following inequalities hold for the indicated ranges of p:*
(a) *for $p \geq 1$*

$$\int_0^b |f^{*\prime}(x)|^p \, dx \leq \int_0^b \left| \frac{f'(x)}{n(f(x))} \right|^p dx, \tag{6.5}$$

(b) *for $0 < p \leq 1$*

$$\int_0^b \left| \frac{f'(x)}{n(f(x))} \right|^p dx \leq \int_0^b |f^{*\prime}|^p \, dx \leq \int_0^b |f'(x)|^p \, dx, \tag{6.6}$$

(c) *for $p < 0$*

$$\int_0^b \frac{1}{|f'(x)|^{|p|}} \, dx \leq \int_0^b \frac{1}{|f^{*\prime}(x)|^{|p|}} \, dx \leq \int \frac{n(f(x))^{|p|}}{|f'(x)|^{|p|}} \, dx. \tag{6.7}$$

Equality holds on the right in (a) and (c), and on the left in (b), if and only if the values of $|f'(x_k)|$ are independent of k, $k = 1, 2, \ldots, n$ for almost all x_k.

Multidimensional generalizations for spherical symmetric equimeasurable decreasing rearrangement are also given.

The following generalizations were given in [**39**]:

THEOREM 6. *Let f be differentiable almost everywhere in $[0,b]$ and let $G(y)$ be a function convex for $y \geq 0$. Then with n finite a.e.*

$$\int_0^b G\left(|f^{*\prime}(x)|\right) \, dx \leq \int_0^b G\left(\frac{|f'(x)|}{n(f(x))} \right) dx. \tag{6.8}$$

The multidimensional generalization is:

THEOREM 7. *Let $G(x)$ be convex for $x > 0$, and let S be the level surface area. Then*

$$\int G\left(|f^{*\prime}(x)|\right) dV \leq \int G\left(\frac{|\nabla f|}{S} \right) dV. \tag{6.9}$$

For an application of the above result see [**40**].

7. The following result is indicated in [**4**, pp. 291–295]:

THEOREM 1. *Suppose that $f(x)$ is nonnegative and integrable in a finite interval $(0,a)$, that $f^*(x)$ is the rearrangement of $f(x)$ in decreasing order, that*

$$\theta(x) = \sup_{0\le\delta<x}\left(\frac{1}{x-\delta}\int_{\delta}^{x} f(t)\,dt\right).$$

and that $\theta^*(x)$ is the rearrangement of $\theta(x)$ in decreasing order.

(i) *For $0 < x \le a$, $\theta^*(x) \le \frac{1}{x}\int_0^x f^*(t)\,dt$.*

(ii) *If s is any increasing function defined for $y \ge 0$, then*

$$\int_0^a s\,(\theta(x))\,dx \le \int_0^a s\left\{\frac{1}{x}\int_0^x f^*(t)\,dt\right\}dx.$$

If f is defined on $[a,b]$ we can in the definition of θ replace $0 \le \delta < x$ by $a \le \delta < x$. This function is the well-known Hardy-Littlewood maximal function.

Generalizations of Theorem 1 for functions of two variables are given in [**41**].

The following results for maximal function are also valid:

(iii) $\int_a^b \theta(x)^p\,dx \le \left(\frac{p}{p-1}\right)^p \int_a^b f(x)^p\,dx, \qquad p>1;$

(iv) *If $\phi(x)/x^{1+\delta}$ is increasing in $(0,\pi)$ and $\phi(x)/x^k$ is decreasing in $(0,\pi)$ for some positive k, where $\phi(x)$ is absolutely continuous, then*

$$\int_a^b \phi\,(\theta(x))\,dx \le A\int_a^b \phi\,(f(x))\,dx,$$

where A is a positive constant, depending on ϕ only.

(v) *Under the conditions of* (iv),

$$\int_a^b x^{\alpha}\theta(x)^p\,dx \le A\int_a^b x^{\alpha} f(x)^p\,dx, \qquad p>1,$$

where $-1 < \alpha < p-1$.

These results are given, for example, in [**42**], and it is noted that (i) is due to Hardy and Littlewood, (ii) is due to Marcinkiewicz and Zygmund, and (iii) is due to Babenko [**43**]. Note that in [**42**] there are 77 references on these results. Chen in [**41**] also gave many new results. For example, he proved a general theorem which includes both (iv) and (v).

A generalization of (iii) to C^r-valued functions is given in [**44**] where the following theorem is proved: *For* $f = (f_1, f_2, \dots)$ *a sequence of functions on* R^n *form the sequence* $\theta = (\theta_1, \theta_2, \dots)$, *where* θ_k *is the classical maximal function of* f_k. *Writing* $\|f(x)\|_r$ *for the* C^r*-norm of the sequence* $(f_1(x), f_2(x), \dots)$, *then*

$$\int_{R^n} \|\theta(x)\|_r^p \, dx \le A_{p,r} \int_{R^n} \|f(x)\|_r^p \, dx$$

provided $1 < p, r < \infty$.

For other generalizations see [**45**],[**46**],[**47**],[**48**],[**49**],[**50**], [**51**]. Here we shall give the following result from [**52**].

Suppose that $U(x) \ge 0$, $V(x) \ge 0$, $1 < p < +\infty$, $f(x) = 0$ for $x < 0$ and for every interval $I \subset (-\infty, +\infty)$ of length $|I|$ we have

$$\left(\int_{-\infty}^{+\infty} \frac{|I|^{p-1} U(x)}{(|I| + |x - x_I|)^p} \, dx \right) \left(\frac{1}{|I|} \int_I (V(x))^{-1/(p-1)} \, dx \right)^{p-1} \le B,$$

where B does not depend on I, and x_I denotes the center of the interval I.

Suppose also that any of the following conditions is valid:

(a) There exists $A > 0$ such that $U(y) \le AU(x)$ and $V(y) \le AV(x)$ for $x \le y \le 2x$ and $x > 0$.

(b) There exists $A > 0$ such that $U(y) \ge AU(x)$ and $V(y) \ge AV(x)$ for $x \le y \le 2x$ and $x > 0$.

(c) There exists $A > 0$ and $d > 1$ such that $U(x) \le AV(y)$ for $x/d \le y \le dx$ and $x > 0$.

(d) $f(x)$ is a monotonic function on $[0, +\infty)$.

Then there exists a constant C, independent of f, such that

$$\int_{-\infty}^{+\infty} \theta(x)^p U(x) \, dx \le C \int_0^{+\infty} |f(x)|^p V(x) dx.$$

Some results on maximal function are also given in [**53**]. For example the following result is valid:

Let f *be a decreasing function such that* $f' \in L^p(0, +\infty)$ $(p > 1)$, *and let the function* $F(x) = \frac{1}{x} \int_0^x f(t) \, dt$ *be defined for* $x \ge 0$. *Then*

$$\|F'(x)\|_p \le \int_0^x \left| \frac{s}{x} \right|^{2-1/p} \|f'(s)\|_p \frac{ds}{s},$$

where the p-norm is defined by $\|g(s)\|_p^p = \int_0^s |f(t)|^p\, dt$.

8. Let f be a real-valued 2π-periodic function of class $L^1(0,2\pi) = L$, and let $\tilde{f}$ be its conjugate function given by the Cauchy-Lebesgue integral

$$\tilde{f}(x) = -\frac{1}{\pi}\int_{-\pi}^{\pi} \frac{f(t)}{2\tan\frac{1}{2}(t-x)}\, dt$$

$$= \lim_{h\to 0}\left\{-\frac{1}{\pi}\int_{h}^{\pi} \frac{h(x+t) - f(x-t)}{2\tan\frac{1}{2}t}\, dt\right\}.$$

The following results are valid:

(i) (M. Riesz).

$$\int_{-\pi}^{\pi} |\tilde{f}(x)|^p\, dx \le A_p \int_{-\pi}^{\pi} |f(x)|^p\, dx, \qquad p > 1.$$

(ii) (M. Riesz, A. Zygmund and S. K. Pichorides) [**54**]. A necessary and sufficient condition for $\tilde{f} \in L$ is that $f \log^+ f \in L$ (i.e. $f \in L \log^+ L$). If the later condition is satisfied, then

$$\int_{-\pi}^{\pi} |\tilde{f}(x)|\, dx \le A \int_{-\pi}^{\pi} |f(x)| \log^+ |f(x)|\, dx + B.$$

(iii) (K. I. Babenko) Under the conditions of (i)

$$\int_{-\pi}^{\pi} |x^\alpha |\tilde{f}(x)|^p\, dx \le A \int_{-\pi}^{\pi} x^\alpha |f(x)|^p\, dx,$$

$$p > 1, \quad -1 < \alpha < p - 1.$$

For some generalizations see [**42**]. Now, the symbol K will be used for the constant

$$K = \frac{1^{-2} - 3^{-2} + 5^{-2} - 7^{-2} + \dots}{1^{-2} + 3^{-2} + 5^{-2} + 7^{-2} + \dots} = 0.7424537\dots .$$

(iv) (Kolmogorov, Davis) [**55**]. When $f \in L$ the conjugate function $\tilde{f}$ satisfies the inequality

$$ym\{x\colon |\tilde{f}(x)| > y\} \le K^{-1} \int_0^{2\pi} |f(x)|\, dx, \qquad y > 0,$$

where $m(E)$ denotes the Lebesgue measure of a measurable subset E of $(0, 2\pi)$. The constant K^{-1} is best-possible.

(v) (O'Neil and Weiss) [**56**]. When $f \in L\log^+ L$ the conjugate function $\tilde{f}$ satisfies the inequality

$$\int_0^{2\pi} |\tilde{f}(x)|\, dx \leq \frac{2}{\pi} \int_0^{2\pi} f^*(t) \log\cot\left(\frac{t}{8}\right)\, dt,$$

where f^* denotes the decreasing rearrangement on $(0, 2\pi)$ of f. The constant $2/\pi$ is best-possible.

Now, set

$$f^{**}(t) = t^{-1} \int_0^t \int_0^t f^*(s)\, ds, \qquad 0 < t < 2\pi.$$

The following results are proved by C. Bennett [**57**]:

(vi) When $f \in L\log^* L$ the conjugate function satisfies the inequality

$$\int_0^{2\pi} |\tilde{f}(x)|\, dx \leq K \int_0^{2\pi} f^{**}(t)\, dt,$$

and the constant K is best-possible.

(vii) If f is (essentially) bounded on $(0, 2\pi)$, then its conjugate function $\tilde{f}$ satisfies the inequality

$$\sup_{0<t<2\pi} \frac{\tilde{f}^{**}(t)}{1 + \log(2\pi/t)} \leq K \operatorname*{ess\,sup}_{0<x<2\pi} |f(x)|,$$

and the constant K is best-possible.

REMARK: Some related results can be given for the Hilbert transformation given by the Cauchy principal value integral

$$(Hf)(x) = \frac{1}{\pi} \int_{-\infty}^{+\infty} \frac{f(t)}{t - x}\, dx, \qquad x \in (-\infty, +\infty).$$

Hilbert transforms and conjugate functions of even and odd functions are considered in [**58**].

9. Let f be real-valued, Lebesgue measurable, and defined in R^d. For $0 < p < \infty$ we write

$$\|f\|_p = \left(\int_{R^d} |f(x)|^p \, dx \right)^{1/p} .$$

For $0 < a < d$ the Riesz potentials $I_a(f)$ are defined by

$$I_a(f)(x) = \int_{R^d} f(y)|x-y|^{a-d} \, dy.$$

The maximal function $M(f)$ is defined by

$$M(f)(x) = \sup_{r>0} r^{-d} \int_{|y|<r} |f(x+y)| \, dy.$$

We will denote various constants, independent of f, by A.

The Hardy-Littlewood-Wiener maximal theorem (see also [**59**, I.1]) is

$$\|M(f)\|_p \leq A\|f\|_p, \qquad p > 1,$$

If f is supported by a finite ball B, then [**59**, I.5.2]

$$\int_B M(f) \, dx \leq A \int_B \left(1 + |f| \log^+ |f|\right) \, dx.$$

The following theorem is due to Hardy and Littlewood [**60**] for $d = 1$, and to Sobolev [**61**] in the general case (a simple proof is given in [**59**, V.1.2]):

(i) *Let $0 < a < d$, $1 < p < q < \infty$, and $1/q = 1/p - a/d$. Then*

$$\|I_a(f)\|_Q \leq A\|f\|_p.$$

If f is supported by a ball B, and $1/q = 1 - a/d$, then $I_a(f) \in L^q(B)$ if $\int_B |f| \log^+ |f| \, dx < \infty$.

Now, we shall give two results from [**62**]:

(ii) *Let $f \geq 0$ be measurable on R^d. Then*

$$\|I_{a\theta}(f)\|_r \leq A\|f\|_p^{1-\theta} \|I_a(f)\|_q^{\theta},$$

with $0 < a < d$, $0 < \theta < 1$, $1 < p < \infty$, $p < q \leq \infty$, and $1/r = (1-\theta)/p + \theta/q$.

(iii) *Let f be as in* (ii). *Then*

$$\|I_{a\theta}(f^t)\|_r \leq A\|f\|_p^{t-\theta} \|I_a(f)\|_q^{\theta},$$

with $0 < a < d$, $0 < \theta < 1$, $0 < p < \infty$, $0 < q \leq \infty$, $\theta < t < \theta + (1-\theta)p$, and $1/r = (t-\theta)/p + \theta/q$.

10. Monotonic rearrangements can be extended to the case when the elements a_j of n-tuple a are from any partial ordered set (see [**63**]–[**65**] and [**15**]). So, a generalization of Theorem 3 from 3. is given in [**15**], as well as an integral analogue.

11. Let $f(\theta)$ be a measurable complex function defined on $(-\pi,\pi)$. If r is a real number ≥ 1, denote by J_r the corresponding potential mean of order r, i.e.

$$J_r(f) = \left(\frac{1}{2\pi}\int_{-\pi}^{\pi} |f(\theta)|^r \, d\theta\right)^{1/r}.$$

Suppose that $c_0, c_2, c_{-1}, c_2, c_{-2}, \ldots$ is a sequence of complex numbers which tend to zero, and let $c_0^* \geq c_1^* \geq c_{-1}^* \geq c_2^* \geq c_{-2}^* \geq \ldots$ be the sequence $|c_0|, |c_1|, |c_{-1}|, |c_2|, |c_{-2}|, \ldots$ arranged in nonincreasing order.

Hardy and Littlewood [**66**],[**67**] proved the following result:

Suppose that $1 < r \leq 2$ and that $\sum c_n e^{n\theta i}$ is the Fourier series of a function $f \in L^r(-\pi,\pi)$. Then $\sum c_n^* e^{n\theta i}$ is the Fourier series of the function $f^\dagger$ (see 5.) from $L^r(-\pi,\pi)$ and the following inequality is valid

$$J_r(f^\dagger) \leq A(r) J_r(f),$$

where $A(r)$ is a constant which depends only on r.

This and many other results are considered in papers [**68**]–[**77**].

12. Convolution operators are considerd in [**78**].

DEFINITION: Given three measure spaces (X,μ), $(\overline{X},\bar{\mu})$, and (Y,ν), a bilinear operator, T, which maps measurable functions on X and $\overline{X}$ into measurable functions on Y is called a convolution operator if

1. $T(f_1+f_2,g) = T(f_1,g) + T(f_2,g)$,
2. $T(f,g_1+g_2) = T(f,g_1) + T(f,g_2)$,
3. $\|T(f,g)\|_1 \leq \|f\|_1 \|g\|_1$,
4. $\|T(f,g)\|_\infty \leq \|f\|_1 \|g\|_\infty$,
5. $\|T(f,g)\|_\infty \leq \|f\|_\infty \|g\|_1$

($\|\cdot\|_1$ denotes L^1 norm, $\|\cdot|_\infty$ the L^∞ norm on any of the three spaces $X, \overline{X}, Y$).

The following results are valid:

(i) A bilinear operator T is a convolution operator if and only if for $h = T(f,g)$,

$$h^{**}(x) \leq x f^{**}(x) g^{**}(x) + \int_x^\infty f^8(t) g^*(t) \, dt.$$

(ii) If T is a convolution operator $h = T(f,g)$ and $p \geq 1$, $q \geq 1$, $r \geq 1$ satisfy the equation $1/p + 1/q - 1 = 1/r$, then

$$\|h\|_r \leq \|f\|_p \|g\|_q.$$

13. P. W.Day [**15**] has noticed that if $a, b \in R^+$ with $a_i, b_i \geq 0$ for $i = 1, \ldots, n$, then for $k = 1, \ldots, n$

$$\sum_{i=1}^{k} \bar{a}_i \underline{b}_i \leq \sum_{i=1}^{k} a_i b_i \leq \sum_{i=1}^{k} bua_i \underline{b}_i.$$

N. Komaroff [**79**] used these inequalities to prove that if $a_i \geq 0$, then for $k = 1, \ldots, n$

$$\sum_{i=1}^{k} \underline{a}_i b_i \leq \sum_{i=1}^{k} \underline{a}_i \underline{b}_i, \quad \text{and}$$

$$\sum_{i=1}^{k} \underline{a}_i \bar{b}_i \leq \sum_{i=1}^{k} a_i b_i.$$

14. S. Saithoh [**80**] shows that if q is a positive integer and $F_j \in L^2(R^+)$, then

$$\int_0^\infty \left| \prod_{j=1}^{2q} {}^* F_j(t) \right|^2 t^{1-2q} dt \leq \frac{1}{(2q-1)!} \prod_{j=1}^{2q} \int_0^\infty |F_j(t)|^2 dt$$

with equality if and only if every $F_j(t) = C_j e^{-tu}$, for some fixed u independent of j. Here $\prod_{j=1}^{2q} *$ means the iterated convolution product. A related result is given in Saithoh [**81**].

References

1. YOUNG, W. H., *On the multiplication of successions of Fourier constants*, Proc. Royal Soc. (A) **87** (1912), 331–339.
2. __________, *Sur la généralisation du théorème de Parseval*, C. R. Acad. Sci. Paris **155** (1912), 30–33.

3. ________, *On the determination of the summability of a function by means of its Fourier constant*, Proc. London Math. Soc. **12** 2 (1913), 71--88.
4. HARDY, G. H., J. E. LITTLEWOOD and G. PÓLYA, *The maximum of a certain bilinear form*, Proc. Lond. Math. Soc. **25** 2 (1926), 265–282.
5. BECKNER, H., *Inequalities in Fourier analysis*, Ann. of Math. **102** (1975), 159–182.
6. BRASCAMP, H. J. and E. H. LIEB, *Best constants in Young's inequality, its converse, and its generlaization to more than three functions*, Advances Math. **20** (1976), 151- -173.
7. BRASCAMP, H. J., E. H. LIEB and J. M. LUTTINGER, *A general rearrangement inequality for multiple integrals*, J. Func. Anal. **17** (1974), 227–237.
8. SUBRAHMANYAM, M. B., *An extremal problem for convolution inequalities*, J. Math. Anal. Appl. **87** (1982), 509–516.
9. BOAS, P. R. and C. O. IMORU, *Elementary convolution inequalities*, SIAM J. Math. Anal. **6** (1975), 457–471.
10. BEESACK, P. R. and J. E. PEČARIĆ, *Inequalities involving nth-order means*, J. Math. Anal. Appl. **99** (1984), 127–149.
11. BEESACK, P. R., *Inequalities involving iterated kernels and convolutions*, Univ. Beograd. Publ. Elektrotehn. Fak. Ser. Mat. Fiz. No. **247–301** (1969), 11- -16.
12. PEČARIĆ, J. E., *On some inequalities analogous to Grüss' inequality*, Mat. Vesnik **4** 17 32 (1980), 197–202.
13. RUDERMAN, H. D., *Two new inequalities*, Amer. Math. Monthly **59** (1952), 29–32.
14. LORENTZ, G. G., *An inequality for rearrangements*, Amer. Math. Monthly **60** (1953), 176–179.
15. DAY, P. W., *Rearrangement inequalities*, Canad. J. Math. **24** (1972), 930–943.
16. BECK, E., *Ein Satz über umerdnungs- ungleichungen*, Univ. Beograd. Publ. Elektrotehn. Fak. Ser. Mat. Fiz. No. **388–409** (1972), 31–35.
17. DERMAN, C., G. J. LIEBERMAN and S. M. ROSS, *Assembly of systems having maximum reliability*, Naval. Res. Logist. Quart. **21** (1974), 1–13.
18. RINOTT, Y., *Multivariate majorization and rearrangement inequalities with some applications to probability and statistics*, Israel J. Math. **15** (1973), 60–77.
19. JURKAT, W. B. and H. J. RYSER, *Term ranks and permanents of nonnegative matrices*, J. Algebra **5** (1967), 342–357.
20. MINC, H. *Rearrangements*, Trans. Amer. Math. Soc. **159** (1971), 497–504.
21. LONDON, D., *Rearrangement inequalities involving convex function*, Pacific J. Math. **34** (1970), 749–753.
22. BORELL, G., *A note on an inequality for rearrangements*, Pacific J. Math. **47** (1973), 39–41.

23. ABRAMOVICH, S., *The increase of sums and products dependent on* $(y_1, \ldots, y_2)$ *by rearangement of this set*, Israel J. Math. **5** (1967), 177–181.
24. OPPENHEIM, A., *Inequalities connected with definite Hermitian forms, II*, Amer. Math. Monthly **61** (1954), 463–466.
25. KOVAČEC, A., *A note on Popoviciu's inequality for bilinear forms*, Anal. Numer. Th. Approx. **12** 2 (1983), 151–155.
26. ________, *On positive definite differences of multilinear forms and their representation*, Proc. Roy. Soc. Edinburg **96** (1984), 97- -108.
27. DARÓCZY, Z., *Rearrangement inequalities*, Publ. Math., Debrecen **20** (1973), 273–279.
28. METCALF, F. T., *A theorem concerning rearrangements*, Amer. Math. Monthly **71** (1964), 172–174.
29. GABRIEL, R. M., *The rearrangement of positive Fourier coefficients*, Proc. Lond. Math. Soc. **33** 2 (1932), 32–51.
30. HARDY, G. H., J. E. LITTLEWOOD and G. PÓLYA, *The maximum of a certain bilinear form*, Proc. Lond. Math. Soc. **25** 2 (1926), 265–282.
31. LEHMAN, A. L., *A result on rearrangements*, Israel J. Math. **1** (1963), 22–28.
32. DUFFIN, R. J. and A. G. SCHAEFFER, *A redefinement of an inequality of the brothers Markoff*, Trans. Amer. Math. Soc. **50** (1941), 517–528.
33. BURKILL, H., *A note on rearrangements of functions*, Amer. Math. Monthly **71** (1964), 887–888.
34. LUTTINGER, J. M. and R. FRIEDBERG, *A new rearrangement inequality for multiple integrals*, Arch. Rat. Mech. Anal. **61** (1976), 45–64.
35. DUFF, G. F. D., *Differences, derivatives and decreasing rearrangements*, Canad. J. Math. **19** (1967), 1153–1178.
36. RYFF, J. V., *Measure preserving transformations and rearrangements*, J. Math. Anal. Appl. **31** (1970), 449–458.
37. DUFF, G. F. D., *Integral inequalities for equimeasurable rearrangements*, Can. J. Math. **22** (1970), 408–430.
38. MITRINOVIĆ, D. S. and J. E. PEČARIĆ, *A general integral inequality for the derivative of an equimeasurable rearrangement*, C. R. Math. Rep. Acad. Sci. Canada **11** (1989), 201–205.
39. DUFF, G. F. D., *A general integral inequality for the derivative of an equimeasurable rearrangement*, Can. J. Math. **28** (1976), 793–804.
40. ________, *Navier Stokes derivative estimates in three space dimensions with boundary values and body forces*, C. R. Math. Rep. Acad. Sci. Canada **11** (1989), 195–204.
41. BURKILL, J. C., *Rearrangements of functions*, J. London Math. Soc. **27** (1952), 393–401.
42. CHEN, Y. M., *Theorems of asymptotoic approximation*, Math. Annalen **140** (1960), 360–407.
43. BABENKO, K. I., *On conjugate functions* (Russian), Doklady Akad. Nauk SSSR, N. S. **62** (1948), 157–160.

44. FEFFERMAN, C. and E. M. STEIN, *Some maximal inequalities*, Amer. J. Math. **93** (1971), 107–115.
45. ALONSO, I. P., *A general result on the equivalence between derivation of integrals and weak inequalities for the Hardy-Littlewood maximal operator*, Studia Math. **60** (1977), 91–96.
46. KURTZ, D. S., *Weighted norm inequalities for the Hardy-Littlewood maximal function for one parameter rectangles*, Studia Math. **53** (1975), 39–54.
47. CAFFARELLI, L. A. and C. P. CALDERÓN, *Weak type estimates for the Hardy-Littlewood maximal functions*, Studia Math. **49** (1974), 217–223.
48. HEINIG, H., *A weighted norm inequality for a maximal function*, Notices AMS **26** (1979), A–206.
49. IMORU, C. O., *Variations of integral means and Hardy-Littlewood maximal functions*, Ann. Soc. Sci. Bruxelles **96** (1983), 165–173.
50. MUCKENHOUPT, B., *Weighted norm inequalities for the Hardy maximal function*, Trans. Amer. Math. Soc. **165** (1972), 207–226.
51. CHEN, Y. M., *On conjugate functions*, Can. J. Math. **15** (1963), 486–494.
52. MUCKENHOUPT, B. and R. L. WHEEDEN, *Two weight function norm inequalities for the Hardy-Littlewood maximal function and the Hilbert transform*, Studia Math. **55** (1976), 279–294.
53. DUFF, G. F. D. *The variation of maximal function*, Bull. Malaysian Math. Soc. **2** 2 (1979), 75–88.
54. PICHORIDES, S. K., *On the best values of the constants in the theorems of M. Riesz, Zygmund and Kolmogorov*, Studia Math. **44** (1972), 154–179.
55. DAVIS, B., *On the weak-type* (1,1) *inequality for conjugate functions*, Proc. Amer. Math. Soc. **44** (1974), 307–311.
56. O'NEIL, R. and G. WEISS, *The Hilbert transform and rearrangement of functions*, Studia Math. **23** (1963), 189–198.
57. BENNETT, C., *A best constant for Zygmund's conjugate function inequality*, Proc. Amer. Math. Soc. **56** (1976), 256–260.
58. ANDERSEN, K. F., *Weighted norm inequalities for Hilbert transforms and conjugate functions of even and odd functions*, Proc. Amer. Math. Soc. **56** (1976), 99–107.
59. STEIN, E. M., "Singular Integrals and Differentiability Properties of Functions," Princeton, 1970.
60. HARDY, G. H. and J. E. LITTLEWOOD, *Some properties of fractional integrals*, Math. Z. **27** (1928), 565–606.
61. SOBOLEV, S. L., *On a theorem of functional analysis*, Mat. Sb. **4** 46 (1938), 471–497. English transl., Amer. Math. Soc. Transl. (2) **34** (1963), 39–68.
62. HEDBERG, L. I., *On certain convolution inequalities*, Proc. Amer. Math. Soc. **36** (1972), 505–510.
63. LUXEMBURG, W. A. J., *Rearrangement invariant Banach function spaces*, Queen's Papers in Pure and Applied Math. **10** (1967), 83–144.

64. LORENTZ, G. G. and T. SHIMOGAKI, *Interpolation theorems for operators in function spaces*, J. Func. Anal. **2** (1968), 31–51.
65. CHONG, K. M. and N. M. RICE, *Equimeasurable rearrangements of functions*, Queen's Papers in Pure and Applied Mathematics **28** (1971).
66. HARDY, G. H. and J. E. LITTLEWOOD, *Notes on the theory of series (XIII): Some new properties of Fourier constants*, J. London Math. Soc. **6** (1931), 3–9.
67. __________, *A new proof of a theorem on rearrangements*, J. London Math. Soc. **23** (1949), 163–168.
68. LITTLEWOOD, J. E., *On a theorem of Paley*, J. Lon. Math. Soc. **29** (1954), 387–395.
69. __________, *On ineuqlities between f and f^**, J. London Math. Soc. **35** (1960), 352–365.
70. PALEY, R. E. A. C., *Some theorems on orthogonal functions*, Studia Math. **3** (1931), 336–238.
71. KEOGH, F. R., *On a problem of Hardy and Littlewood concerning rearrangements*, J. London Math. Soc. **36** (1961), 353–361.
72. __________, *On some inequalities of Littlewood and a problem on rearrangements*, J. London Math. Soc. **36** (1961), 362–276.
73. BACHELIS, G. A., *On the upper and lower majorant properties of L^p*, Quart. J. Math. (Oxford) **24** 2 (1973), 119–128.
74. MONTGOMERY, H. L., *Topics in multiplicative number theory*, in "Lecture Notes in Mathematics," Vol. 227, 1971, 187 pp.
75. __________, *A notc on rearrangements of Fourier coefficients*, Ann. Inst. Fourier, Grenoble **26** 2 (1976), 29–34.
76. LEHMER, D. H., *On a problem of Hardy and Littlewood*, J. London Math. Soc. **34** (1959), 395–396.
77. KAHANE, J. P., *Sur un problème de Littlewood*, Proc. Kon. Ned. Ak. v. Wet., Amsterdam **60** (1957), 268–271.
78. O'NEIL, R., *Convolution operators and $L(p,q)$ spaces*, Duke Math. J. **30** (1963), 129–142.
79. KOMAROFF, N., *Rearrangement and matrix product inequalities*, Linear Alg. and Appl. **140** (1990), 155–161.
80. SAITHOH, S., *A fundamental inequality in the convolution of L_2 functions on the half line*, Proc. Amer. Math. Soc. **91** (1984), 285–286.
81. __________, *On the convolution of L_2 functions*, Kodai Math. J. **9** (1986), 50–57.

CHAPTER XI

INEQUALITIES OF ČAPLYGIN TYPE

1. In quantitative analysis of differential equations the use of differential inequalities is very wide. In this Chapter we shall give various theorems on differential inequalities of the first order, of higher orders, as well as of systems of differential inequalities.

2. Let us consider the initial value problem

$$\frac{dx}{dt} = F(t,x), \qquad x(a) = x_0, \tag{2.1}$$

where F is a continuous function on an open set D which contains the point $(0, x_0)$.

Define a rectangle R by

$$R = \{(t,x) \,|\, a \le t \le b_x,\ a < b_x,\ |x - x_0| \le c,\ c > 0\} \subset D.$$

For differentiable functions x, defined on the interval $a \le t \le b_x$, we say that x is a solution of (2.1) if $(t, x(t)) \in D$ for $a \le t \le b_x$ and

$$x'(t) = F(t, x(t)), \quad a \le t \le b_x \text{ with } x(a) = x_0.$$

For a solution r, we say that it is a maximal or an upper solution (minimal or lower) of problem (2.1) on $[a, b_r)$ if $x(t) \le r(t)$ $(x(t) > r(t))$ holds on $a < t < \min(b_x, b_r)$ for every solution x of problem (2.1).

In connection with problem (2.1) we give two important results.

LEMMA 1. *If the above conditions are valid, then the initial value problem* (2.1) *has both maximal and minimal solutions. If* $|F(t,x)| \le M$ *for* $(t,x) \in R$, *then these solutions, as well as all others, are defined at least on the interval* $[a, b_1)$ *where* $b_1 - a = \min[b-a, bM^{-1}]$. *If* $[a, b_r)$ *is the maximal domain of existence of the maximal solution* r, *then there exists an* $\varepsilon_0 > 0$ *such that for* $0 < \varepsilon < \varepsilon_0$, *there exists a maximal (minimal) solution of the system of*

$$\frac{dx}{dt} = F(t,x) \pm \varepsilon, \quad x(a) = x_0 \pm \varepsilon, \tag{2.2}$$

on $[a, b_r)$ where the sign + or − depends on whether we consider maximal or minimal solutions, respectively.

REMARK 1: The first part of the Lemma is proven in Hille [**1**, p. 46] and Reid [**2**, Th 4]. The second part of the Lemma appears in Szarski [**3**, Lemma 10.1] and Lakshmikantham and Leela [**4**, Lemma 1.3.1].

LEMMA 2. *Let $F(t,x)$ be a continuous function on an open set D which contains the point (a, x_0), and let $\{\varepsilon_n\}$ be a strictly monotonic sequence with $\lim \varepsilon_n = 0$. Consider the initial value problems*

$$\frac{dx}{dt} = F(t,x) + \varepsilon_n, \quad x(a) = x_0 + \varepsilon_n. \tag{2.3}$$

Let $x = r_n(t)$ be the maximal or minimal solution of this problem on $[a, b_1)$ according to whether $\varepsilon_n > 0$ or $\varepsilon_n < 0$.

Then the sequence $\{r_n\}$ converges uniformly on $[a, b - \delta]$ for any $\delta > 0$, to a function r which is the maximal solution of (2.1) if $\varepsilon_n > 0$, or the minimal solution if $\varepsilon_n < 0$.

REMARK 2: For a proof, see Beesack [**5**, pp. 38-39], for example.

THEOREM 1. *Let $F(t,x)$ be a continuous function on an open set D which contains the point (a, x_0), and let y be any solution of the initial value problem (1). If x is any differentiable function on $[a, b_1]$ such that $(t, x(t)) \in D$ for $t \in [a, b_1]$ and*

$$x'(t) < F\,(t, x(t))\,, \; t \in [a, b_1], \text{ with } x(a) < x_0, \tag{2.4}$$

then also

$$x(t) < y(t), \; t \in [a, b_1]. \tag{2.5}$$

If all of the inequalities in (2.4) and (2.5) are reversed then the result also holds.

PROOF: We prove the case with <. Suppose that for some $t \in (a, b_1)$ we have $y(t) \le x(t)$. Let $\tau = \inf\{t \in (a, b_1) \,|\, y(t) \le x(t)\}$. Then $a < \tau < b_1$ and by continuity $y(\tau) = x(\tau)$ while $x(t) < y(t)$ on $[a, \tau)$. However $x'(\tau) < F\,(\tau, x(\tau)) = F\,(\tau, y(\tau)) = y'(\tau)$. Since $x(\tau) = y(\tau)$ this implies that $x(t) < y(t)$ on $(\tau - \varepsilon, \tau)$ which is contrary to the definition of τ. The proof for > is analogous.

THEOREM 2. *Let $F(t,x)$ be defined as in* Theorem 1 *and let the initial value problem* (2.1) *have maximal solution $x = r(t)$ on $[a, b_1)$. If x is any differentiable function on $[a, b_1]$ such that $(t, x(t)) \in D$ for $t \in [a, b_1)$ and*

$$x'(t) \leq F(t, x(t)), \; x(a) = x_0, \; t \in [a, b_1), \tag{2.6}$$

then

$$x(t) \leq r(t), \; t \in (a, b_1). \tag{2.7}$$

If "maximal" is replaced by "minimal" and the signs are reversed, then we have a valid statement.

PROOF: We prove only the maximal case. Let $\{\varepsilon_n\}$ and $\{r_n\}$ be defined as in Lemma 2. Then we have $x'(t) \leq F(t, x(t)) < F(t, x(t)) + \varepsilon_n$, $t \in [a, b_1)$ and $x(a) < x_0 + \varepsilon_n$. By Theorem 1, $x(t) < r_n(t)$ for $t \in [a, b)$. Be letting $n \to \infty$ we get using Lemma 2 that $x(t) \leq r(t)$ for $t \in [a, b_1)$.

REMARK 3: Very often we find that the previous Theorem is ascribed to S.A. Čaplygin [**6**]–[**7**]. However, these results from 1919 were first proved in 1886 by G. Peano [**8**] and in 1899 by M. Petrovitch [**9**]. The question of priorities of these results was considered in a paper of Bertolino and Trifunović [**10**] (see also Bertolino [**11**, pp.220–249]).

REMARK 4: Theorem 2 is well-known as a Comparison Theorem.

The following result is given in Beesack [**5**, pp. 39–40].

THEOREM 3. *Let the functions F and G be continuous and of one sign in a domain D containing the point (a, r_0), and suppose*

$$\frac{dr_1}{dt} = G(t, r_1), \quad r_1(a) = r_0 \tag{2.8}$$

has maximal (if $G \geq 0$) or minimal (if $G \leq 0$) solution $r_1(t)$ on $[a, b_1)$ such that

$$S_+ = \{(t,r) \,|\, a \leq t \leq b_1, \; r_0 \leq r \leq r_1(t)\} \subset D \quad \text{if } G \geq 0,$$

or

$$S_- = \{(t,r) \,|\, a \leq t \leq b_1, \; r_0 \geq r \geq r_1(t)\} \subset D \quad \text{if } G \leq 0.$$

If $0 \leq (\pm F(t,r)) \leq (\pm G(t,r))$ for $(t,r) \in S_\pm$, then all solutions r of the initial value problem

$$\frac{dr}{dt} = F(t, r), \quad r(a) = r_0 \tag{2.9}$$

exist on $[a, b_1)$. Moreover

$$r_0 \leq r(t) \leq r_1(t) \quad \text{on } [a, b_1) \text{ if } G \geq 0, \quad \text{or}$$
$$r_1(t) \leq r(t) \leq r_0 \quad \text{on } [a, b_1) \text{ if } G \leq 0.$$

P. R. Beesack [**5**, pp. 40–48] used the previous result in a proof of the following result.

COROLLARY 1. *Let σ and k be continuous and of one sign on an interval $J = [a, b]$, and let g be continuous, monotonic, and never zero on an interval I containing the point r_0. Suppose that either g is nondecreasing and $k \geq 0$ or g is nonincreasing and $k \leq 0$. If $\sigma \geq 0$ ($\sigma \leq 0$), let the maximal (minimal) solution of the initial value problem*

$$\frac{dr}{dt} = k(t)g(t) + \sigma(t), \quad r(a) = r_0 \tag{2.10}$$

exist on the interval $[a, b_1)$. Let

$$u_1 = \sup\left\{u \in J \,|\, r_0 + \int_a^u \sigma\, ds \in I\right\}.$$

$$u_2 = \sup\left\{u \in J \,|\, G\left(r_0 + \int_a^t \sigma\, ds\right) + \int_a^t k\, ds \in G(I),\ a \leq t \leq u\right\},$$

with $G(u) = \int_{u_0}^{u} dy/g(y)$, $u \in I$ ($u_0 \in I$). Let $b_2 = \min(u_1, u_2)$, and $\bar{b}_1 = \min(b_1, b_2)$. Then for $t \in [a, \bar{b}_1)$ we have

$$r(t) \leq (\geq) G^{-1}\left[\int_a^t k\, ds + G\left(r_0 + \int_a^t \sigma\, ds\right)\right] \tag{2.11}$$

if $\sigma \geq 0$ ($\sigma \leq 0$). Moreover, if $\sigma \geq 0$ and k and g have the same sign ($\sigma \leq 0$ and k and g have opposite signs), then $b_1 \geq b_2$ and we also have

$$G^{-1}\left(\int_a^t k\, ds + G(r_0)\right) \leq (\geq) r(t), \quad a \leq t \leq b_2. \tag{2.12}$$

3. It is natural to generalize to Dini derivates. For example,

$$D^+(t_0) = \limsup_{t \to t_0^+} \frac{\phi(t) - \phi(t_0)}{t - t_0}$$

is the right upper derivative, and the right lower (D_+ as lim inf). The left upper D^- and left lower D_- derivatives are defined analogously. We have $D_+(t_0) \leq D^+\phi(t_0)$ and $D_-\phi(t_0) \leq D^-\phi(t_0)$ and $\phi'_+(t_0)$ exists if and only if $D^+\phi(t_0)$ and $D_+\phi(t_0)$ are finite and equal, and analogously for $\phi'_-(t_0)$.

Proofs are similar to the previous ones so we shall give only a proof of the following theorem:

THEOREM 1. *Let the function $\phi(t,u)$ be continuous for $t \in [0,T]$ and $|u| < \delta$, and let the continuous function v satisfy*

$$D_-v(t) < \phi(t, v(t)) \quad \text{on } (0,T] \tag{3.1}$$

with $v(0) \leq u_0$ and if $v(0) = u_0$,

$$D^+v(0) < \phi(0, u_0). \tag{3.2}$$

Then

$$v(t) < u(t) \quad \text{on } (0,T] \tag{3.3}$$

where $u(t)$ is the minimal solution of the initial value problem

$$\frac{du}{dt} = \phi(t,u), \quad u(0) = u_0 \tag{3.4}$$

defined on $[0,T]$.

PROOF: It follows from (3.2) that the inequality (3.3) is valid on some neighborhood of the point $t = 0$. Suppose that (3.3) is not valid and let t^* denote the first point where $u(t) = v(t)$. Therefore we have $v(t) < u(t)$ for $0 < t < t^*$, $v(t^*) = u(t^*)$ so that $D_-v(t^*) = \liminf_{t \to t^*-} \frac{v(t)-v(t^*)}{t-t^*} > \lim_{t \to t^*-} \frac{u(t)-u(t^*)}{t-t^*} = u'(t^*) = \phi(t^*, u(t^*)) = \phi(t^*, v(t^*))$ contrary to the assumption (3.1). This proves the theorem.

This Theorem as well as the following are given in Mamedov, Aširov, and Atdaev [**12**, pp. 7–10].

THEOREM 2. *Let $\phi(t,u)$ be a continuous function on $[0,T] \times [-\delta,\delta]$ and let v be a continuous function which satisfies one of the following two conditions on $(0,T]$,*

$$D_-v(t) \leq \phi(t, v(t)), \quad v(0) \leq u_0 \qquad \text{or} \tag{3.5}$$

$$D_+v(t) \leq \phi(t, v(t)), \quad v(0) \leq u_0. \tag{3.6}$$

Then

$$v(t) \leq u(t) \quad \text{on } [0,T], \tag{3.7}$$

where $u(t)$ is the maximal solution of (3.4) defined on $[0,T]$.

COROLLARY 2. *(Flett [**13**, pp. 94–95]). Let $J = [t_0, t_0 + \alpha]$ and let p, q, and ϕ be real valued continuous functions defined on J such that*

$$D_+\phi(t) < p(t)\phi(t) + q(t) \qquad \text{a.e. on } J.$$

Then ϕ is dominated by the solution of the scalar linear differential equation $y' = p(t)y + q(t)$ that agrees with ϕ at t_0. That is

$$\phi(t) \leq \phi(t_0) \exp\left(\int_{t_0}^{t} p(v)dv\right) + \int_{t_0}^{t} q(u) \exp\left(\int_{u}^{t} p(v)dv\right) du.$$

The following Theorem is also proved in Mamedov,Aširov and Atdaev [**12**, pp. 12–13]:

THEOREM 3. *Let $\phi(t,u)$ be a continuous function on $[0,T] \times [-\sigma,\sigma]$. Suppose $v_1(t)$ is a solution of the differential inequality*

$$\frac{dv_1(t)}{dt} \leq \phi(t, v_1(t)) \quad \text{on } (0,T]. \tag{3.8}$$

Then there exists a solution $v_2(t)$ of the differential inequality

$$\frac{dv_2(t)}{dt} \geq \phi(t, v_2(t)) \quad \text{on } (0,T], \quad v_1(0) \leq v_2(0), \tag{3.9}$$

such that on this interval $v_1(t) \leq v_2(t)$.

PROOF: Let $\psi(t,v_2) = \phi(t, \max(v_2, v_1(t)))$. Obviously, ψ is a continuous function on $[0,T] \times [-\sigma,\sigma]$. First we shall prove the inequality

$$v_1(t) \leq \omega(t) \quad \text{on } [0,T] \tag{3.10}$$

where $\omega(t)$ satisfies the inequality

$$\frac{d\omega(t)}{dt} \geq \psi(t, \omega(t)), \quad \omega(0) = v_2(0). \tag{3.11}$$

Suppose that this is not true, i.e. that for some value τ, $\omega(\tau) < v_1(\tau)$. Let τ_0 be the lower bound of numbers s for which we have $\omega(t) < v_1(t)$ for $s \leq t \leq \tau$. Then $\omega(\tau_0) = v_1(\tau_0)$ and $\omega(t) < v_1(t)$ for $\tau_0 < t < \tau$. Therefore on the segment $[\tau_0, \tau]$ we get

$$\psi(t, \omega(t)) = \phi(t, v_1(t)).$$

Using inequality (3.8) we get

$$\begin{aligned} v_1(\tau) &\leq v_1(\tau_0) + \int_{\tau_0}^{\tau} \phi(s, v_1(s))\, ds \\ &= \omega(\tau_0) + \int_{\tau_0}^{\tau} \psi(s, \omega(s))\, ds, \end{aligned} \tag{3.12}$$

and from (3.11) and (3.12) we get $v_1(\tau) \leq \omega(\tau)$, which is in contradiction with our supposition. This proves the inequality (3.10). But now because (3.10) implies that a solution of inequality (3.11) is also a solution of inequality (3.9), we see that that result follows from (3.10).

4. The question of obtaining the best possible bounds in differential inequalities is considered in Perov [**14**] and Adamov and Sabirov [**15**] (see also Mamedov, Aširov and Atdaev [**12**, pp. 14-23]). Here we shall cite some of these results.

Let us consider the differential equation

$$\frac{du}{dt} = \phi(t,u) \tag{4.1}$$

with initial condition

$$u(0) = u_0. \tag{4.2}$$

Let $\underline{u}(t,t_1,u_1)$ denote a minimal solution of (4.1) which passes through the point (t_1,u_1). We say that the a solution $u(t)$ has property E_* at the point $t=t_1$ if there exists a $T_1 > 0$, such that $u(t) = \underline{u}\,(t,t_1,u(t_1))$ on $[t_1,T_1]$. A solution of the problem (4.1)–(4.2) defined on an interval $[0,T]$ which at all points of this interval has property E_* is call an E_* solution. The set of all E_* solutions U is not empty since it contains the minimal solution. The formula

$$\omega(t) = \sup_{u(t)\in U} \{u(t)\} \quad \text{on } (0,T] \tag{4.3}$$

defines a maximal E_* solution of problem (4.1)–(4.2). It is not the case that $\omega(t)$ is a maximal or minimal solution of (4.1)–(4.2). Indeed, the maximal E_* solution of the equation

$$\frac{du}{dt} = \phi(t,u) \equiv \begin{cases} 2t, & 0 \leq t \leq 1, \quad t^2 \leq u, \\ 2\left(t - \sqrt{t^2-u}\right), & 0 \leq t \leq 1, \quad 0 \leq u \leq t^2, \\ 2u/t, & 0 < t \leq 1, \quad -t^2 \leq u \leq 0, \\ -2t, & 0 \leq t \leq 1, \quad u \leq -t^2 \end{cases}$$

which passes through the point $(0,0)$ is the function $\omega(t) = 0$, but the maximal solution is $\bar{u}(t) = t^2$, and the minimal solution is $\underline{u}(t) = -t^2$.

THEOREM 1. *If the differentiable function $v(t)$ satisfies*

$$v(0) \leq u_0, \quad v'(t) < \phi\,(t,v(t)) \quad \text{on } (0,T], \tag{4.4}$$

then

$$v(t) < \omega(t) \quad on\ (0, T], \tag{4.5}$$

where ω is defined by (4.3).

THEOREM 2. *If a continuous function $v(t)$ satisfies*

$$v(0) < u_0, \quad D_{-}v(t) < \phi(t, v(t)) \quad on\ (0, T] \tag{4.6}$$

then (4.5) is valid.

THEOREM 3. *If a continuous function $v(t)$ satisfies*

$$v(0) < u_0, \quad D_{+}v(t) < \phi(t, v(t)) \quad on\ (0, T], \tag{4.7}$$

then

$$v(t) < \omega(t) \quad on\ (0, T]. \tag{4.8}$$

REMARK 1: In all cases $\omega(t)$ can be obtained as $\sup\{v(t)\}$ over all functions $v(t)$ which satisfy the corresponding theorem, i.e. satisfy the inequalities (4.4), (4.6) and (4.7), respectively.

5. It is clear that differential inequalities are very important for numerical analysis, especially for numerical solution of differential equations. Of course, here it is of interest to discuss comparison theorems with the unique solution of a differential equation as was proved by Čaplygin [**6**] (see also Berezin and Žitkov [**16**, pp. 261–264] or Rabczuk [**17**, pp. 11–16]:

THEOREM 1. *Let functions $f(t, y)$ and $F(t, y)$ be continuous in a domain*

$$R = \{(t, y)\,|\, t_0 \le t \le t_0 + a;\ |y - y_0| \le b;\ a > 0,\ b > 0\} \tag{5.1}$$

and satisfy

$$f(t, y) \le F(t, y). \tag{5.2}$$

Further let $y = y(t)$ and $X = X(t)$ be solutions of the differential equations

$$y' = f(t, y), \quad X' = F(t, X) \tag{5.3}$$

which pass through the point (t_0, y_0), defined on $[t_0, t_0 + a]$, and which lie between $y_0 - b$ and $y_0 + b$. If the function $f(t, y)$ satisfies a Lipschitz condition in y on R then

$$X(t) \ge y(t). \tag{5.4}$$

In fact, if at any point t_1 inequality (5.4) is strict then it is strict on $[t_1, t_1 + a]$.

In practice, when solving the equation $y' = f(t, y)$ we look for functions F and ϕ such that $F(t, y) \geq f(t, y) \geq \phi(t, y)$, and that the equations $X' = F(t, X)$ and $x' = \phi(t, x)$ are easily solved. For then we have $x \leq y \leq X$.

EXAMPLE: Let us consider the initial value problem

$$y' = t^2 + y^2, \quad y(0) = 0, \quad t \in [0, 1].$$

Choose $F(t, y) = 1 + y^2$ and $\phi(t, y) = t^2$. We find that $X = \tan t$ and $x = t^3/3$. It is clear that

$$\frac{dX}{dt} - f(t, X) \geq 0, \quad \frac{dx}{dt} - f(t, x) \leq 0 \tag{5.5}$$

and

$$X \geq y \geq x. \tag{5.6}$$

REMARK 1: Inequalities (5.5) and (5.6) are satisfied if

$$X = y_0 + Y \quad \text{and} \quad x = y_0 - Y, \tag{5.7}$$

where $Y = \int_{t_0}^{t} e^{L(t-s)} |f(s, y_0)| ds$ and L is the Lipschitz constant for the function f. Indeed

$$\begin{aligned} X' - f(t, X) &= |f(t, y_0)| + LY - f(t, y_0 + Y) \\ &= |f(t, y_0)| - f(t, y_0) + LY - \{f(t, y_0 + Y) - f(t, y_0)\} \\ &> |f(t, y_0)| - f(t, y_0) \geq 0. \end{aligned}$$

The second inequality in (5.5) follows in the same way.

REMARK 2: The functions X and x are called "frame curves".

REMARK 3: Even if Čaplygin does not have priority in proving diffential inequalities, he was the first to systematically use them (using basically comparison theorems of the first order for finding one pair of frame curves $X_1(t)$ and $x_1(t)$), to obtain convergent sequences of pairs of frame curves

$$\begin{aligned} x_1(t) < x_2(t) < \cdots < x_n(t) < \cdots < y(t) < \cdots < X_n(t) \\ < X_{n-1}(t) < X_1(t), \end{aligned}$$

in which we only solve linear differential equations.

We now exposit the previous remark and give Čaplygin's method. Suppose that f_{yy} $(= \partial^2 f/\partial y^2)$ has constant sign is a domain bounded by the curves $y = X(t)$ and $y = x(t)$ and the lines $t = t_0$, and $t = t_0 + a$, i.e. that sections of the surface $z = f(x, y)$ cut by planes $t =$ constant are either always convex or always concave, (fig.1).

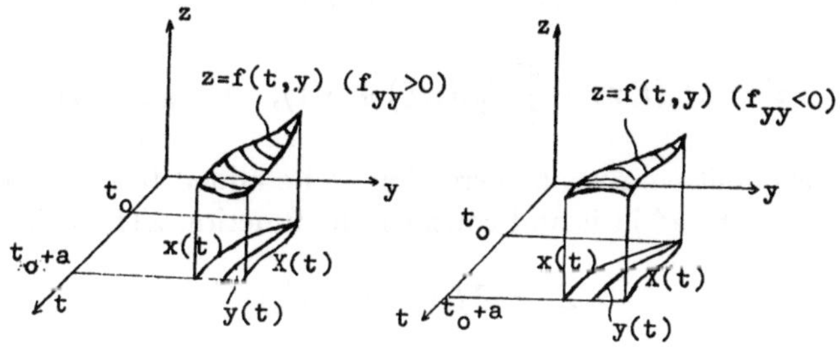

fig 1

Let us consider any section of this surface cut by $t =$ constant (fig 2) with x, y, X on this section. Construct the chord AB and the tangent AT, if $f_{yy} > 0$, or BT if $f_{yy} < 0$. In this way we get two surfaces $z = F(t, y)$ and $z = \phi(t, y)$ with $F(t, y) \geq f(t, y) \geq \phi(t, y)$. We have that F and ϕ are linear with respect to y so the differential equations

$$\frac{dx_1}{dt} = F(t, X_1) \quad \text{and} \quad \frac{dx_1}{dt} = \phi(t, x_1)$$

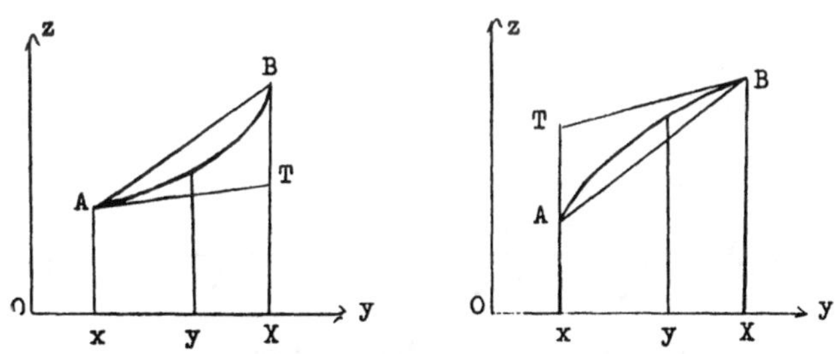

fig 2

are linear. Let X_1 and x_1 be the solutions of these equations which satisfy the same initial conditions as above. We can now show that these functions form a pair of frame curves. Indeed, along the curve $y = X(t)$ we have

$$\frac{dX}{dt} - f(t, X) \geq 0$$

and also $f(t,X) = F(t,X)$. So we have $\frac{dX}{dt} - F(t,X) \geq 0$ and $X \geq X_1$, and from the inequality $f(t,y) \leq F(t,y)$ we have $X_1 \geq y$. There is an analogous proof for x.

Now we formulate Čaplygin's Method analytically. Suppose that $f_{yy} > 0$. Then $y' = f(t,y)$, $x' = f(t,x) - \alpha(t)$ $(\alpha(t) \geq 0)$. Let $z = y - x$. Using the Taylor formula for the difference $f(t,y) - f(t,x)$ we get

$$z' = zf_y(t,x) + \frac{z^2}{2} f_{yy}(t,\xi) + \alpha(t), \quad (x \leq \xi \leq y).$$

Solution of this equation gives a correction to be added to x to obtain y. If we neglect the term with z^2 in it and consider the function $Z(t)$ which satisfies

$$Z' = Zf_y(t,x) + \alpha(t) \tag{5.8}$$

with initial condition $Z(t_0) = 0$, then we find that $Z \leq z$. On the other hand, if we set $\alpha(t) \equiv 0$, the solution of (5.8) with zero initial condition will be identically zero. Thus $Z \geq 0$ and the function $x_1 = x + Z$ satisfies $x \leq x_1 \leq y$, as well as

$$\begin{aligned} x' = x' + Z' &= f(t,x) - \alpha(t) + Zf_y(t,x) + \alpha(t) \\ &= f(t,x) - (x_1 - x)f_y(t,x) = \phi(t,x_1). \end{aligned}$$

If we also want to improve the upper estimate we consider the equations $X' = f(t,X) + \theta(t)$ $(\theta(t) \geq 0)$ and $y' = f(t,y)$. Set $u = X - y$ and write $u' = f(t,X) - f(t,y) + \theta(t)$. Introduce the notation

$$F(t) = \frac{f(t,X) - f(t,x)}{X - x}.$$

Then $f(t,X) - f(t,y) = uF(t) + \beta(t)$, where

$$\begin{aligned} \beta(t) &= f(t,X) - f(t,y) - (X-y)F(t) \\ &= (X-y)\left[\frac{f(t,X) - f(t,y)}{X-y} - \frac{f(t,X) - f(t,x)}{X-x}\right]. \end{aligned}$$

Since $f(t,y)$ is convex with respect to y, the function

$$\lambda(\tau) = \frac{f(t,X) - f(t,\tau)}{X - \tau} \qquad (\tau \leq X)$$

is nondecreasing (see for example Mitrinović [**AI**] p. 18), so $\lambda(y) > \lambda(x)$, that is $\beta(t) \geq 0$. In this way we argue that the solution of the initial value problem

$$U' = UF(t) + \theta(t), \quad U(t_0) = 0,$$

for $t_0 < t$, satisfies $U \leq u$. As for Z we get $U \geq 0$. If we set $X_1 = X - U$ we improve the upper frame curve and

$$X_1' = X' - U' = f(t,X) + \theta(t) - UF(t) - \theta(t)$$
$$= f(t,X) + (X_1 - X)F(t) = F(t,X_1).$$

The case when $f_{yy} < 0$ is handled similarly.

In the previously considered example we have $f_{yy} = 2 > 0$, so we can use Čaplygin's Method. Then Z is the solution of the initial problem

$$Z' = Z\frac{2t^3}{3} + \frac{t^6}{9}, \quad Z(0) = 0,$$

so

$$Z = e^{t^4/6} \int_0^t e^{-s^4/6} \frac{s^6}{9}\, ds,$$

and

$$x_1 = \frac{t^3}{3} + \frac{1}{9} e^{t^4/6} \int_0^t s^6 e^{-s^4/6}\, ds.$$

The equation for U is

$$U' = \left(\tan t + \frac{t^3}{3}\right) U + (1 - t^2), \quad \text{so}$$

$$U = (\sec t) e^{t^4/12} \int_0^t (1 - s^2) \cos s\, e^{-s^4/12}\, ds,$$

and

$$X_1 = \tan t - (\sec t) e^{t^4/12} \int_0^t (1 - s^2) \cos s\, e^{-s^2/12}\, ds.$$

N. N. Luziń [**18**] showed that if the starting frame curve is sufficiently close to the solution y, then if $f_{yy} \geq 0$ the sequence $y - x_n$ (if $f_{yy} \leq 0$ then $X_n - y$) converges to zero as $C/2^{2^n}$, where C is a constant and 2^{2^n} are Fermat's numbers. This is a rapidly convergent scheme.

A deficiency is Čaplygin's method is that f_{yy} must have the same sign throughout the considered domain. This deficiency does not exist, for example, in Quade's method [**19**] (see also Berezin and Žitkov [**16**, pp. 269–273] or Rabczuk [**17**, pp. 21–24].

THEOREM 2. *Let functions $f(t,y)$, $F(t,y)$, $y(t)$ and $X(t)$ be defined as in Theorem 1. If we set*

$$h(t) = X'(t) - f(t, X(t)),$$

then the function

$$X_1(t) = X(t) - \int_{t_0}^{t} e^{-L(t-s)} h(s)\, ds,$$

where L is the Lipschitz constant for the function $f(t,y)$, is an upper frame curve on the interval $[t_0, t_0 + a]$ and on that interval we have the inequalities

$$y(t) \le X_1(t) \le X(t).$$

PROOF: The inequality $X_1(t) \le X(t)$ is obvious. On the other hand we have

$$X'(t) - f(t, X_1(t)) = X'(t) - h(t) + L \int_{t_0}^{t} e^{-L(t-s)} h(s) ds - f(t, X_1(t))$$
$$= f(t, X_1(t)) - f(t, X(t)) + L(X(t) - X_1(t)) \ge 0.$$

Thus $y(t) \le X_1(t)$.

REMARK 4: If $x(t)$ is a lower frame curve and if we set

$$h_1(t) = x'(t) - f(t, x(t)),$$

then

$$x_1(t) = x(t) - \int_{t_0}^{t} e^{-L(t-s)} h_1(s) ds$$

is also a lower frame curve and we have $x(t) \le x_1(t) \le y(t)$.

In this case we can continue the process, so we have the following

THEOREM 3. *Let $f(t,y)$, $F(t,y)$, $y(t)$ and $X(t)$ be defined as in Theorem 1. Set*

$$h_0(t) = X'(t) - f(t, X(t))$$

and define the sequence

$$X_n(t) = X_{n-1}(t) - \int_{t_0}^{t} e^{-L(t-s)} h_{n-1}(s) ds \quad (n = 1, 2, \ldots)$$
$$h_n(t) = X_n'(t) - f(t, X_n(t)) \quad (n = 1, 2, \ldots).$$

Then the sequence $\{X_n(t)\}$ converges to $y(t)$ uniformly on $[t_0, t_0 + a]$.

REMARK 5: For some other methods for improving frame curves see Rabczuk [**17**, pp. 24–38] and references therein.

6. In this Section we shall consider differential inequalities of higher order. We begin with linear differential inequalities of second order.

THEOREM 1. *Let $p_0(t)$ and $g(t)$ be continuous on $[0,T]$, and $p_1(t)$ be continuously differentiable on the same interval. If a continuous function $v(t)$ satisfies on $[0,T]$*

$$v''(t) - p_1(t)v'(t) - p_0(t)v(t) - g(t) < 0, \;\; v(0) = u_0, \;\; v'(0) = u_0', \tag{6.1}$$

then

$$v(t) < u(t) \quad on\ (0, T_1) \quad with\ T_1 \le T, \tag{6.2}$$

where $u(t)$ is a solution (on $[0,T_1]$) of

$$u''(t) - p_1(t)u'(t) - p_0(t)u(t) - g(t) = 0, \;\; u(0) = u_0, \;\; u'(0) = u_0'. \tag{6.3}$$

PROOF: By using inequality (6.1) and equation (6.3) we can write

$$\begin{aligned} u''(t) - v''(t) + \sigma(t)\,(u'(t) - v'(t)) - (\sigma(t) + p_1(t))\,(u'(t) - v'(t)) \\ - p_0(t)\,(u(t) - v(t)) > 0, \end{aligned} \tag{6.4}$$

where the function $\sigma(t)$ is to be determined by the equation

$$\frac{d}{dt}\left(\left(\sigma(t) + p_1(t)\exp\int_0^t \sigma(s)ds\right)\right) = p_0(t)\exp\left(\int_0^t \sigma(s)ds\right). \tag{6.5}$$

Multiplying the inequality (6.4) by $\exp\int_0^t \sigma(s)ds$ and using (6.5) we get

$$\frac{d}{dt}\left((u'(t) - v'(t))\exp\int_0^t \sigma(s)ds\right) - \frac{d}{dt}\left((\sigma(t) + p_1(t))\,(u(t) - v(t))\exp\int_0^t \sigma(s)ds\right) > 0.$$

By integration of this inequality from 0 to t we obtain

$$u'(t) - v'(t) - (\sigma(t) + p_1(t))\,(u(t) - v(t)) \geq 0,$$

and in turn

$$u(t) - v(t) > 0 \quad \text{on } (0, T_1).$$

It is obvious that the inequality (6.2) holds on any interval $(0, T_1)$ close to the point $t = 0$, since we have $v(0) = u(0)$ and $v'(0) = u'(0)$ and $v''(0) < u''(0)$. There is the problem to decide what the largest such interval $(0, T^*)$ might be on which (6.2) holds. Call T^* the *bound of application* of the differential inequality.

7. S. A. Čaplygin [20] gave a method for estimation of the bound of application. This is the method of using the Riccati equation of a linear equation. The above proof works on any interval where σ is a continuous solution of (6.5).

In differential form it is

$$\sigma' + \sigma^2 + p_1\sigma + (p_1' - p_0) = 0. \tag{7.1}$$

This is the Riccati equation associated with the Lagrange adjoint of the operator

$$Lu \equiv u'' - pu' - pu.$$

The maximal interval over which a solution of (7.1) is continuous is the interval of disconjugacy of the equation $L^+u = 0$, i.e. the largest interval over which $L^+u = 0$ has a solution with just one zero. If u is a solution of $L^+u = 0$, then $\sigma = u'/u$ satisfies (7.1). The operator L is given explicitly by

$$L^+u = u'' + (p_1u)' - p_0u.$$

EXAMPLE: Let the functions p_1, and p_0 be constants. Then a solution of (7.1) is given by

$$-\int \frac{d\sigma}{\sigma^2 + p_1\sigma - p_0} = t + C. \tag{7.2}$$

There are the three possible cases.

1. The roots of the equation

$$\sigma^2 + p_1\sigma - p_0 = 0 \tag{7.3}$$

are equal so that (7.3) is $(\sigma - a)^2 = 0$, and an integral of (7.2) is given be $\sigma(t) = a + (t + C)^{-1}$. In this case we can choose C so that σ is defined on $[0, \infty)$, so T is arbitrarily large.

2. The roots are real and distinct, say a and b. Then an integral of (7.1) is given by

$$\sigma(t) = \frac{aC - c\exp(b-a)t}{C - \exp(b-a)t}$$

and again we may choose C so that T is arbitrarily large.
3. The roots are complex, say $a \pm bi$. Now the solution of (7.1) is $\sigma(t) = a + b\tan(bt + c)$ and the maximum interval is π/b. The method of Riccati equations gives only $T \geq \pi/b$.

8. Čaplygin also gives a generalization of this method to linear differential inequalities of the nth order.

THEOREM 1. *Let $p_0(t)$ and $g(t)$ be continuous on $[0,T)$, and $p_i(t)$ be continuously differentiable functions on $[0,T)$, $i = 1, ..., n-1$. If $v(t)$ is a function that satisfies*

$$v^{(n)}(t) - \sum_{k=1}^{n} p_{n-k} v^{(n-k)}(t) - g(t) < 0, \qquad t \in (0,T), \tag{8.1}$$
$$v^{(k)}(0) = u_0^{(k)}, \qquad k = 0, 1, ..., n-1,$$

then also

$$v(t) < u(t) \quad \text{on } (0, T_1) \quad (T_1 \leq T) \tag{8.1$'$}$$

where $u(t)$ is a solution of the problem

$$u^{(n)}(t) - \sum_{k=1}^{n} p_{n-k}(t) u^{(n-k)}(t) - g(t) = 0, \tag{8.2}$$
$$u^{(k)}(0) = u_0^{(k)}, \qquad k = 0, 1, \ldots, n-1.$$

The proof is similar to that of Theorem 1 of 6., and one can estimate the bound on application by the following related system of equations,

$$\sigma_1'(t) + \sigma_1^2(t) + p_{n-1}(t)\sigma_1(t) + p_{n-1}'(t) = -\sigma_2(t),$$
$$\sigma_2'(t) + \sigma_1(t)\sigma_2(t) + p_{n-2}(t)\sigma_1(t) + p_{n-2}'(t) = -\sigma_3(t),$$
$$\vdots$$
$$\sigma_{n-1}'(t) + \sigma_1(t)\sigma_{n-1}(t) + p_1(t)\sigma_1(t) + p_1'(t) = -p_0(t).$$

REMARK 1: Čaplygin in a similar way considered a nonlinear differential inequality of the second order

$$v''(t) - \phi(t, v(t), v'(t)) < 0, \quad v(0) = u_0, \; v'(0) = u_0'.$$

In such problems there also exists an effective radius of application.

9. The method of associated equations give only sufficient conditions for the corresponding inequalities of functions to follow from the differential inequalities. Here we shall give necessary and sufficient conditions for linear differential inequalities of the nth order.

Consider the linear differential equation

$$L[y] = y^{(n)} + p_1(t)y^{(n-1)} + \cdots + p_{n-1}y' + p_n(t)y = f(t)$$

with initial conditions $y(t_0) = y_0, \ldots, y^{(n-1)}(t_0) = y_0^{(n-1)}$.

Further, let $v(t)$ be a given function on $[t_0, t_1)$ such that $L[v] - f(t) > 0$ and

$$v^{(k)}(t_0) = y_0^{(k)}, \quad k = 0, 1, ..., n-1.$$

The question is whether $v > y$ in (t_0, t_1). By the substitution $w = v - y$ the problem is reduced to the differential inequality

$$L[w] > 0 \tag{9.1}$$

with zero initial conditions at t_0, and the question is whether $w > 0$ on (t_0, t_1).

In the theory of differential equations it is well known that the general solution of $L[y] = f(t)$ is given by $y = Y + y_1$ where Y is a particular solution of the nonhomogeneous equation and y_1 is the general solution of the corresponding homogeneous equation $L[y] = 0$. A particular solution of the nonhomogeneous equation can be found in the following way:

Let $K(t, \alpha)$ be the Cauchy function, i.e. it is a solution of the homogeneous equation $L[y] = 0$ (as a function of t) which satisfies the conditions ($' = \frac{d}{dt}$)

$$K^{(k)}(\alpha, \alpha) = 0, \quad k = 0, 1, ..., n-2, \quad K^{(n-1)}(\alpha, \alpha) = 1.$$

Then

$$Y(t) = \int_{t_0}^{t} k(s, \alpha)\phi(s)ds.$$

In our case we have $L[w] > 0$, so we can write

$$L[w] = \phi(t) > 0.$$

So

$$w = \int_{t_0}^{t} K(s, \alpha)\phi(s)ds,$$

is a solution of the equation with the correct initial conditions.

We pose the problem of finding necessary and sufficient conditions that $w > 0$ follows from $L[w] > 0$ on an interval $[t_0, t_1)$. We claim that such a condition is $K(t,\alpha) \geq 0$ for $t_0 \leq \alpha \leq t \leq t_1$. That the condition is sufficient is obvious from the equation for w above. Moreover, it is also necessary since, if for some α, $K(t,\alpha)$ changes sign, then a ϕ can be chosen so that w is negative at α. Note that this condition on K depends only on the coefficients of the operator L.

REMARK 1: N. A. Kaščeev [**21**],[**22**] was the first to formulate the above result(see also Bertolino [**11**, pp. 228–229]).

REMARK 2: Note that the nonnegativity of the corresponding derivative of the Cauchy function is necessary and sufficient for the validity of Čaplygin's Theorem in the wider sense of obtaining the inequalities $v^{(k)}(t) \geq y^{(k)}(t)$.

REMARK 3: Let $y_1(t), ..., y_n(t)$ be a system of linearly independent solutions of the homogeneous equation. The Cauchy function can be written in the form

$$K(t,\alpha) = \frac{\begin{vmatrix} y_1(\alpha) & \dots & y_n(\alpha) \\ y_1'(\alpha) & \dots & y_n'(\alpha) \\ \vdots & & \vdots \\ y_1^{(n-2)}(\alpha) & \dots & y_n^{(n-2)}(\alpha) \\ y_1(t) & \dots & y_n(t) \end{vmatrix}}{\begin{vmatrix} y_1(\alpha) & \dots & y_n(\alpha) \\ y_1'(\alpha) & \dots & y_n'(\alpha) \\ \vdots & & \vdots \\ y_1^{(n-2)}(\alpha) & \dots & y_n^{(n-2)}(\alpha) \\ y_1^{(n-1)}(\alpha) & \dots & y_n^{(n-1)}(\alpha) \end{vmatrix}}. \tag{9.2}$$

REMARK 4: In the case of a linear differential inequality with constant coefficients, if the roots of the characteristic equation are all real, then we have an infinite bound of application since $K(t,\alpha) \geq 0$ for $t \geq \alpha$ (see Bertolino [**11**, pp. 251–252]). He gave a very simple proof in the cases when the roots are either all equal or all distinct (ibid pp. 229–232) that the Cauchy function given by (9.2) is positive.

Similarly for an equation of the second order $y'' + py' + qy = 0$ with constant coefficients p, q, whose roots of the characteristic equation are $a \pm bi$, we have that the Cauchy function is $(1/b)\exp(a(t-\alpha))\sin b(t-\alpha)$ and the bound of application is $T^* = \pi/b$. This is the same result we obtained by using the Riccati equation.

The case of complex roots of the characteristic equation for linear differential equations of the third and fourth order with constant coefficients was considered by Z. Radišin (Mat.Vesnik 10 (25)(1973), 75-81). For this see also Bertolino [**11**, pp. 233–234].

REMARK 5: As in Remark 3, let $y_0(t), ..., y_n(t)$ be a system of linearly independent solutions of $L[y] = 0$, and let $W_m(t)$ denote the Wronskian

$$W_m = w(y_1, \ldots, y_m) = \begin{vmatrix} y_1 & \cdots & y_m \\ y_1' & \cdots & y_m' \\ \vdots & & \vdots \\ y_1^{(m-1)} & & y_m^{(m-1)} \end{vmatrix}.$$

Then (see Pólya [**23**], Mammana [**24**], or Rabczuk [**17**, pp. 101])

$$L[y] = \frac{W_n}{W_{n-1}} \frac{d}{dt} \frac{W_{n-1}^2}{W_n W_{n-2}} \cdots \frac{d}{dt} \frac{W_2^2}{W_3 W_1} \frac{d}{dt} \frac{W_1^2}{W_2 W_0} \frac{d}{dt} \left(\frac{y}{W_1}\right) \tag{9.3}$$

where $W_0(t) \equiv 1$.

A simple consequence of this result is that the differential inequality

$$L[y] \geq 0, \quad y^{(k)}(t_0) = 0, \quad k = 0, 1, ..., n-1, \text{ on } [t_0, t_1] \tag{9.4}$$

implies that $y(t) > 0$ on $[t_0, t_1]$ provided there exists a linear independent system of solutions $y_i(t)$ of $L[y] = 0$ such that

$$W_i'(t) > 0, \; i = 1, ..., n-1, \text{ on } [t_0, t_1]. \tag{9.5}$$

On the other hand, the inequality $L[y] > 0$ is a generalization of the inequality $y'' \geq 0$. This latter inequality is the differential definition of convex functions, see for example Mitrinović [**AI**, pp 17–18]. The inequality $L[y] \geq 0$ has a close relation with convex functions with respect to ECT- systems of functions, see Karlin and Studden [**25**] or Pečarić [**26**]. Note also that the inequality (9.5) is a necessary and sufficient condition that the system of functions $y_1(t), ..., y_n(t)$ is an ECT-system(Extended Complete Tchebycheff System) of functions (see Karlin and Studden [**25**, pp. 375–466, especially pp. 376 and 379–381].

10. In this Section we shall first give two theorems of B. N. Babkin [**27**],[**28**]. Both this theorem and another result on differential inequalities of higher order are given in Mamedov, Aširov, and Atdaev [**12**].

THEOREM 1. *Let $\omega(t)$ be an arbitrary function defined on $[0, T]$ which satisfies the conditions*

$$\omega^{(k)}(0) = u_0^{(k)}, \quad k = 0, 1, ..., n-1, \tag{10.1}$$

and the differential inequality

$$L[\omega] - g(t) < (>) 0.$$

If a function v satisfies the conditions (10.1) and the inequality

$$L[v] - g(t) > (<) \sum_{k=0}^{n-2} (M_k + |M_k|)(v - \omega)^{(k)} \quad \text{on } [0,T],$$

where $M_k = \max_{0 \le t \le T} p_k(t)$, and a function u is a solution of the equation $L[u] = g(t)$ with initial conditions (10.1), then

$$v^{(k)}(t) > (<) u^{(k)}(t), \qquad k = 0, \ldots n-1, \text{ on } [0,T].$$

THEOREM 2. *Let the function $\phi(t,u)$ be continuous on $[0,T] \times R$, and have continuous nonnegative derivative with respect to u. If v is a function that satisfies*

$$v(0) = u_0, \quad v(T) = u_T \tag{10.2}$$

and the differental inequality

$$v'' - \phi(t,v) \le (\ge) 0 \quad \text{on } [0,T],$$

then

$$v(t) \ge (\le) u(t) \quad \text{on } (0,T), \tag{10.3}$$

where u is a solution of

$$u'' - \phi(t,u) = 0 \tag{10.4}$$

with the boundary conditions (10.2).

PROOF: Note first that the initial value problem (10.4) and (10.2) has a unique solution for if u and z are two solutions, then the difference w satisfies

$$w'' - \phi_u(t, z + \theta w)) = 0, \quad w(0) = w(T) = 0, \quad 0 \le \theta \le 1. \tag{10.5}$$

Multiplying the equation in (10.5) by w and integrating by parts we arrive at

$$-\int_0^T (w')^2 dt - \int_0^T \phi_u w^2 = 0. \tag{10.6}$$

From this equation we infer that $w = 0$, since $\phi_u \geq 0$. For the existence of a solution of the boundary value (10.4)–(10.2) we refer to Babkin [**27**]. Now to prove the inequality (10.3) we let $z = u - v$ and we get

$$z'' - \phi_u(t, u + \theta z) = \delta(t) \tag{10.7}$$

where $\delta(t) = v''(t) - \phi(t, v(t)) \leq 0$.

As above we multiply by z and integrate to arrive at

$$-\int_0^T (z')^2 - \int_0^T \phi_u z^2 = \int_0^T \delta(t) z(t) dt. \tag{10.8}$$

Since $\phi_u \geq 0$, we get $\int_0^T \delta(t)z(t)dt \leq 0$. If in some interval z is negative then we can find a subinterval (ξ_1, ξ_2) on which z is negative and zero at the ends. The above argument with $(0, T)$ replaced by (ξ_1, ξ_2) now results in

$$\int_{\xi_1}^{\xi_2} \delta(t) z(t) dt \leq 0 \tag{10.9}$$

which is a contradiction. So we have $z(t) \geq 0$ as is required.

REMARK 1: A function that satisfies the differential inequality $v'' - \phi(t, v) \leq (\geq) 0$ and the boundary conditions (10.2) is called an upper (lower) function for the differential equation (10.4). There is a simple construction of such functions. Let $z(t)$ be an arbitrary function belonging to the class $C^2[0, T]$ which satisfies the boundary conditions (10.2). Let $\lambda(t) = z''(t) - \phi(t, z)$ and $\sigma(t)$ be a solution of the boundary value problem

$$\sigma'' + |\lambda(t)| = 0, \quad \sigma(0) = \sigma(T) = 0.$$

The function σ is nonnegative on the interval $(0, T)$. We shall show that the function $v(t) = z(t) + \sigma(t)$ is an upper function for the equation (10.4). Indeed

$$\begin{aligned} v'' - \phi(t, v) &= z'' + \sigma'' - \phi(t, z + \sigma) \\ &= z'' - \phi(t, z) - |\lambda(t)| + \phi(t, z) - \phi(t, z + \sigma) \\ &= \lambda(t) - |\lambda(t)| - [\phi(t, z + \sigma) - \phi(t, z)], \end{aligned}$$

so it is obvious that $v'' - \phi(t, v) \leq 0$. Similarly if σ is replaced by δ where δ satisfies

$$\delta'' - |\lambda(t)| = 0, \quad \delta(0) = \delta(T) = 0,$$

then $y(t) = z(t) + \delta(t)$ is a lower function.

11. The following result given in Grace and Lalli [**29**] is a generalization of a result of Kiguradze [**30**]:

THEOREM 1. *Suppose n is an even integer and let*

$$L_0x(t) = x(t),\ L_kx(t) = a_k(t)\frac{d}{dt}\left(L_{k-1}x(t)\right), \quad k = 1,\dots,n,$$

where the a_i, $i = 1,2,\dots,n-1$, are positive continuous functions defined on $[t_0,\infty)$ with $a_0 = a_n = 1$ and $\int_0^\infty \frac{ds}{a_i(s)} = +\infty$, $1 \le i \le n-1$. Let $D(L_n)$ be the set of all real valued functions defined on $[t_0,\infty)$ such that $L_kx(t)$ exist for $0 \le k \le n$ and are continuous.

If $x \in D(L_n)$ and $x(t)L_n(x(t)) < 0$ on $[t_0,\infty)$, then there exists an even m, $1 \le m \le n-1$ and $T > t_0$ such that $x(t)L_jx(t) > 0$ on $[T,\infty)$ for $0 \le j \le m$ and

$$(-1)^{j-1}x(t)L_jx(t) > 0 \text{ on } [T,\infty) \text{ for } m+1 \le j \le n.$$

REMARK 1: For some other resuts on nth order differential inequalities see Atkinson [**31**], Teufel [**32**], Kartsatos [**33**].

REMARK 2: For related results see Muldowney [**34**] and Muldowney and Willett [**35**], [**36**].

12. For systems of differential inequalities, we let $u, v \in R^m$. We write $u \le v$ to mean that $u_i \le v_i$, for $i = 1,\dots,m$.

THEOREM 1. *Let the vector function $\phi(t,u)$ be continuous and have continuous partial derivatives with respect to u_i on $[0,T]\times(-\delta,\delta]$ such that $\frac{\partial\phi_i}{\partial u_j} > 0$. Suppose the continuously differentiable vector function v satisfies*

$$v'(t) < \phi(t,v(t)), \quad v(0) = u_0. \tag{12.1}$$

Then

$$v(t) < u(t) \quad \text{on } [0,T], \tag{12.2}$$

where $u(t)$ is the solution of the problem

$$u'(t) = \phi(t,u(t)), \quad u(0) = u_0. \tag{12.3}$$

PROOF: Since $v'(0) < u'(0)$ we have $v'(t) < u'(t)$ on some interval $[0,h]$ by continuity. On this interval we have $v(t) < u(t)$. The above argument may be done coordinate wise. Conversely, if at some point $v(t) < u(t)$ (or on some interval) then $v'(t) < u'(t)$. This is an easy consequence of the condition that $\frac{\partial\phi_i}{\partial u_j} > 0$. Let $[0,t_0)$ be the maximal interval on which $v(t) < u(t)$, $(t_0 < T)$. If $t_0 < T$ we proceed as follows. If $v_i(t_0) = u_i(t_0)$, then we have $v(t) < u(t)$ on $[0,t_0)$ and thus $v'(t) < u'(t)$ on the same interval. We thus argue that $v_i(t_0) < u_i(t_0)$, contrary to assumption and thus $t_0 = T$.

THEOREM 2. *Let $\phi(t,u)$ satisfy the hypotheses of* Theorem 1 *and suppose the continuously differential function v satisfies*

$$v'(t) \leq \phi(t, v(t)), \quad v(0) = u_0. \tag{12.4}$$

Then

$$v(t) \leq u(t) \quad \text{on } [0,T], \tag{12.5}$$

where u is the solution of (12.3).

PROOF: The function v satisfies the diferential equation

$$v'(t) = \phi(t, v(t)) + p(t)$$

where $p(t) \leq 0$ according to (12.4). Consider the system of equations

$$v_n'(t) = \phi(t, v_n(t)) + p(t) + \varepsilon_n(t), \quad v_n(0) = u(0),$$

where $\varepsilon_n < 0$ and $\lim \varepsilon_n = 0$. On the basis of Theorem 1, we have $v_n(t) < u(t)$. By taking limits we get that $v_n(t) \to v(t)$ so that v satisfies (12.5).

REMARK 1: The previous two theorems are given in Gel'fand [**37**] (see also Mamedov, Aširov, and Atdaev [**12**, pp. 58–61].

In the same book [pp. 61–63] the following two theorems are also given.

THEOREM 3. *Let $\phi(t,u)$ satisfy the conditions of* Theorem 1 *and let v be a continuously differentiable function that satisfies*

$$v'(t) < \phi(t, v(t)), \quad v(0) \leq u_0 \quad \text{on } (0,T], \quad \text{and}$$

if $v_i(0) = u_{0i}$, then $v_i'(0) < \phi_i(0, u_0)$. Then

$$v(t) < u(t) \quad \text{on } (0,T]$$

where u is the minimal solution of (12.3) *defined on* $[0,T]$.

PROOF: As in the previous proof, there is a small interval $(0, t_0)$ where the conclusion holds because of continuity and the hypotheses. The completion of the argument is exactly the same as for the proof of Theorem 1. There is a maximal t_1 such that $v(t) < u(t)$ on $(0, t_1)$ but some component v_i has equality at t_1. Then $v_i'(t_1) < \phi(t_1, v(t_1)) = \phi(t_1, u(t_1)) = u'(t_1)$, a contradiction.

THEOREM 4. *Let* $\phi(t,u)$ *satisfy the conditions of* Theorem 1 *and let* v *be a continuously differentiable function that satisfies*

$$v'(t) \le \phi(t, v(t)), \quad v(0) \le u_0 \quad \text{on } (0,T],$$

then

$$v(t) \le u(t) \quad \text{on } [0,T]$$

where u *is the maximal solution of* (12.3) *defined on* $[0,T]$.

The previous two theorems are not direct generalizations of the related results for functions of one variable since here we do not assume the monotonicity of ϕ with respect to u. Moreover the conditions on the function ϕ can be relaxed.

13. Let $\min(u,v)$ and $\max(u,v)$ be defined as the coordinate wise minimun and maximum respectively, and let $[u,v]$ denote the interval $[u_1,v_1]\times\cdots\times[u_m,v_m]$. We say that the function satisfies the Kamke-Ważewski condition (K–W) in u if ϕ_i is nondecreasing in each u_j for each $j \ne i$. The following Theorem was proved in Perov [**38**] (see also Ważewski [**39**], [**40**] or Mamedov, Aširov and Atdaev [**12**, pp. 63–64]).

THEOREM 1. *Let* $\phi(t,u)$ *be a continuous vector function defined on* $[0,T]\times(-\delta,\delta)$ *and satisfy the* K-W *condition in* u. *If a continuously differentiable function satisfies the inequality*

$$v'(t) \le \phi(t, v(t)), \quad v(0) \le u_0 \quad \text{on } [0,T]. \tag{13.1}$$

then

$$v(t) \le u(t) \quad \text{on } [0,T], \tag{13.2}$$

where $u(t)$ *is a solution of* (12.3) *defined on* $[0,T]$.

PROOF: If $\psi(t,u) = \phi(t, \max(u, v(t)))$, then satisfies the K–W condition in u. We shall show that for every solution $\omega(t)$ of the problem $u' = \psi(t,u)$, $u(0) = u_0$, we have

$$v(t) \le \omega(t). \tag{13.3}$$

Suppose the contrary. Then for some i and some τ we have $\omega_i(\tau) < v_i(\tau)$. Let τ_0 be the smallest value of s (>0) for which $\omega_i(t) < v(t)$ for $s \le t \le \tau$. Then $v_i(\tau_0) = \omega_i(\tau_0)$. On $[\tau_0, \tau]$ we have $\psi_i(t, \omega(t)) = \phi_i(t, \max(\omega(t), v(t))) \ge$

$\phi_i(t, v(t))$ since $v_j(t) \leq \max(v_j(t), \omega_j(t))$ for $j \neq i$ and $v_i(t) = \max(v_i(t), \omega(t))$ by the K–W condition. From this condition we derive the inequality

$$\omega_i(\tau) - \omega_i(\tau_0) = \int_{\tau_0}^{\tau} \psi_i(s, \omega(s))\, ds \geq \int_{\tau_0}^{\tau} \phi_i(s, v(s))\, ds. \tag{13.4}$$

On the other hand we get from (13.1) that

$$v_i(\tau) - v_i(\tau_0) \leq \int_{\tau_0}^{\tau} \phi_i(s, v(s))\, ds \tag{13.5}$$

and hence $v_i(\tau) \leq \omega_i(\tau)$ contrary to our assumption. We thus have proved (13.3). But now ω is also a solution of (12.3) so the Theorem follows.

REMARK 1: If the function $\phi(t, u)$ satisfies the K–W condition in u then the solutions u of (12.3) lie between the minimal solution $\underline{u}$ and the maximal solution $\bar{u}$, so that we may replace the inequality (13.2) by

$$v(t) \leq \bar{u}(t) \quad \text{on } [0, T].$$

(see Mamedov, Aširov and Atdaev [**12**,p.65]).

14. The K–W condition for the vector function $\phi(t, u)$ in u is essential for the validity of Theorem 4 of 12. We shall give an example. Consider the problem

$$\begin{aligned} u_1' &= u_1 - u_2 + 2t + 1, \qquad u_1(0) = 0, \\ u_2' &= -u_1 + u_2 + 2t - 1, \quad u_2(0) = 1. \end{aligned}$$

This problem does not satisfy the K–W condition in u. The solution of the problem is $u_1 = t^2$, and $u_2 = 1 + t^2$. For a function to compare this with we take $v_1 = t/5$, and $v_2 = 1 + 3t$. Then we have that for $0 < t < 5/8$

$$\begin{aligned} v_1' - v_1 + v_2 - 2t - 1 &= (1 + 4t)/5 > 0, \\ v_2' + v_1 - v_2 - 2t + 1 &= 3 - 24t/5 > 0. \end{aligned}$$

But $v_1 - u_1 = t(1/5 - t) < 0$ for $t > 1/5$, i.e. the differential inequality is not satisfied.

15. Perov [**38**] gives the following generalization of Theorem 1 of 13.

THEOREM 1. *Let $\phi(t,u)$ satisfy the hypotheses of* Theorem 1 *of* 13. *If the continuous function v satisfies*

$$\min\left(D^-v(t), D_+v(t)\right) \le \phi\left(t, v(t)\right), \quad v(0) < u_0, \quad \text{on } (0,T] \text{ or}$$
$$\min\left(D_-v(t), D^+v(t)\right) \le \phi\left(t, v(t)\right), \quad v(0) < u_0, \quad \text{on } (0,T],$$

then

$$v(t) \le u(t) \quad \text{on } [0,T],$$

where u is the maximal solution of the problem (12.3) *on* $(0,T]$.

16. We now consider integro-differential inequalities. First we shall give some comparison theorems for integro-differential inequalities of first order (see Mamedov, Aširov, and Atdaev [**12**, pp. 147–157]):

THEOREM 1. *Let the functions $\phi(t,u,v)$ and $\psi(t,s,u)$ be continuous for $0 \le t, s \le T$, $|u| < \delta_1 < \infty$, $|v| < \delta_2 < \infty$, such that $\phi(t,u,v)$ is decreasing in v and $\psi(t,s,u)$ is decreasing in u. Let $v(t)$ be a continuous function on $[0,T)$ such that*

$$(16.1) \qquad v'(t) < (>)\phi\left\{t, v(t), \int_0^t \psi\left[t,s,v(s)\right] ds\right\} \quad \text{on } (0,t],$$

$$v(0) \le (\ge) u_0, \text{ or } v(0) = u_0 \text{ with } v'(0) < (>)\phi(0,u_0,0).$$

If $u(t)$ is any solution of the problem

$$(16.2) \qquad u'(t) = \phi\left\{t, u(t), \int_0^t \psi\left[t,x,u(s)\right] ds\right\}, \quad u(0) = u_0, \quad \text{on } [0,T],$$

then

$$(16.3) \qquad v(t) < (>) u(t) \quad \text{on } (0,T].$$

PROOF: Since $v(0) < u_0$ or $v'(0) < \phi(0,u_0,0) = u'(0)$ we have the conclusion (16.3) on some interval containing zero. If (16.3) is not true then there is a first point t^* where $u(t^*) = v(t^*)$. Then we have

$$\begin{aligned} v'(t^*) &< \phi\left\{t^*, v(t^*), \int_0^{t^*} \psi\left[t^*,s,v(s)\right] ds\right\} \\ &= \phi\left\{t^*, u(t^*), \int_0^{t^*} \psi\left[t^*,s,v(s)\right] ds\right\} \\ &< \phi\left\{t^*, u(t^*), \int_0^{t^*} \psi\left[t^*,s,u(s)\right] ds\right\} = u'(t^*). \end{aligned}$$

Therefore to the left of t^* we have $v(t) > u(t)$ which is a contradiction. Thus (16.3) holds on $(0, T]$.

THEOREM 2. *Let the functions $\phi(t,u,v)$ and $\psi(t,s,u)$ satisfy the conditions of* Theorem 1 *and $v(t)$ be a continuous function that satisfies the integro-differential inequality*

$$(16.4) \qquad v'(t) \leq (\geq)\phi\left\{t, v(t), \int_0^t \psi\left[t, s, v(s)\right] ds\right\} \text{ on } (0,T], \quad v(0) \leq (\geq) u.$$

Then

$$(16.5) \qquad v(t) \leq (\geq) u(t) \quad on [0,T],$$

where $u(t)$ is the maximal (minimal) solution of (16.2) *on* $[0,T]$.

PROOF: For any fixed n let $u_n(t)$ be a solution of the problem

$$u_n'(t) = \phi\left\{t, u_n(t), \int_0^t \psi\left[t, s, u_n(s)\right] ds\right\} + \frac{1}{n}, \quad u_n(0) = u_0 + \frac{1}{n}.$$

These solutions exist on some interval $[0,h)$, $h \leq T$. We have

(16.6)
$$u_n'(t) > \phi\left\{t, u_n(t), \int_0^t \psi\left[t, s, u_n(s)\right] ds\right\},$$

(16.7)
$$u_{n+1}'(t) < \phi\left\{t, u_{n+1}(t), \int_0^t \psi\left[t, s, u_{n+1}(s)\right] ds\right\} + \frac{1}{n}, \quad \text{and}$$

(16.8)
$$v(t) < u_n(t) \quad \text{on } (0,h].$$

The last inequality is by the previous Theorem. Also by the previous Theorem we have $u_{n+1}(t) < u_n(t)$. It follows that $\lim u_n(t) = u(t)$ exists for $[0,h]$. By writing the defining equation for $u_n(t)$ in integral form we get the limiting function $u(t)$ to satisfy (16.2). In fact it is the maximal solution and by (16.8) we get (16.5) on $[0,h]$. If $[0,T_1]$ is the maximal interval on which (16.5) holds and $T_1 < T$, we have $v(T_1) = u(T_1)$ and

$$v'(t) < \phi\left\{t, v(t), \int_0^{T_1} \psi\left[t, s, u(s)\right] ds + \int_{T_1}^t \psi\left[t, s, v(s)\right] ds\right\}$$

on $[T_1, T]$. We may repeat the above argument to get (16.5) on an interval $[T_1, T_1 + h]$ contradicting the maximality of T. Thus (16.5) is proved for $[0, T]$.

17. The previous two theorems considered integro-differential inequalities of the Volterra type. In the book Mamedov, Aširov, and Atdaev [**12**, pp 154–156] have also considered inequalities of the Fredholm type such as in the following theorem:

THEOREM 1. *Let the functions $\phi(t, u, v)$ and $\psi(t, s, u)$ satisfy the conditions of* Theorem 1. *Suppose that for any given continuous function $\lambda(t)$ on $[0, T]$ with $|\lambda(t)| \le \delta_2$, the problem*

$$u'(t) = \phi\left(t, u(t), \lambda(t)\right), \quad u(0) = u_0$$

has a continuous solution $u(t)$ on $[0, T]$ with $|u(t)| \le \delta_1$. If $v(t)$ is a continuous solution of

$$v'(t) \le \phi\left\{t, v(t), \int_0^T \psi\left[t, x, v(s)\right] ds\right\} \quad \text{on } (0, T], \ v(0) < u_0,$$

then $v(t) \le u(t)$ on $[0, T]$, where $u(t)$ is the maximal solution of

$$u'(t) = \phi\left\{t, u(t), \int_0^T \psi\left[t, x, u(s)\right] ds\right\}, \quad u(0) = u_0 \text{ on } [0, T].$$

18. In Mamedov, Aširov, and Atdaev [**12**, pp. 157–163] the following system of integro-differential equations was considered:

$$u'(t) = \phi\left\{t, u(t), \int_0^t \psi\left[t, s, u(s)\right] ds\right\}, \quad u(0) = u_0, \tag{18.1}$$

where u, u_0, ϕ and ψ are vector functions. Then they considered the corresponding inequalities:

THEOREM 1. *Let the vector functions $\phi(t, u, v)$ and $\psi(t, s, u)$ be continuous for $t, s \in [0, T]$ and $|u_i| \le \delta_1^i$, $|v_i| \le \delta_2^i$. Suppose that the function $\phi(t, u, v)$ satisfies the Kamke–Ważewski* (K–W) *condition,* see 12.1.3) *in u and be nondecreasing in v while $\psi(t, s, u)$ is nondecreasing in u. If v is a continuous solution of*

$$v'(t) < (>) \phi\left\{t, v(t), \int_0^t \psi\left[t, s, v(s)\right] ds\right\} \quad \text{on } (0, T],$$

$$\text{and} \quad v(0) \le (\ge) u_0, \quad \text{if } v_i(0) = (u_0)_i \text{ then } v_i'(0) < (>) \phi_i(0, u_0, 0),$$

then any solution $u(t)$ of (18.1) defined on $[0,T]$ satisfies

$$v(t) < (>) u(t) \quad \text{on } (0,T].$$

THEOREM 2. *Let ϕ and ψ be defined as in* Theorem 1 *and let $v(t)$ be a continuous solution of*

$$v'(t) \le \phi\left\{t, v(t), \int_0^t \psi[t,s,v(s)]\,ds\right\} \quad \text{on } (0,T],\ v(0) \le u_0.$$

Then $v(t) \le u(t)$ on $[0,T]$ where $u(t)$ is the maximal solution of (18.1) defined on $[0,T]$.

REMARK 1: Generalizations of the previous results are also given in the same book.

19. Not all integro-differential inequalities are included in (16.1) or (16.4). For example, P. G. Pachpatte [**41**] proved the following result:

THEOREM 1. *Let $x(t)$, $x'(t)$, $a(t)$ and $b(t)$ be nonnegative continuous functions defined on $I = [0,\infty)$ such that*

$$x'(t) \le a(t) + b(t)\int_0^t c(s)\,(x(s) + x'(s))\,ds, \quad t \in I. \tag{19.1}$$

Then on I we have

$$x'(t) \le a(t) + b(t)\int_0^t c(s)\,[A(s) + B(s)]\exp\left(\int_s^t c(r)\,[b(r) - 1]\,dr\right) ds \tag{19.2}$$

where

$$A(t) = x(0) + a(t) + \int_0^t a(s)ds, \quad t \in I, \text{ and}$$

$$B(t) = \int_0^t c(s)A(s)\exp\left(\int_s^t [b(t) + c(r) + b(r)c(r)]\,dr\right) ds, \quad t \in I.$$

PROOF: Let $m(t) = \int_0^t c(s)\,(x(s) + x'(s))\,ds$, $m(0) = 0$, so that (19.1) becomes

$$x'(t) \le a(t) + b(t)m(t). \tag{19.3}$$

Differentiation of (19.3) and using (19.3) we get

$$m'(t) \le c(t)\,[x(t) + a(t) + b(t)m(t)]\,. \tag{19.4}$$

On the other hand, integration of (19.3) from 0 to t leads to

$$x(t) \le x(0) + \int_0^t a(s)ds + \int_0^t b(s)m(s)ds, \tag{19.5}$$

so that (19.4) becomes

$$m'(t) \le c(t)\left[x(0) + a(t) + \int_0^t a(s)ds + b(t)m(t) + \int_0^t b(s)m(s)ds\right].$$

Introducing the function $v(t) = m(t) + \int_0^t b(s)m(s)ds$ this inequality may be rewritten

$$v'(t) \le c(t)A(t) + b(t)c(t)m(t) + c(t)v(t). \tag{19.6}$$

Using the inequality $v(t) \ge m(t)$ we get

$$v'(t) \le [b(t) + c(t) + b(t)c(t)]\,v(t) + c(t)A(t)$$

from which it follows that

$$v(t) \le \int_0^t c(s)A(s)\exp\left(\int_s^t [b(r) + c(r) + b(r)c(r)]\,dr\right)ds = B(t).$$

By observing that $m'(t) = v'(t) - b(t)m(t)$ we combine the latter inequality with (19.6) to arrive at

$$m'(t) \le c(t)\,[b(t) - 1]\,m(t) + c(t)\,[A(t) + B(t)]\,,$$

and in turn

$$m(t) \le \int_0^t c(s)\,[A(s) + B(s)]\exp\left(\int_s^t c(r)\,[b(r) - 1]\,dr\right)ds.$$

From (19.3) we get (19.2).

In the same paper Pachpatte also proved the following theorem:

THEOREM 2. *Let $x(t)$ and $x'(t)$ be nonnegative continuous functions defined on I with $b(t) \geq 1$ and continuous on I. Let $W(u)$ be a positive continuous strictly increasing function for $u > 0$, with $W(0) = 0$. If*

$$x'(t) \leq a + \int_0^t b(s)W(x(s) + x'(s))\, ds \quad \text{on } I,$$

where a is a nonnegative constant, then on I ,

$$x'(t) \leq a + \int_0^t b(s)W\left\{G^{-1}\left[G(x(0) + a) + \int_0^s b(r)dr\right]\right\} ds,$$

where $G(r) = \int_{r_0}^{r} ds/(r_0 s + W(s))$, $r \geq r_0 > 0$, and I_1 is chosen so that $G(x(0) + a) + \int_0^t b(r)dr$ is in the domain of G^{-1} for $t \in I_1$.

There are many other results for integro-differential inequalities of the first order as well as some for higher order. For example, Pachpatte [**42**] considered the following inequality in connection with the previous results,

$$x''(t) \leq f(t) + \int_{t_0}^t g(s)(x(s) + x'(s))\, ds,$$

while in [**43**] he proved the following (among others):

THEOREM 3. *Let $x(t)$, $x'(t)$, $x''(t)$, and $a(t)$ be continuous nonnegative functions defined on $I = [0, \infty)$ such that on I*

$$x''(t) \leq x(0) + x'(0) + \int_0^t a(s)(x(s) + x'(s) + x''(s))\, ds.$$

Then on I

$$x''(t) \leq x(0) + x'(0) + \int_0^t a(s) \left[2\left(x(0) + x'(0)\right) \exp\left(\int_0^s a(r)dr \right) \right.$$
$$+ \int_0^s \left[x'(0) + 2\left(x(0) + x'(0)\right)\right]$$
$$\left. \times \exp\left(\int_0^r \left(2 + a(u)\right) du \right) \exp\left(\int_r^s a(u)du \right) dr \right] ds.$$

REMARK 1: Note that the various multidimensional integral inequalities were also considered for example in Simeonov and Bainov [**44**], Singare and Pachpatte [**45**], Shastri and Kasture [**46**], Rab [**47**], Margolis [**48**], Lu and Wu [**49**], Lu [**50**], and Abramovich [**51**].

20. Further references may be found in [**52**]–[**106**].

REFERENCES

1. HILLE, E., "Lectures on ordinary differential equations," Reading, Mass., 1969.
2. REID, W., *Properties of solutions of an infinite system of ordinary linear differential equations of the first order with auxiliary boundary conditions*, Trans. Amer. Math. Soc. **32** (1930), 284–318.
3. SZARSKI, J., "Differential Inequalities," Warsaw, 1965.
4. LAKSHMIKANTHAM, V., *Upper and lower bounds of the norm of a solution of differential equations*, Proc. Amer. Math. Soc. **13** (1962), 615–616.
5. BEESACK, P. R., *Gronwall Inequalities*, Carleton Un. Math Notes No. **11**, (1975).
6. ČAPLYGIN, S.A., *Osnovanija novoga sposoba približennogo integrirovanija differencial'nyh uravnenii̋*, Moskva 1919. Sobr. Sočin. Vol. I. Moskva, 1948, 348–368.
7. __________, *Približennoe integrirovanie obyknovennogo differencial'nogo uravnenija pervogo porjadka*, Moskva, 1920. Sobr. Sočin. Vol. I. Moskva, 1948, 402–419.

8. PEANO, G. *Sull' integrabilità delle equazioni differenziali del primo ordine,* Atti. R. Accad. Sc. Torino **21** (1885-1886), 677–685. English translation in "Selected Works of Giuseppe Peano," pp. 51-57, translated and edited by H.C. Kennedy, Toronto, 1973.
9. PETROVITCH, M., *Sur une manière d'étendre le théorème de la moyenne aux équations différentielles du premier ordre,* Math. Annalen, **54** (1899), 417–436.
10. BERTOLINO, M. and D. TRIFUNOVIĆ, *Sur le théorème fondamental de S.A. Čaplygin sur l'inégalité différentielle du premier ordre,* Math. Balkanica **1** (1971), 11–18.
11. BERTOLINO, M., "Diferencijalne Jednačine," Beograd, 1980.
12. MAMEDOV, Ja.D., S. AŠIROV and S. ATDAEV, "Teoremy o Neravenstvah," Ašhabad, YLYM, 1980.
13. FLETT, T. M., "Differential analysis," Cambridge-London-New York-New Rochelle-Melbourne-Sydney, 1980.
14. PEROV, A.I., *K voprosu o strukture integral'noĭ voronki,* Nauč. Dokl. Vysšeĭ Školy. Ser FMN, 1959, No. 2.
15. ADAMOV, V.G. and T. SABIROV, *O differencial'nyh neravenstvah s proizvodnymi čislami,* Sb. Rabot. St. I Asp. Po Mat. Voronež. Voron. Un-t 1965.
16. BEREZIN, I.S. and N.P. ŽITKOV, "Metodi vyčisleniĭ II," Moskva, 1960.
17. RABCZUK, R., "Elementy nierówności vózniczkowych," Warszawa, 1976.
18. LUZIŃ, N.N., *O metode S.A. Čaplygina s analitičeskoi točki zrenija,* Sobr. sočin Vol. **3**, Moskva 1953, str. 146–167.
19. QUADE, W., *Ein neues Verfahren der schrittweisen Naherungen zur Lösung von* $y' = f(x, y)$, Math. Zeit. **48** (1942), 324–368.
20. ČAPLYGIN, S. A., "Novyĭ metod približennogo integrirovanija differencial'nyh uravneniĭ," Moskva-Leningrad, 1950.
21. KAŠČEEV, N.A., *K voprosu o granice primenimosti teoremy S.A. Čaplygina o diff. ner. k lineĭnnym diff. ur. vysših porjadkov,* Uč. zap. Kuib. gos. ped. in-ta 1956, vyp. 14, 53–56.
22. __________, *Točnaja granica primenimosti teoremi Čaplygina dlja lineĭnogo uravnenija,* DAN SSSR **111** (1956), No. 5, 937–940.
23. PÓLYA, G., *On the mean-value theorem corresponding to a given linear homogeneous differential equation,* Trans. Amer. Math. Soc. **24** (1922), 312–324.
24. MAMMANA, G., *Decompositione delle espressioni differenzialli omogene in prodotto di fattori symbolici e applicazione relativa allo studio delle equazioni differenzialli lineari,* Math. Zeit. **33** (1931).

25. KARLIN, S. and W.J. STUDDEN, "Tchebycheff systems: with applications in analysis and statistics," New York-London-Sydney, 1966.
26. PEČARIĆ, J.E., "Konveksne funkcije: Nejednakosti," Beograd, 1987.
27. BABKIN, B., *On a generalization of a theorem of Academician S.A. Čaplygin on a differential inequality* (Russian). Molotov. Gos. Univ. Uč. **8** (1953), 3–6.
28. ________, *Rešenie odnoǐ kraevoǐ zadači dlja obyknovennogo differencial'nogo uravnenija vtorogo porjadka metodom Čaplygina*, Prikladnaja Mat. I Meh. **18** (1954).
29. GRACE, S.R. and B.S. LALLI, *A comparison theorem for general nonlinear ordinary differential equations*, J. Math. Anal. Appl. **129** (1986), 39–43.
30. KIGURADZE, I.T., *Oscillation properties of certain ordinary differential equations* (Russian), Soviet Math. (Iz.VUZ) **3** (1962), 649-652.
31. ATKINSON, F. V., *On second order differential inequalities*, Proc. Roy. Soc. Edinburgh Sect. A **72** (1974), 109–127.
32. TEUFEL, M., *A note on second order differential inequalities and functional differential equations*, Pacific J. Math. **41** (1972), 537-541.
33. KARTSATOS, A.G., *On nth-order differential inequalities*, J. Math. Anal. Appl. **52** (1975), 1–9.
34. MULDOWNEY, J.S., *On an inequality of Peano*, Canad. Math. Bull. **16** (1) (1973), 79–81.
35. MULDOWNEY, J.S. and D. WILLETT, *An elementary proof of the existence of solutions to second order nonlinear boundary value problems*, SIAM J. Math. Anal. **5** (1979), 701–707.
36. ________, *An intermediate value property for operators with applications to integral and differential equations*, Can. J. Math. **26** (1979), 27–41.
37. GEL'FAND, A.V., *Približennoe integrirovanie sistemy obiknovennyh differencial'nyh uravneniǐ pervogo porjadka*, Izv. AN SSSR. Otd. M i EN, (**1938**), No. 5–6.
38. PEROV, A.I., *Neskoljko zamečanii otnostel'no differencial'nyh neravenstv*, Izv. Vuzov. Matematika, (**1965**), No. 4.
39. WAŻEWSKI, T., *Systémés des équations et des inegalites differentielles ordinaires aux deuxiémes membres monotones et leurs applications*, Ann. Sci. Polon. Math. **23** (1950), 112–166.
40. ________, *Certaines propositions de caractere "epidermique" relatives aux inégalitès differéntielles*, Ann. Sci. Polon. Math. **24** (1952), 1–12.
41. PACHPATTE, B.G., *On some fundamental integrodifferential and integral inequalities*, An. Ştinţ. Univ."Al. I. Cuza" Iaşi. Sect Ia Mat. (N.S.) **23** (1977), 77–86.

42. __________, *A note on some fundamental integro-differential inequalities*, Tamkang J. Math. **13** (1982), 63–67.
43. __________, *A note of second order integro-differential inequalities of the Gronwall-Bellman type*, Univ. Beograd. Publ. Elektrotehn. Fak. Ser. Mat. Fiz. No. **678–715** (1980), 35–41.
44. SIMEONOV, P.S. and D.D. BAĬNOV, *On some generalizations of the Gronwall-Bellman integral inequality for scalar functions of many variables*, Rend. Circ. Math. Palermo, II Ser. **32** (1983), 26–38.
45. SINGARE, V.M. and B.G. PACHPATTE, *On some discrete generalizations of an inequality of Gronwall*, Chinese J. Math. **6** (1978), 121–135.
46. SHASTRI, R.P. and D. Y. KASTURE, *Wendroff type inequalities*, Proc. Amer. Math. Soc. **72** (1978), 248–250.
47. RAB, M., *Linear integral inequalities*, Arch. Math. Brno **15** (1979), 37–46.
48. MARGOLIS, B., *An extension of Gronwall inequality*, Rev. Un. Math. Argentina **25** (1970/71), 247- -251.
49. LU, W. and Y. WU, *Integral inequalities of Gronwall type in n independent variables and their applications to partial differential equations*, J. Sichuan Univ. Nat. Sci. Ed. 1982, No. 1, 1–19 (Chinese).
50. LU, W., *Generalizations of integral inequalities of Gronwall-Wendroff type*, J. Sichuan Univ. Nat. Sci. Ed. 1989, No. 2, 12–29 (Chinese).
51. ABRAMOVICH, J., *On Gronwall and Wendroff type inequalities*, Proc. Amer. Math. Soc. **87** (1983), 481–486.48.
52. APARCIN, A.S. and Ten Men JAN, *Unimprovable estimates of the solutions of certain integral inequalities* (Russian), Sibirsk. Mat. Ž. **20** (1979), 192-195, 207.
53. AZBELEV, N.V., *O granicah primenimosti teoremi S.A. Čaplygina*, DAN SSSR, **89** (1953), No 4.
54. __________, *Nekotorye uslovija razrešimosti zadači Čaplygina dlja sistemy obyknovečnyh differencial'nyh uravneniĭ*, Naučn. dokl. vyssei školy. Ser. FMN, 1958, No 6.
55. AZBELEV, N.V. and Z.B. ÇALJUK, *O zadače Čaplygina*, Ukrain. mat. Ž. **10** (1958), No 1.
56. __________, *K voprosu ob integral'nyh neravenstvah*, Tr. Iževskogo mat. sem. Iževsk. vyp. 1, 1963.
57. __________, *K voprosu ob differencial'nom neravenstve*, Differencial'nye uravnenija 1, 1965, No 4.
58. __________, *Ob odnom metode ocenok rešenii uravnenii*, Volžskii mat. sb. 1966, vyp. **5**.

59. AZBELEV, N.V., A.Ja. HOHRJAKOV and Z.B. CALJUK, *Teoremy o differencial'nom neravenstve dlja kraevyh zadac*, Mat. sbornik **59** (101), 1962.
60. BABKIN, B. *K teoreme S.A. Čaplygina o differencial'nih neravenstvah*, Mat. sb. **46** (88), 1958, No 4.
61. BAIADA, E., *Confronto e dipendenze dai parametri degli integrali delle equazioni differenziali, I*, Rend. Accad. Lincei, (8) **3**, 258-263 (1974).
62. BALUEV, A.N., *O metode Čaplygina*, Vestnik Lening. Univ., **13**, (1956) vyp. 3.
63. BITTNER, L. *Die elementaren differential-und integralgleichungen mit einem allgemeinen Ungleichungsbegriff*, Math. Nach. **38** (1968), 1–17.
64. BRAUER, F., *Bounds for solutions of ordinary differential equations*, Proc. Amer. Math. Soc. **14** (1963), 538–545.
65. CAFIERO, F., *Su due teoremi di confronto relativi ad un'equazioni differenziale ordinaria del primo ordine*, Boll. Un. Mat. Ital. **3** (1948), 124–128.
66. CALJUK, Z.B., *Volterra integral equations* (Russian), Math. Analysis Vol. 15, Akad. Nauk SSSR Vsesojuz. Inst. Naučn. i Tehn. Informacii, Moscow, 1977.
67. ČANDIROV, G.I., *O suščestvovanii* (ε, B)*-rešenija smešannoi zadači dlja kvazilineĭnogo giperboličeskogo uravnenija*, Uč. zap. AGU, Ser. Fiz.-Mat.nauk, 1970, No 2.
68. ČAPLYGIN, S.A., *Približennoe integrirovanie sistemy dvuh differencial'nyh uravnenii*, Sobr. sočin. Vol. I. Moskva, 1948, 427–444.
69. CONLAN, J. and J. DIAZ, *Existence of solutions of an nth order hyperbolic partial differential equations*, Contrib. to Diff. Equations (1963), 277–289.
70. DAS, P. and R. SHARMA, *On optimal controls for measure delay differential equations*, SIAM J. Control. **9** (1971), 43-61.
71. GRAMMATIKOPOULIS, M.K., *On the effect of deviating argument on the behaviour of bounded solutions of nonlinear differential equations* (Russian), Ukrain. Mat. Ž. **30** (1978), 462–473.
72. GRONWALL, T., *On the existence and properties of the solutions of a certain differential equation of the second order*, Ann. Math. (2) **28** (1927), 355–364.
73. GUTOWSKI, R., *Nekotorye voprosy ustoĭčivosti differencial'nyh uravneniĭ v častnyh proizvodnyh, opisyvajuščih dviženie mehaniceskih sistem v teorii kolebanii*, VII über. nichtlineare Schwing. B.1. 1977, 289–305.

74. GUTOWSKI, R. and B. RADZISZEWSKI, *Asymptotic behaviour and properties of solutions of a system of non-linear second order ordinary differential equations describing motion of mechanical systems*, Arch. Mech. Stos. **22** (1970), 675–694.
75. _________, *Asymptotic behaviour and properties of solutions of nonlinear ordinary differential equations of first order describing the motion of a mechanical system*, Arch. Mech. Stos. **23** (1971), 17–25.
76. JACKSON, L. and K. SCHRADER, *On second order differential inequalities*, Proc. Amer. Math. Soc. **17** (1966), No. 5.
77. KNOBLOCH, H.W., *Second order differential inequalities and a nonlinear boundary value problem*, J. Different. Equat. **5** (1969), No. 1.
78. KRASNOSEL'SKIĬ, M.A. and Ja D. MAMEDOV, *Zamečanie o primenenii differencial'nyh i integral'nyh neravenstv v voprosah o korrektnosti zadači Koši dlja obyknovennyh differencial'nyh uravneniĭ v banahovyh prostranstvah*, Naučn. dokl. vysš. šk. ser. FNN, 1959, No. 2.
79. KRASNOSEL'SKIĬ, M.A., A. Ju. LEVIN and Ja.D. MAMEDOV, *Ob ocenkah rešeniĭ differencial'nyh uravnenii vtorogo porjadka*, Ukrain. mat. ž. **18** (1966), No. 1.
80. KURPEL', N.S. and V.I. OHRONCUK, *O differencial'nyh neravenstvah v banahavyh prostranstvah*, In: "Metody približennogo rešenija differenc. i integral'n. uravneniĭ," Kiev. Izd. AN USSR, 1973.
81. LUZIŃ, N.N., *O metode približennogo integrirovanija akademika S.A. Čaplygina*, Trudy CAGI, vyp. 141 (1932). Sobr. sočin.T.e, Moskva 1953, str. 145–180.
82. _________, *O metode približennogo integrirovanija akademika S.A. Čaplygina*, Uspehi mat. nauk. 6, vyp. **6** Sobr. sočin. T. e, Moskva 1953, str. 181–208.
83. MAMEDOV, Ja.D. and P.E. SOBOLEVSKIĬ, *Differencial'noe uravnenie s neograničennympreremennym operatorom v banahovom pronstranstve*, Tr. sem. po funkc. anal. Voronež. Voronežki in-t., 1963, vyp. 7.
84. MARTYNYUK, A.A. and R. GUTOVSKI, "Integral inequalities and stability of motion" (Russian), Kiev 1979, 272 pages.
85. MARTYNYUK, A.A., V. LAKSHMIKANTHAM, and S. LEELA, "Motion Stability, the Method of Integral Inequalities" (Russian), Kiev, 1989.
86. MLAK, W., *Differential inequalities in linear spaces*, Ann. Polon. Math. **5**, 1958, No. 1.
87. _________, *Differential inequalities with unbounded operators in Banach spaces*, Ann. Polon. Math. **9**, 1960, No. 1.
88. _________, *Note on abstract linear differential inequalities*, Rev. math. pures. appl. **6**, 1961, No. 4.

89. MULDOWNEY, I. S. *On an inequality of Čaplygin and Polya*, Proc. Roy. Irish Acad.

90. MYŠKIS, A.D., *Ob odnom differencial'no-funcional'nom neravenstve*, Uspehi mat. nauk **15**, 1960, vyp. 4 (94).

91. ________, *Žamečanije k stat'e P.M. Ždanova "O približennom rešenii sistemy differencial'nyh uravnenii pervogo porjadka s zapazdyvajuščim argumentom"*. Uspehi mat. nauk **16**, 1961, vyp. 2 (98).

92. OHRONCUK, V., *O differencial'nyh neravenstvah vtorogo porjadka v banahovom prostranstve*, Ukrain. mat. ž. **27**, 1975, No. 1.

93. ________, *O differencijal'nyh neravenstvah dlja nelineiyh differencijal'nyh uravnenija s zapazdyvajuščim argumentom*, Ukrain. mat. ž. **27** (1975), 256–262.

94. OPOĬCEV, V.I. and T. A. HURODZE, *Some theorems on differential and integral inequalities* (Russian), Sakharth. SSR Mecn. Akad. Moambe **87** (1977), 565–568.

95. PAK, S.A. *Teoremy o differencial'nyh neravenstvah dlja mnogotočečnyh kraevyh zadač*. In: "Priblizennye metody rešenija differencial'nyh uravnenii," Kiev, 1964, vyp. 2.

96. PETROV, B.N., *Granica primenimosti teoremi S.A. Čaplygina o differencial'nyh neravenstvah k lineinym uravnenijam s obyknovenym proizvodnymi vtorogo porjadka*, DAN SSSR, **51**, 1946, No. 4.

97. ________, *Neprimenimost' teoremy o differencial'nom neravenstve S.A. Čaplygina k nekotorym uravnenijam s obyknovennymi proizvodnym vtoroga porjadka*, DAN SSSR, **51**, 1946, No. 7.

98. RAHMATULLINA, L.F., *Ob odnom primenenii uslovii razresimosti zadači Čaplygina k voprosam ograničennosti i ustoičivosti rešenii differencial'nyh uravnenii*, Izv. VUZov Matematika, 1959, No. 2.

99. SADOVSKIĬ, B.N., *Nekotorye teoremy o differencial'nyh neravenstvah*, V. sb.: Problemy mat. analiza sloznih sistem. Voronež. Voronežkii Un-t, 1967, vyp. **1**.

100. SCHRADER, K., *A note on second order differential inequalities*, Proc. Amer. Math. Soc.**19** (1968), 1007–1012.

101. SLUGIN, S.N., Približennoe resenie operatornyh uravnenii na osnove metoda S.A. Čaplygina. DAN SSSR, **103**, 1955, No. 4.

102. SUGUJAMA, Shohei, *On comparison theorems of nonlinear Volterra integral equations*, Kodai Math. Semin. Repts. **27** (1976), 147–154.

103. VED', Ju.A., *O vozmuščenijah lineinyh odnorodnyh differencial'nyh uravnenii s preremennymi koefficientami*. In: "*Issledovanija po integrodifferencial'nym uravnenijam v Kirgizii*," vyp. 3, Frunze, "Ilim", 1965.

104. VENCKOVA, M., *On the boundedness of solutions of higher order differential equations*, Arch. Math. 4, Scripta Fac. Sci. Nat. Ujep Brunensis **13** (1977), 235–242.

105. WESTPHAL, H., *Zur Abschätzung der Lösungen nichtlinearer parabplischer Differentialgleichungen.* Math. Zeit. **51** (1949), 690–695.

106. YUE-SHENG, L., *The bound, stability and error estimates for the solution of nonlinear differential equations*, Chinese Math. Acta **3** (1963), 34-41.

CHAPTER XII

INEQUALITIES OF GRONWALL TYPE OF A SINGLE VARIABLE

1. T. Gronwall [**1**] in 1919 proved the following result:

THEOREM 1. *Let $x(t)$ be a continuous real valued function on $J = [\alpha, \alpha + h]$ and let*

$$0 \leq x(t) \leq \int_\alpha^t (a + bx(s))\,ds \quad \text{on } J, \tag{1.1}$$

where a and b are nonnegative constants. Then $x(t) \leq ah\exp(bh)$ on J.

Another inequality of this type was proved in 1943 by R. Bellman [**2**].

THEOREM 2. *Let $x(t)$ and $k(t)$ be real valued nonnegative continuous functions for $t \geq \alpha$. If a is a constant, $a \geq 0$, and*

$$x(t) \leq a + \int_\alpha^t k(s)x(s)ds, \quad t \geq \alpha,$$

then

$$x(t) \leq a\exp\left(\int_\alpha^t k(s)ds\right), \quad \text{for } t \geq \alpha. \tag{1.2}$$

It is clear that Bellman's result contains that of Gronwall. Inequalities of this type were called "Gronwall inequalities" or "Inequalities of Gronwall type" (see Beckenbach and Bellman [**3**], Mitrinović [**4**], Reid [**5**], Hille [**6**],[**7**], Lakshmikantham and Lella [**8**], Walter [**9**]), "Bellman's lemma or inequality" (see Bihari [**10**], Viswanatham [**11**], Győri [**12**]) or sometimes "Bellman-Gronwall" or "Gronwall-Bellman" inequalities (see for example Ziebur [**13**], Lakshmikantham [**14**], Lees [**15**]).

Moreover, the essential idea for such inequalities (together with their applications to differential equations) dates as do related differential inequalities to 1885. For example, Peano [**16**] proved a special case of (1.1) for $a = 0$. In 1930 E. Kamke [**17**] proved and made use of an inequality which is essentially that of Gronwall. Similarly, Reid [**18**] before Bellman's result, (see also [**5**]) proved a result which is essentially the same as that of Bellman except that the hypothesis on k was relaxed to be Lebesgue integrable.

2. These results are related to those of Theorems 1 and 2 of 2. in Chapter XI for linear equations. In fact Bellman's inequality (1.2) implies the corresponding result, Theorem 2 in 2. of Chapter XI, when f is linear in y. The converse implication is also true. For example, consider the inequality

$$x(t) \leq a + \int_{\alpha}^{t} k(s)x(s)ds.$$

Define the function $R(t)$ by the right hand side of this inequality. Then $R(t)$ satisfies the differential inequality

$$R'(t) = k(t)x(t) \leq k(t)R(t), \quad R(\alpha) = a.$$

By Theorem 2 of 2. in Chapter XI $R(t) \leq y(t)$ where y is a solution to the differential system $y' = ky$, $y(\alpha) = a$. This gives Bellman's inequality (without a or x being nonnegative).

3. Inequalities of this type, which give explicit bounds for functions which satisfy a Volterra integral type inequality, have numerous applications to boundedness, uniqueness, and stability theorems in differential equations and to Volterra type integral equations. Some of these applications appear in Hille [**6**], Reid [**5**], Filatov and Šarova [**19**], Martynjuk and Gutovski [**20**], and Demidovič [**21**].

Both of the Theorems cited above have hypotheses which are not necessary. In this section we give various generalizations of Gronwall's inequality involving an unknown function of a single variable. Later we will consider other generalizations.

A. Filatov [**22**] proved the following linear generalization of Gronwall's inequality.

THEOREM 1. *Let $x(t)$ be a continuous nonnegative function such that*

$$x(t) \leq a + \int_{t_0}^{t} (b + cx(s))\, ds, \quad \text{on } t \geq t_0,$$

where $a \geq 0$, $b \geq 0$, $c > 0$. Then for $t \geq t_0$, $x(t)$ satisfies

$$x(t) \leq \left(\frac{b}{c}\right)(\exp(c(t-t_0)) - 1) + a \exp c(t-t_0).$$

REMARK 1: As the above proof for Bellman's inequality shows (modified appropriately here), the hypotheses that x, a, and b are positive are irrelevant.

K. V. Zadiraka [**23**] (see also Filatov and Šarova [**14**, p. 15]) proved the following:

THEOREM 4. *Let the continuous function $x(t)$ satisfy*

$$|x(t)| \leq |x(t_0)| \exp(-\alpha(t-t_0)) + \int_{t_0}^{t} (a|x(s)| + b)\, e^{-\alpha(t-s)} ds,$$

where a, b, and α are positive constants. Then

$$|x(t)| \leq |x(t_0)| \exp(-\alpha(t-t_0)) + b(\alpha - a)^{-1}(1 - \exp(-(\alpha - a)(t-t_0))).$$

In the book [**24**] R. Bellman cites the following result (see also Filatov and Šarova [**14**, pp. 10–11]):

THEOREM 3. *Let $x(t)$ be a continuous function that satisifies*

$$x(t) \leq x(\tau) + \int_{\tau}^{t} a(s)x(s)ds,$$

for all t and τ in (a, b) where $a(t) \geq 0$ and continuous. Then

$$x(t_0) \exp\left(-\int_{t_0}^{t} a(s)ds\right) \leq x(t) \leq x(t_0) \exp\left(\int_{t_0}^{t} a(s)ds\right)$$

for all $t \geq t_0$.

The following two theorems were given in the book of Filatov and Šarova [**14**, pp. 8–9 and 18–20] and are due to G. I. Čandirov [**25**]:

THEOREM 4. *Let $x(t)$ be continuous and nonnegative on $[0,h]$ and satisfy*

$$x(t) \le a(t) + \int_0^t (a_1(s)x(s) + b(s))\,ds,$$

where $a_1(t)$ and $b(t)$ are nonnegative integrable functions on the same interval with $a(t)$ bounded there. Then on $[0,h]$

$$x(t) \le \int_0^t b(s)ds + \sup_{0\le t\le h} |a(t)| \exp\left(\int_0^t a_1(s)ds\right).$$

THEOREM 5. *Let $x(t)$ be a nonnegative continuous function on $[0,\infty)$ such that*

$$x(t) \le ct^\alpha + mt^\beta \int_0^t \frac{x(s)}{s} ds,$$

where $c > 0$, $\alpha \ge 0$, $\beta \ge 0$. Then

$$x(t) \le ct^\alpha \left(1 + \sum_{n=1}^{\infty} \frac{m^n t^{n\beta}}{\alpha(\alpha+\beta) + \cdots + (\alpha + (n-1)\beta)}\right).$$

4. A more general result was given in Willett [**26**] and Harlamov [**27**]. Here we shall give an extended version due to Beesack [**28**, pp. 3–4].

THEOREM 1. *Let x and k be continuous and a and b Riemann integrable functions on $J = [\alpha,\beta]$ with b and k nonnegative on J.*

(i) *If*

$$x(t) \le a(t) + b(t) \int_\alpha^t k(s)x(s)ds, \quad t \in J, \tag{4.1}$$

then

$$x(t) \le a(t) + b(t) \int_\alpha^t a(s)k(s) \exp\left(\int_s^t b(r)k(r)dr\right) ds, \quad t \in J. \tag{4.1)'$$

Moreover, equality holds in (4.1)′ for a subinterval $J_1 = [\alpha,\beta_1]$ of J if equality holds in (4.1) for $t \in J_1$.

(ii) *The result remains valid if $\le$ is replaced by $\ge$ in both (4.1) and (4.1)′.*

(iii) *Both (i) and (ii) remain valid if $\int_\alpha^t$ is replaced by $\int_t^\beta$ and $\int_s^t$ by $\int_t^s$ throuhgout.*

PROOF: This proof is typical of those for inequalities of the Gronwall type, we set

$$U(t) = \int_{\alpha}^{t} k(s)x(s)ds \text{ so that } U(\alpha) = 0,$$

and

$$U'(t) = k(t)x(t).$$

Hence $U'(s) \le a(s)k(s) + b(s)k(s)U(s)$. Multiplying by the integrating factor $\exp\left(\int_s^t b(r)k(r)dr\right)$ and integrating from α to t gives

$$(4.1)'' \qquad U(t) \le \int_{\alpha}^{t} a(s)k(s)\exp\left(\int_s^t b(r)k(r)dr\right) ds, \quad t \in J.$$

Since $b \ge 0$ substitution of (4.1)$''$ into (4.1) gives (4.1)$'$. The equality conditions are obvious and the proof of (ii) is similar or can be done by the change of variables $t \to -t$.

REMARK 1: Observe that the stated hypotheses assure the existence of all the integrals that appear. In case the integrals are interpreted to be Lebesgue integrals (Willett [**26**]) the hypotheses can be relaxed to: x, a, b, k are measurable with kx, ka, kb being integrable. The equality and inequality conditions then are to be interpreted as a.e. as is usual. In this sense the stated conditions for equality are necessary and sufficient. Similar remarks apply to the other inequalities considered in this Chapter.

REMARK 2: B. Pachpatte [**29**] proved an analogous result on R^+ and $(-\infty, 0]$.

REMARK 3: Willett's paper [**26**] also contains a linear generalization in which $b(t)k(s)$ is replaced by $\sum_{i=1}^{n} b_i(t)k_i(s)$. Such a result is also given by Š. T. Gamidov [**30**] (see also Filatov and Šarova [**19**, pp. 19–20]). Moreover this result can be derived from the case of $n = 1$ by observing that $\sum b_i k_i < \sup_{1\le i\le n}\{b_i\} \sum k_i$ (note that the functions are positive).

5. Š. T. Gamidov [**30**] also gave the following theorem:

THEOREM 1. *If*

$$x(t) \le a(t) + a_1(t)\int_{t_1}^{t} b_1(s)x(s)ds + a_2(t)\sum_{i=2}^{n} c_i \int_{t_1}^{t_i} b_i(s)x(s)ds,$$

for $t \in [a,b]$, where $a = t_0 < \cdots < t_n = b$, c_i are constants and the functions appearing are all real, continuous and nonnegative, and if

$$\sum_{i=2}^{n} c_i \int_{t_1}^{t_i} b_i(t)\left[a_2(t) + a_1(t)\int_{t_1}^{t} b_1(s)a_2(s)\exp\left(\int_{t_1}^{t} a_1(r)b_1(r)dr\right)ds\right]dt < 1,$$

then

$$x(t) \le K_1(t) + MK_2(t),$$

where

$$K_1(t) = a(t) + a_1(t)\int_{t_1}^{t} b_i(s)a(s)\exp\left(\int_{s}^{t} a_1(r)b_1(r)dr\right)ds,$$

$$K_2(t) = a_2(t) + a_1(t)\int_{t_1}^{t} b_1(s)a_2(s)\exp\left(\int_{s}^{t} a_1(r)b_1(r)dr\right)ds,$$

and

$$M = \left(\sum_{i=2}^{n} c_i \int_{t_1}^{t_i} b_i(s)K_1(s)ds\right)\left(1 - \sum_{i=2}^{n} c_i \int_{t_1}^{t_i} b_i(s)K_2(s)ds\right)^{-1}.$$

6. H. Movljankulov and A. Filatov [**31**] proved the following result:

THEOREM 1. *Let $x(t)$ be real, continuous, and nonnegative such that for $t > t_0$*

$$x(t) \le c + \int_{t_0}^{t} k(t,s)x(s)ds, \quad c > 0,$$

where $k(t,s)$ is a continuously differentiable function in t and continuous in s with $k(t,s) \ge 0$ for $t \ge s \ge t_0$. Then

$$x(t) \le c\exp\left(\int_{t_0}^{t}\left(k(s,s) + \int_{t_0}^{s}\frac{\partial k}{\partial s}(s,r)dr\right)ds\right).$$

In the same paper the following result appears:

THEOREM 2. *Let $x(t)$ be real, continuous, and nonnegative on $[c, d]$ such that*

$$x(t) \leq a(t) + b(t) \int_0^t k(t,s)x(s)ds,$$

where $a(t) \geq 0$, $b(t) \geq 0$, $k(t,s) \geq 0$ and are continuous functions for $c \leq s \leq t \leq d$. Then

$$x(t) \leq A(t) \exp\left(B(t) \int_c^t K(t,s)ds \right),$$

where $A(t) = \sup_{c \leq s \leq t} a(s)$, $B(t) = \sup_{c \leq s \leq t} b(s)$, $K(t,s) = \sup_{s \leq \sigma \leq t} k(\sigma, s)$.

REMARK 1: The inequality

$$x(t) \leq a(t) + \int_{t_0}^t (a_1(t)a_2(s)x(s) + b(t,s))\, ds$$

was considered by S.M. Sardar'ly [**32**]. Using the substitution $a(t) + \int_{t_0}^t b(t,s)$ is $a(t)$ then we see that this is the same as (4.1) so that this result follows from Theorem 1 of 4.

7. S. Chu and F. Metcalf [**33**] proved the following linear generalization of Gronwall's inequality:

THEOREM 1. *Let x and a be real continuous functions on $J = [\alpha, \beta]$ and let k be a continuous nonnegative function on $T : \alpha \leq s \leq t \leq \beta$. If*

$$x(t) \leq a(t) + \int_\alpha^t k(t,s)x(s)ds, \quad t \in J, \tag{7.1}$$

then

$$x(t) \leq a(t) + \int_\alpha^t R(t,s)a(s)ds, \quad t \in J, \tag{7.1$'$}$$

where $R(t,s) = \sum_{i=1}^{\infty} K_i(t,s)$, with $(t,s) \in T$, is the resolvent kernel of $k(t,s)$ and $K_i(t,s)$ are iterated kernels of $k(t,s)$.

REMARK 1: P. Beesack [**34**] extended this result for the case when $x, a \in L^2(J)$ and $k \in L^2(T)$, and he noted that the result remains valid if $\geq$ is substituted for $\leq$ in both (7.1) and (7.1)$'$.

REMARK 2: If we put $k(t,s) = b(t)k(s)$ and $k(t,s) = \sum_{i=1}^{n} b_i(t)k_i(s)$ we get the results of D. Willett [**26**].

8. G. Jones [**35**] extended Willett's result in the case of Riemann-Stieltjes integrals. This result is related to discrete results so it will be given in Chapter XIV. For some analogous results see Wright, Klasi, and Kennebeck [**36**], Schmaedeke and Sell [**37**], Herod [**38**], and B. Helton [**39**],[**40**].

9. We can consider various nonlinear generalizations of Gronwall's inequality. The following Theorem is proved in Perov [**41**]:

THEOREM 1. *Let $u(t)$ be a nonnegative function that satisfies the integral inequality*

$$u(t) \le c + \int_{t_0}^{t} (a(s)u(s) + b(s)u^{\alpha}(s))\,ds, \quad c \ge 0, \quad \alpha \ge 0, \tag{9.1}$$

where $a(t)$ and $b(t)$ are continuous nonnegative functions for $t \ge t_0$.
For $0 \le \alpha < 1$ we have

(9.1′)

$$u(t) \le \left\{ c^{1-\alpha} \exp\left[(1-\alpha)\int_{t_0}^{t} a(s)ds\right] + (1-\alpha)\int_{t_0}^{t} b(s)\exp\left[(1-\alpha)\int_{s}^{t} a(r)dr\right] ds \right\}^{\frac{1}{1-\alpha}};$$

for $\alpha = 1$,

$$u(t) \le c\exp\left\{\int_{t_0}^{t} [a(s)+b(s)]\,ds\right\}; \tag{9.1″}$$

and for $\alpha > 1$ with the additional hypothesis

$$c < \left\{\exp\left[(1-\alpha)\int_{t_0}^{t_0+h} a(s)ds\right]\right\}^{\frac{1}{\alpha-1}} \left\{(\alpha-1)\int_{t_0}^{t_0+h} b(s)ds\right\}^{-\frac{1}{\alpha-1}} \tag{9.2}$$

we also get for $t_0 \le t \le t_0 + h$, *for* $h > 0$,

(9.1‴)

$$u(t) \le c \left\{ \exp\left[(1-\alpha)\int_{t_0}^{t} a(s)ds\right] - c^{-1}(\alpha-1)\int_{t_0}^{t} b(s)\exp\left[(1-\alpha)\int_{s}^{t} a(r)dr\right] ds \right\}^{\frac{1}{\alpha-1}}.$$

PROOF: For $\alpha = 1$ we get the usual linear inequality so that (9.1′) is valid. Assume now that $0 < \alpha < 1$. Denote by v a solution of the integral equation

$$v(t) = c + \int_{t_0}^{t} a(s)v(s) + b(s)v^{\alpha}(s)ds, \quad t \ge t_0.$$

In differential form this is the Bernoulli equation

$$v'(t) = a(t)v(t) + b(t)v^{\alpha}(t), \quad v(0) = c.$$

This is linear in the variable $v^{1-\alpha}$ so can readily be integrated to produce $v(t) =$ the right hand side of (9.1′). Since $u(t) \le v(t)$ we have completed the proof of (9.1′).

For $\alpha > 1$ we again get a Bernouilli equation and an analogous proof where we need the extra condition (9.2) if this condition is to hold on the bounded interval $t_0 \le t \le t_0 + h$.

REMARK 1: Inequality (9.1) is also considered in Willett [**42**] and Willett and Wong [**43**]. For a related result, see Ho [**44**].

10. The following theorem is a modified version of a theorem proved in Gamidov [**30**]:

THEOREM 1. *If*

$$u(t) \le f(t) + c\int_{0}^{t} \phi(s)u^{\alpha}(s)ds,$$

where all functions are continuous and nonnegative on $[0, h]$, $0 < \alpha < 1$, $c \ge 0$, *then*

$$u(t) \le f(t) + c\xi_0^{\alpha}\left(\int_{0}^{t} \phi^{\frac{1}{1-\alpha}}(s)ds\right)^{1-\alpha}$$

where ξ_0 *is the unique root of* $\xi = a + b\xi^{\alpha}$.

11. Gamidov [**30**] also proved the following result:

THEOREM 1. *If*

$$u(t) \leq c_1 + c_2 \int_0^t \phi(s)u^\alpha(s)ds + c_3 \int_0^h \phi(s)u^\alpha(s)ds,$$

$c_1 \geq 0$, $c_2 \geq 0$, $c_3 > 0$, *and the functions* $u(t)$ *and* $\phi(t)$ *are continuous and nonnegative on* $[0, h]$, *then for* $0 < \alpha < 1$ *we have*

$$u(t) \leq \left(\xi_0^{1-\alpha} + c_2(1-\alpha) \int_0^t \phi(s)ds \right)^{\frac{1}{1-\alpha}},$$

where ξ_0 *is the unique root of the equation*

$$\left[\frac{c_2 + c_3}{c_3} \xi - \frac{c_1 c_2}{c_3} \right]^{1-\alpha} - \xi^{1-\alpha} - c_2(1-\alpha) \int_0^h \phi(s)ds = 0.$$

If $\alpha > 1$ *and* $c_2(\alpha - 1) \int_0^t \phi(s)ds < c_1^{1-\alpha}$, *there exists an interval* $[0, \delta] \subset [0, h]$ *where*

$$u(t) \leq \left(c_1^{1-\alpha} - c_2(\alpha - 1) \int_0^t \phi(s)ds \right)^{\frac{1}{1-\alpha}}.$$

A related result was proved by B. Stachurska **[45]**:

THEOREM 2. *Let the functions* x, a, b, *and* k *be continuous and nonnegative on* $J = [\alpha, \beta]$, *and* n *be a positive integer* ($n \geq 2$) *and* a/b *be a nondecreasing function. If*

$$x(t) \leq a(t) + b(t) \int_\alpha^t k(s)x^n(s)ds, \quad t \in J, \tag{11.1}$$

then

$$x(t) \leq a(t) \left\{ 1 - (n-1) \int_\alpha^t k(s)b(s)a^{n-1}(s)ds \right\}^{\frac{1}{1-n}}, \quad \alpha \leq t \leq \beta_n, \tag{11.1'}$$

where $\beta_n = \sup\left\{t \in J: (n-1)\int\limits_{\alpha}^{t} kba^{n-1}ds < 1\right\}$.

REMARK 1: The inequality (11.1) was considered by P. Maroni [**46**], but without the assumption of the monotonicity of the ratio a/b. He obtained two estimates, one for $n = 2$ and another for $n \geq 3$. Both are more complicated than (11.1′). For $n = 2$ and a/b nondecreasing, Starchurska's result can be better than Maroni's on longer intervals.

12. One of the more important nonlinear generalizations of Gronwall's inequality is the well-known one of Bihari [**10**].

This result was proved seven years earlier by J. P. Lasalle [**47**].

THEOREM 1. *Let $u(t)$ and $k(t)$ be positive continuous functions on $[c, d]$ and let a and b be nonnegative constants. Further let $g(z)$ be a positive nondecreasing function for $z \geq 0$. If*

$$u(t) \leq a + b\int_{c}^{t} k(s)g\left(u(s)\right)ds, \quad t \in [c, d],$$

then

$$u(t) \leq G^{-1}\left(G(a) + b\int_{c}^{t} k(s)ds\right), \quad c \leq t \leq d_1 \leq d,$$

where $G(\lambda) = \int\limits_{\xi}^{\lambda} ds/g(s)$ $(\xi > 0,\ \lambda > 0)$ and d_1 is defined such that $G(a) + b\int\limits_{c}^{t} k(s)ds$ belongs to the domain of G^{-1} for $t \in [c, d_1]$.

13. The following generalization of the Bihari-Lasalle inequality was given by I. Győri [**12**]:

THEOREM 1. *Suppose that $u(t)$ and $\beta(t)$ are continuous and nonnegative on $[t_0, \infty)$. Let $f(t)$, $g(u)$, and $\alpha(t)$ be differentiable functions with f nonnegative, g positive and nondecreasing, and α nonegative and nonincreasing. Suppose that*

$$u(t) \leq f(t) + \alpha(t)\int_{t_0}^{t} \beta(s)g\left(u(s)\right)ds. \tag{13.1}$$

If

$$f'(t)\left\{\frac{1}{g(\eta(t))}-1\right\}\leq 0 \quad \text{on } [t_0,\infty) \tag{13.2}$$

for every nonnegative continuous function η, then

$$u(t)\leq G^{-1}\left\{G(f(t_0))+\int_{t_0}^{t}[\alpha(s)\beta(s)+f'(s)]\,ds\right\} \tag{13.1$'$}$$

where

$$G(\delta)=\int_{\varepsilon}^{\delta}\frac{ds}{g(s)},\quad \varepsilon>0,\ \delta>0, \tag{13.3}$$

and (13.1′) *holds for all values of t for which the function*

$$\delta(t)=G[f(t_0)]+\int_{t_0}^{t}[\alpha(s)\beta(s)+f'(s)]\,ds$$

belongs to the domain of the inverse function G^{-1}.

PROOF: Let $V(t)=f(t)+\int_{t_0}^{t}\alpha(s)\beta(s)g[u(s)]\,ds$. Since g is nondecreasing and α is nonincreasing we get from (13.1) that $g(u(t))\leq g(V(t))$. From this we obtain

$$f'(t)+\alpha(t)\beta(t)g[u(t)]\leq\alpha(t)\beta(t)g[V(t)]+f'(t)$$

which may be written as

$$\frac{V'(t)}{g[V(t)]}\leq\alpha(t)\beta(t)+\frac{f'(t)}{g[V(t)]}.$$

Using (13.2) we get

$$\frac{V'(t)}{g[V(t)]}\leq\alpha(t)\beta(t)+f'(t).$$

Upon integration we get

$$G[V(t)]\leq G[f(t_0)]+\int_{t_0}^{t}[\alpha(s)\beta(s)+f'(s)]\,ds.$$

If we suppose that $\delta(t)$ is in the domain of G^{-1} then we get the result (13.1′) since $u(t)\leq V(t)$.

REMARK 1: Note that the special case $\alpha(t)\equiv 1$ already implies the general result since we substitute for β the product $\alpha\beta$ and observe that $\alpha(t)\beta(s)\leq\alpha(s)\beta(s)$ for $s\leq t$.

14. K. Ahmedov, A. Jakubov and A. Veisov [**48**] proved the following theorem:

THEOREM 1. *Let $u(t)$ be a continuous function on $[t_0, T]$ such that*

$$0 \le u(t) \le f(t) + \int_{t_0}^{\phi(t)} k(t,s) g\left(u(s)\right) ds,$$

where

1) $f(t)$ *is continuous, nonnegative, and nonincreasing;*
2) $\phi(t)$ *is differentiable,* $\phi'(t) \ge 0$, $\phi(t) \le t$, $\phi(t_0) = t_0$;
3) $g(u)$ *is positive and nondecreasing on* R; *and*
4) $k(t,s)$ *is nonnegative and continuous on* $[t_0, T] \times [t_0, T]$ *with* $\frac{\partial k}{\partial t}(t,s)$ *nonnegative and continuous.*

Then for G *defined by (13.3) we have*

$$u(t) \le f(t) - f(t_0) + G^{-1}\left\{ G\left(f(t_0)\right) + \int_{t_0}^{t} F(s) ds \right\},$$

where

$$F(t) = k\left(t, \phi(t)\right) \phi'(t) + \int_{t_0}^{\phi(t)} \frac{\partial k}{\partial t}(t,s) ds.$$

15. C. E. Langenhop [49] proved the following result:

THEOREM 1. *Let the functions $u(t)$ and $\alpha(t)$ be nonnegative and continuous on $[a, b]$, $g(u)$ be positive and nondecreasing for $u > 0$, and suppose that for every y in $[a, x]$*

$$u(y) \le u(x) + \int_{y}^{x} \alpha(r) g\left(u(r)\right) dr.$$

Then for every x in $[a, b]$ we have

$$u(x) \ge G^{-1}\left\{ G\left(u(a)\right) - \int_{a}^{x} \alpha(r) dr \right\},$$

where G is defined by (13.3) and we assume that the term in the $\{\ \}$ is in the domain of G^{-1}.

PROOF: We define $R(y) = \int_y^x \alpha(r) g\left(u(r)\right) dr$ so that the inequality becomes

$$u(y) \le u(x) + R(y).$$

Since g is nondecreasing we have

$$g[u(y)] \leq g[u(x) + R(y)]$$

which may be written as

$$\frac{d(u(x) + R(y))}{g[u(x) + R(y)]} \geq -\alpha(y)dy.$$

An integration from y to x $(y \leq x)$ yields

$$-G(u(x) + R(y)) + G(u(x)) \geq -\int_y^x \alpha(s)ds.$$

But G is nondecreasing so $G(u(y)) \leq G(u(x) + R(y))$. Combining the last two inequalities and rearranging we arrive at

$$G(u(x)) > G(u(y)) - \int_y^x \alpha(s)ds.$$

Applying G^{-1} we get the result when y is set to a.

16. The following result is proved in Ahmedov, Jakubov, and Veisov [48].

THEOREM 1. *Suppose that*

$$u(t) < f(t) + \sum_{i=1}^{n} \alpha_i(t) \int_{t_0}^{t} \beta_i(s) g[u(s)]\, ds, \quad t \geq t_0,$$

where $u(t)$, $f(t)$, and $\beta_i(t)$ are positive and continuous on $[t_0, \infty)$, $\alpha_i(t) > 0$ while $\alpha_i'(t) \geq 0$; g is a nondecreasing function that satisfies $g(z) \geq z$ for $z > 0$. Then

$$u(t) \leq f(t) - \bar{f} + G^{-1}\left[G(\bar{f}) + \log \prod_{i=1}^{n} \alpha_i(t) - \ln \prod_{i=1}^{n} \alpha_i(t_0) + \int_{t_0}^{t} \sum_{i=1}^{n} \alpha_i(s)\beta_i(s)ds\right]$$

where $\bar{f} = \max f(t)$ and G is defined by (13.3).

THEOREM 2. *Let the positive continuous function $u(t)$ satisfy*

$$u(t) \leq f(t) + \int_{t_0}^{\phi_1(t)} a_1(s)F_1(u(s))\,ds + \int_{t_0}^{\phi_2(t)} a_2(s)F_2(u(s))\,ds,$$

with the following conditions:

1) $f(t)$ *is a nonincreasing function on* $[t_0, T]$;
2) *the functions* $a_1(t)$ *and* $a_2(t)$ *are continuous and nonnegative on* $[t_0, T]$;
3) $\phi_1(t)$, *and* $\phi_2(t)$ *are continuously differentiable and nondecreasing functions with* $\phi_i(t_0) = t_0$, $i = 1, 2$, *and* $\phi_1(t) \leq t$;
4) *the functions* $F_1(z)$ *and* $F_2(z)$ *are continuous, nondecreasing and satisfy* $F_2(z) > 0$ *for all* z *and*

$$\frac{d}{dz}\left[\frac{F_1(z)}{F_2(z)}\right] = \frac{C}{F_2(z)}$$

for some constant C.

Then

$$u(t) \leq f(t) - f(t_0) + z(t),$$

where

$$G(z) = \frac{F_1(z)}{F_2(z)} = G(z_0) + \int_{z_0}^{z} \frac{C}{F_2(s)}ds,$$

G^{-1} *is its inverse function and*

$$z(t) = G^{-1}\left\{\exp\left(C\int_{t_0}^{\phi_1(t)} a_1(s)ds\right)\left[G(f(t_0))\right.\right.$$
$$\left.\left.+C\int_{t_0}^{t} a_2(\phi_2(s))\,\phi_2'(s)\exp\left(-C\int_{t_0}^{\phi_1(s)} a_1(r)dr\right)ds\right]\right\}$$

is a continuous solution of the initial value problem

$$z'(t) = a_1(\phi_1(t))\,\phi_1'(t)F_1(z) + a_2(\phi_2(t))\,\phi_2'(t)F_2(z),$$
$$z(t_0) = f(t_0).$$

THEOREM 3. *Let the continuous function $u(t)$ satisfy*

$$u(t) \le f(t) + \sum_{i=1}^{n} a_i(t) \int_{t_0}^{\phi_i(t)} b_i(s) f(u(s))\, ds \text{ on } [t_0, T] \text{ with}$$

1) *the $a_i(t)$ bounded nonnegative nonincreasing functions;*
2) *the $b_i(t)$ continuous nonnegative functions;*
3) *the $\phi_i(t)$ continuous with $\phi_i'(t) > 0$, $\phi_i(t) \le t$, $\phi_i(t_0) = t_0$;*
4) *$f(t)$ a continuous nonincreasing function;*
5) *$g(z)$ a nondecreasing positive function defined on R.*

Then

$$u(t) \le f(t) - f(t_0) + z(t)$$

where G is defined by (13.3) and

$$z(t) = G^{-1}\left[G(f(t_0)) + \sum_{i=1}^{n} \int_{t_0}^{\phi_i(t)} a_i(s) b_i(s) ds\right]$$

is a continuous solution of the initial value problem $z(t_0) = f(t_0)$ and

$$z'(t) = \sum_{i=1}^{n} a_i(\phi_i(t))\, b_i(\phi_i(t))\, \phi_i'(t) g(z).$$

17. The following result appears in S. S. Dragomir [**50**].

THEOREM 1. *Let the functions $A, B\colon [\alpha, \beta] \to R^+$, $L\colon [\alpha, \beta] \times R^+ \to R^+$ be continuous and*

(17.1) $$0 \le L(t, u) - L(t, v) \le M(t, v)(u - v), \ t \in [\alpha, \beta], \ u \ge v \ge 0,$$

where M is a nonnegative continuous function on $[\alpha, \beta) \times R^+$. Then for every nonnegative continuous solution of the integral inequality

(17.1') $$x(t) \le A(t) + B(t) \int_{\alpha}^{t} L(s, x(s))\, ds, \quad t \in [\alpha, \beta),$$

we also have on $[\alpha, \beta)$

(17.1'') $$x(t) \le A(t) + B(t) \int_{\alpha}^{t} L(u, A(u)) \exp\left(\int_{u}^{t} M(s, A(s))\, B(s) ds\right) du.$$

PROOF: Let $y(t) = \int_{\alpha}^{t} L(s, x(s))\, ds$. Then $y'(t) = L(t, x(t))$ and $y(\alpha) = 0$. Using the integral inequality that x satisfies and then condition (17.1) we have

$$y'(t) \le L(t, A(t) + B(t)y(t)) \le L(t, A(t)) + M(t, A(t))\, B(t)y(t)$$

which gives the above estimate.

Some related results are given in Pečarić and Dragomir [**51**].

18. In the previous results we have a nonlinearity in the unknown function only under the integral sign. In the next few results we allow the nonlinearity to appear everywhere.

THEOREM 1. *Let the positive functions $u(t)$, $a(t)$, and $b(t)$ be bounded on $[c, d]$; $k(t, s)$ be a bounded nonnegative function for $c \le s \le t \le d$; $u(t)$ is a measurable function and $k(., t)$ is a measurable function. Suppose that $f(u)$ is strictly increasing and $g(u)$ is nondecreasing. If $A(t) = \sup_{c \le s \le t} a(s)$, $B(t) = \sup_{c \le s \le t} b(s)$, and $K(t, s) = \sup_{s \le \sigma \le t} k(\sigma, s)$, then from*

$$f(u(t)) \le a(t) + b(t) \int_{c}^{t} k(t, s) g(u(s))\, ds, \quad t \in [c, d],$$

follows

$$u(t) \le f^{-1}\left[G^{-1}\left\{G(A(t)) + B(t) \int_{c}^{t} K(t, s) ds\right\}\right] \quad t \in [c, d'],$$

where $G(u) = \int_{\xi}^{u} dw/g\left(f^{-1}(w)\right)$ $(\xi > 0,\ u > 0)$ and

$$d' = \max\left[c \le r \le d\colon G\left(A(r) + B(r) \int_{c}^{r} K(r, s) ds\right) \le G(f(\infty))\right].$$

This theorem was proved in Butler and Rogers [**52**] and the following one in Györi [**12**]:

THEOREM 2. *Suppose*

$$F(u(t)) \leq f(t) + \alpha(t) \int_{t_0}^{t} \beta(s) g[u(s)]\, ds,$$

where the functions $f(t)$, $\alpha(t)$, $\beta(t)$, and $g(u)$ satisfy the conditions of Theorem 1 *in* 13., *and the function $F(u)$ is monotone decreasing and positive for $u > 0$. Then on $[t_0, d']$*

$$u(t) \leq F^{-1}\left\{G^{-1}\left[G(f(t_0)) + \int_{t_0}^{t} [\alpha(s)\beta(s) + f'(s)]\, ds\right]\right\}$$

where $G(z) = \int_{\varepsilon}^{t} ds/g\left[F^{-1}(s)\right]$, $z > \varepsilon \geq 0$ and d' is defined such that the function $\delta(t)$ defined in Theorem 1 *of* 13. *belongs to the domain of definition of the function $F^{-1} \circ G^{-1}$.*

19. The next result allows the integral to appear in the nonlinearity. It is due to Willett and Wong [**43**].

THEOREM 1. *Let the functions x, a, b, and k be continuous and nonnegative on $J = [\alpha, \beta]$, $1 \leq p < \infty$, and*

$$x(t) \leq a(t) + b(t)\left(\int_{\alpha}^{t} k(s) x^p(s) ds\right)^{1/p}, \qquad t \in J.$$

Then

$$x(t) \leq a(t) + b(t)\frac{\left(\int_{\alpha}^{t} k(s)e(s)a^p(s)ds\right)^{1/p}}{1 - [1 - e(t)]^{1/p}}, \qquad t \in J,$$

where $e(t) = \exp\left(-\int_{\alpha}^{t} k(s)b^p(s)ds\right)$.

Gollwitzer [**53**] replaces x and x^p by g and g^p respectively.

20. Generalizations of this result were given in Beesack [**28**, pp. 20-30]. Here we shall give some results obtained in Filatov and Šarova [**19**, pp. 34-37] and from Deo and Murdeshwar [**54**].

THEOREM 1. *Suppose*

1) $u(t)$, $f(t)$ *and* $F(t,s)$ *are positive continuous functions on* R, *and* $s \le t$;
2) $\partial F(t,s)/\partial t$ *is nonnegative and continuous;*
3) $g(u)$ *is positive, continuous, additive, and nondecreasing on* $(0,\infty)$;
4) $h(z) > 0$ *and is nondecreasing and continuous on* $(0,\infty)$.

If

$$u(t) \le f(t) + h\left(\int_0^t F(t,s)g(u(s))\,ds\right),$$

then, for $t \in I$, *we have*

$$u(t) \le f(t) + h\left\{G^{-1}\left[G\left(\int_0^t F(t,s)g(f(s))\,ds\right) + \int_0^t \phi(s)ds\right]\right\},$$

where $G(u) = \int_\varepsilon^u ds/g(h(s))$, $u > 0$, $\varepsilon > 0$,

$$\phi(t) = F(t,t) + \int_0^t \frac{\partial F}{\partial t}(t,s)ds,$$

and

$$I = \left\{t \in (0,\infty)\colon G(\infty) \ge G\left(\int_0^t F(t,s)g(f(s))\,ds\right) + \int_0^t \phi(s)ds\right\}.$$

PROOF: Using the additivity of the function g and that nondecreasing nature of $F(t,s)$ in t, we have

$$u(t) - f(t) \le h(v(t)),$$

where

$$v(t) = \int_0^t F(t,s)g(u(s) - f(s))\,ds + \int_0^T F(T,s)g(f(s))\,ds,$$

$t \in (0,T)$ and $T \in (0,\infty)$. Since g is nondecreasing we find that

$$g(u(t) - f(t)) \le g(h(v(t))). \tag{20.1}$$

Multiplying this inequality by $\partial F(t,s)/\partial t$ and integrating from 0 to t we arrive at

$$\int_0^t \frac{\partial F}{\partial t}(t,s)g\left(u(s)-f(s)\right)ds \le \int_0^t \frac{\partial F}{\partial t}(t,s)g\left(h\left(v(s)\right)\right)ds.$$

On the other hand if we multiply (20.1) by $F(t,t)$ and using this last inequality we get

$$v'(t) \le F(t,t)g\left(h\left(v(t)\right)\right) + \int_0^t \frac{\partial F(t,s)}{\partial t}g\left(h\left(v(s)\right)\right)ds,$$

i.e.

$$\frac{d}{dt}G\left(v(t)\right) \le F(t,t) + \int_0^t \frac{\partial F(t,s)}{\partial t}ds.$$

Now by integrating from 0 to T we get

$$G\left(v(T)\right) - G\left(v(0)\right) \le \int_0^T \phi(t)dt,$$

and since $u(T)-f(T) < h\left(v(T)\right)$ we have

$$u(T)-f(T) \le h\left\{G^{-1}\left(G\left[\int_0^T F(T,s)g\left(f(s)\right)ds\right] + \int_0^T \phi(t)dt\right)\right\}.$$

Since T is arbitrary we have the result.

A simple consequence of this result is:

THEOREM 2. *Suppose that $J=(0,\infty)$ and*

1) *$u(t)$, $f(t)$, and $F(t)$ are positive and continuous on J;*
2) *$g(u)$ is positive, continuous, additive and nondecreasing on J;*
3) *$h(z)>0$, nondecreasing and continuous.*

If

$$u(t) \le f(t) + h\left(\int_0^t F(s)g\left(u(s)\right)ds\right), \quad t \in J,$$

then for $t \in J_1$, we have

$$u(t) \le f(t) + h\left\{G^{-1}\left[G\left(\int_0^t F(s)g\left(f(s)\right)ds\right) + \int_0^t F(s)ds\right]\right\},$$

where G is defined as in Theorem 1 *and*

$$J_1 = \left\{ t \in J \colon G(\infty) \geq G\left(\int_0^t F(s) g\left(f(s)\right) ds \right) + \int_0^t F(s) ds \right\}.$$

Deo and Murdeshwar [**54**] also proved the following theorem:

THEOREM 3. *Let the conditions* 1), 2), *and* 3) *of the previous theorem hold and let $g(u)$ be an even function on R. If*

$$u(t) \geq f(t) - h\left(\int_0^t F(s) g\left(u(s)\right) ds \right), \quad t \in (0, \infty),$$

then for $t \in J_1$ and G as defined in Theorem 2 *we have*

$$u(t) \geq f(t) - h\left\{ G^{-1} \left[G\left(\int_0^t F(s) g\left(f(s)\right) ds \right) + F(s) ds \right] \right\}.$$

21. Further generalizations of this result are given in Deo and Dhongade [**55**], [**56**] and Beesack [**28**, pp. 65-86]. Bessack has also given corrections of some results from Deo and Dhongade [**55**],[**56**]. Here we give only one result of P.R. Beesack (this result for $f(x) = x$, $h(u) = u$, and $a(t) = a$ becomes Theorem 2 in Deo and Dhongade).

THEOREM 1. *Let x, a, k, and k_1 be nonnegative continuous functions on $J = [\alpha, \beta)$, and let $a(t)$ be nondecreasing on J. Let g and h be continuous nondecreasing functions on $[0, \infty)$ such that g is positive, subadditive and submultiplicative on $[0, \infty)$, and $h(u) > 0$ for $u > 0$. Suppose f is a continuous strictly increasing function on $[0, \infty)$ with $f(u) \geq u$ for $u \geq 0$ and $f(0) = 0$. If*

$$f\left(x(t)\right) \leq a(t) + h\left(\int_\alpha^t k(s) g\left(x(s)\right) ds \right) + \int_\alpha^t k_1(s) x(s) ds, \quad t \in J,$$

then

$$x(t) \leq \left(f^{-1} \circ F^{-1}\right) \left\{ \int_\alpha^t k_1(s) ds + F\left[a(t) + \left(h \circ G^{-1}\right) \left\{ \int_\alpha^t k(s) g\left(E(s)\right) ds \right.\right.\right.$$
$$\left.\left.\left. + G\left(\int_\alpha^t k(s) g\left(a(s) E(s)\right) ds \right) \right\} \right] \right\}, \quad \text{for } \alpha \leq t \leq \beta_1,$$

where

$$E(t) = \exp\left(\int_{\alpha}^{t} k_1 ds\right), \quad F(u) = \int_{y_0}^{u} dy/f^{-1}(y), \quad y > 0, \quad (y_0 > 0)$$

and

$$G(u) = \int_{u_0}^{u} dy/g(h(y)), \quad u > 0, \quad (u_0 > 0)$$

with

$$\beta_1 = \sup\left\{t \in J: G\left(\int_{\alpha}^{t} k(s)g(a(s)E(s))\,ds\right) + \int_{\alpha}^{t} k(s)g(E(s))\,ds \in G(R^+)\right\}.$$

If $a(t) = a$ then we may omit the requirement that g be subadditive and then for $\alpha \leq t \leq \beta_2$ we have

$$x(t) \leq (f^{-1} \circ F^{-1})\left\{\int_{\alpha}^{t} k_1 ds + F\left[a + (h \circ G_a^{-1})\left(\int_{\alpha}^{t} k(s)g(E(s))\,ds\right)\right]\right\},$$

where $G_a(u) = \int_0^u dy/g(|a + h(y)|)$, $u > 0$ and

$$\beta_2 = \sup\left\{t \in J: \int_{\alpha}^{t} k(s)g(E(s))\,ds \in G_a(R^+)\right\}.$$

22. In previous sections we have given explicit estimates for unknown functions which satisfy integral inequalities. Several of these results may be given in terms of solutions of some differential or integral equation and are similar to results for differential inequalities exposed in Chapter XI. The first one we give is due to B. N. Babkin [**57**].

THEOREM 1. *Let $\phi(t,u)$ be continuous and nondecreasing in u on $[0,T] \times (-\delta,\delta)$ with $\delta \leq \infty$. If $v(t)$ is continuous and satisfies*

$$v(t) \leq u_0 + \int_0^T \phi(t, v(s))\,ds,$$

where u_0 is a constant, then

$$v(t) \leq u(t)$$

where $u(t)$ is the maximal solution of the problem

$$u'(t) = \phi(t, u), \quad u(0) = u_0$$

defined on $[0, T]$.

PROOF: Introduce the function $w(t) = u_0 + \int_0^t \phi(s, v(s))\, ds$. Then $v(t) \leq w(t)$ and

$$w'(t) = \phi(t, v(t)) \leq \phi(t, w(t)), \quad \text{with } w(0) = u_0.$$

By Theorem 2 from 2. of Chapter XI we have $w(t) \leq u(t)$ so that the Theorem is proved.

23. The following two theorems are generalizations of the preceding result.

THEOREM 1. *Let $\phi(t, s, u)$ be continuous and nondecreasing in u for $0 \leq t, s \leq T$ and $|u| \leq \delta$. Let $u_0(t)$ be a continuous function on $[0, T]$. If $v(t)$ is a continuous function that satisfies the integral inequality (on $[0, T]$)*

$$v(t) < u_0(t) + \int_0^t \phi(t, s, v(s))\, ds \tag{23.1}$$

then

$$v(t) < u(t) \quad \text{on } [0, T], \tag{23.1'}$$

where $u(t)$ is a solution of the equation

$$u(t) = u_0(t) + \int_0^t \phi(t, s, u(s))\, ds \quad \text{on } [0, T]. \tag{23.2}$$

PROOF: From (23.1) and (23.2) we get that (23.1′) is valid at $t = 0$. By continuity of the functions involved we get (23.1′) holding on some nontrivial interval. If the result does not hold on $[0, T]$ then there is a t_0 such that $v(t) < u(t)$ on $[0, t_0)$ but $v(t_0) = u(t_0)$. From (23.1) and (23.2) we have

$$\begin{aligned} v(t_0) &< u_0(t_0) + \int_0^{t_0} \phi(t_0, s, v(s))\, ds \\ &\leq u_0(t_0) + \int_0^{t_0} \phi(t_0, s, u(s))\, ds = u(t_0). \end{aligned}$$

This contradiction proves the theorem.

We shall say that the function $\phi(t,s,u)$ satisfies the condition (μ) if the equation

$$W(t) = u_0(t) + \lambda + \int_0^t \phi(t,s,W(s))\,ds$$

has a solution defined on $[0,T]$ for every constant $\lambda \in [0,\mu]$.

THEOREM 2. *Let $\phi(t,s,u)$ be defined for $0 \le t,s \le T$, $|u| < \delta$, and be continuous and nondecreasing in u and satisfying conditon (μ). If the continuous function $v(t)$ satisfies*

$$v(t) \le u_0(t) + \int_0^t \phi(t,s,v(s))\,ds \tag{23.3}$$

on $[0,T]$, then

$$v(t) \le u(t)$$

on $[0,T]$ where $u(t)$ satisfies (23.2) on the same interval.

PROOF: For every fixed n we denote by $W_n(t)$ a solution of the integral equation

$$W_n(t) = \frac{\varepsilon}{n} + u_0(t) + \int_0^t \phi(t,s,W_n(s))\,ds$$

defined on $[0,T]$. For ε sufficiently small we may employ Theorem 1 to conclude that

$$u(t) < W_{n+1}(t) < W_n(t) < W_1(t)$$

as well as $v(t) < W_n(t)$. Letting n tend to ∞ we get the result.

REMARK 1: It can be shown (see Mamedov, Aširov, and Atdaev [**58**, pp. 96–98]) that the condition of monotonicity of the function ϕ in u is sufficient for the validity of the theorem on integral inequalities but is not necessary.

24. A generalization of this result has a Fredholm term.

THEOREM 1. *Let the functions $\phi(t,s,u)$ be continuous and nondecreasing in u for $0 \le t,s \le T$, $|u| < \delta$. Suppose that $u_0(t)$ is continuous on $[0,T]$ and either*

a) *for every fixed continuous function $W_0(t)$ with values in $|u| < \delta$ on $[0,T]$ and every sufficiently small positive number λ we have*

$$W(t) = u_0(t) + \lambda + \int_0^t \phi_1(t,s,W(s))\,ds + \int_0^T \phi(t,s,W_0(s))\,ds$$

has a continuous solution on $[0,T]$; *or*

b) $|u(t)| + \int_0^T \max_{0\le t\le T} |\phi_1(t,s,\delta) + \phi_2(t,s,\delta)| ds \le \delta.$

If a continuous function $v(t)$ *satisfies*

$$v(t) \le u_0(t) + \int_0^t \phi_1(t,s,v(s))\,ds + \int_0^T \phi_2(t,s,v(s))\,ds,$$

then

$$v(t) \le u(t) \quad \text{on } [0,T],,$$

where

$$u(t) = u_0(t) + \int_0^t \phi_1(t,s,u(s))\,ds + \int_0^T \phi_2(t,s,u(s))\,ds.$$

REMARK 1: The special cases when $\phi_1 \equiv 0$ and $\phi_2 \equiv 0$ in the previous theorem gives integral inequalities of Fredholm and Volterra respectively.

REMARK 2: The previous theorem is given in Mamedov, Aširov and Atdaev [58]. This book also contains the following result of Ja. D. Mamedov [**59**].

THEOREM 2. *Let the function* $\phi(t,s,u)$ *be continuous in* t *(in* $[0,\infty)$*) for almost all* $s \in [0,\infty)$ *and* u *with* $|u| < \infty$. *Suppose that for fixed* t *and every continuous function* $u(s)$ *on* $[0,\infty)$ *the function* $\phi(t,s,u(s))$ *is mesurable in* s *on* $[0,\infty)$. *Further let* ϕ *be nondecreasing in* u *and* $u_0(t)$ *be a continuous function on* $[0,\infty)$

.

If the continuous function $v(t)$ *satisfies*

$$v(t) < u_0(t) + \int_t^\infty \phi(t,s,v(s))\,ds \quad \text{on } [0,\infty), \tag{24.1}$$

then

$$v(t) < u(t) \quad \text{on } [0,\infty) \tag{24.1'}$$

where $u(t)$ *is a solution of*

$$u(t) = u_0(t) + \int_t^\infty \phi(t,s,u(s))\,ds \quad \text{on } [0,\infty).$$

REMARK 3: Mamedov [59] (see also Mamedov, Aširov, and Atdaev [58]) also considered (24.1) with "$\leq$" instead of " $<$ ", as well as the inequality

$$v(t) \leq u_0(t) + \int_0^t \phi_1(t, s, v(s))\, ds + \int_t^\infty \phi_2(t, s, u(s))\, ds.$$

25. Theorem 1 of 22. was generalized in another way by V. Lakshmikantham [14]. His results were extended by P. R. Beesack [28]. For this result we shall use the following special notation and terminology. If the function $F_1(t, s, u)$ is defined on $I_1 \times J_1 \times K_1$ where each of the symbols represents an interval and if for every $(s, u) \in J_1 \times K_1$ the function $f(t) = F_1(t, s, u)$ is monotone on I_1, we say that $F_1(\cdot, s, u)$ is monotone. We say that F_1 and F_2 are monotone in the same sense of they are both nondecreasing or both nonincreasing, and monotone in the opposite sense if one is nondecreasing and the other nonincreasing.

THEOREM 1. *Let $f(x)$ be continuous and strictly monotone on an interval I, and let $H(t, v)$ be continuous on $J \times K$ where $J = [\alpha, \beta]$ and K is an interval containing zero and H monotonic in v. Let $T_1 = \{(t, s) \colon \alpha \leq s \leq t \leq \beta\}$ and assume that $W(t, s, u)$ is continuous and of one sign on $T_1 \times I$, monotone in u, and monotone in t. Suppose also that the functions x and a are continuous on J with $x(J) \subset I$ and*

$$a(t) + H(t, v) \in f(I) \text{ for } t \in J \text{ and } |v| \leq b, \tag{25.1}$$

for some constant $b > 0$. Let

$$f(x(t)) \leq a(t) + H\left(t, \int_\alpha^t W(t, s, x(s))\, ds\right), \quad t \in J, \tag{25.2}$$

and let $r = r(t, T, \alpha)$ be the maximal (minimal) solution of the system

$$\begin{aligned} &r' = W\left(T, t, f^{-1}\left[a(t) + H(t, r)\right]\right), \\ &r(\alpha) = 0, \quad \alpha \leq t \leq T \leq \beta_1 \ (\beta_1 < \beta), \end{aligned} \tag{25.3}$$

if $W(t, s, .)$ and f are monotonic in the same (opposite) sense, where $\beta_1 > a$ is chosen so that the maximal (minimal) solution exists for the indicated interval. Then if $W(., s, u)$ and $H(t, .)$ are monotonic in the same sense,

$$x(t) \leq \ (\geq)\ f^{-1}\left[a(t) + H(t, \tilde{r}(t))\right], \quad \alpha \leq t \leq \beta_1, \tag{25.2'}$$

where $\tilde{r}(t) = r(t,t,\alpha)$ if

i) $H(t,.)$ and $W(t,s,.)$ *are monotonic in the same sense and* f *is increasing; if*

ii) f *is decreasing and* $H(t,.)$ *and* $W(t,s,.)$ *are monotonic in the opposite sense (and the second reading of the other hypotheses), then the inequality is reversed in* (25.2′).

PROOF: The function $F(T,t,r) = W\left(T,t,f^{-1}\left[a(t)+H(t,r)\right]\right)$ is by (25.1) continuous on the compact set $T_1\times[-b,b]$, so it is bounded there, say by the constant M. It follows from Lemma 1 in 2. of Chapter XI that there exists, independent of T, a $\alpha<\beta$ (in fact $\beta_1>\alpha+\min(\beta-\alpha, bM^{-1})$) such that the maximal (minimal) solution of the system (25.3) exists on $[\alpha,\beta]$. Now fix $T\in(\alpha,\beta]$ and let $t\in[\alpha,T]$. Then define $v(t,u)=\int_\alpha^t W\left(u,s,x(s)\right)ds$ and we have

$$(25.4)\qquad v(t,t)=\int_\alpha^t W\left(t,s,x(s)\right)ds \leq\ (\geq)\ \int_\alpha^t W\left(T,s,x(s)\right)ds = v(t,T)$$

if $W(.,s,u)$ is increasing (decreasing). Observe that (25.2) implies that $v(t,t)\in K$ for $t\in J$. Since K is an interval containing zero it follows that $v(t,T)\in K$ in both cases of (25.4) regardless of the sign of W.

By (25.2) we obtain

$$(25.5)\qquad x(t)\leq\ (\geq)\ f^{-1}\left[a(t)+H\left(t,v(t,t)\right)\right],$$

if f is increasing (decreasing). Since $v'(t,T)=W\left(T,t,x(t)\right)$ for $\alpha\leq t\leq T\leq\beta$, we have

$$(25.6)\qquad v'(t,T)\leq\ (\geq)\ W\left(T,t,f^{-1}\left[a(t)+H\left(t,v(t,t)\right)\right]\right),$$

if the functions $W(t,s,.)$ and f are monotonic is the same (opposite) sense. On the other hand by (25.4) we have

$$(25.4')\qquad H\left(t,v(t,t)\right)\leq\ (\geq)\ H\left(t,v(t,T)\right),\quad \alpha\leq t\leq T,$$

if (a) $W(t,s,.)$ and $H(t,.)$ are monotonic in the same ((b) opposite) sense. Thus

$$f^{-1}\left[a(t)+H\left(t,v(t,t)\right)\right]\leq\ (\geq)\ f^{-1}\left[a(t)+H\left(t,v(t,T)\right)\right],$$

on $\alpha\leq t\leq T$ if (a′): f is increasing and (a) or f is decreasing and (b) ((b′) f is increasing and (b) or f is decreasing and (a)). This in turn implies that

$$(25.7)\qquad \begin{aligned} &W\left(T,t,f^{-1}\left[a(t)+H\left(t,v(t,t)\right)\right]\right)\\ &\qquad\leq\ (\geq) W\left(T,t,f^{-1}\left[a(t)+H\left(t,v(t,T)\right)\right]\right),\end{aligned}$$

if (a''): $W(t,s,.)$ is increasing and (a') or $W(t,s,.)$ is decreasing and (b') ((b''): $W(t,s,.)$ is increasing and (b') or $W(t,s,.)$ is decreasing and (a')). Combining this with (25.6) we see that if $W(t,s,.)$ and $H(t,.)$ are monotonic in the same sense then

$$(25.8) \quad v'(t,T) \leq \ (\geq)\ W\left(T,t,f^{-1}\left[a(t)+H\left(t,v(t,T)\right)\right]\right), \quad \alpha \leq t \leq T < \beta,$$

if $W(t,s,.)$ and f are monotonic in the same (opposite) sense.

Since $v(\alpha,T)=0$, Theorem 2 from 2. of Chapter XI shows that if $W(t,s,.)$ and $H(t,.)$ are monotonic in the same sense and if $r(t,T,\alpha)$ is the maximal or the minimal solution of (25.3) as specified, then

$$v(t,T) \leq \ (\geq)\ r(t,T,\alpha) \quad \text{for } \alpha \leq t \leq T \leq \beta_1$$

from which we get in particular that this holds when $t=T$. Since T is an arbitrary element of $(\alpha,\beta_1]$ we have

$$(25.9) \qquad v(t,t) \leq \ (\geq)\ \tilde{r}(t) \quad \text{on } [\alpha,\beta_1]$$

provided that (A): $W(t,s,.)$ and f are monotonic in the same sense ((B): $W(t,s,.)$ and f are montonic in the opposite sense).

As in the analysis of (25.4) and (25.4') we have on $[\alpha,\beta_1]$

$$H\left(t,v(t,t)\right) \leq (\geq)\ H\left(t,\tilde{r}(t)\right)$$

if (A'): $H(t,.)$ is increasing and (A) or $H(t,.)$ is decreasing and (B), ((B'): $H(t,.)$ is increasing and (B) or $H(t,.)$ is decreasing and (A)). Now if (A''): f is increasing and (A') or f is decreasing and (B') ((B''): f is increasing and (B') or f is decreasing and (A')) then

$$f^{-1}\left[a(t)+H\left(t,v(t,t)\right)\right] \leq (\geq)\ f^{-1}\left[a(t)+H\left(t,\tilde{r}(t)\right)\right].$$

Analyzing the various cases we see that if $W(.,s,u)$ and $H(t,.)$ are monotonic in the same sense and $W(t,s,.)$ and $H(t,.)$ are monotonic in the same (opposite) sense then

$$(25.10) \qquad f^{-1}\left[a(t)+H\left(t,v(t,t)\right)\right] \leq \ (\geq) f^{-1}\left[a(t)+H\left(t,\tilde{r}(t)\right)\right]$$

on $[\alpha,\beta_1]$. The conclusion (25.2') now follows in cases (i) or (ii) from (25.5) and (25.1).

In the same way one can prove:

THEOREM 2. *Under the hypotheses of* Theorem 1 *suppose that*

$$f(x(t)) \geq a(t) + H\left(t, \int_\alpha^t W(t, s, x(s))\, ds\right), \quad t \in J.$$

and that $W(., s, u)$ and $H(t, .)$ are monotonic in the opposite sense. Let $\tilde{r}(t) = r(t, t, \alpha)$ where $r(t, T, \alpha)$ is the maximal (minimal) solution of problem (25.3) and suppose that $W(t, s, .)$ and f are monotonic in the opposite (same) sense. Then

$$x(t) \geq (\leq)\ f^{-1}[a(t) + H(t, r(t))] \quad \text{on } [\alpha, \beta_1]$$

provided that conditions (i) or (ii) of Theorem 1 hold.

REMARK 1: Similar results with $\int_t^\beta$ instead of $\int_\alpha^t$ can be obtained from the previous theorems, see P. R. Beesack [**28**, p. 52].

REMARK 2: In case $K = [0, t_0]$, then $W > 0$ holds (since $v(t,t) \in K$), so in the condition (25.1) $|v| \leq b$ can be changed to $0 \leq v \leq b$.

REMARK 3: The assumption that W has one sign can be replaced by the condition that $v(t, T) \in K$ for $\alpha \leq t \leq T \leq \beta$.

26. The following theorem is a consequence of Theorem 1 of 25. (Lakshmikantham [**14**, Th.3.1 (ii)]):

THEOREM 1. *Let x, a, b, and c be continuous nonnegative functions on $J = [\alpha, \beta]$ and f and h be continuous nonegative functions on R^+, with f strictly increasing and h nondecreasing. In addition, suppose that $k(t, s)$ is continuous and nonnegative on $T = \{(t, s) \colon \alpha \leq s \leq t \leq \beta\}$, and $w(t, u)$ is continuous and nonnegative on $J \times R^+$, with $w(t, .)$ nondecreasing on R^+. Define $C(t) = \max_{\alpha \leq s \leq t} c(s)$, and $K(t, s) = \max_{s \leq \sigma \leq t} k(\sigma, s)$, for $\alpha \leq s \leq t \leq \beta$. If*

$$f(x(t)) \leq a(t) + b(t) h\left(c(t) + \int_\alpha^t k(t, s) w(s, x(s))\, ds\right), \quad t \in J, \tag{26.1}$$

then

$$x(t) \leq f^{-1}[a(t) + b(t) h(\tilde{r}_1(t, C(t)))], \quad t \in J_0, \tag{26.1'}$$

with $\tilde{r}_1(t, c(\alpha)) = r(t, t, c(\alpha))$, where $r = r(t, \beta, c(\alpha))$ is the maximal solution on $J = [\alpha, \beta_0]$ of

$$r' = K(\beta_0, t) w\left(t, f^{-1}[a(t) + b(t) h(r)]\right), \ r(\alpha) = c(\alpha).$$

27. P. R. Beesack [**2**, pp. 56–65] showed that a sequence of well-known results can be simply obtained by using the previous results. Here we shall give one example. We consider the following inequality of Gollwitzer [**53**]:

$$x(t) \le a + g^{-1}\left(\int_\alpha^t k(s)g\left(x(s)\right)ds\right), \quad t \in J = [\alpha, \beta].$$

We use Theorem 1 of 25. with: $f(x) = x$, $H(t,v) = g^{-1}(v)$, $W(t,s,u) = k(s)g(u)$, $K = g(I)$ and I is an interval such that $x(J) \subset I$. The comparison equation is

$$r' = k(t)g\left(a + g^{-1}(r)\right), \quad r(\alpha) = 0. \tag{27.1}$$

By Theorem 1 of 25. we have

$$x(t) \le a + g^{-1}\left(r(t)\right), \quad \alpha \le t \le \beta_1, \tag{27.2}$$

where $r(t)$ is the unique solution of problem (27.1) on $[\alpha, \beta_1]$. If we define G by

$$G(u) = \int_0^u dr/g\left[a + g^{-1}(r)\right], \quad u \in g(I) = K,$$

then by (27.2) we obtain

$$x(t) \le a + g^{-1}\left[G^{-1}\left(\int_\alpha^t kds\right)\right], \quad \alpha \le t \le \beta, \tag{27.2'}$$

where $\beta_1 = \sup\left\{t\colon \int_\alpha^t kds \in G(K)\right\}$.

In fact, Beesack showed that this result is better than that of Gollwitzer so he formulated the following more general result:

Let x and k be continuous functions with $k \ge 0$ on $J = [\alpha, \beta]$, and let g be a continuous and monotonic on an interval I such that $x(J) \subset I$ and g is nonzero on I except perhaps at an endpoint of I. Let h be continuous and monotone on an interval K such that $0 \in K$, and let a and b be constants such that $a + h(v) \in I^0$ for $v \in K$, $|v| \le b$.

If g and h are monotone in the same sense and

$$x(t) \le a + h\left(\int_\alpha^t k(s)g\left(x(s)\right)ds\right), \quad t \in J,$$

then

$$x(t) \le a + h\left[G^{-1}\left(\int_\alpha^t k\,ds\right)\right], \quad \alpha \le t \le \beta_1,$$

where

$$G(u) = \int_0^u dr/g\,[(a + h(r)], \quad u \in K, \text{ and}$$

$$\beta_1 = \sup\left\{t \in J:\ \int_\alpha^t k\,ds \in G(K)\right\}.$$

28. It is interesting that there exist a sequence of integral inequalities with an unknown function of one variable in which we have several integrals.

B. Pachpatte [**60**] proved the following result:

THEOREM 1. *Let $x(t)$, $a(t)$, $b(t)$, $c(t)$, and $d(t)$ be real, nonnegative, and continuous functions defined on R^+ such that for $t \in R^+$,*

$$x(t) \le a(t) + b(t)\left(\int_0^t c(s)x(s)ds + \int_0^t c(s)b(s)\left(\int_0^s d(u)x(u)du\right)ds\right).$$

Then on the same interval we have

$$x(t) \le a(t) + b(t)\left(\int_0^t c(s)\left(a(s) + b(s)\exp\left(-\int_0^s b(r)\left(c(r) + d(r)\right)dr\right)\right.\right.$$

$$\left.\left.\times \int_0^s \left(a(r)\left(c(r) + d(r)\right)\exp\left(-\int_0^t \left(b(u)\left(c(u) + d(u)\right)du\right)dr\right)ds\right)\right).$$

29. The following three theorems from Bykov and Salpagarov [**61**] are given in the book Filatov and Šarova [**19**].

THEOREM 1. *Let $u(t)$, $v(t)$, $h(t,r)$, and $H(t,r,x)$ be nonnegative functions for $t \ge r \ge x \ge a$ and c_1, c_2, and c_3 be nonnegative constants not all zero. If*

$$u(t) \le c_1 + c_2\int_a^t \left[v(s)u(s) + \int_a^s h(s,r)u(r)dr\right]ds \tag{29.1}$$

$$+ c_3\int_a^t\int_a^r\int_a^s H(s,r,x)u(x)dsdrds,$$

then for $t \geq a$
(29.1′)

$$u(x) \leq c_1 \exp\left\{c_2 \int_a^t \left[v(s) + \int_a^s h(s,r)dr\right] ds + c_3 \int_a^t \int_a^s \int_a^r H(s,r,x)dxdrds\right\}.$$

PROOF: Let the right hand side of (29.1) be denoted by $b(t)$. Then $b(s) \leq b(t)$ for $s \leq t$ since all the terms are nonnegative. We have

$$\frac{b'(t)}{b(t)} = c_2 v(t)\frac{u(t)}{b(t)} + c_2 \int_a^t \frac{h(t,r)u(r)}{b(t)}dr$$

$$+ c_3 \int_a^t \int_a^r \frac{H(t,r,x)u(x)}{b(t)}dxdr$$

$$\leq c_2 v(t) + c_2 \int_a^t h(t,r)dr + c_3 \int_a^t \int_a^r H(t,r,x)dxdr.$$

Integration from a to t yields

$$\log b(t) - \log c_1 \leq c_2 \int_0^b \left[v(s) + \int_a^s h(s,r)dr\right] ds$$

$$+ c_3 \int_a^t \int_a^s \int_a^r H(s,r,x)dxdrds.$$

Writing this in terms of $b(t)$ and using $u(t) \leq b(t)$ completes the proof.

THEOREM 2. *Let the nonnegative function* $u(t)$ *defined on* $[t_0, \infty)$ *satisfy the inequality*

$$u(t) \leq c + \int_{t_0}^t k(t,s)u(s)ds + \int_{t_0}^t \int_{t_0}^s G(t,s,\sigma)\, u(\sigma)d\sigma ds,$$

where $k(t,s)$ *and* $G(t,s,\sigma)$ *are continuously differentiable nonnegative functions for* $t \geq s \geq \sigma \geq t_0$, *and* $c > 0$. *Then*

$$u(t) \leq c \exp\left\{\int_{t_0}^t \left[k(s,s) + \int_{t_0}^s \left(\frac{\partial k(s,\sigma)}{\partial s} + G(s,s,\sigma)\right) d\sigma\right.\right.$$

$$\left.\left. + \int_{t_0}^s \int_{t_0}^\sigma \frac{\partial G(s,\sigma,r)}{\partial s} dr d\sigma\right] ds\right\}.$$

THEOREM 3. *Let the functions $u(t)$, $\sigma(t)$, $v(t)$, and $w(t,r)$ be nonnegative and continuous for $a \le r \le t$, and let c_1, c_2, and c_3 be nonnegative. If for $t \in I = [a,\infty)$*

$$u(t) \le c_1 + \sigma(t)\left\{c_2 + c_3\int_a^t\left[v(s)u(s) + \int_a^s w(s,r)u(r)dr\right]ds\right\},$$

then for $t \in I$,

$$u(t) \le c_1 + \sigma(t)\left\{c_2 \exp\left[c_3\int_a^t\left(v(s)\sigma(s) + \int_a^s w(s,r)\sigma(r)dr\right)ds\right]\right.$$
$$+ c_1c_3\int_a^t\left(v(s) + \int_a^s w(s,r)dr\right)$$
$$\left.\times \exp\left[c_3\int_s^t\left(v(\delta)\sigma(\delta) + \int_a^\delta w(\delta,r)\sigma(r)dr\right)d\delta\right]ds\right\}.$$

30. Other related results exist. Here we shall mention one which appears in E. H. Yang [**62**].

THEOREM 1. *Let $x(t)$ be continuous and nonnegative on $I = [0,h)$ and let $p(t)$ be continuous, positive, and nondecreasing on I. Suppose that $f_i(t,s)$, $i = 1,\dots,n$, are continuous nonnegative functions on $I \times I$, and nondecreasing in t. If for $t \in I$*

$$x(t) \le p(t) + \int_0^t f_1(t,t_1)\int_0^{t_1} f_2(t_1,t_2)\dots\int_0^{t_{n-1}} f_n(t_{n-1},t_n)x(t_n)dt_n\dots dt_1$$

then

$$x(t) \le p(t)U(t), \quad t \in I,$$

where $U(t) = V_n(t,t)$ and $V_n(T,t)$ is defined successively by

$$V_1(T,t) = \exp\left\{\int_0^t \sum_{j=1}^t f_j(T,s)ds\right\}$$

$$V_k(T,t) = F_{n-k+1}(T,t)\left\{1 + \int_0^t f_{n-k+1}(T,s)\frac{V_{k-1}(T_1,s)}{F_{n-k+1}(T,s)}ds\right\},$$

where t, and T are in I and $k = 2, 3, \ldots, n$ and for $i = 1, \ldots, n-1$

$$F_i(T,t) = \exp\left\{\int_0^t \left[\sum_{j=1}^{i-1} f_j(T,s) - f_i(T,s)\right] ds\right\}.$$

REMARK 1: In the special case of this theorem for $n = 2$, one can get the conclusion for $t \in I$

$$x(t) \le p(t)\exp\left(-\int_0^t f_1(t,s)ds\right)$$
$$\times\left\{1 + \int_0^t f_1(t,s)\left\{\exp\int_0^s [2f_1(s,r) + f_2(s,r)]\, dr\right\} ds\right\}.$$

31. In the previous theorems we have had linear inequalities. We now turn to some nonlinear inequalities. B. Pachpatte in **[63]** and **[29]** proved the following two theorems.

THEOREM 1. *Let $x(t)$, $a(t)$, and $b(t)$ be real nonnegative and continuous functions on $I = [0, \infty)$ such that*

$$x(t) \le x_0 + \int_0^t a(s)x(s)ds + \int_0^t a(s)\left(\int_0^s b(r)x^p(r)dr\right) ds, \quad t \in I,$$

where x_0 is a nonnegative constant and $0 \le p \le 1$. Then for $t \in I$

$$x(t) \le x_0 + \int_0^t a(s)\exp\left(\int_0^s a(r)dr\right)\left\{x_0^{1-p} + (1-p)\int_0^s b(r)\right.$$
$$\left.\times \exp\left(-(1-p)\int_0^r a(u)du\right) dr\right\}^{\frac{1}{1-p}} ds.$$

THEOREM 2. *Let $x(t)$, $a(t)$, $b(t)$, and $c(t)$ be nonnegative and continuous on R such that for $t \in I$*

$$x(t) \le x_0 + \int_0^t a(s)\left(x(s) + \int_0^s a(r)\left(\int_0^r b(u)x(u) + c(u)x^p(u)du\right) dr\right) ds$$

where x_0 is a nonnegative constant and $0 \le p \le 1$. Then for $t \in I$

$$x(t) \le x_0 + \int_0^t a(s) \left(x_0 + \int_0^s a(r) \exp\left(\int_0^r (a(u)+b(u))\,du \right) \right.$$

$$\left. \times \left\{ x_0^{1-p} + (1-p) \int_0^r c(u) \exp\left(-(1-p) \int_0^r (a(v)+b(v))\,dv \right) du \right)^{\frac{1}{1-p}} dr \right\} ds.$$

32. E. H. Yang [**64**] proved the following result.

THEOREM 1. *Let $u(t)$, $a(t)$, $f(t,s)$, $g_i(t,s)$, and $h_i(t,s)$, $i = 1,\dots,n$, be nonnegative continuous functions defined on $I = [0,h)$ and $I \times I$. Let $a(t)$ be nondecreasing and $f(t,s)$, $g_i(t,s)$ and $h_i(t,s)$ be nondecreasing in t. If $0 < p \le 1$ and*

$$u(t) \le a(t) + \int_0^t f(t,s)u(s)ds + \sum_{i=1}^n \int_0^t g_i(t,s) \left[\int_0^s h_i(s,r)u^p(r)dr \right] ds,$$

then

(A) *for $0 < p < 1$ and $t \in I$ we have*

$$u(t) \le \left\{ [a(t)F(t)]^{1-p} + (1-p) \sum_{i=1}^n G_i(t)F(t) \int_0^t h_i(t,s)ds \right\}^{\frac{1}{1-p}}.$$

(B) *for $p = 1$ and $t \in I$ we have*

$$u(t) \le a(t) \exp\left\{ \int_0^t \left[f(t,s) + \sum_{i=1}^n G_i(t)F(t)h_i(t,s) \right] ds \right\},$$

where $F(t) = \exp \int_0^t f(t,s)ds$, and $G_i(t) = \int_0^t g_i(t,s)ds$, $i = 1,\dots,n$.

33. The nonlinearity x^p in the previous results may be replaced by a more general nonlinearity and we give two results of Pachpatte [**65**] and [**66**].

THEOREM 1. *Let $x(t)$, $a(t)$, and $b(t)$ be nonnegative continuous functions defined on $I = [a, b]$, and let $g(u)$ be a positive continuous strictly increasing subadditive function for $u > 0$ with $g(0) = 0$. If for $t \in I$*

$$x(t) \leq a(t) + \int_a^t b(s)\left(x(s) + \int_a^s b(r) g(x(r))\, dr\right) ds,$$

then for $t \in I_0$ we have

$$x(t) \leq a(t) + \int_a^b b(s)\left(a(s) + \int_a^s b(r) g(a(r))\, dr\right) ds$$
$$+ \int_a^t b(s) G^{-1}\left(G\left(\int_a^b b(r)\left(a(r) + b(u) g(a(u))\, du\right) dr\right) + \int_a^s b(r) dr\right) ds,$$

where

$$G(u) = \int_{u_0}^u \frac{ds}{(s + g(s))}, \quad u \geq u_0 > 0,$$

and

$$I_0 = \left\{t \in I \colon G(\infty) \geq G\left(\int_a^b b(s)\left(a(s) + b(r) g(a(r))\, dr\right) ds + \int_a^t b(s) ds\right)\right\}.$$

THEOREM 2. *Let $x(t)$, $a(t)$, $b(t)$, $c(t)$, and $k(t)$ be continuous on $I = [a, b]$ and $f(u)$ be positive, continuous, strictly increasing, submultiplictive, and subadditive function for $u > 0$, with $f(0) = 0$. If for $t \in I$ we have*

$$x(t) \leq a(t) + b(t)\left(\int_a^t c(s) f(x(s)) + b(s)\int_0^s k(r) f(x(r))\, dr\right) ds,$$

then we also have for $t \in I_0$

$$x(t) \leq a(t) + b(t)\left(d + \int_a^t c(s) f\left(b(s) G^{-1}\left(G(d)\right.\right.\right.$$
$$\left.\left.\left. + \int_a^s f(b(r))(c(r) + k(r))\, dr\right)\right) ds\right),$$

where

$$d = \int_a^b c(s) f\left(a(s) + b(s)\int_a^s k(r) f(a(r))\,dr\right) ds,$$

$$G(u) = \int_{u_0}^u \frac{ds}{f(s)}, \quad u \geq u_0 > 0,$$

and

$$I_0 = \left\{ t \in I\colon G(\infty) \geq G(d) + \int_a^b f(b(r))(c(r) + k(r))\,dr \right\}.$$

THEOREM 3. *Let the functions x, a, b, c, k, and f satisfy the hypotheses of the previous theorem. If for $t \in I$ we have*

$$x(t) \leq a(t) + b(t) f^{-1}\left(\int_a^t c(s) f(x(s))\,ds + \int_a^t c(s) f(b(s))\left(\int_a^s k(r) f(x(r))\,dr\right) ds\right),$$

then we also have for $t \in I$

$$\begin{aligned} x(t) \leq a(t) + b(t) f^{-1}\Bigg(\int_a^t & c(s)\left(f(a(s)) + f(b(s))\right. \\ & \times \left(\exp\left(\int_a^s f(b(r))(c(r) + k(r))\,dr\right)\int_a^s f(a(r))(c(r) - k(r))\right. \\ & \times \left.\left.\exp\left(-\int_a^r f(b(u))(c(u) + k(u))\,du\right) dr\right)\right) ds. \end{aligned}$$

THEOREM 4. *Let the functions be defined as in Theorem 2, while for $a \leq s \leq t \leq b$ we have*

$$x(t) \geq x(s) - b(t) f^{-1}\left(\int_s^t c(r) f(x(r))\,dr + \int_s^t c(r)\left(\int_r^t k(u) f(x(u))\,du\right) dr\right),$$

then for the same range of values we also have

$$x(t) \geq x(s)\left(f^{-1}\left(1 + f(b(t))\int_s^t c(r)\exp\left(\int_r^t (c(u) f(b(t)) + k(u))\,du\right) dr\right)\right)^{-1}$$

34. In Bykov and Salpagarov [61] the following theorem was proved:

THEOREM 1. *Suppose that the functions $u(t)$, $\alpha(t)$, and $\beta(t,s)$ are nonnegative for $0 < s < t < b$ and $\phi(s)$ is positive, nondecreasing and continuous for $s > 0$. If*

$$u(t) \leq c + \int_a^t \left\{ \alpha(r)\phi(u(r)) + \int_a^r \beta(r,s)\phi(u(s))\,ds \right\} dr$$
$$\equiv c + \int_a^t L\phi(u)dr = b(t),$$

where c is a positive number, then for $t \in (a,b)$ we have

$$\int_a^{u(t)} \frac{dx}{\phi(x)} \leq \int_a^t \left[\alpha(r) + \int_a^r \beta(r,s)ds \right] dr = P(t).$$

PROOF: From the hypotheses it follows that

$$\frac{b'(t)}{\phi(b(t))} = \alpha(t)\frac{\phi(u(t))}{\phi(b(t))} + \int_a^t \beta(t,s)\frac{\phi(u(s))}{\phi(b(t))}ds$$
$$\leq \alpha(t) + \int_a^t \beta(t,s)ds.$$

By integration from a to t we get the result.

REMARK 1: For $\beta \equiv 0$ we get the Bihari inequality.

REMARK 2: If the function $P(t)$ belongs to the domain of definition of the inverse function H^{-1}, where $H(h) = \int_0^h dx/\phi(x)$, then $u(t) \leq H^{-1}[P(t)]$.

REMARK 3: Analogously one can prove the corresponding theorem in which the integral $\int_a^t$ is replaced by the integral $\int_t^\infty$.

35. The following generalization of the previous theorem was given by H. M. Salpagarov [**67**]:

THEOREM 1. *Suppose that*

1) $\phi(u)$ *is a nonnegative, continuous nondecreasing function on* $[0,\infty)$;
2) $H(u) = \int_{u_0}^{u} ds/\phi(s)$ $(0 < u < \infty,\ u_0 \in (0,\infty)$ *is fixed), and* H^{-1} *is the inverse function of* H;
3) $u(t)$ *is continuous on* $[0,\infty)$; *and*
4) $M(t)$ *is a nondecreasing nonnegative function.*

If

$$u(t) \le M(t) + \int_a^t L\phi(u)dr,$$

where the operator L *is defined in the previous theorem, then*

$$u(t) \le H^{-1}\{H[M(t) + P(t)]\}$$

where $P(t)$ *is also defined as in* Theorem 1 *of* 34.

PROOF: Let $T > a$ be fixed. Then for $t \in (a,T]$ we have

$$u(t) \le M(T) + \int_0^t L\phi(u)dr,$$

since $M(T) \ge M(t)$. On the basis of Theorem 1 of 34. we have

$$u(t) \le H^{-1}\{H[M(T) + P(t)]\} \quad \text{for } t \le T.$$

Setting $t = T$ we get the conclusion.

36. Further bibliography for this Chapter can be found as items [**68**]–[**151**].

REFERENCES

1. GRONWALL, T, *Note on the derivatives with respect to a parameter of the solutions of a system of differential equations*, Ann. Math. (2)20 (1919), 292–296.
2. BELLMAN, R., *The stability of solutions of linear differential equations*, Duke Math. J. **10** (1943), 643–647.
3. BECKENBACH, E. and R. BELLMAN, "Inequalities," Berlin-Heidelberg-New York, 1965.
4. MITRINOVIĆ, D. S., "Analytic Inequalities," Berlin-Heidelberg-New York, 1970.
5. REID, W., "Ordinary Differential Equations," New York-London, 1971.
6. HILLE, E., "Lectures on Ordinary Differential Equations," Reading, Mass., 1969.
7. __________, "Methods in Classical and Functional Analysis," Reading, Mass., 1972.
8. LAKSHMIKANTHAM, V. and S. LEELA, "Differential and Integral Inequalities," Vol. 1, New York-London, 1969.
9. WALTER, W., "Differential and Integral Inequalities," New York-Heidelberg-Berlin, 1970.
10. BIHARI, I., *A generalization of a lemma of Bellman and its application to uniqueness problems of differential equations*, Acta Math. Acad. Sci. Hungar. **7** (1956), 81–94.
11. VISWANATHAM, B., *A generalization of Bellman's lemma*, Proc. Amer. Math. Soc. **14** (1963), 15–18.
12. GYÖRI, I., *A generalization of Bellman's inequality for Stieltjes integrals and a uniqueness theorem*, Studia Sci. Math. Hungar. **6** (1971), 137–145.
13. ZIEBUR, A., *On the Gronwall-Bellman lemma*, J. Math. Anal. Appl. **22** (1968), 92–95.
14. LAKSMIKANTHAN, V., *Upper and lower bounds of the norm of a solutions of differential equations*, Proc. Amer. Math. Soc. **13** (1962), 615–616.
15. LEES, M., *Energy inequalities for the solutions of ordinary differential equations*, Trans. Amer. Math. Soc. **94** (1960), 58–73.
16. PEANO, G., *Sull' integrabilità delle equazioni differenziali del primo ordine*, Att. R. Accad. Sc. Torino **21** (1885–1886), 677–685.
17. KAMKE, E., "Differentialgleichungen reeller Funktionen," Leipzig, 1930.
18. REID, W., *Properties of solutions of an infinite system of ordinary linear differential equations of the first order with auxiliary boundary conditions*, Trans. Amer. Math. Soc. **32** (1930), 284–318.
19. FILATOV, A. and L. ŠAROVA, "Integral'nye neravenstva i teorija nelineinyh kolebanii," Moskva, 1976.

20. MARTYNJUK, A. and R. GUTOWSKI, "Integral'nye neravenstva i ustoičivost' dviženija," Kiev, 1979.
21. DEMIDOVIČ, B., "Lekcii po mathematičeskoi teorii ustoičivosti," Moskva, 1967.
22. FILATOV, A., "Metody usrednenija v differencial'nyh i integro-differencial'nyh uravnenijah," Taškent, 1971.
23. ZADIRAKA, K. V., *Issledovanie nereguljarnogo vozmuščennyh differencial'nyh uravnenii.* In: "Voprosy teorii i istorii differencial'nyh uravnenii," Kiev, 1968.
24. BELLMAN, R., "Stability theory of differential equations," New York, 1953.
25. ČANDIROV, G. I., *Obodnom obobščenii neravenstva Gronuolla i ego priložcnijah*, Učen. Zap. Azerb. Univ., Ser. Fiz-Mat. Nauk, **1970**, No. **2**, 9–13.
26. WILLETT, D., *A linear generalization of Grownwall's inequality*, Proc. Amer. Math. Soc. **16** (1965), 774–778.
27. HARLAMOV, P., *Ob ocenke rešenii sistemy differencial'nyh uravnenii*, Ukrain. Mat. Ž. (1955), 471–473.
28. BEESACK, P. R., *Gronwall inequalities*, Carleton Univ. Math. Notes No. **11**, 1975.
29. PACHPATTE, B. G., *A note on some integral inequalities*, Math. Student **42** (1974), 409–411.
30. GAMIDOV, S., *Nekotorye integral'nye neravenstva dlja kraevyh zadač differencial'nyh uravnenii*, Differencial'nye Uravnenia **5** (1969).
31. MOVLJANKULOV, H. and A. FILATOV, *Ob odnom približennom metode postroenija rešenii integral'nyh uravnenii*, Tr. In'ta Kibern. AN UzSSR, Taškent, 1972, No. **12**, 11–18.
32. SARDAR'LY, S. M., *Issledovanie rešenii nelineinyh integral'nyh uravnenii na poluoisi po parametry*, Učen. Zap. Azerb. Univ. **1965**, No. **6**.
33. CHU, S. and F. METCALF, *On Gronwall's inequality*, Proc. Amer. Math. Soc. **18** (1967), 439–440.
34. BEESACK, P. R., *Comparison theorems and integral inequalities for Volterra integral equations*, Proc. Amer. Math. Soc. **20** (1969), 61–66.
35. JONES, G., *Fundamental inequalities for discrete and discontinuous functional equations*, J. Soc. Ind. Appl. Math. **12** (1964), 43–57.
36. WRIGHT, F., M. KLASI, and D. KENNEBECK, *The Gronwall inequality for weighted integrals*, Proc. Amer. Math. Soc. **30** (1971), 504–510.
37. SCHMAEDEKE, W. and G. SELL, *The Gronwall inequality for modified Stieltjes integrals*, Proc. Amer. Math. Soc. **19** (1968), 1217–1222.
38. HEROD, J., *A Gronwall inequality for linear Stieltjes integrals*, Proc. Amer. Math. Soc. **23** (1969), 34–36.

39. HELTON, B., *A product integral representation for a Gronwall inequality*, Proc. Amer. Math. Soc. **23** (1969), 493–500.
40. ________, *The solution of a nonlinear Gronwall inequality*, Proc. Amer. Math. Soc. **38** (1973), 337–342.
41. PEROV, A. I., *K voprosu o strukture integral'noĭ voronki*, Nauč. Dokl. Vysšeii Školy. Ser FMN, 1959, No. 2.
42. WILLETT, D. *Nonlinear vector integral equations as contraction mappings*, Arch. Rational Mech. Anal. **15** (1964) 79–86.
43. WILLETT, D. and J. WONG, *On the discrete analogues of some generalizations of Gronwall's inequality*, Monatsh. Math. **69** (1965), 362–367.
44. HO, T. K., A note on Gronwall-Bellman inequality, Tamkang J. Math. **11** (1980) 249–255.
45. STACHURSKA, B., *On a nonlinear integral inequality*, Zeszyzy Nauk Univ. Jagiellonskiego 252, Prace. Mat. **15** (1971), 151–157.
46. MARONI, P., *Une généralisation non linéaire de l'inégalité de Gronwall*, Bull. Acad. Polon. Sc. Ser. Sc. Math. Astr. Phys. **16** (1968), 703–709.
47. LASALLE, J. P., *Uniqueness theorems and successive approximations*, Ann. of Math. **50** (1949), 722–730.
48. AHMEDOV, K., M. JAKUBOV and I. VEISOV, *Nokotorye integral'nye neravenstva*, Izv. AN UzSSR **1972**.
49. LANGENHOP, C. E., *Bounds on the norm of a solution of a general differential equation*, Proc. Amer. Math. Soc. **11** (1960), 795–799.
50. DRAGOMIR, S. S., *The Gronwall type lemmas and applications*, Monografii Matematice, No. **29**. Univ. Timişoara, 1987.
51. PEČARIĆ, J. E. and S. S. DRAGOMIR, *On some Gronwall type lemmas*, Sem. Ecual. Func. Univ. Timisoara No. 86(1988), 1–10.
52. BUTLER, G. and T. ROGERS, *A generalization of a lemma of Bihari and applications to pointwise estimates for integral equations*, J. Math. Anal. Appl. **33** (1971), 77–81.
53. GOLLWITZER, H., *A note on a functional inequality*, Proc. Amer. Math. Soc. **23** (1969), 642–647.
54. DEO, S. and M. MURDESHWAR, *On a system of integral inequalities*, Proc. Amer. Math. Soc. **26** (1970), 141–144.
55. DEO, S. and U. DHONGADE, *Some generalizations of Bellman-Bihari integral inequalities*, J. Math. Anal. Appl. **44** (1973), 218–226.
56. ________, *Pointwise estimates of solutions of some Voltera integral equations*, J. Math. Anal. Appl. **45** (1974), 615–628.
57. BABKIN, B., *On a generalization of a theorem of Academician S. A. Čaplygin on a differential inequality* (Russian), Molotov. Gos. Univ. Uč. **8** (1953), 3–6.

58. MAMEDOV, Ja. D., S. AŠIROV and S. ATDAEV, "Teoremy on Neravenstvah," Ašhabad, YLYM, 1980.
59. MAMEDOV, Ja. D., *Odnostrannie ocenki v uslovijah suščestvovanija i edinstvennosti rešenija predel'noi zadači Koši v Banahovom prostranstve*, Sib. Mat. Ž. **6** (1965), No. 5.
60. PACHPATTE, B. G., *Integral inequalities of Gronwall-Bellamn type and their applications*, J. math. phys. Sci., Madras **8** (1974), 309–318.
61. BYKOV, Ja. V. and H. M. SALPAGAROV, *K theorii integrodifferencial'nyh uravnenijam v Kirgizii*, Izd-vo AN Kirg.SSR, **1962**, No. 2.
62. YANG, E. H., *On the most general form of Bellman-type linear inequalities involving multiple-fold integral functionals*, J. Math. Anal. Appl. **103** (1984), 184–197.
63. PACHPATTE, B. G., *A note on Gronwall-Bellman inequality*, J. Math. Anal. Appl. **44** (1973), 758–762.
64. YANG, E. H., *On Asymptotic behavior of certain integrodifferential equations*, Proc. Amer. Math. Soc. **90** (1984), 271–276.
65. PACHPATTE, B. G., *An integral inequality similar to Bellman-Bihari inequality*, Bull. Soc. math. Grèce. N.S. **15** (1974), 7–12.
66. __________, *A note on integral inequalities of the Bellman-Bihari type*, J. Math. Anal. Appl. **49** (1975), 295–301.
67. SALPAGAROV, H. M., *O suščestvovanii rešeniĭ odnogo klassa integrodifferencial'nyh uravneniĭ tipa Vol'terra s osobennost'ju.* In: "Issledovanija po integrodifferencial'nym uravnenijam v Kirgizii," vyp. **6**. Frunze, "Ilim", 1969.
68. AGARWAL, R. P. and E. THANDAPANI, *Remarks on generalizations of Gronwall's inequality*, Chinese J. Math. **9** (1981), 1–22.
69. __________, *Some inequalities of Gronwall type*, An. Ştint. Univ. "Al. I. Cuza" Iaşi Sec. I. a Mat. (N.S.) **28** (1982), 71–75.
70. ALIEV, R.M., *Some integral inequalities* (Russian), Akad. Nauk. Azerbaĭdzan SSR Dokl. **32** (1976), No. 9, 12–18.
71. ALIEV, R. M. and G. DINOVRUZOV, *Some theorems on integral inequalities* (Russian), Akad. Nauk. Azerbaidzan SSR Dokl. **25** (1969), No. 6, 3–9.
72. AŠIROV, S. and S. ATDAEV, *Ob odnom obobščenii neravenstva Bellmana.* Izv. Akad. Nauk Turkmen SSR Ser. Fiz'Tehn. Him. Geol. Nauk., **1973**, 102–104.
73. __________, *Ob odnom integral'nom neravenstve.* In: Matematika. Ašhabad, **1973**.
74. __________, *O nekotoryh integrad'hyn neravenstvah tipa Bellmana-Gronoulla.* Izv. An TSSR, Ser. TFN i. GN, **1974**, No. 6.

75. ATDAEV, S. and S. AŠIROV, *Ob odnom obobščenii neravenstva Bellmana.* Izvestija AN TSSR, Ser. FTN i GN, **1973**, No. 3
76. ______________, *Certain integral inequalities of Bellman-Gronwall type,* Izv. AN Turkmen SSR Ser Fiz-Tehn Kim. Geol. Nauk 1979, 3–8.
77. BAINOV, D.D. and A. J. LAHARIEV, *A generalization of the Bellman-Gronwall inequality*, Godišnik Visš Učebn. Zaved. Priložna Mat. **12** (1976), 207–210.
78. BANAŚ, J. and A. HAJNOSZ, *On some integral inequalities* (Russian), Zeszyty Nauk. Politech. Rzeszowskiej. Mat. Fiz. No 1 (1984), 105–115.
79. BEATRIZ, M., *An extension of Gronwall's inequality*, Rev. Unión. Mat. Argent. **25** (1971), 247-251.
80. BURLAČENKO, V.P. and N.I. SIDENKO, *On improving the solution of integral inequalities of the Gronwall-Bellman type* (Russian), Ukrain. Mat. Ž. **32** (1980), 538–545.
81. BUSTOZ, J. and D. WRIGHT, *A note on Gronwall transforms.* Applicable Anal. **6** (1976/77), No 1, 39–45.
82. BYKOV, Ja.V., *O nekotoryh zadačah teorii integrodifferencial'nyh uravneniĭ*, Frunze Kirg. G.U., **1967**.
83. CHANDIROV, G. A., *On a generalization of an inequality of Gronwall and its applications*, Proc. Indian Acad. Sci. Sect. A (1978), 27–31.
84. CINQUINI, S., *Un' osservazione sopra un'estensions del lemma di Gronwall*, Ist. Lombardo Accad. Sci. Lett. Rend. **A 97** (1963), 64-67.
85. CLARK, D., *Generalization of an inequality of Gronwall-Reid*, J. Approx. Theory **58** (1989), 12–14.
86. DANNAN, F.M., *Integral inequalities of Gronwall-Bellman-Bihari type and asymptotic behavior of certain second order nonlinear differential equations*, J. Math. Anal. Appl. **108** (1985), 151-164.
87. DEO, S. and M. MURDESHWAR, *A note on Gronwall's inequality*, Bull. London Math. Soc. **3** (1971), 34-36.
88. DHONGADE, U. and S. G. DEO, *A nonlinear generalization of Bihari's inequality*, Proc. Amer. Math. Soc. **54** (1976), 211–216.
89. DŽABAROV, S., *O nekotoryh integral'nyh neravenstvah i primenenii ih k voprosam ustoičivosti, ograničennosti, stremlenija k nulju i edinstvennosti rešenii opredelennyh klassov differencial'nyh uravnenii*, Kand. diss., Baku, 1966.
90. FAGBOHUN, A.B. and C.O. IMORU, *A new class of integrl-differential inequalities*, Simon Stevin, A. Quart. J. Pure. Appl. Math. **60** (1986), 301–311.

91. GRUDO, Ė. I. and L. F. JANČUK, *Inequalities of Gronwall, Bihari, and Langenhop type for Pfaffian systems* (Russian), Differencial'nye Uravnenija **13** (1977), 749–752, 775.
92. GUILLIANO, L., *Generalizzazione di un lemma di Gronwall e di una diseguaglianza de Peano*, Rend. Accad. Lincei (8) **1** (1964), 1264–1271.
93. GYÖRI, I. and L. PINTÉR, *On some Stieltjes integral inequalities*, Kozl.-MTA Śzamitastech. Automat. Kutató. Inst. Budapest. No **26** (1982), 79–84.
94. HELTON, J. C., *A special integral and a Gronwall inequality*, Trans. Amer. Math. Soc. **217** (1976), 163–181.
95. HRISTOVA, S.G. and D.D. BAĬNOV, *A generalization of Bihari's inequality* (Russian). Godišnik Visš. Učebn. Zaved. Prilŏzna. Mat. **13** (1977), 87–92.
96. ___________, *A generalization of an inequality of Bihari* (Russian), Izv. AN KazSSR, Ser. FM **1978**, 88–89.
97. KONSTANTINOV, M., *Ob odnom integral'nom neravenstve*, Arch. Mat. 2, Scr. Fac. Sci. Nat. Ujer Runesis **12** (1976), 81–86 (Russian).
98. KRECMETOV, G.S., *On an integral inequality* (Russian), Differ. Uravn. **29**, No 5 (1988), 894–896.
99. LAKSHMIKANTHAN, V., *A variation of constants formula and Bellman-Gronwall-Reid inequalities*, J. Math. Appl. **41** (1973), 199–204.
100. LAPTINSKIĬ, V.N., *Ob odnom neravenstve*, Differencial'nye Uravnenija **4**, No 2, (1968).
101. LIN, C.T., *A note on an inequality of Gronwall type*, Soochow J. Math. **9** (1983), 167–177.
102. LIN, C.T. and C.M. WU, *A note on an inequality ascribed to Pachpatte*, Tamkang J. Math. **20** (1989), 221–225.
103. LIU, W.B., *Gronwall's equation and Gronwall's inequalities* (Chinese), J. Math. (Wuhan) **5** (1985), 89–97.
104. MAMIKONYAN, F.U. and A.B. NERSESYAN, *Inversion of some integral inequalities* (Russian), Izv. AN Armyan.SSR. Ser. Math. **17** (1982), 62–79.
105. MANOUGIAN, M.N., *Sur l'inégalité de Gronwall*, Bull. Soc. Math. Grèce, n. Ser. **10** (1969), 45–47.
106. MULDOWNEY, J. and J. WONG, *Bounds for solutions of nonlinear integro-differential equations*, J. Math. Anal. Appl. **23** (1968), 487–499.
107. NARSIMHA REDDY, K., *Integral inequalities and applications*, Bull. Austral. Math. Soc. **21** (1986), 13–20.
108. OSTASZEWSKI, K. and J. SOHACKI, *Gronwall's inequality and the Henstock integral*, J. Math. Anal. Appl. 128 (1987), 370-374.

109. PACHPATTE, B.G., *On some generalization of Bellman's lemma*, J. Math. Anal. Appl. **51** (1975), 141–150.

110. __________, *On some inequalities similar to Bellman-Bihari inequalities*, J. Math. Anal. Appl. **49** (1975), 794–802.

111. __________, *A note on an integral inequality*, J. Math. Phys. Sc. **9** (1975), 11–14.

112. __________, *On an integral inequality of Gronwall-Bellman*, J. Math. Phys. Sc. **9** (1975), 405–416.

113. __________, *On some new integral inequalities for differential and integral equations*, J. Math. Phys. Sc. **10** (1976), 101–116.

114. __________, *A new integral inequality for differential and integral equations*, Proc. Nat. Acad. Sci. India Sect. A **46** (1976), 21–26.

115. __________, *On some integral inequalities and their applications to integrodifferential equations*, Indian J. Pure Appl. Math. **8** (1977), 1157–1175.

116. __________, *A note on Gronwall type integral and integrodifferential inequalities*, Tamkang J. Math. **8** (1977), 53–59.

118. __________, *On some fundamental integral inequalities of the Bellman-Bihari type*, Chin. J. Math. **5** (1977), 71–80.

119. __________, *On the generalization of Bellman's lemma*, Math. Student **45** (1977), no. 4, 33–36.

120. __________, *A note on Gronwall type integrodifferential inequalities*, Tamkang J. Math. **9** (1978), 151–156.

121. __________, *On some fundamental integral inequalities with applications*, Bul. Ştinţ. Tehn. Insst. Politehn. "Traian Vuia" Timişoara **23** (37) (1978), No 1, 25–33.

122. __________, *On some new integral inequalities involving functions and their derivatives*, Indian J. Pure Appl. Math. **11** (1980), 1039–1044.

123. __________, *On some fundamental partial integral inequalities*, J. Math. Anal. Appl. **73** (1980), 238–251.

124. __________, *Some inequalities between functions and their derivatives*, Acta Cinc. Indica Math. **7** (1981), 223–234.

125. __________, *A note on some nth order integr-differential inequalities*, Acta Cinc. Indica Math. **19** (1984), 8–12.

126. __________, *On some fundamental integrodifferential inequalities for differential equations*, Chin. J. Math. **6** (1978), 17–23.

127. __________, *On some integral inequalities of the Gronwall-Bellman type*, Indian J. Pure and Appl. Math. **6** (1975), 169–172.

128. __________, *On some nonlinear generalizations of Gronwall's inequality*, Proc. Indian Acad. Sci. Sect. A. (1976), 1–19.

129. PERESTYUK, N.A., S.G. HRISTOVA and D.D. BAĬNOV, *A linear integral inequality for piecewise continuous functions* (Bulgarian), Godišnik Visš. Učebn Zaved. Priložna Mat. **16** (1980), 221–226.
130. RAGHAVENDRA, *A remark on an integral inequality*, Bull. Austral. Math. Soc.**23** (1981), 195–197.
131. ROGERS, T.D., *An integral inequality*, J. Math. Anal. Appl. **82** (1981), 470–472.
132. SATO, T., *Sur l'équation intégrale non linéaire de Volterra.* Compositio Math. **11** (1953), 271–289.
133. SOBEIH, M. and G. HAMAD, *On a generalization to Gronwall's integral inequality.* Afrika Mat. **6** (1984), 5–12.
134. STORCHI, E., *Su una generalizzazione dela lemma di Peano-Gronwall.* Matematiche **16** (1961), 8–26.
135. THANDAPANI, E., *A note on Bellman-Bihari type integral inequalities.* J. Math. Phys. Sci. **21** (1987), 1–3.
136. __________, *On some new integrodifferential inequalities theory and applications.* Tamkang J. Math. **11** (1981), 169-184.
137. VERLAN', A.F. and V.S. GODLEVSKIĬ, *Some integral estimates of the type of Gronwall-Bellman and Bihari inequalities* (Russian), Točn. i Nadežn. Kibernnet. Sistem. No 2 (1974), 3–8, 140.
138. VOLTERRA, V., "Theory of Functionals and of Integral and Integrodifferential Equations," New York, 1959.
139. WANG, C.L., *A short proof of a Greene theorem*, Proc. Amer. Math. Soc. **69** (1978), 357–358.
140. WANG, C.L. and L.C. ZHONG, *Some remarks on integral inequalities of Bellman-Gronwall type* (Chinese), Sichuan. Daxue Xuebao 1984, No 1, 18–27.
141. WATTAMWAR, M.J., *On some integral inequalities of Gronwall-Bellman type*, J. Maulaua Azad. College. Tech. **15** (1982), 13–20.
142. YANG, C.C., *On some nonlinear generalizations of Gronwall-Bellman's inequality*, Bull. Inst. Math. Acad. Sinica **7** (1979), 15–19.
143. YANG, E.H., *A generalization of Bihari's inequality and its applications to nonlinear Volterra integral equations* (Chinese), Chinese Ann. Math. **3** (1982), 209–216.
144. YANG, G.S., *A note on an integro-differential inequality*, Tamkang J. Math. **12** (1981), 257–264.
145. __________, *A note on some integro-differential inequalities.* Soochow J. Math. **9** (1983), 231–236.

146. YOUNG, E.C., *On Bellman-Bihari integral inequalities*, Intern. J. Math. Sci. **5** (1982), 97–103.

147. __________, *On integral inequalities of Gronwall-Bellman type*, Proc. Amer. Math. Soc. **94** (1985), 636–640.

148. ZAGHROUT, A.A.S. and I.A. Aly, *On integral inequalities of Gronwall-Bellman-Reid type I*, Kyuugpook Math. J. **27** (1987), 145-152.

149. __________, *On integral inequalities of Gronwall-Bellman-Reid type II*, Kyuugpook Math. J. **27** (1987), 153–161.

150. ZHANG, B.G., *A generalization of the Bellman-Bihari integral inequality* (Chinese), J. Math. Res. Exposition **2** (1982), 27–32.

151. __________, *On integral inequalities of Gronwall-Bellman-Reid type I*, Tamkang J. Math. **18** (1987), 1–11.

CHAPTER XIII

GRONWALL INEQUALITIES IN HIGHER DIMENSIONS

1. Several authors generalized inequalities of Gronwall type to the case of functions of two or more variables. Of course, such results have application in the theory of partial differential equations and Volterra integral equations.

We begin with some results for two variables. In the book by Beckenbach and Bellman [**1**] the following unpublished Wendroff result was given:

THEOREM 1.

1) *If*

$$u(x,y) \le a(x) + b(y) + \int_0^x \int_0^y v(r,s)u(r,s)\,dr\,ds \tag{1.1}$$

where $a(x), b(y) > 0$, $a'(x), b'(y) \ge 0$, $u(x,y), v(x,y) \ge 0$, *then*

$$u(x,y) \le \frac{(a(0)+b(y))\,(a(x)+b(0))}{a(0)+b(0)} \exp\left(\int_0^x \int_0^y v(r,s)\,dr\,ds\right).$$

2) *If*

$$u(x,y) \le c + a\int_0^x u(s,y)\,ds + b\int_0^y u(x,s)\,ds$$

then $u(x,y) \le c\exp(ax+by+abxy)$, *where* a, b, c *are constants.*

3) *If*

$$u(x,y) \le a(x) + b(y) + a\int_0^x u(s,y)\,ds + b\int_0^y u(x,s)\,ds,$$

then $u(x,y) \le Q(x,y)$ *where*

$$Q(x,y) = \left[a(0)+b(0)+\int_0^x e^{-as}a'(s)\,ds\right]\left[a(0)+b(0)+\int_0^y e^{-bs}b'(s)\,ds\right] \times \frac{(\exp(ax+by+abxy))}{(a(0)+b(0))}.$$

PROOF:: We shall only prove the first conclusion. Denote the expression on the right hand side of (1.1) by $b(x,y)$. Then we have

$$\frac{b_x(x,y)}{b(x,y)} \le \frac{a'(x)}{b(x,y)} + \int_0^x v(x,s)ds \tag{1.2}$$

$$\le \frac{a'(x)}{b(x,0)} + \int_0^y v(x,s)ds.$$

By integrating with respect to x from zero to x we get

$$\ln b(x,y) - \ln b(0,y) \le \int_0^x \frac{a'(x)}{(a(x)+a(0))}dx + \int_0^x\int_0^y v(r,s)drds.$$

Performing the first integration and using $u(x,y) \le b(x,y)$ this rearranges into (1.1′).

2. The Wendroff inequality (1.1) was generalized by T. Nurimov [**2**]:

THEOREM 1. *Let $u(x,y)$ be a nonnegative continuous function on $D = [0,x_0] \times [0,y_0]$ such that*

$$u(x,y) \le a(x,y) + b(x,y)\int_0^x\int_0^y c(z,t)u(z,t)dtdz, \tag{2.1}$$

where $a(x,y)$, $b(x,y)$, and $c(x,y)$ are nonnegative continuous functions on D. Then in D we have

(2.1′)

$$u(x,y) \le a(x,y) + b(x,y)\int_0^x\int_0^y \exp\left(\int_z^x\int_t^y c(r,s)b(r,s)drds\right) \times a(z,t)c(z,t)dtdz.$$

REMARK 1: In fact, Nurimov proved a complicated result which can be obtained from the above result by using the subsitution

$$a(x,y) \to a(x,y) + b(x,y)\int_0^x\int_0^y d(x,y,z,t)dtdz.$$

REMARK 2: Note that in the case when $a(x,y) \equiv C$ and $b(x,y) \equiv 1$, (2.1′) becomes $u(x,y) \leq Cw(x,y)$ where

$$w(x,y) = 1 + \int_0^x \int_0^y \exp\left(\int_z^x \int_t^y c(r,s)drds\right) c(z,t)dtdz.$$

This inequality is better than Wendroff's obtained from (1.1′) by taking $a(x,y) \equiv C$ and $b(x,y) \equiv 0$, since

$$w(x,y) \leq \exp\left(\int_0^x \int_0^y c(z,t)dtdz\right).$$

T. Nurimov [3] also proved the following:

THEOREM 2. *Let $u(x,y)$ be continuous and nonnegative on $[0,\infty) \times [0,\infty)$ such that*

$$u(x,y) \leq c + \int_x^\infty \int_y^\infty a(z,t)u(z,t)dtdz < \infty.$$

Then if $c > 0$, and $a(x,y)$ is nonnegative, continuous and $\int_x^\infty \int_y^\infty a(z,t)dz\,dt < \infty$, we have

$$u(x,y) \leq v(x,y)$$

where

$$v(x,y) = c\left(1 + \int_x^\infty \int_y^\infty a(z,t)\exp\left(\int_z^\infty \int_t^\infty a(r,s)drds\right)dtdz\right).$$

REMARK 3: Note that this is better than the corresponding Wendroff inequality since

$$v(x,y) \leq c\exp\left(\int_x^\infty \int_y^\infty a(z,t)dtdz\right).$$

REMARK 4: Nurimov in [2] gave an analogous result where the integrals $\int_x^\infty \int_y^\infty$ and $\int_z^\infty \int_t^\infty$ are replaced by $\int_0^x \int_y^\infty$ and $\int_0^z \int_t^\infty$ respectively.

In the paper [3] Nurimov also proved the following result:

THEOREM 3. *Let $u(x,y)$ be nonnegative and continuous on $D=[0,x_0]\times[0,y_0]$ and satisfy*

$$u(x,y)\le f(x,y)+a(x)b(y)\int_0^x\int_0^y \phi(z)\psi(t)u(z,t)dtdz$$
$$+a(x)\int_0^x \phi(z)u(z,y)dz+b(y)\int_0^y \psi(t)u(x,t)dt,$$

where f, a, b, ϕ, and ψ are nonnegative and satisfy

$$\int_0^x\int_0^y f(z,t)dtdz<\infty,\quad \int_0^x\int_0^y f(z,t)\phi(z)\psi(t)dtdz<\infty,$$
$$\int_0^x a(z)\phi(z)dz<\infty,\ \text{and}\ \int_0^y b(t)\psi(t)dt<\infty,$$

then

$$u(x,y)\le f(x,y)+\sum_{n=1}^{\infty}G_n(x,y),$$

where

$$G_n(x,y)=\frac{a(x)}{(n-1)!}\int_0^x f(z,y)\phi(z)\left[\int_z^x a(\sigma)\phi(\sigma)d\sigma\right]^{n-1}dz$$
$$+\frac{b(y)}{(n-1)!}\int_0^y f(x,t)\psi(t)\left[\int_t^y b(\delta)\psi(\delta)d\delta\right]^{n-1}dt$$
$$+\sum_{\nu=1}^{n-1}c_{n,\nu}\frac{a(x)b(y)}{(n-\nu+1)!(\nu-1)!}\int_0^x\int_0^y f(z,t)\phi(z)\psi(t)$$
$$\times\left[\int_z^x a(\sigma)\phi(\sigma)d\sigma\right]^{n-\nu-1}\left[\int_t^y b(\delta)\psi(\delta)d\delta\right]^{\nu-1}dtdz;$$

and

$$c_{n+1,\nu} = c_{n,\nu} + c_{n,\nu-1} + c_{n-1,\nu-1}, \quad c_{1,0} = c_{1,1} = 1,$$
$$c_{2,0} = c_{2,2} = 1, \quad c_{2,1} = 3.$$

REMARK 5: Previously in [2] Nurimov had proved the special case: $f(x,y) = cx^\alpha y^\beta$, $a(x) = ax^\gamma$, $b(y) = by^\delta$, $\phi(z) = z^{-r}$, $\psi(t) = t^{-s}$, with $c, a, b \geq 0$, $\alpha, \beta > 0$, $\gamma, \delta \geq 0$, $r - 1 < \alpha$, $s - 1 < \beta$, $r - 1 < \gamma$, $s - 1 < \delta$. In this case one obtains the result with

$$G_n(x,y) = cx^\alpha y^\beta \sum_{\nu=0}^{\infty} c_{n,\nu} \left[ax^{\gamma-r+1}\right]^{n-\nu} \left[by^{\delta-s+1}\right]^{\nu}$$
$$\times \{[\alpha - (r-1)] \ldots [\alpha + (n-\nu+1)\gamma - (n-\nu)(r-1)]$$
$$\times [\beta - (s-1)] \ldots [\beta + (\nu-1)\delta - \nu(s-1)]\}^{-1}.$$

3. H. Ešmatov [4] gave the following result:

THEOREM 1. *Suppose that for $0 \leq s \leq z \leq x$, and $0 \leq r \leq t \leq y$ the functions $u(x,y)$, $\phi(z,t)$, and $f(z,t,s,r)$ are nonnegative and continuous such that*

$$u(x,y) \leq a + b \int_0^x \int_0^y \left[\phi(z,t)u(z,t) + \int_0^z \int_0^t f(z,t,s,r)u(s,r)dsdr\right] dzdt,$$

where $a > 0$, and $b \geq 0$. Then

$$u(x,y) \leq a \exp\left\{b \int_0^x \int_0^y \left[\phi(z,t) + \int_0^z \int_0^t f(z,t,s,r)dsdr\right] dzdt\right\}.$$

4. The final result that we give for functions of two variables is from Bykov and Salpagarov [5]:

THEOREM 1. *Let the functions $u(t,x)$, $f_1(\sigma,\delta)$, $f_2(\sigma,\delta,s)$, $f_3(\sigma,\delta,r)$, and $f_4(\sigma,\delta,s,r)$*

be nonnegative on $a \le s \le \delta \le x < a_1$, $b \le r \le \sigma \le t < b_1$ *and satisfy*

$$u(t,x) \le c(x) + \int_a^x \int_b^t \{f_1(\sigma,\delta)\phi[u(\sigma,\delta)] + \int_a^\delta f_2(\sigma,\delta,s)\phi[u(\sigma,s)]\,ds + \int_b^\sigma f_3(\sigma,\delta,r)\phi[u(r,\delta)]\,dr + \int_a^\delta \int_b^\sigma f_4(\sigma,\delta,s,r)\phi[u(r,s)]\,drds\}\,d\sigma d\delta, \quad a \le x \le a_1, \quad b \le t \le b_1,$$

where $c(x)$ *is a positive function. Then*

$$\int_{c(x)}^{u(t,x)} \frac{d\sigma}{\phi(\sigma)} \le g(t,x) \equiv \int_a^x \int_b^t \left\{ f_1(\sigma,\delta) + \int_a^\delta f_2(\sigma,\delta,s)ds + \int_b^\sigma f_3(\sigma,\delta,r)dr + \int_a^\delta \int_b^\sigma f_4(\sigma,\delta,s,r)drds \right\} d\sigma d\delta.$$

REMARK 1: If $g(t,x)$ belongs to the domain of definition of H^{-1}, where $H(h) = \int_{c(x)}^{h} ds/\phi(s)$, then

$$u(t,x) \le H^{-1}[g(t,x)].$$

REMARK 2: For some related results see Bondge, Pachpatte, and Walter [**6**], Dragomir and Ionescu [**7**],[**8**], Pachpatte [**9**]. See also Nurimov [**10**].

5. Let x and y be vectors in Euclidean n-space. We say that $x < y$ ($x \le y$) if and only if for each $i = 1,\dots,n$, $x_i < y_i$ ($x_i \le y_i$). Let ([0,T]) denote the n-dimensional parallelepiped, and the integral $\int_{x_1}^{y_1} \dots \int_{x_n}^{y_n} \cdot ds_n \dots ds_1$ is written as $\int_x^y \cdot ds$ while $D_i = \frac{\partial}{\partial x_i}$ and $D = D_1 D_2 \dots D_n$.

The following theorem is a direct generalization of Theorem 1 from 9. in Chapter XII (see for example Mamedov, Aširov and Atdaev [**11**, pp. 129]:

THEOREM 1. *Let the nonnegative function $v(t)$ satisfy*

$$v(t) \le u_0(t) + \int_0^t a_1(s)v(s)ds + \int_0^t a_2(s)v^p(s)ds$$

on $([0,T])$ where $u_o(t)$, $a_i(t)$, $i = 1,2$ are nonnegative and continuous and $p > 0$.
If $p > 1$ then

$$\bar{u}_0^{1-p} + (1-p)\int_0^t a_2(s)\exp\left(-(1-p)\int_0^{s_1}\int_0^{t_2}\cdots\int_0^{t_n} a_1(r)dr\right)ds > 0,$$

where $\bar{u}_0 = \max_{0\le t\le T} u_0(t)$. If $0 < p \ne 1$ we have

$$v(t) \le u_0(t) - \bar{u}_0 + \exp\left(\int_0^t a_1(s)ds\right)$$
$$\left(\bar{u}_0^{1-p} + (1-p)\int_0^t a_2(s)\exp\left(-(1-p)\int_0^{s_1}\int_0^{t_2}\cdots\int_0^{t_n} a_1(r)dr\right)ds\right)^{\frac{1}{1-p}},$$

and for $p = 1$ we have

$$v(t) \le u_0(t) - \bar{u}_0 + \bar{u}_0\exp\left(\int_0^t (a_1(s) + a_2(s))\,ds\right).$$

Related to Theorem 3 from 16. in Chapter XII is the following theorem of S. Atdaev and S. Aširov (see Mamedov, Aširov, and Atdaev [**11**, pp. 123-127]):

THEOREM 2. *Let $u_0(t)$, $g(t)$, and $a_i(t)$, $i = 1,2$, on $([0,T])$ be nonnegative and continuous functions. Suppose further that $w_i(t)$, $i = 1,2$, are continuous nondecreasing functions on $[0,\infty)$ such that $w_i(0) \ge 0$, $w_i(u) > 0$ for $u > 0$, and*

$$\frac{d}{du}\left[\frac{w_1(u)}{w_2(u)}\right] = \frac{C}{w_2(u)}, \quad (C \ne 0 \text{ constant}).$$

If $v(t)$ is a continuous nonnegative function such that

$$v(t) \le u_0(t) + g(t)\sum_{i=1}^{2}\int_0^t a_i(s)w_i(v(s))\,ds \quad \text{on } ([0,T]),$$

then

$$v(t) \leq u_0(t) - g(t)\frac{\bar{u}_0}{\bar{g}_0} + \frac{g(t)}{\bar{g}_0} G^{-1} \left\{ \exp\left(C\bar{g}_o \int_0^t a_1(s)ds \right) \left[G(\bar{u}_0) \right.\right.$$

$$\left.\left. + C\bar{g}_0 \int_0^t a_2(s) \exp\left(-C\bar{g}_0 \int_0^{s_1}\int_0^{t_2} \cdots \int_0^{t_n} a_1(r)dr \right) ds \right] \right\}$$

where $\bar{u}_0 = \max_{0 \leq t \leq T} u_0(t)$, $\bar{g}_0 = \max_{0 \leq t \leq T} g(t)$, *and* $G(u) = w_1(t)/w_2(t) = G(v_0) + \int_{v_0}^{u} C/w_2(s)ds$, *for every t for which the value of the function given in the* $\{\ \ \}$ *is in the domain of* G^{-1}.

Atdaev and Aširov also gave the following result:

THEOREM 3. *Let $a(t)$ on $([0,T])$ be continuous and nonnegative and suppose that $u_0(t)$, on $[0,T]$, is nondecreasing and continuous in each variable with the partial derivative existing in at least one variable. Let $w(u)$ be a positive nondecreasing function for $u > 0$ with $w(0) \geq 0$. If $v(t)$ is continuous and satisfies*

$$v(t) \leq u_0(t) + \int_0^t a(s)w(v(s))\,ds, \tag{5.1}$$

then

$$v(t) \leq G^{-1}\left[G(u_0(t)) + \int_0^t a(s)ds \right] \tag{5.1'}$$

where $G(u) = \int_{\sigma}^{u} du/w(u)$ $(0 < \sigma \leq u)$.

PROOF: Denote the right hand side of (5.1) by $z(t) - \sigma$ for σ a sufficiently small positive number. Assume that u_0 is differentiable in t_1. Then

$$D_1 z(t) = D_1 u_0(t) + \int_0^{t_2} \cdots \int_0^{t_n} a(t_1, s_2, \ldots, s_n) w(v(t_1, s_2, \ldots, s_n))\, ds_2 \ldots ds_n.$$

Since $w(u)$is nondecreasing we get

$$D_1 G(z(t)) \leq D_1 G(u_0(t) + \sigma) + \int_0^{t_2} \cdots \int_0^{t_n} a(t_1, s_2, \ldots, s_n) ds_2 \ldots ds_n,$$

and by integration

$$G(z(t)) \leq G(u_0(t)+\sigma) + \int_0^t a(s)ds.$$

Taking inverse functions and letting σ go to zero, we get the result (5.1′).
REMARK 1: Atdaev and Aširov also gave analogous results for Theorems 2 and 3 in which the integral $\int_0^t$ is replaced by integral of the form $\int_{t_1}^{\infty} \dots \int_{t_n}^{\infty}$ (see Mamedov, Aširov, and Atdaev [**11**, pp. 144-146]).

6. M. H. Shih and C. C. Yeh [**12**] proved a similar generalization of Theorem 1 of 15. in Chapter XII:

THEOREM 1. *Let $a(s)$ and $b(s)$ be nonnegative continuous functions defined on $([0,\infty))^n$ and $u(s)$ be a positive continuous function on the same set. Further let $w(r)$ be a positive, continuous, and nondecreasing function on $[0,\infty)$ with $w(0)=0$ and $w'(r)$ continuous and nonnegative. If*

$$u(s) \geq u(x) - a(s)\int_x^s b(t)w(u(t))\,dt \quad \text{for } 0 < x < s,$$

then for $0 \leq x \leq s \leq s_0$,

$$u(s) \geq G^{-1}\left[G(u(x)) - a(s)\int_x^s b(t)dt\right]$$

where G is defined as in the previous Theorem and the term in the [ ] is in the domain of G for $0 \leq x \leq s \leq s_0$.

7. Note also that there exist various generalizations of Wendroff's inequality Theorem 1 of 1. (see for example Bondge and Pachpatte [**13**] or Yeh[**14**]). Here we shall give only the following result of Yeh [**14**]:

THEOREM 1. *Let $u(x)$ and $p(x)$ be nonnegative continuous functions defined for $x \geq x_0$ and $H(u)$ a positive differentiable function for $u > 0$ with $H(0)=0$ and $H'(u) \geq 0$. Suppose that*

$$u(x) \leq \sum_{i=1}^n a_i(x_i) + \int_{x_0}^x p(t)\left(u(t) + \int_{x_0}^t p(s)H(u(s))\,ds\right)dt,$$

where $a_i(x_i) > 0$ and $a_i'(x) \geq 0$, $i = 1, \ldots, n$. Then for $x_0 \leq x \leq \bar{x}$ we have

$$u(x) \leq \sum_{i=1}^{n} a_i(x_i) + \int_{x_0}^{x} p(t) G^{-1}\Bigg\{ G\left(\sum_{i=2}^{n} a_i(t_i) + a_1(x_{01})\right) + \int_{x_{01}}^{t_1} \frac{a_1'(s_1)ds_1}{\sum_{i=3}^{n} a_i(x_i) + a_2(x_{02}) + a_1(s_1) + H\left(\sum_{i=3}^{n} a_i(x_i) + a_2(x_{02}) + a_1(s_1)\right)} + \int_{x_0}^{t} p(s)ds \Bigg\} dt$$

where $G(r) = \int_{r_0}^{r} ds/(s + H(s))$, $0 < r_0 \leq r$, and $\bar{x}$ is taken such that the expression in $\{\ \}$ is in the domain of G^{-1}.

8. Returning for a moment to linear integral inequalities, the following theorem is a simple form of a result from Mamedov, Aširov, and Atdaev [**11**, pp. 131–136]:

THEOREM 1. *Let $u_0(t)$, $g(t)$, and $a(t)$ be nonnegative continuous functions on $([0,T])$ and let the nonnegative continuous function $v(t)$ satisfy*

$$v(t) \leq u_0(t) + g(t) \int_0^t a(s)v(s)ds.$$

Then

$$v(t) \leq u_0(t) + g(t) \int_0^t \exp\left(\int_s^t a(r)g(r)dr\right) u_0(s)a(s)ds.$$

The case $g(t) \equiv 1$ was considered in Young [**15**]:

THEOREM 2. *Let $u(x)$, $a(x)$ and $b(x) \geq 0$ be continuous functions defined on an open bounded set Q in R^n. Let $v(s;x)$ be the solution of the characteristic initial value problem*

$$(8.1) \qquad (-1)^n \frac{\partial^n v(s;x)}{\partial s_1 \ldots \partial s_n} - b(s)v(s;x) = 0 \quad \text{on } Q,$$
$$v(s;s) = 1,$$

and let D^+ be the corresponding subdomain of Q which contains all x such that $v \geq 0$ for all $s \in D^+$. If $D \subset D^+$ is a parallelepiped defined by $x_0 < t < x$ and

$$u(x) \leq a(x) + \int_{x_0}^{x} b(s)u(s)ds, \tag{8.2}$$

then

$$u(x) \leq a(x) + \int_{x_0}^{t} a(s)b(s)v(s;x)ds. \tag{8.2'}$$

REMARK 1: In Snow [**16**] it was noted that $v(t;x)$ is a solution of the integral equation

$$v(t;x) = 1 + \int_{t}^{x} b(s)v(s;x)ds. \tag{8.3}$$

This can be shown either by integration of (8.1) or by differentiation of (8.3). Moreover, $v(t;x)$ has a representation in the form of a Neumann series

$$v(t;x) = \sum_{0}^{\infty} u_k(t;x), \tag{8.4}$$

where $u_0(t;x) = 1$ and for $k \geq 1$,

$$u_k(t;x) = \int_{t}^{x} b(t)u_{k-1}(t_1;x)dt_1 = \cdots = \int_{t}^{x}\int_{t_1}^{x}\cdots\int_{t_{k-1}}^{x} \prod_{1}^{k} b(t_i)dt_k \ldots dt_1 \tag{8.5}$$

(see Beesack [**17**], Young [**15**], or Walter [**18**, pp. 143–144]).

A simple proof of the inequality

$$v(t;x) \leq \exp\left(\int_{t}^{x} b(r)dr\right)$$

is given by Beesack [**17**]. This inequality is also established in Fink [**19**] without the continuity hypotheses. He merely requires that u be bounded and measurable on bounded sets and that b is integrable. We give the proof.

PROOF: Let $C(y) = ([y, x])$. We iterate (8.3) to get

$$v(t;x) = 1 + \int\limits_{C(t)} b(s)ds + \int\limits_{C(t)} b(s) \int\limits_{C(s)} b(u)v(u;x)du\,ds$$
$$= 1 + \int\limits_{C(t)} b(s)ds + \int\limits_{C(t)} b(s) \int\limits_{C(s)} b(u)du\,ds$$
$$+ \int\limits_{C(t)} b(s) \int\limits_{C(s)} b(u) \int\limits_{C(u)} b(y)v(y;x)dy\,du\,ds$$
$$\vdots$$
$$= 1 + \int\limits_{C(t)} b(t_1)dt_1 + \int\limits_{C(t)} t(t_1) \int\limits_{C(t_1)} b(t_2)dt_2\,dt_1 + \dots$$
$$+ \int\limits_{C(t)} b(t_1) \int\limits_{C(t_1)} b(t_2) \int\limits_{C(t_2)} b(t_3)\dots \int\limits_{C(t_{n-1})} b(t_n)dt_n \dots dt_1 + R_n,$$

where

$$R_n = \int\limits_{C(t)} b(t_1) \int\limits_{C(t_1)} b(t_2) \int\limits_{C(t_2)} b(t_3)\dots \int\limits_{C(t_n)} b(u)v(u;x)du\,dt_n \dots dt_1.$$

Now

$$I_k = \int\limits_{C(t)} b(t_1) \int\limits_{C(t_1)} b(t_2)\dots \int\limits_{C(t_k)} b(t_{k+1})dt_{k+1}\dots dt_1$$
$$= \int\limits_{G} b(t_1)b(t_2)\dots b(t_{k+1})dt_{k+1}\dots dt_1,$$

where $G = \{(t_1, \dots, t_{k+1}) \,|\, t \le t_1 \le \dots \le t_{k+1} \le x\}$.

If σ is a permutation of the integers $1, \dots, k+1$, let $\sigma(G) = \{(t_1, \dots, t_{k+1} \,|\, t \le t_{\sigma(t_1)} \le t_{\sigma(t_2)} \le \dots \le t_{\sigma(k+1)} \le x\}$. Then $\int\limits_{\sigma(G)} b(t_1)\dots b(t_{k+1})dt_{k+1}\dots dt_1$ is invariant under σ by symmetry. If $\sigma_1 \ne \sigma_2$ then $\sigma_1(G) \cap \sigma_2(G)$ is a measure zero set. Since $C(t) = \bigcup\limits_{\sigma\in\pi_{k+1}} \sigma(G)$ we have

$$I_k = \frac{1}{(k+1)!} \int\limits_{C(t)} b(t_1)\dots b(t_{k+1})dt_1\dots dt_{k+1} = \frac{1}{(k+1)!} \left(\int\limits_{C(t)} b(s)ds \right)^{k+1}.$$

Since $b \geq 0$ and $v(t,x)$ is bounded (being continuous) say by M, then $R_n \leq \frac{M}{(n+1)!}\left(\int_{C(t)} b(s)ds\right)^{n+1}$ and so $R_n \to 0$ as $n \to \infty$ and

$$v(t;x) \leq 1 + \int_{C(t)} b(s)ds + \frac{1}{2!}\left(\int_{C(t)} b(s)ds\right)^2 + \dots$$
$$= e^{\int_{C(t)} b(s)ds}.$$

9. Similar results for various other integral inequalities are given in Yeh [**20**]. Here we shall give the following one:

THEOREM 1. *Suppose that $u(x)$, $a(x)$, $b(x)$, and $c(x)$ are nonnegative continuous functions defined on Q with D and D^+ defined as in* Theorem 2 of 8. *Let $v(s;x)$ be the solution of the characteristic initial value problem*

$$(-1)^n \frac{\partial^n v(s;x)}{\partial s_1 \dots \partial s_n} - [b(s) + c(s)]\, v(s;x) = 0 \quad \text{on } Q, \quad v(s;x) = 1 \text{ for } s = x.$$

If

$$u(x) \leq a(x) + \int_{x_0}^{x} b(s)\left[u(s) + \int_{x_0}^{s} c(t)u(t)dt\right] ds, \quad \text{on } Q,$$

then

$$u(x) \leq a(x) + \int_{x_0}^{x} b(s)\left[a(s) + \int_{x_0}^{s} a(t)\,[b(t) + c(t)]\, v(t;s)dt\right] ds.$$

10. We have previously given comparison theorems for integral inequalities in the one dimensional case. We now give two such theorems for the case of several variables, both come from the book by Mamedov, Aširov, and Atdaev [**11**, pp. 118–121]:

THEOREM 1. *Let the function $\phi(t,s,u)$ be continuous on $0 \leq t, s \leq T$, $|u| \leq \delta < \infty$ and nondecreasing in u. Let $u_0(t)$ be a continuous function on $([0,T])$ and $v(t)$ a continuous solution of*

$$v(t) < u_0(t) + \int_0^t \phi\,(t,s,v(s))\, ds, \quad \text{on } ([0,T]), \tag{10.1}$$

then

$$v(t) < u(t) \quad on \quad ([0,T]) \tag{10.1'}$$

where $u(t)$ is a solution of the integral equation

$$u(t) = u_0(t) + \int_0^t \phi(t,s,u(s))\, ds \quad on \quad ([0,T]). \tag{10.2}$$

PROOF: The parallelepiped $([0,T])$ is a convex set and (10.1) and (10.2) imply (10.1′) on any face of dimension $r \leq n-1$ that has one coordinate zero. Denote by G the set of points in in $([0,T])$ for which $v(t) \not< u(t)$. This set is closed and by the above remark does not contain any points of the faces mentioned above. Let t denote a point which is the upper limit of values α such that the interval $0 \leq t \leq (\alpha, \alpha^2, \ldots, \alpha^n)$ does not intersect G. There exists a point τ such that $v(\tau) \not< u(\tau)$ and $\tau \leq \alpha$. On the other hand, for $0 \leq t \leq (\tau_1, \alpha^2, \ldots, \alpha^n)$ we have the $v(t) < u(t)$ so as a consequence of (10.1) and (10.2) we have

$$v(\tau) < u_0(\tau) + \int_0^\tau \phi(\tau,s,v(s))\, ds \leq u_0(\tau) + \int_0^\tau \phi(\tau,s,u(s))\, ds = u(\tau).$$

This contradiction proves that G is empty and the theorem is proved.

We say that the function $\phi(t,s,u)$ satisfies the condition (μ) if the n-dimensional equation

$$W(t) = u_0(t) + \lambda + \int_0^t \phi(t,s,W(s))\, ds$$

has a solution on $([0,T])$ for every $\lambda \in [0,\mu]$.

THEOREM 2. *Let $\phi(t,s,u)$ be continuous for $0 \leq t, s \leq T$; $|u| < \delta$, and nondecreasing in u and satisfy condition (μ) for some small μ. Further assume that $u_0(t)$ is continuous on $([0,T])$. If $v(t)$ is a continuous solution of*

$$v(t) \leq u_0(t) + \int_0^t \phi(t,s,v(s))\, ds \quad on \quad ([0,T]),$$

then $v(t) \leq u(t)$ where $u(t)$ is the maximal solution of (10.2) defined on $([0,T])$.

The proof is similar to that of Theorem 2 from 23. of Chapter XII.

REMARK 1: With respect to condition (μ) note that the following result holds (see Beesack [**21**, pp. 88–89]):

Let G be an open set in R^n, and for $x, y \in G$ let $G(x,y)$ denote the parallelepiped with diagonal joining x to y. Let the points $x_0, y \in G$ such that $G_0 = G(x_0, y) \subset G$ and suppose the functions $a(x)$ and $k(x,t,z)$ be continuous on G_0 and $G_T \times R$ respectively where $G_T = \{(x,t)\colon x \in G_0,\ t \in G(x_0,x)\}$. Suppose further that k is nondecreasing in z for $(x,t) \in G_T$. If

$$|k(x,t,z)| < h(t)g(|z|) \quad \text{for } (x,t,z) \in G_T \times R, \tag{10.3}$$

where $h \in L(G_0)$ and g is continuous and nondecreasing on $[0,\infty)$ with $\int\limits_1^\infty ds/g(s) = \infty$, then

$$u(x) = a(x) + \int\limits_{x_0}^{x} k(x,t,u(t))\,dt \tag{10.4}$$

has a solution which is continuous in G_0. Moreover, if $\{\varepsilon_n\}$ is a strictly decreasing sequence which converges to zero and if u_n is a corresponding solution of the integral equation

$$u_n(x) = a(x) + \varepsilon_n + \int\limits_{x_0}^{x} k(x,t,u_n(t))\,dt, \tag{10.5}$$

then $U(x) = \lim u_n(x)$ exists uniformly on G_0, and $U(x)$ is the maximal solution of (10.4).

It is clear that using this and Theorem 2 we have: if a continuous function $v(t)$ satisfies

$$v(x) \le a(x) + \int\limits_{x_0}^{x} k(x,t,v(t))\,dt, \quad x \in G_0,$$

then

$$v(x) \le U(x), \quad x \in G_0,$$

where $U(x)$ is the maximal solution of (10.4) on G_0 (see Headley [**22**]).

11. K. M. Das [**23**] in 1979 gave a short and elegant proof of an inequality of Gronwall type for systems of two integral inequalities

$$\begin{aligned} f(t) &\le K_1 + \int\limits_0^t h_1(s)f(s)ds + \int\limits_0^t e^{\mu s}h_2(s)g(s)ds \\ g(t) &\le K_2 + \int\limits_0^t e^{-\mu s}h_3(s)f(s)ds + \int\limits_0^t h_4(s)g(s)ds \end{aligned} \tag{11.1}$$

where K_1, K_2, and μ are nonnegative constants and f, g, and h_i are nonnegative continuous functions on $[0, \infty)$.

This result is due to D. E. Greene [**24**] and says that there are constants C_i and M_i such that for $t \geq 0$

$$f(t) \leq M_1 \exp(C_1 t), \quad g(t) \leq M_2 \exp(C_2 t) \tag{11.2}$$

if in addition the functions h are bounded on $[0, \infty)$.

P. R. Beesack [**25**] gave a simple proof of a result with explicit values for the constants in this estimate using the following result of Z. Opial [**26**] (see also Beesack [**21**, Th. 4.1]) for a system (possibly nonlinear) of integral inequalities:

THEOREM 1. *Let $J = [\alpha, \beta)$ and f be continuous on $J \times R^n$ into R^n, and satisfy $f(t, x) \leq f(t, y)$ whenever $x \leq y$. If x is a continuous vector function such that on J*

$$x(t) \leq c + \int_\alpha^t f(s, x(s))\, ds,$$

for some c in R^n and if r is the maximal solution on $J_1 = [\alpha, \beta_1)$ with $\beta_1 < \beta$) of

$$r(t) = c + \int_\alpha^t f(s, r(s))\, ds,$$

then $x(t) \leq r(t)$ for $t \in J_1$.

12. Mamedov, Aširov and Atdaev [**11**, pp. 101–102] proved the same result by replacing f by a more general function:

THEOREM 1. *Let $\phi(t, s, u)$ be continuous and nondecreasing in u on $0 \leq s, t \leq T$, $|u_i| < \delta_i$, and let $w(t)$ be a continuous vector function on $[0, T]$. If $v(t)$ is a continuous vector function such that*

$$v(t) < w(t) + \int_0^t \phi(t, s, v(s))\, ds, \tag{12.1}$$

then $v(t) < u(t)$ on $(0, T]$ where $u(t)$ is the minimal solution of

$$u(t) = w(t) + \int_0^t \phi(t, s, u(s))\, ds \tag{12.2}$$

defined on $[0, T]$.

Continuity of the function ϕ may be somewhat relaxed to the conditions of Caratheodory:

We say that the vector function $\phi(t, s, u)$ satisfies the conditon of Caratheodory (K) for $t, s \in [0, T]$ and $\|u\| = \max_{1 \le i \le n} |u_i| < \delta$ if

a) $\phi(t, s, u)$ is continuous in u for every t and for almost all s, and measurable in s for every t and u;

b) for any positive number $\delta_1 < \delta$ there exist functions $u_1(t, s)$ and $v_1(t_1, t, s)$ $(0 \le s \le t \le t_1 \le T)$ measurable in s such that

$$\sup_{\|u\| \le \delta_1} \|\phi(t, s, u)\| \le u_1(t, s),$$

$$\sup_{\|u\| \le \delta_1} \|\phi(t_1, s, u) - \phi(t, s, u)\| \le v_1(t_1, t, s), \quad \text{with}$$

$$\lim_{t_1 - t \to \infty} \left[\int_t^{t_1} u_1(t_1, s) ds + \int_0^t v_1(t_1, t, s) ds \right] = 0$$

for fixed t or t_1.

Denote by $u_\tau(t)$ the solution of the system (12.2) defined on $[0, \tau) \subset [0, T)$. If $\tau < \xi$ and $u_\xi(t) = u_\tau(t)$ for $t \in [0, \tau)$ then we say that $u_\xi(t)$ is an extension of $u_\tau(t)$ on $[0, \xi)$, and $u_\tau(t)$ is part of $u_\xi(t)$. Finally, if $u_\tau(t)$ is not part of another solution, we say that it cannot be extended.

The following theorem is also given in Mamedov, Aširov, and Atdaev [**11**, pp. 103]:

THEOREM 2. *Let the vector function* $\phi(t, s, u)$ *satisfy the condition* (K) *for* $t, s \in [0, T]$ $(T \le \infty)$ *and* $|u| \le \delta \le \infty$. *Moreover assume that it is nondecreasing in* u *and that* $w(t)$ *is continuous and* $\|w(0)\| < \delta$. *Let* $u_\alpha(t)$ *be the minimal (maximal) solution of the system (12.2) which cannot be extended. If* $v(t)$ *is a continuous vector function on* $[0, d) \subset [0, T)$ *with* $\|v(t)\| < \delta$, *and satisfying*

$$v(t) \ge \ (\le)\ w(t) + \int_0^t \phi(t, s, v(s))\, ds,$$

then $v(t) \ge \ (\le)\ u_\alpha(t)$ *on* $[0, d)$.

13. Let us return to inequality (11.1). B.B. Shinde and B.G. Pachpatte [**27**, Th.1] dealt with a generalization of (11.1) where the independent variable is n-dimensional. Their result is in a certain sense included in a general theorem proved in 1976 by J. Chandra and P. W. Davis [**28**], which we give in our next theorem. Moreover, this latter result is turn included in an abstract version of a general linear Gronwall inequality formulated at least as early as 1969 by E. Hille [**29**, pp. 18–20],and [**30**, pp. 364–376]. Of course, comparison theorems give the best possible results for the considered inequalities. From the practical point of view the problem of solving a system of inequalities has only been replace by the possibly more difficult problem of solving a system of equations. This is not of much use to someone who is seeking explicit upper bounds for the functions satisfying the inequalities.

THEOREM 1. *Let $G(x)$ and $H(x)$ be $N \times N$ matrices and $a(x)$ and $u(x)$ be N-vectors that are continuous functions for $x_0 \le x$. Let all the components of H and G be nonnegative. If*

$$u(x) \le a(x) + G(x) \int_{x_0}^{x} H(y)u(y)dy, \quad x_0 \le x, \tag{13.1}$$

then

$$u(x) \le a(x) + G(x) \int_{x_0}^{x} V(x,y)H(y)a(y)dy, \quad x_0 \le x, \tag{13.1'}$$

where

$$V(x,y) = I + \int_{y}^{x} H(z)G(z)V(z,y)ds, \quad x_0 \le y \le x. \tag{13.2}$$

(I is the identity matrix and the integral is a multiple integral as before.)

14. Using this best possible estimate for u, we can obtain various explicit extimates (see Chandra and Davis [**28**], and especially Beesack [**31**, pp. 11–17]). Of course, some explicit estimates can be obtained directly. Here we shall give some results of P. R. Beesack [**31**, pp. 4–11]:

THEOREM 1. *Suppose u_i, f_i, g_{ij}, and h_{ij} are all continuous functions with u_i, g_{ij}, and h_{ij} all nonnegative for $x \ge 0$. If $G = [g_{ij}]$ and $H = [h_{ij}]$ and u satisfy the inequality* (13.1) *for $x \ge 0$ then for such x and $i = 1, \ldots, N$* we have

$$u_i(x) \le f_i(x) + R_0(x)G_i(x), \tag{14.1}$$

where

$$G_i(x) = \max_{1\le k\le N} g_{ik}(x), \quad 1 \le i \le N, \tag{14.2}$$

$$R_0(x) = \exp\left(\int_0^{x_1} P(s_1,x)ds_1\right)\int_0^{x_1} A(t_1,\tilde{x})\exp\left(-\int_0^{t_1} P(s_1,\tilde{x})ds_1\right)dt_1 \tag{14.3}$$

with $\tilde{x} = (x_2,\dots,x_n)$, $x = (x_1,\tilde{x})$, $\tilde{t} = (t_2,\dots,t_n)$, and

$$A(x_1,\tilde{x}) = \sum_{j,k=1}^{N}\int_0^{\tilde{x}} h_{kj}(x_1,\tilde{t})f_j(x_1,\tilde{t})d\tilde{t}, \tag{14.4}$$

and

$$P(x_1,\tilde{x}) = \max_{1\le\sigma\le N}\sum_{j,k=1}^{N}\int_0^{\tilde{x}} h_{kj}(x_1,\tilde{t})g_{j\sigma}(x_1,\tilde{t})d\tilde{t}. \tag{14.5}$$

REMARK 1: In the one dimensional case (14.3–14.5) become

$$R_0(x) = \exp\left(\int_0^{x} P(s)ds\right)\int_0^{x} A(t)\exp\left(-\int_0^{t} P(s)ds\right)dt,$$

$$A(x) = \sum_{j,k=1}^{N} h_{kj}(x)f_j(x), \text{ and } P(x) = \max_{1\le\sigma\le N}\sum_{j,k=1}^{N} h_{kj}(x)g_{j\sigma}(x).$$

REMARK 2: Using the above notation, the system (11.1) can be rewritten in the form

$$u_1 \le K_1 + \int_0^x h_1u_1dt + \int_0^x e^{\mu t}h_2u_2dt,$$

$$u \le K_2 + \int_0^x e^{-\mu t}h_3dt + \int_0^x h_4u_2dt,$$

so that $G(x) = I$, $h_{11} = h_1$, $h_{12} = e^{\mu t} h_2$, $h_{21} = e^{-\mu t} h_3$, $h_{22} = h_4$, so by Remark 1 we have

$$A(x) = K_1(h_1 + e^{-\mu s} h_3) + K_2(e^{\mu s} h_2 + h_4),$$
$$P(s) = \max\{h_1 + e^{-\mu s} h_3,\ e^{\mu s} h_2 + h_4\}, \text{ and}$$

$$u_i(x) \le K_i + \int_0^x A(t) \exp\left(\int_t^x P(s)ds\right) dt, \quad x \ge 0. \tag{14.6}$$

For a crude estimate note that $A(s) \le (K_1 + K_2)P(s)$, so (14.6) becomes

$$\begin{aligned} u_1(x) &\le -K_2 + (K_1 + K_2) \exp\left(\int_0^x P ds\right), \\ u_2(x) &\le -K_1 + (K_1 + K_2) \exp\left(\int_0^x P ds\right). \end{aligned} \tag{14.7}$$

The bounds given by Das [23] are

$$\begin{aligned} u_1(x) &\le (K_1 + K_2) \exp\left(\mu x + \int_0^x h ds\right), \\ u_2(x) &\le (K_1 + K_2) \exp\left(\int_0^x h ds\right), \end{aligned} \tag{14.8}$$

where $h = \max(h_1 + h_3, h_2 + h_4)$. The crude bound (14.7) is thus better than (14.8) for small x. However in case $h_2 + h_4 \ge h_1 + h_3$ then $h \le P$ and Das's bounds are better for large x.

THEOREM 2. *If u_i, f_i, g_{ij}, and h_{ij} are defined as in Theorem 1. and u satisfies the inequality (13.1) for $x \ge 0$ then*

$$\tilde{u}(x) \le \tilde{f}(x) + \tilde{G}(x) \int_0^x \tilde{f}(t)\tilde{H}(t) \exp\left(\int_t^x \tilde{G}(s)\tilde{H}(s)ds\right) dt, \tag{14.9}$$

where $\tilde{u}(x) = \max_{1 \le i \le N} u_i(x)$, $\tilde{f}(x) = \max_{1 \le i \le N} f_i(x)$, and

$$\tilde{G}(x) = \sum_{k=1}^{N} \max_{1 \le i \le N} g_{ik}(x), \quad \tilde{H}(x) = \max_{1 \le k \le N} \sum_{i=1}^{N} h_{kj}(x). \tag{14.10}$$

PROOF: Consider a component of u in (13.1),

$$u_i(x) \leq f_i(x) + \sum_{k=1}^{N} g_{ik}(x) R_k(x),$$

where

$$R_k(x) = \sum_{j=1}^{N} \int_0^x h_{kj}(t) u_j(t) dt, \quad 1 \leq k \leq N.$$

From this we get

$$\tilde{u}(x) \leq \tilde{f}(x) + \sum_{k=1}^{N} \tilde{G}_k(x) R_k(x), \qquad \left(\tilde{G}_k(x) = \max_{1 \leq i \leq N} g_{ik}(x)\right)$$

as well as

$$R_k(x) \leq \sum_{j=1}^{N} \int_0^x h_{kj}(t) \tilde{u}(t) dt = \int_0^x H_k(t) \tilde{u}(t) dt,$$

where $H_k = \sum_{j=1}^{N} h_{kj}$; that is we get

$$\text{(14.11)} \qquad \tilde{u}(x) \leq \tilde{f}(x) + \tilde{G}(x) \int_0^x \tilde{H}(t) \tilde{u}(t) dt, \quad x \geq 0.$$

Using Theorem 1 from 8. we obtain (14.9).

15. Some generalizations of Greene's Theorem are also given in G. S. Yang [**32**] and Pachpatte [**33**].

16. The method used in either Theorems of 14. can also be applied to nonlinear systems of inequalities. Beesack illustrated this by using the method of Theorem 2 of 14. for the system

$$\text{(16.1)} \qquad u_i(x) \leq f_i(x) + \int_0^x k_i(t) g_i\left(u_1(t), \ldots, u_N(t)\right) dt,$$
$$x \geq 0, \quad 1 \leq i \leq N,$$

to prove the following result:

THEOREM 1. *Let u_i, f_i, and k_i be continuous nonnegative functions on $x \geq 0$ in R^N, and suppose that $g_i(u)$ are continuous and nondecreasing in every u, and $1 \leq i \leq N$. If $\tilde{f}(0) > 0$ and $\tilde{g}(u) > 0$ for $u > 0$ and (16.1) holds, then*

$$u(x) \leq \phi^{-1}\left[\phi\left(\hat{f}(x)\right) + \int_0^x \tilde{k}(t)dt\right], \quad 0 \leq x \leq x_1,$$

where $\tilde{k}(t) = \max_{1\leq i\leq N} k_i(t)$, $\tilde{u}(x) = \max_{1\leq i\leq N} u_i(x)$, and $\tilde{f}(x) = \max_{1\leq i\leq N} f_i(x)$, and the functions $\hat{f}$, g and ϕ are defined by $\hat{f}(x) = \max_{0\leq t\leq x} f(t)$, $\phi(r) = \int_{r_0}^{r} ds/\tilde{g}(s)$ $(r > 0,\ r_0 > 0)$ and $\tilde{g}(u) = \max_{1\leq i\leq N} g_i(u,\ldots,u)$. Finally x_1 is chosen so that $\phi\left(\hat{f}(x)\right) + \int_0^x k(t)dt \in \operatorname{dom}(\phi^{-1})$ for $0 \leq x \leq x_1$.

17. As is usual, we say that a matrix is continuous or bounded or measurable on ([0, T]) if and only if each component satisfies the requirement and $|\ \ |$ denotes the usual Euclidean 2-norm.

Theorem 1 of 13. was generalized by P. R. Beesack [**34**]. In fact he firstly considered a system of integral equations with unknown functions of several variables and proved the following results:

THEOREM 1. *Let $a, b \in R^n$, with $a_j < b_j$ for each j. Let $J = [a, b]$ and $T = \{(x,t) : a \leq t \leq x \leq b\}$.*

Suppose that $f \in L^2(J)$ and $K \in L^2(T)$. Then the systems of integral equations

$$u_i(x) = f_i(x) + \sum_{j=1}^{N} \int_0^x k_{ij}(x,t)u_j(t)dt, \quad 1 \leq i \leq N, \tag{17.1}$$

or in matrix form

$$u(x) = f(x) + \int_a^x K(x,t)u(t)dt \tag{17.2}$$

has a unique solution $u \in L^2(J)$. This solution is given be the Neumann series

$$u(x) = \sum_{r=0}^{\infty} v_r(x), \tag{17.3}$$

where

(17.4) $$v_0(x) = f(x), \quad v_r(x) = \int_a^x K^{(r)}(x,t)f(t)dt, \quad r \geq 1,$$

and $K^{(r)}$ *are the iterated kernels of* K *defined by*

(17.5) $$K^{(1)}(x,t) = K(x,t), \quad K^{(r)}(x,t) = \int_a^x K(x,s)K^{(r-1)}(s,t)ds, \quad r \geq 2.$$

The series $\sum_0^\infty |v_r(x)|$ *is convergent in* $L^2(J)$ *as well as pointwise convergent on* J*. The solution is also given be the representation*

(17.6) $$u(x) = f(x) + \int_a^x M(x,s)f(s)ds, \quad x \in J,$$

where $M \in L^2(T)$ *and denotes the resolvent kernel in* L^2 *defined by the* L^2 *convergent series*

(17.7) $$M(x,t) = \sum_{r=1}^\infty K^{(r)}(x,t).$$

THEOREM 2. *Let* $J = [a,b]$ *be compact in* R^n*, and suppose that* f *and* K *are bounded and measurable (or continuous) on* J *and* T *respectively (as in Theorem 1). Then the system (17.1) or (17.2) has a unique solution which is bounded and measurable (continuous) on* J*. The solution is given by the Neumann series (17.3) which is uniformly absolutely convergent on* J*, and is also given be the representation (17.6) where is bounded and measurable (continuous) on* T*. The Neumann series (17.7) for is uniformly convergent on* T*.*

Note also that

$$\begin{aligned} K^{(r)}(x,t) &= \int_t^x \int_t^{t_1} \cdots \int_t^{t_{r-2}} K(x,t_1)K(t_1,t_2)\ldots K(t_{r-1},t)dt_{r-1}\ldots dt_1 \\ &= \int_t^x \int_{t_{r-1}}^x \cdots \int_{t_2}^x K(x,t_1)K(t_1,t_2)\ldots K(t_{r-1},t)dt_1\ldots dt_{r-1} \end{aligned}$$

and

$$K^{(r)}(x,t) = \int_t^x K^{(i)}(x,s)K^{(j)}(s,t)ds, \quad r \geq 2, \quad i+j=r,$$

$$\text{and } 1 \leq j \leq r-1.$$

Also we have for $(s,t) \in T$ that

$$M(x,t) = K(x,t) + \int_t^x K(x,s)M(s,t)ds = K(x,t) + \int_t^x M(s,t)K(x,s)ds.$$

Thus, Beesack proved:

THEOREM 3. *Let J, T, f, and K be as in* Theorem 1 *or* 2, *and suppose that $K \geq 0$ on T. If u satisfies the same conditions as f on J and*

$$u(x) \leq f(x) + \int_a^x K(x,t)u(t)dt, \quad x \in J, \tag{17.8}$$

then

$$u(x) \leq f(x) + \int_a^x M(x,t)f(t)dt, \quad \text{for } x \in J, \tag{17.8'}$$

where M is the resolvent kernel of K defined by (17.7).

PROOF: It is easy to prove that $K \geq 0$ implies that $K^{(r)} \geq 0$ on T and consequently that M is also nonnegative on T. Now define the function $\hat{f}$ on J by

$$\hat{f}(x) = f(x) + \int_a^x K(x,t)u(t)dt - u(x).$$

Thus $\hat{f}(x) \geq 0$ on J by (17.8). Also

$$u(x) = \{f(x) - \hat{f}(x)\} + \int_0^x K(x,t)u(t)dt, \quad x \in J.$$

Since $\hat{f} - f$ is a function of the same class as f and u, the basic Theorems 1 and 2 yield

$$u(x) = f(x) - \hat{f}(x) + \int_0^x \Gamma(x,t)\{f(t) - \hat{f}(t)\}dt, \quad x \in J.$$

Since $\hat{f} \geq 0$, and $M \geq 0$, we arrive at (17.8′).

REMARK 1: The case when $K(x,t) = G(x)H(t)$ gives Theorem 1 of 13. since $M(x,t) = G(x)V(x,t)H(t)$ where $V(x,t) = I + \int_t^x H(x)G(s)V(s,t)ds = I + \int_t^x V(s,t)H(s)G(s)ds$.

REMARK 2: Using Theorem 1, Beesack gave a series of explicit bounds for solution of the inequality (17.8).

REMARK 3: Beesack in [**34**] considered also the analogous nonlinear systems of inequalities.

18. Further references are [**35**]–[**124**].

REFERENCES

1. BECKENBACH, E. and R. BELLMAN, "Inequalities," Berlin-Heidelberg-New York, 1965.
2. NURIMOV, T., *Issledovanie rešenija dvumernogo nelineinogo integral'nogo uravnenija Vol'terra,* Doklady AN UzSSR, **1971**, 6–8.
3. __________, *Issledovanie rešenii sistem dvumernyh nelineinyh integral'nyh uravnenii tipa Vol'terra po parametram*, Sb. Kraevie zadači dlja differencial'nih uravnenii s častnymi proizvodymi. Taškent, **1972**, 129–148.
4. EŠMATOV, H., *Ob odnom integral'nom neravenstve*, Trudi In-ta kiber. s VC AN UzSSR. "Voprosy vyčisl. i prikl. matematiki", vyp **18** Taškent, 1973.
5. BYKOV, Ja.V. and H.M. SALPAGAROV, *K teorii integrodifferencial'nyh uravnenii.* In: "Issledovanija po integrodifferencial'nym uravnenijam v Kirgizii," vyp. 2. Izd-vo AN Kirg. SSR, 1962.
6. BONDGE, B.K., B.G. PACHPATTE, and W. WALTER, *On generalized Wendroff type inequalities and their applications,* Nonlinear Anal. Th. and Method. Appl. **4** (1980), 491–495.
7. DRAGOMIR, S.S. and N.M. IONESCU, *The Wendroff type lemmas and applications (I).* In: Simp. of Funct. Eg. Approx. and Conv. Cluj-Napoli, 1981.

8. __________, *The Wendroff type lemmas and applications (II)*, Simp. Math. Applic. Timmoşoara, 1987, 65–68.
9. PACHPATTE, B.G., *On Wendroff type partial integral inequalities.* Tamkang J. Math. **10** (1979), 141–150.
10. NURIMOV, T., *Some certain two-dimensional integral inequalities* (Russian), Dokl. Akad. Nauk AzSSR **1977**, no 1.
11. MAMEDOV, Ja. D., S. AŠIROV and S. ATDAEV, "Teoremy on Neravenstvah," Ašhabad, YLYM, 1980.
12. SHIH, M.N. and C.C. YEH, *Some integral inequalities in n independent variables*, J. Math. Anal. Appl. **84** (1981), 569–583.
13. BONDGE, B.K. and B.G PACHPATTE, *On two independent variable generalizations of certain integral inequalities*, Tamkang J. Math. **11** (1980), 37–47.
14. YEH, C.C., *Wendroff's inequality in n independent variables*, J. Math. Anal. Appl. **78** (1980), 78–87.
15. YOUNG, E.C., *Gronwall's inequality in n independent variables*, Proc. Amer. Math. Soc. **41** (1973), 241–244.
16. SNOW, D., *Gronwall's inequality for systems of partial differential equations in two independent variables*, Proc. Amer. Math. Soc. **33** (1972), 46–54.
17. BEESACK, P.R. *Riemann's function has an exponential bound*, Proc. Amer. Math. Soc. **88** (1983), 313- -316.
18. WALTER, W. "Differential and Integral Inequalities," Berlin–Heidelberg–New York, 1970.
19. FINK, A.M., *Wendroff's inequality*, Nonlinear Anal. **5** (1981), no 3, 813–819.
20. YEH, C.C., *On some integral inequalities in n independent variables and their applications*, J. Math. Anal. Appl. **86** (1982), 387–410.
21. BEESACK, P.R., *Gronwall inequalities*, Carleton Un. Math. Notes, No 11, 1975.
22. HEADLEY, V., *A multidimensional nonlinear Gronwall inequality*, J. Math. Anal. Appl. **47** (1974), 250–255.
23. DAS, K., *A note on a inequality due to Greene*, Proc. Amer. Math. Soc. **77** (1979), 424–425.
24. GREENE, D., *An inequality for a class of integral systems*, Proc. Amer. Math. Soc. **62** (1977), 101–104.
25. BEESACK, P.R., Review 26015. Zentralbl. für Math. **351**, 1978.
26. OPIAL, Z., *Sur un système d'inéqualities intégrales*, Ann. Polon. Math. **7** (1960), 247–254.
27. SHINDE, B.B. and B.G. PACHPATTE, *On multidimentional simultaneous integral inequalities*, J. Math. Phys. Sci. **17** (1983), 219–243.

28. CHANDRA, J. and P.W. DAVIS, *Linear generatizations of Gronwall's inequality*, Proc. Amer. Math. Soc. **60** (1976), 157–160.
29. HILLE, E., "Lectures on Ordinary Differential Equations," Reading, Mass., 1969.
30. __________, "Methods in Classical and Functional Analysis," Reading, Mass., 1972.
31. BEESACK, P.R., *Systems of multidimensional Gronwall integral inequalities*, Carleton Math. Ser. No. **206**, 1984.
32. YANG, G.S., *A note on a theorem of Green.* Tamkang J. Math. **19** (1983), 211–214.
33. PACHPATTE, B.G., *A note on Greene's inequality.* Tamkang J. Math **15** (1989), 49–59.
34. BEESACK, P.R., *Systems of multidimensional Volterra integral equations and inequalities*, Carleton Math. Ser. No. **209**, 1984.
35. AGARWAL, R.P., *On an integral inequality in n independent variables*, J. Math. Anal. Appl. **85** (1982), 192–196.
36. __________, *On integro-differential inequalities in two independent variables*, J. Math. Anal. Appl. **89** (1982), 581–597.
37. __________, *Inequalities involving partial derivatives*, J. Math. Anal. Appl. **89** (1982), 628–638.
38. AHMEDOV, K. and T. NURIMOV, *O dvuhmernyh integral'nyh uravnenijah i neravenstvah tipa Vol'terra*, Doklady Akademii Nauk Uz SSR, **1972**, 3–4.
39. AKINYELE, O., *On an integral inequality in n independent variables*, Internt. J. Math. Sci. **7** (1984), 455–467.
40. __________, *On Gronwall-Bellman-Bihari-type integral inequalities in several variables with retardation*, J. Math. Anal. Appl. **104** (1984), 1–26.
41. ALEKSEEV, V.M., *Teorema ob integralnom neravenstve i nekotorye ee priloženija*, Mat. Sbornik **68** (110) (1965), 251–273.
42. ATDAEV, S. and S. AŠIROV, *Ob odnom metode dvustoronnih približeniǐ dlja resenija integro-differencial'nyh uraveniǐ Vol'terra.* In: "Difference. i integral'n uravnenija Teorija verojatnostei," Ašhabad, 1976.
43. __________, *Issledovanie rešeniǐ nelineinyh mnogomernyh operatornyh uravnenii Vol'terra*, Ukr. Mat. Ž. **29** (1977), No 5.
44. __________, *K teorii mnogomernyh integral'nyh neravenstvah.* In: "Diferenc. i integral'n. uravnenija," Teorija verojatnostei. Ašhabad, 1976.
45. AZBELEV, N.V. and Z.B. DALJUK, *Ob integral'nyh neravenstvah*, Mat. Sbornik **56**, 1962, No 3.

46. BAĬNOV, D.D. and S.G. KHRISTOVA, *Integral inequalities for functions of several variables* (Russian), Bull. Math. Soc. Sci. Math. R. S. Roumaine (N.S.) **26** 74 (1982), 15–20.
47. BEESACK, P.R., *On some Gronwall-type integral inequalities in n independent variables*, J. Math. Anal. Appl. **100** (1984), 393–408.
48. BOBROWSKI, D., J. POPENDA and J. WERBOWSKI, *On the systems integral inequalities with delay of Gronwall-Bellman type*, Fasc. Math. **10** (1978), 97–104.
49. BONDGE, B.K. and B.G. PACHPATTE, *On nonlinear integral inequalities of the Wendroff type*, J. Math. Anal. Appl. **70** (1979), 161–169.
50. __________, *On some fundamental integral inequalities in two independent variables*, J. Math. Anal. Appl. **72** (1979), 533–544.
51. __________, *On Wendroff type integral inequalities in n independent variables*, Chinese J. Math. **7** (1979), 37–46.
52. __________, *On generalized Wendroff-type inequalities and their applications*, Nonlinear Anal. **4** (1980), 491–495.
53. __________, *On some fundamental integral inequalities in n independent variables*, Bull. Inst. Math. Acad. Sinica **8** (1980), 553–560.
54. __________, *On some partial integral inequalities in two independent variables*, Funkcion. Ekvac. **23** (1980), 327–334.
55. BURLAČENKO, V.P. and N.I. SIDENKO, *A generalization of the Wendroff inequality* (Russian), Question of the theory of the approximation of functions (Russian), pp. 14-21, 193. Akad. Nauk USSR, Inst. Mat. Kiev, 1980.
56. CHEN, L.S. and C.C. YEH, *Some integro-differential inequalities in n independent variables*, Proc. Roy. Soc. Edinburg. Sect. A. **89** (1981), 347–353.
57. CHOSHAL, S.K. and M.A. MASOOD, *Generalized Gronwall's inequality and its applications to a class of non-selfadjoint linear and non-linear hyperbolic partial differential equations, to establish their uniqueness continuous dependence and comparison theorems*, Indian J. Pure Appl. Math. **5**, (1974), No 3.
58. CORLAN, J. and C.L. WANG, *Gronwall-Bellman type inequalities for systems*, J. Math. Anal. Appl. **120** (1986), 622–630.
59. __________, *Higher dimensional Gronwall-Bellman type inequalities*, Applicable Analysis **18** (1984), 1–12.
60. CONSTANTIN, A. *A Gronwall-like inequality and its applications*, Rend. Mat. Acc. Lincei Ser IX. vol. I (1990), 111–115.
61. CORDUNEANU, A., *A note on the Gronwall inequality in two independent variables*, J. Integral. Equations **4** (1982), 271–276.

62. __________, *An integral inequality in two independent variables*, An. Ştinţ. Univ. "Al. I. Cuza" Iaşi Sect. Ia Mat. (N.S.) **29** (1983) 69–74.
63. __________, *A nonlinear integral inequality in two independent variables*, An. Ştinţ. Univ. "Al. I. Cuza" Iaşi Sect. Ia Mat (N.S.) **30** (1984), 15–18.
64. __________, *Nonlinear integral inequalities in n independent variables*, An. Ştinţ. Univ. "Al. I. Cuza" Iaşi Sect. Ia Mat. (N.S.) **31** (1985), 281–288.
65. __________, *Integral inequalities in two independent variables*, Rev. Roum. Math. Pures Appl. **32** (1987), 331–341.
66. DANNAN, F.M., *Submultiplative and subadditive functions and integral inequalities of Bellman-Bihari type*, J. Math. Anal. Appl. **120** (1986), 631–646.
67. DAS, P. and R. SHARMA, *Existence and stability of measure differential equations*, Chechoslavak Math. J. **22** (97) (1972), 145–158.
68. DE FRANCO, R., *Gronwall's inequality for systems of multiple Volterra integral equations*, Funkcial. Ekvac. **19** (1976), 1–9.
69. DRAGOMIR, S.S. and N. M. IONESCU, *On nonlinear integral inequalities of two independent variables*, Studia Univ. Babes-Bolyai, Math. **34** 2 (1989), 11–17.
70. DŽABAROV, S.T. and G.K. MAMEDOV, *An integral inequality for functions of two variables and its applications* (Russian), Izv. Akad. Nauk Azerb. SSR, Ser. FTMN **5** (1984), no. 4, 32–37.
71. GOSHAL, A., *Generalized n-dimensional Gronwall's inequality and its application to non-selfadjoint and nonlinear hyperbolic equations*, Ann. Polon. Math. **33** (1976/77), 223–233.
72. GUTOWSKI, R., *Étude d'une inégalité intégrale non linéaire en deux variables*, Ann. Polon. Math. **35** (1978), 247–252.
73. HELTON, J., *Two generalizations of the Gronwall inequality by product integration*, Rocky Mount. J. Math. **7** (1977), 733–749.
74. HRISTOVA, S.G. and D.D. BAĬNOV, *Some integral inequalities* (Russian), Glas. Mat. Ser. III **13** (33) (1978), 249–253.
75. __________, *An integral inequality for a scalar function of a vector argument* (Russian), Godišnik Visš. Učebn. Zaved. Priložna. Mat. **16** (1980), 217–220.
76. __________, *Some generalizations of integral inequalities of Bellman-Bihari type for functions of several variables*, Rend. Circ. Mat. Palermo (2) **30** (1981), 378–382.
77. __________, *On some integral inequalities for a scalar function of a vector argument*, Godišnik. Visš. Učebn. Zaved. Priložna. Mat. **17** (1981), 67–70.

78. __________, *Integral inequalities for functions of many variables*, Godišnik. Visš. Učebn. Zaved. Priložna. Mat. **17** (1981), 163–170.
79. __________, *On some Bihari type inequalities for a scalar function of a vector argument*, Godišnik. Visš. Učebn. Zaved. Priložna. Mat. **17** (1981), 171–176.
80. __________, *Integral inequalities for functions of many variables*, Mathematica (Cluj) **23** (46) (1981), 205–210.
81. KASATKINA, N.V., *K voprosu ob integral'nyh neravenstvah dlja mnogomernyh sistem tipa Vol'terra*, Tr. Iževskogo Mat. Sem. Izevsk, vyp. 1, 1963.
82. __________, *Integral'nye neravenstva dlja mnogomernyh integral'nyh uravneniĭ tipa Vol'terra*, Izv. Vuz-ov. Matematika, **1965**, No. 2.
83. KASATURE, D. and S. DEO, *Inequalities of Gronwall type in two independent variables*, J. Math. Anal. Appl. **58** (1977), 361–372.
84. KERIMOV, Dž.S., *Multipoint integral inequalities* (Russian), Investigations on some questions of the constructive theory of functions and differential equations, pp. 116–121. Azerb. Inst. Nefti i Khimii, Baku, 1981.
85. LIN, C.T., *A note on Pachpatte's inequality*, Tamkang J. Math. **9** (1978), 129–133.
86. __________, *On Wendroff's inequality in n independent variables*, Tamkang J. Math. **14** (1983), 167–177.
87. __________, *A note on some fundamental partial integral inequality*, Tamkang J. Math **18** (1987), 91–101.
88. MAO, X.R., *Lebesgue-Stieltjes integral inequalities of the Wendroff type in N-independent variables*, Chinese J. Math. **17** (1989), 29–50.
89. MAHMUDOV, A.N. and V.M. MUSAEV, *K teorii nelineĭnyh integral'nyh uravneniĭ tipa Vol'terra-Uryson*, Uč. zap. AGU. Ser. FMN, **1969**, No. 2.
90. NORMURADOV, H. and T. NURIMOV, *Nekotorye obobščenija neravenstv Vendroffa.* In: "Voprossy vyčeslitel'noĭ i prikladnoĭ matematiki," Taškent, vyp. **49**, 1977.
91. OLECH, C., *On a system of integral inequalities*, Colloq. Math. **16** (1967), 137–139.
92. ONESTI, N.B., *Sopra un'estensione del lemma di Gronwall*, Ist. Lombardo Sci. Lett. Red. Sc. **91** (1957), 643–649.
93. __________, *A proposito del lemma di Gronwall per le funzioni di due variabli*, Ist. Lombardo Sci. Lett. Red. Sc. A **95** (1961), 119–126.
94. ORNELAS, A., *A continuous version of the Filippov-Šarova inequality for differential inclusions*, Rend. Mat. Acc. Lincei Ser IX. Vol I. (1990), 105–110.

95. PACHPATTE, B.G., *On some integro-differential inequalities similar to Wendroff's inequality*, Ann. Ştinţ. Univ. "Al.I. Cuza" Iaşi Sec. Ia Mat. (N.S.) **25** (1979), no. 1, 127–136.
96. __________, *On Wendroff type partial integral inequalities*, Tamkang J. Math. **10** (1979), 141–150.
97. __________, *On some new integral and integrodifferential inequalities in two independent variables and their applications*, J. Differential Equations **33** (1979), 249–272.
98. __________, *A note on some new partial integral inequalities*, Chinese J. Math. **8** (1980), 13–17.
99. __________, *On some new integro-differential inequalities of the Wendroff type*, J. Math. Anal. Appl. **73** (1980), 491–500.
100. __________, *On some partial integral inequalities in n independent variables*, J. Math. Anal. Appl. **79** (1981), 256–272.
101. __________, *On certain integral equations*, An. Ştinţ. Univ. "Al. I. Cuza" Iaşi Sect. Ia Mat. (N.S.) **27** (1981), 365–374.
102. __________, *On comparison method for some fundamental partial integral inequalities*, J. Math. Phys. Sci. **15** (1981), 341–352.
103. __________, *On some fundamental hyperbolic partial integral inequalities*, J. Math. Phys. Sci. **16** (1982), 243–252.
104. PENG, G. *Some generalizations of Bellman-Bihari integral inequalities* (Chinese), J. Sichuan Univ. Nat. Sci. Ed. **24** (1987), 376–381.
105. RASMUSSEN, D., *Gronwall's inequality for functions of two independent variables*, J. Math. Anal. Appl. **55** (1976), 407–417.
106. REDHEFFER, R., *Differential and integral inequalities*, Proc. Amer. Math. Soc. **15** (1964), 715–716.
107. SHASTRI, R.P. and D.Y. KASTURE, *Wendroff type inequalities*, Proc. Amer. Math. Soc. **72** (1978), 248–250; **75** (1979), 374.
108. SHUVAR, B.A., *Integral inequalities of Bihari and Wendroff type* (Russian), Ukrain. Mat. Ž. **36** (1984), 532–536.
109. SIMEONOV, P.S. and D.D. BAĬNOV, *Integral and differential inequalities for a class of piecewise continuous functions* (Russian), Ann. Pol. Math. **98** (1988), 207–216.
110. SINGARE, V.M., *On some integral inequalities in two independent variables*. J. Math. Phys. Sci. **17** (1983), 301–311.
111. SINGARE, V.M. and B.G. PACHPATTE, *Lower bounds on some integral inequalities in n independent variables*, Indian J. Pure Appl. Math. **12** (1981), 318–331.
112. __________, *On some basic partial integral inequalities and their applications*, J. Math. Phys. Sci. **16** (1982), 301–317.

113. SNOW, D., *A two independent variable Gronwall-type inequality.* In: "Inequalities," Vol. III, 333–340, New York 1972.
114. THANDAPANI, E. and R.P. AGARWAL, *On some new inequalities in n independent variables*, J. Math. Anal. Appl. **86** (1982), 542–561.
115. WAŻEWSKI, T., *Remarque sur un système d'inégalités intégrales*, Ann. Polon. Math. **3** (1957), 210–212.
116. YANG, E.H., *On some new integral inequalities in n independent variables*, J. Math. Anal. Appl. **109** (1985), 171–181.
117. __________, *On some new n-independent-variable discrete inequalities of the Gronwall type*, Int. J. Math. Sci. **9** (1986), 485–495.
118. YANG, Y.S., *On Bellman-Gronwall's inequality*, J. Math. (Wuhan) **2** (1982), 269–273.
119. YEH, C.C., *Bellman-Bihari integral inequalities in several independent variables*, J. Math. Anal. Appl. **87** (1982), 311–321; Errata. **90** (1982), 583.
120. YEH, C.C. and M.N. SHIH, *The Gronwall-Bellman inequality in several variables*, J. Math. Anal. Appl. **86** (1982), 157–167.
121. YOUNG, E.C., *Functional integral inequalities in n variables*, Chinese J. Math. **10** (1982), 1–7.
122. ZAHARIEV, A.I. and D.D. BAĬNOV, *A generalization of the Bellman-Gronwall inequality* (Bulgarian), Godišnik Visš. Učebn. Zaved. Priložna Mat. **12** (1976), 207–210.
123. __________, *A note on Bellman-Gronwall's inequality*, J. Math. Anal. Appl. **80** (1981), 147–149.
124. PECH, Pavel, *Inequality between sides and diagonals of a space n-gon and its integral analog*, Časopis Pěst. Mat. **115** (1990), 343–350.

CHAPTER XIV

GRONWALL INEQUALITIES ON OTHER SPACES: DISCRETE, FUNCTIONAL AND ABSTRACT

1. There are Gronwall type inequalities in which the unknown function is not a function on R^n, rather in some other space. This Chapter is devoted to these kinds; on discrete, functional and then abstract spaces.

2. We first consider comparison theorems.

Let the functions $f(x,y)$ and $g(x,y)$ be defined for $x = x_0 + ih$, $h > 0$, $i = 1,\ldots,n$, and $a(x) \leq y \leq b(x)$. Let $y = \phi(x)$ and $z = \psi(x)$ be solutions of the difference equations

(2.1) $$\Delta y = hf(x,y) \text{ and } \Delta z = hg(x,z)$$

with the same initial conditions

(2.2) $$\phi(x_0) = y_0 = \psi(x_0), \quad a(x_0) \leq y_0 \leq b(x_0),$$

with $a(x) \leq \phi(x)$, $\psi(x) \leq b(x)$ for $x = x_0 + jh$, $j \leq N$ (N is a natural number).

The following Theorem is valid, see Rabczuk [**1**, pp. 200-202]:

THEOREM 1. *If for $x = x_0 + jh$, $j \leq N - 1$,*

(2.3) $$f(x,y) < g(x,z) \text{ for } y \leq z,$$

then for $x = x_0 + jh$, $j \leq N$, we have

(2.3′) $$\phi(x) < \psi(x).$$

PROOF: We use mathematical induction. Using (2.3) and (2.2) we have

$$f\left(x_0, \phi(x_0)\right) < g\left(x_0, \psi(x_0)\right),$$

so on the basis of (2.1) we have $\Delta\phi(x_0) < \Delta\psi(x_0)$ and thus $\phi(x_0+h) < \psi(x_0+h)$ and the assertion is true for $n = 0$. Suppose by way of proof that the result is valid for an index k, i.e. that $\phi(x_k) < \psi(x_k)$ for $k < N$. Combining the inequalities $f(x_k, \phi(x_k)) < g(x_k, \psi(x_k))$ and $\Delta\phi(x_k) < \Delta\psi(x_k)$ we get $\phi(x_k+h) < \psi(x_k+h)$ which is the assertion for $k+1$. The Theorem 1 is proven.

Further let the functions $u(x)$ and $v(x)$ be solutions of the difference equations

$$\Delta y = hf_1(x,y), \quad \Delta z = hf_2(x,z), \quad x = x_0 + jh, \ j \leq N,$$

with $u(x_0) = y_0 = v(x_0)$ and $a(x) \leq u(x)$, $v(x) \leq b(x)$ for the same x. ($\phi(x)$ will be defined as above.)

Corollary 1. *Let the functions $f_1(x,y)$ and $f_2(x,y)$ be defined for $x = x_0+jh$, $j \leq N-1$, and $a(x) \leq y \leq b(x)$ and suppose they are increasing in y. If*

$$f_1(x,y) < f(x,y) < f_2(x,y) \tag{2.4}$$

then for $x = x_0 + hj$, $j \leq N$, we have

$$u(x) < \phi(x) < v(x).$$

Corollary 2. *If instead of (2.4) we have*

$$f_1(x,y) \leq f(x,y) \leq f_2(x,y) \tag{2.5}$$

then $u(x) \leq \phi(x) \leq v(x)$.

Corollary 3. *If for $x = x_0 + jh$, $j \leq N-1$, the function $f(x,y)$ is increasing in y and*

$$\Delta v - hf(x,y) > 0, \tag{2.6}$$

then for $x = x_0 + jh$, $j \leq N$, we have

$$v(x) > y(x), \tag{2.7}$$

where $y = y(x)$ is the solution of the difference equation $\Delta y = hf(x,y)$, $y(x_0) = y_0$. The assertion is also valid if both (2.6) and (2.7) reverse the inequalities.

3. In the following result let $h = 1$ so that $\Delta f(x) = f(x+1) - f(x)$. In Grace, Lalli, and Yeh [**2**] the following result is proved:

THEOREM 1. *Let the function f be a real valued function on $R^+ \times R$, and v and w be real valued functions on R^+ which satisfy*

(A) *$f(t,u)$ is increasing in u and*

$$\Delta v(t) - f(t, v(t)) \leq \Delta w(t) - f(t, w(t)) \quad \text{for } t \in R^+,$$

or

(B) *$f(t,u)$ is decreasing in u and*

$$\Delta v(t) - f(t, v(t+1)) \leq \Delta w(t) - f(t, w(t+1)) \quad \text{for } t \in R^+,$$

then from $v(t) \leq w(t)$ on $[0,1)$ we conclude the same inequality on R^+.

4. Now we shall consider the difference equation

$$x(t) = g(t) + \sum_{s=0}^{t-1} f(t, s, x(s)), \quad t = 1, 2, \ldots, \tag{4.1}$$

with the initial condition $x(0) = g(0)$, where x, g, and f can be scalar or vector and $f(t,s,x)$ is defined for $t, s = 0, 1, 2, \ldots$ and for all x.

The following theorem was proved in Pachpatte and Singare [**3**]:

THEOREM 1. *Let $f(t,s,x)$ be defined as above with f nondecreasing in x. If $u(t)$ satisfies the inequality*

$$u(t) \leq g(t) + \sum_{s=0}^{t-1} f(t, s, u(s)), \quad u(0) \leq g(0), \tag{4.2}$$

then $u(t) \leq x(t)$, $t = 0, 1, \ldots$, where $x(t)$ is the solution of the difference equation (4.1).

PROOF: First $u(0) \leq g(0) = x(0)$. Suppose that the inequality $u(t) \leq x(t)$ is not valid for for some t_0, and suppose it is the first one for which the inequality fails. Then

$$x(t_0) - u(t_0) \geq \sum_{s=0}^{t-1} [f(t_0, s, x(s)) - f(t_0, x, u(s))] \geq 0,$$

which is a contradiction so the Theorem is proved.

REMARK 1: For analogous results see Rabczuk [**1**, pp. 206-220], Redheffer and Walter [**4**], and Grace, Lalli, and Yeh [**2**].

5. We observe first that discrete inequalities of Gronwall type may be obtained from the continuous case by appropriate specialization. For example, consider the inequality

$$x_i \le a_i + b_i \sum_{j=0}^{i-1} k_j x_j, \quad i = 0, 1, \ldots, N, \tag{5.1}$$

where b_j and k_j are nonnegative.(Here and elsewhere we set $\sum_{j\in\varphi} A_j = 0$ and $\prod_{j\in\varphi} A_j = 1$ if φ is the empty set.)

Set

$$x(t) = \begin{cases} x_i, & i \le t < i+1,\ 0 \le i \le N-1, \\ x_N, & t = N, \end{cases}$$

and analogously for a, b, and k, we get (5.1) from

$$x(t) \le a(t) + b(t) \int_0^t k(s)x(s)ds, \quad 0 \le t \le N,$$

if $t = i$. Similarly from

$$x(t) \le a(t) + b(t) \int_0^t a(s)k(s) \exp\left(\int_s^t b(r)k(r)dr \right) ds,\ 0 \le t \le N,$$

we get (see Beesack [**5**, pp.92-93]):

$$x_i \le a_i + b_i \sum_{j=0}^{i-1} a_j k_j \exp\left(\sum_{v=j}^{i-1} b_v k_v \right),\ 0 \le i \le N. \tag{5.1'}$$

6. None of the above results is best possible. The best possible results for the inequality (5.1) was given by G. S. Jones [**6**]:

THEOREM 1. *Let x, a, b, and k be real-valued functions on $J = [\alpha, \beta]$, with $b \ge 0$ and $k \ge 0$, and let $\{t_0, t_1, \ldots, t_N\}$ be a strictly increasing sequence in (α, β). If*

$$x(t) \le a(t) + b(t) \sum_{t_j < t} k(t_j)x(t_j), \quad t \in J, \tag{6.1}$$

then for all $t \in J$

(6.1′) $$x(t) \le a(t) + b(t) \sum_{t_i<t} a(t_j)k(t_j) \prod_{t_j<t_k<t} [1 + b(t_k)k(t_k)]\,.$$

PROOF: Denote the right hand side of (6.1′) by $y(t)$. Then the function y is a solution of the equation

(6.2) $$y(t) = a(t) + b(t) \sum_{t_j<t} k(t_j)y(t_j), \quad t \in J.$$

This is clearly the case for $\alpha \le t \le t_0$, and for all t such that $b(t) = 0$. If $t_0 \le t \le t_1$, then

$$y(t) = a(t) + b(t)a(t_0)k(t_0) = a(t) + b(t) \sum_{t_j<t} k(t_j)y(t_j),$$

since $y(t_0) = a(t_0)$. One can prove that if $y(t)$ satisfies (6.2) for $\alpha \le t \le t_j$, then $y(t)$ also satisfies (6.2) for $t_i \le t \le t_{i+1}$ if $i < N$, or for $t_i < t \le \beta$ if $i = N$. The desired conclusion now follows by mathematical induction.

A similar induction shows that if $w = x - y$, then $w(t) \le 0$ for $\alpha \le t \le t_i$, $0 \le i \le N+1$ ($t_{N+1} = \beta$). This proves the Theorem 1 and shows that (6.1′) is best possible.

REMARK 1: To reduce (6.1) to (5.1) it suffices to set

$$x(t) = x_i \text{ if } t = t_i,\ 0 \le i \le N;\ x(t) = 0 \text{ else},$$

with a corresponding definition of a, b, and k. Then (5.1) implies (6.1) and hence also (6.1′). Setting $t = t_i$ we obtain the best possible result

(6.3) $$x_i \le a_i + b_i \sum_{j=0}^{i-1} a_j k_j \prod_{j<m<i} [1 + b_m k_m]\,, \quad 0 \le i \le N.$$

REMARK 2: Note that the previous result gives

$$\prod_{j=0}^{i-1}(1 + k_j) \le \exp\left(\sum_{j=0}^{i-1} k_j\right).$$

7. T. E. Hull and W. A. J. Luxemburg [7] (see also Filatov and Šarova [8, pp. 42-45]) obtained the following discrete version of Bihari's inequality but with all $k_j = k > 0$.

THEOREM 1. *Let the function g be continuous, positive, and nondecreasing on an interval $I = [u_0, \infty)$ and let $a \in I$ and $k_j \geq 0$ for $j = 0, 1, \ldots$. If $x_i \in I$ for $i = 0, 1, \ldots$, and*

$$x_i \leq a + \sum_{j=0}^{i-1} k_j g(x_j), \quad i \geq 0, \tag{7.1}$$

then

$$x_i \leq G^{-1}\left[G(a) + \sum_{j=0}^{i-1} k_j\right], \quad 0 \leq i \leq N, \tag{7.1'}$$

where $G(u) = \int_{u_0}^{u} dy/g(y)$, $u \geq u_0$, and $N = \sup\{i\colon G(a) + \sum_{j=0}^{i-1} k_j \in G(I)\}$.

8. The following two theorems give discrete analogues of Theorems 1 of 19. and Theorem 1 of 9. in Chapter XII (see Willett and Wong [**9**]):

THEOREM 1. *Let $\{x_n\}$, $\{a_n\}$, $\{b_n\}$, and $\{k_n\}$ be nonnegative sequences and suppose that $1 \leq p < \infty$. If*

$$x_n \leq a_n + b_n \left(\sum_{j=0}^{n-1} k_j x_j^p\right)^{1/p}, \quad n = 0, 1, \ldots, \tag{8.1}$$

then

$$x_n \leq a_n + b_n \left(\sum_{j=0}^{n-1} k_j a_j^p e_{j+1}\right)^{1/p} \Big/ \left[1 - (1 - e_n)^{1/p}\right], \; n = 0, 1, \ldots, \tag{8.1'}$$

where

$$e_i = \prod_{j=0}^{i-1} \left(1 + k_j b_j^p\right)^{-1}, \; i = 0, 1, \ldots. \tag{8.2}$$

PROOF: Note that $\{e_i\}$ satisfies the difference equation

$$e_{i+1} - e_i = -k_i b_i^p e_{i+1}, \quad e_0 = 1. \tag{8.3}$$

Define the sequence $\{c_i\}$ by

$$c_i = e_i \sum_{j=0}^{i-1} k_j x_j^p, \quad i \geq 0, \tag{8.4}$$

and using (8.1–8.3) we obtain

$$c_{i+1} - c_i \leq \left\{ a_i k_i^{1/p} e_{i+1}^{1/p} + \frac{b_i k_i^{1/p} c_i^{1/p}}{(1 + k_i b_i^p)^{1/p}} \right\}^p - \frac{k_i b_i^p c_i}{1 + k_i b_i^p}.$$

We now sum over i noting that $c_0 = 0$ to get

$$c_n + \sum_{i=0}^{n-1} \frac{k_i b_i^p c_i}{1 + k_i b_i^p} \leq \sum_{i=0}^{n-1} \{\cdots + \ldots\}^p.$$

By taking pth roots and applying Minkowsi's inequality we get

$$(c_n + A)^{1/p} - A^{1/p} \leq \left(\sum_{i=0}^{n-1} e_{i+1} k_i a_i^p \right)^{1/p}, \tag{8.5}$$

where $A = \sum\limits_{i=0}^{n-1} k_i b_i^p c_i / (1 + k_i b_i^p)$.

Since the function $f(x) = (c+x)^{1/p} - x^{1/p}$ is decreasing for $x > 0$, (8.5) remains valid when A is increased. By (8.4)

$$\begin{aligned} A &= \sum_{i=0}^{n-1} \frac{k_i b_i^p e_i}{1 + k_i b_i^p} \left(\sum_{j=0}^{n-1} k_j x_j^p \right) \\ &\leq \sum_{i=0}^{n-1} k_i b_i^p e_{i+1} \left(\sum_{j=0}^{n-1} k_j x_j^p \right) \quad \text{by (8.2)} \\ &= \left(\sum_{j=0}^{n-1} k_j x_j^p \right) \sum_{i=0}^{n-1} (e_i - e_{i+1}) \quad \text{by (8.3)}, \end{aligned}$$

so

$$A \leq (1 - e_n) \sum_{j=0}^{n-1} k_j x_j^p.$$

Hence (8.5) implies that

$$\left(c_n + (1-e_n)\sum_{j=0}^{n-1} k_j x_j^p\right)^{1/p} \le (1-e_n)^{1/p}\left(\sum_{j=0}^{n-1} k_j x_j^p\right)^{1/p} + \left(\sum_{i=0}^{n-1} e_{i+1} k_i a_i^p\right)^{1/p}.$$

Using the definition of c_n, we get

$$\left(\sum_{j=0}^{n-1} k_k x_j^p\right)^{1/p}\left(1-(1-e_n)^{1/p}\right) \le \left(\sum_{j=1}^{n-1} k_j a_j^p e_{j+1}\right)^{1/p}.$$

On substituting this into (8.1) we obtain (8.1′).

THEOREM 2. *Let* $\{x_n\}$, $\{k_n\}$, *and* $\{K_n\}$ *be nonnegative sequences and let* $a > 0$, $p \ge 0$, $p \ne 1$. *If*

$$x_n \le a + \sum_{j=0}^{n-1} k_j x_j + \sum_{j=0}^{n-1} K_j x_j^p, \quad n \ge 0, \tag{8.6}$$

then

$$x_n \le E_n^{-1}\left\{a^{1-p} + (1-p)\sum_{j=0}^{n-1} K_j E_{j+1}^{1-p}\right\}^{1/(1-p)}, \quad n \ge 0, \tag{8.6'}$$

where

$$E_i = \prod_{j=0}^{i-1}(1+k_j)^{-1}, \quad i \ge 0. \tag{8.7}$$

In case $p > 1$, *(8.6′) is restricted to those* n *for which* $\{\dots\} > 0$.

PROOF: The sequence $\{E_i\}$ satisfies the equation

$$E_{i+1} - E_i = -k_i E_{i+1}, \quad E_0 = 1. \tag{8.8}$$

Denote the expression on the right hand side of (8.6) by C_n and consider the sequence $\{C_n\}$. Note that $C_0 = a$ and by (8.6) $x_i \le C_i$ for $i \ge 0$. Since $C_{i+1} - C_i = k_i x_i + K_i x_i^p$ and $p \ge 0$, it follows that

$$C_{i+1} - C_i - C_i k_i \le K_i C_i^p, \quad i \ge 0. \tag{8.9}$$

Now multiply (8.9) by E_{i+1} and use (8.8) to obtain

$$E_{i+1}C_{i+1} - E_iC_i \le K_iE_{i+1}C_i^p. \tag{8.10}$$

Set $q = 1 - p$. By the mean value theorem

$$(E_{i+1}C_{i+1})^q - (E_iC_i)^q = qX^{-p}(E_{i+1}C_{i+1} - E_iC_i), \tag{8.11}$$

where X lies between E_iC_i and $E_{i+1}C_{i+1}$. Using the fact that $\{C_i\}$ is nondecreasing while $\{E_i\}$ is nonincreasing, it follows that

$$\begin{aligned} X > E_iC_i \ge C_iE_{i+1} \quad &\text{if} \quad E_iC_i < E_{i+1}C_{i+1}, \quad \text{and} \\ X > E_{i+1}C_{i+1} \ge C_iE_{i+1} \quad &\text{if} \quad E_{i+1}C_{i+1} < E_iC_i. \end{aligned}$$

Hence (8.10) and (8.11) imply that

$$(E_{i+1}C_{i+1})^q - (E_iC_i)^q \le qK_iE_{i+1}^q \tag{8.12}$$

where $\le$ or $\ge$ holds according to whether $q > 0$ or < 0. Now sum (8.12) for $i = 0, 1, \ldots, n-1$ using $C_0 = a$ to obtain

$$(E_nC_n)^q \le a^q + q\sum_{i=0}^{n-1} K_iE_{i+1}^q, \quad (\text{or } \ge).$$

Taking q-th roots it follows that in both cases

$$E_nC_n \le \left(a^q + q\sum_{i=0}^{n-1} K_iE_{i+1}^q\right)^{1/q}.$$

Since $x_n \le C_n$ and $E_n > 0$, (8.6′) now follows.

REMARK 1: A special case of the previous theorem when $k = 0$ and $K_j = K > 0$ is in Lees [10].

9. The following two theorems were proved in Singare and Pachpatte [**11**]:

THEOREM 1. *Let $x(n)$, $k(n)$, $p(n)$, $f(n)$, $g(n)$, and $h(n)$ be non-negative functions on $N = \{1, 2, \ldots\}$ for which*

$$x(n) \le k(n) + p(n)\left[\sum_{s=n_0}^{n-1} f(s)x(s) + \sum_{s=n_0}^{n-1} g(s)\left(\sum_{t=n_0}^{s-1} h(t)x(t)\right)\right]$$

holds for every $n \in N$. If $1+p(n)f(n)-g(n) \geq 0$, then

$$x(n) \leq k(n) + p(n)\Bigg[\sum_{s=n_0}^{n-1} k(s)f(s) \prod_{t=s+1}^{n-1} (1+p(t)f(t)-g(t))$$
$$+ \sum_{s=n_0}^{n-1} g(s)\left(\sum_{t=n_0}^{s-1} k(t)(f(t)+h(t)) \prod_{m=t+1}^{s-1} (1+p(m)(f(m)+h(m))+g(m))\right)$$
$$\times \prod_{t=s+1}^{n-1} (1+p(t)f(t)-g(t))\Bigg],$$

for $n \in N$.

THEOREM 2. *Let $x(n)$, $f(n)$, $g(n)$, and $h(n)$ be nonnegative functions defined on N such that*

$$x(n) \leq x_0 + \sum_{s=n_0}^{n-1} f(s)x(s) + \sum_{s=n_0}^{n-1} g(s)\left(\sum_{t=n_0}^{s-1} h(t)x^p(t)\right), \quad n \in N,$$

where x is a nonnegative constant and $0 \leq p < 1$. If $1+f(n)-g(n) \geq 0$, then for $n \in N$,

$$x(n) \leq x_0 \prod_{s=n_0}^{n-1} (1+f(s)-g(s))$$
$$+ \sum_{s=n_0}^{n-1} g(s) \sum_{m=s+1}^{n-1} (1+f(m)-g(m)) \times \prod_{m=n_0}^{n-1} (1+f(m)-g(m))$$
$$\times \left[x_0^{1-p} + (1-p)\sum_{m=n_0}^{s-1} h(m) \prod_{t=n_0}^{s} (1+f(t)-g(t))^{p-1}\right]^{1/(1-p)}.$$

10. Of course there exist discrete analogues for other known inequalities. So in Singare and Pachpatte [**12**] the following discrete version of a Wendroff type inequality is proved:

THEOREM 1. *Let $u(x,y)$ and $c(x,y)$ be nonnegative functions defined for $x, y \geq 0$ and satisfy*

$$u(x,y) \leq a(x) + b(y) + \sum_{s=0}^{x-1}\sum_{t=0}^{y-1} c(s,t)u(s,t), \quad x, y \geq 0,$$

where $a(x)$ and $b(y)$ are positive functions defined for $x, y > 0$ with $\Delta a(x), \Delta b(y) \geq 0$. Then for $x, y \geq 0$

$$u(x,y) \leq (a(0)+b(y)) \prod_{s=0}^{x-1} \left(1 + \frac{\Delta a(s)}{a(s)+b(0)} + \sum_{t=0}^{y-1} c(s,t)\right).$$

11. The previous theorem gives discrete inequalites for two variables. There are analogous results for more than two variables. Here we shall give a result from Yeh [**13**, II].

Let N be a set of nonnegative integers, and n a positive integer. If $\underline{1} = (1, 1, ..., 1)$ and u is a function defined on N to R^+ define: $x = (x_1, \tilde{x})$ where $\tilde{x} = (x_2, \ldots, x_n)$; and $\sum_{y=0}^{x-\underline{1}} u(y) = \sum_{y_1=0}^{x_1-1} \cdots \sum_{y_n=0}^{x_n-1} u(y_1, ..., y_n)$; $\sum_{y=0}^{x-\underline{1}} u(y) = 0$ whenever any $x_i = 0$.

THEOREM 1. *Let the functions $u(x)$ and $k(x)$ be defined on N^n to R^+ and $f(x;s)$ be defined on N^{2n} to R^+ for $s \leq x$. If*

$$u(x) \leq k(x) + \sum_{s=0}^{x-\underline{1}} f(x,s)u(s) \quad \text{for } x \in N^n,$$

then

$$u(x) \leq K(x) \prod_{s_1=0}^{x_1-1} \left[1 + \sum_{\tilde{s}=0}^{\tilde{x}-\tilde{1}} F(x; s_1, s)\right],$$

where

$$K(x) = \sup\{k(s) \,|\, 0 \leq s \leq x\}, \text{ and } F(x;s) = \sup\{f(t,s) \,|\, 0 \leq t \leq x\}.$$

THEOREM 2. *Let u, k, t, K, and F be defined as in Theorem 1 and $g(x;s)$ be defined on N^{2n} to R^+ for $s \leq x_0$. If*

$$u(x) \leq k(x) + \sum_{s=0}^{x-1} f(x;s)\left[u(s) + \sum_{t=0}^{s-\underline{1}} g(s;t)u(t)\right], \quad x \in N^n \tag{11.1}$$

then

$$u(x) \leq K(x) \prod_{s_1=0}^{x_1-1} \left\{1 + \sum_{\tilde{s}=0}^{\tilde{x}-\tilde{1}} [F(x; s_1, \tilde{s}) + G(x; s_1, \tilde{s})]\right\} \tag{11.1'}$$

or

$$(11.1'') \qquad u(x) \le k(x) + \sum_{s=0}^{x-1} f(x;s)K(s) \times \prod_{t_1=0}^{s_1-1} \left\{ 1 + \sum_{\tilde{t}=0}^{\tilde{s}-\tilde{1}} [F(s;t_1,\tilde{t}) + F(s;t_1,\tilde{t})] \right\},$$

where $G(x;s) = \sup\{g(t,s) \mid 0 \le t \le x\}$.

PROOF: Let $w(x) = u(x) + \sum_{t=0}^{x-1} g(x,t)u(t)$. Then

$$(11.2) \qquad u(x) \le w(x),$$

and by (11.1)

$$(11.3) \qquad u(x) = w(x) - \sum_{s=0}^{x-1} g(x;s)u(s) \le k(x) + \sum_{s=0}^{x-1} f(x;s)w(s).$$

Therefore

$$w(x) \le k(x) + \sum_{s=0}^{x-1} [f(x;s) + g(x;s)]\, w(s).$$

Using Theorem 1 we get

$$(11.4) \qquad w(x) \le K(x) \prod_{s_1=0}^{x_1-1} \left\{ 1 + \sum_{\tilde{s}=0}^{\tilde{x}-\tilde{1}} [F(x;s_1,\tilde{s}) + G(x;s_1,\tilde{s})] \right\}.$$

From (11.2) and (11.4) we get (11.1′) and (11.1″).

12. There are also results which are discrete versions of integro-differential inequalities. For example, Pachpatte [**14**] has the following theorem:

THEOREM 1. *Let $x(n)$, $\Delta x(n)$, $f(n)$ and $g(n)$ be nonnegative functions on N, for which*

$$\Delta x(n) \le f(n) \left(x(n) + \sum_{s=n_0}^{n-1} g(s)\Delta x(s) \right), \qquad n \in N.$$

Then

$$x(n) \le x(n_0) \left\{ 1 + \sum_{s=n_0}^{n-1} f(s) \prod_{t=n_0}^{s-1} [1 + f(t) + f(t)g(t)] \right\}, \qquad n \in N.$$

13. J. Popenda [**15**] proved the following result:

THEOREM 1. *Let E_0 be a subset of nonnegative elements of any vector lattice. Let x, Δx be functions from N to E_0, and a and b be functions of N to $[0,\infty)$. If*

$$\Delta x_n \le x_1 + \sum_{j=1}^{n-1} a_j(x_j + \Delta x_j) + \sum_{j=1}^{n-1} a_j \sum_{i=1}^{j-1} b_i \Delta x_i, \quad n \in N,$$

then

$$\Delta x_n \le \left[1 + 2\sum_{j=1}^{n-1} a_j \prod_{i=1}^{j-1} (2 + a_i + b_i)\right] x_1, \quad n \in N.$$

14. Finally, note that there are results which incorporate both discrete and continuous results. For example, such theorems are given in the following result of G. S. Jones [6]:

THEOREM 1. *Let the functions x, a, b, and k on J to R be either, continuous or of bounded variation and a, b, and k be nonnegative, ($J = [\alpha,\beta]$). Let μ be nondecreasing and left continuous on J and suppose that x, a, b, and k are all continuous from the right at all points of discontinuity of μ. If*

$$x(t) \le a(t) + b(t) \int_\alpha^{t^-} k(s)x(s)d\mu(s), \quad t \in J,$$

then for $t \in J$

$$x(t) \le a(t) + b(t) \int_\alpha^{t^-} a(s)k(s)\exp\left(\int_s^{t^-} b(r)k(r)d\mu(r)\right) d\mu(s).$$

Moreover, if $\sigma_0 = \alpha$ and $\sigma_1, \sigma_2, \ldots$ are the points of discontinuity of μ in $(\alpha,\beta]$ and

$$\lambda(t) = \mu(t) - \sum_{\sigma_i < t} K(\sigma_i), \quad t \in J,$$
$$K(t) = \mu(t^+) - \mu(t^-),$$

then for $t \in J$

$$x(t) \le a(t) + b(t) \int_\alpha^{t^-} a(s)k(s)\exp\Big(\int_s^t b(r)k(r)d\lambda(r)$$
$$\times \prod_{s<\sigma_i<t} \{1 + b(\sigma_i)k(\sigma_i)K(\sigma_i)\}\, d\mu(s)\Big).$$

Finally if $|a(t)| \le A$ *and* $0 \le b(t) \le B$ *on* J *then*

$$x(t) \le A \exp\left(B \int_a^t k d\lambda\right) \prod_{s<\sigma_i<t} [1 + Bk(\sigma_i)K(\sigma_i)], \quad t \in J.$$

15. Various discrete inequalities were considered in E. H. Yang [**16**], Thandapani and Agarwal [**17**], Pachpatte and Singare [**18**], Pachpatte [**19**]-[**24**], Agarwal and Thandapani [**25**], and Novotná [**26**].

16. Similar results for Stieltjes integrals are given in Das and Sharma [**27**], and Mingarelli [**28**].

17. By C_T we denote the space of continuous functions $u(t)$ defined on $[0,T]$ with the norm $\|u(t)\| = \max_{0\le t\le T} |u(t)|$, and by S_T we mean a sphere from C_T with center zero and a radius r, $S_T = \{u(t) \in C_T \mid \|u(t)\| \le r\}$.

Let $\phi(t,u_t)$ be a nonlinear Volterra operator, that is an operator which for fixed $t \in [0,T]$, acts from C_T (or S_T) in R, and which satisfies the following conditions:

a) if $u^{(n)}(s) \in S_T$, $n = 1,2,\ldots$, $u^{(n+1)}(s) \le u^{(n)}(s)$ and $\lim_n u^{(n)}(s) = u(s)$ for every fixed $s \in [0,T]$, then $\lim_n \phi(t, u_t^{(n+1)}) = \phi(t,u_t)$ for each $t \in [0,T]$;

b) if $u(s) \in C_T$, then $\phi(t,u_t) = \tilde{u}(t) \in C_T$.

We say that the Volterra operator $\phi(t,u_t)$ is nondecreasing in the second argument if from $u(s) \le v(s)$, $0 \le s \le t \le T$, it follows that $\phi(t,u_t) \le \phi(t,v_t)$ on $[0,T]$.

THEOREM 1. *Let the Volterra operator* $\phi(t,u_t)$ *be nondecreasing in the second argument and suppose that* $u(t)$ *is the lower solution of the equation*

$$u(t) = \phi(t,\mu_t), \tag{17.1}$$

defined on $[0,T]$. *If the function* $v(t) \in C_T$ *(or* $v(t) \in S_T$*) satisfies*

$$v(t) < \phi(t,v_t) \quad \text{on } (0,T], \quad v(0) < u(0), \tag{17.2}$$

then

$$v(t) < u(t) \quad \text{on } [0,T]. \tag{17.2'}$$

PROOF: The inequality (17.2′) is true for $t = 0$, and hence on some nontrivial interval $(0,t^*)$. If we select t^* to be the largest number for which (17.2′) holds and $t^* < T$ then we must have $v(t) < u(t)$ on $[0, t^*)$ and $v(t^*) = u(t^*)$. But we then have

$$v(t^*) < \phi(t^*, v_{t^*}) \leq \phi(t^*, u_{t^*}) = u(t^*),$$

providing a contradiction, and proving the Theorem.

REMARK 1: Theorem 1 is also valid if in (17.2) and (17.2′) all inequalities are reversed and u is the upper solution of (17.1).

We say that the Volterra operator $\phi(t, u_t)$ satisfies the condition (μ) if the equation

$$W(t) = \phi(t, W_t) + \delta, \quad W(0) = u(0) + \delta,$$

has a solution $W(t) \in C_T$ (or S_T) for every fixed δ in $[0, \mu]$.

THEOREM 2. *Let the Volterra operator $\phi(t, u_t)$ be nondecreasing in the second argument and satisfy the condition (μ) for sufficiently small μ. If the function $v(t) \in C_T$ (or S_T) satisfies*

$$v(t) \leq \phi(t, v_t) \quad \text{on } (0, T], \quad v(0) \leq u(0), \tag{17.3}$$

and if $u(t)$ is the upper solution of (17.1), then

$$v(t) \leq u(t) \quad \text{on } [0, T]. \tag{17.3′}$$

PROOF: For every fixed $n = 1, 2, \ldots$, let $u^{(n)}(t)$ denote the solution of the equation

$$u^{(n)}(t) = \phi(t, u_t^{(n)}) + \varepsilon/n, \quad u^{(n)}(0) = u(0) + \varepsilon/n, \tag{17.4}$$

defined on $[0, T]$, where $\varepsilon > 0$ is a sufficiently small number. From (17.4) we obtain

$$u^{(n)}(t) > \phi(t, u_t^{(n)}) \quad \text{on } (0, T], \quad u^{(n)}(0) = u(0) + \varepsilon/n, \tag{17.5}$$

and

$$\begin{aligned} &u^{(n+1)}(t) < \phi\left(t, u_t^{(n+1)}\right) + \varepsilon/n \quad \text{on } (0, T], \\ &u^{(n+1)}(0) = u(0) + \varepsilon/(n+1). \end{aligned} \tag{17.6}$$

From (17.1), (17.5) and (17.6) and Theorem 1 we get

$$u(t) < u^{(n+1)}(t) < u^{(n)}(t) \leq u^{(1)}(t) \quad \text{on } (0, T].$$

From inequality (17.3) we find that

$$v(t) < u^{(n)}(t) \quad \text{on } [0, T]. \tag{17.7}$$

Now $\lim_n u^{(n)}(t) = \tilde{u}(t)$ exists on $[0, T]$ and $u(t) \leq \tilde{u}(t)$ and by taking limits in (17.7) we get the result of the Theorem.

THEOREM 3. *Let the Volterra operator $\phi(t,u_t)$ on S_T to S_T be nondecreasing in the second argument. If the function $v(t) \in S_T$ satisfies the inequality*

$$v(t) \le \phi(t, v_t) \quad \text{on } [0,T],$$

then

$$v(t) \le u(t) \quad \text{on } [0,T] \tag{17.8}$$

where $u(t)$ is the upper solution of (17.1).

PROOF: For the equation (17.1) define the sequence of functions

$$\begin{aligned} \sigma^{(1)}(t) &= \phi(t, \{r\}_t) \\ \sigma^{(n+1)}(t) &= \phi(t, \sigma_t^{(n)}), \quad n = 1, 2, \ldots. \end{aligned} \tag{17.9}$$

It is obvious that $\sigma^{(n)}(t) \in S_T$. First we shall prove that this sequence is nonincreasing. We have $\sigma^{(2)}(t) = \phi(t, \sigma_t^{(1)}) \le \phi(t, \{r\}_t) = \sigma^{(1)}(t)$ to begin the induction. If we assume that $\sigma^{(n)}(t) \le \sigma^{(n-1)}(t)$ then by (17.9) we have

$$\sigma^{(n+1)}(t) = \phi(t, \sigma_t^{(n)}) \le \phi(t, \sigma_t^{(n-1)}) = \sigma^{(n)}(t),$$

to complete the induction. Next we prove that

$$v(t) \le \sigma^{(n)}(t) \quad \text{on } [0,T],\ n = 1, 2, \ldots. \tag{17.10}$$

Indeed

$$v(t) \le \phi(t, v_t) \le \phi(t, \{r\}_t) = \sigma^{(1)}(t),$$

so that (17.10) is true for $n = 1$. Again by way of an induction argument assume that (17.10) is true for a given n. From (17.9) we obtain $v(t) \le \phi(t, \sigma_t^{(n)}) = \sigma^{(n+1)}(t)$, to complete the induction. Hence the sequence $\sigma^{(n)}(t)$ is nonincreasing and bounded from below and so there is a function $\sigma(t)$ so that $\lim_n \sigma^{(n)}(t) = \sigma(t)$ exists on $[0,T]$. From (17.9) we get that $\sigma(t)$ is a solution of (17.1). Taking limits in (17.10) we arrive at (17.8).

18. We need the following notations to discuss the generalizations of the previous results:

1) $\phi(t, u, v_t)$ is a nonlinear Volterra operator for fixed u in $[-r, r]$, and for fixed $t \in [0,T]$ and $v_t \in S_T$, $\phi(t, u, v_t)$ is a continuous function on $[-r, r]$ and satisfies the conditions:

a) if $u^{(n)}(s), v^{(n)}(s) \in S_T, n = 1, 2, \ldots$ with $u^{(n+1)}(s) \le u^{(n)}(s), v^{(n+1)}(s) \le v^{(n)}(s)$ and $\lim_n u^{(n)}(s) = u(s)$, $\lim_n v^{(n)}(s) = v(s)$ on $[0, T]$, then on $[0, T]$ we have $\lim_n \phi\left[t, u^{(n)}(t), v_t^{(n)}\right] = \phi\left[t, u(t), v_t\right]$.

b) if $u(s), v(s) \in S_T$, then $\phi(t, u(t), v_t) = \tilde{u}(t) \in S_T$.

2) $\phi(t, u_t, v_t)$ is a nonlinear Volterra operator for fixed v in S_T, and for fixed $t \in [0, T]$ and $u_t \in S_T$, ϕ acts from S_T to R and satisfies the conditions:

a) if $u^{(n)}(s), v^{(n)}(s) \in S_T$, $n = 1, 2, \ldots$, $u^{(n+1)}(s) \le u^{(n)}(s)$, $v^{(n+1)}(s) \le v^{(n)}(s)$ and $\lim_n u^{(n)}(s) = u(s)$, and $\lim_n v^{(n)}(s) = v(s)$ exist on $[0, T]$, then $\lim_n \phi\left(t, u_t^{(n)}, v_T^{(n)}\right) = \phi(t, u_t, v_T)$ on $[0, T]$.

b) if $u(s), v(s) \in S_T$, then $\phi(t, u_t, v_T) = \tilde{u}(t) \in S_T$.

We say that the operator $\phi(t, u, v_t)$ is nondecreasing in the second and third arguments if from $u(s) \le \tilde{u}(s)$ and $v(s) \le \tilde{v}(s)$ it follows that $\phi(t, u(t), v_t) \le \phi(t, \tilde{u}(t), \tilde{v}_t)$. Similar properties can be defined for the operator $\phi(t, u_t, v_T)$.

THEOREM 1. *Let the operator-function $\phi(t, u, v_t)$ be nondecreasing in the second and third arguments. If $v(t) \in S_T$ and*

$$v(t) \le \phi(t, v(t), v_t) \quad \text{on } [0, T],$$

then

$$v(t) \le u(t) \qquad \text{on } [0, T]$$

where $u(t)$ is the upper solution of the equation

$$u(t) = \phi(t, u(t), u_t). \tag{18.1}$$

THEOREM 2. *Let the operator $\phi(t, u_t, v_T)$ be nondecreasing in the second and third arguments. If $v(t) \in S_T$ and*

$$v(t) \le \phi(t, v_t, v_T) \quad \text{on } [0, T],$$

then

$$v(t) \le u(t) \qquad \text{on } [0, T],$$

where $u(t)$ is the upper solution of the equation

$$u(t) = \phi(t, u_t, u_T). \tag{18.2}$$

19. The above results in 17. and 18. are obtained in Mamedov and Aširov [**29**] as well as in Mamedov, Aširov, and Atdaev [**30**]. Similar results are obtained in the case when $[0,T]$ is replaced by $[0,\infty)$.

As examples of the previous theorems we might take

$$\phi(t,u_t) = \phi\left(t, \int_0^t f(t,s,u(s))\,ds\right),$$

$$\phi(t,u,v_t) = \phi\left(t,u, \int_0^t f(t,s,v(s))\,ds\right),$$

or

$$\phi(t,u_t,u_T) = \phi\left(t, \int_0^t f(t,s,u(s))\,ds, \int_0^T g(t,s,u(s))\,ds\right).$$

20. Mamedov and Novruzov [**31**] (see also Mamedov, Aširov, and Atdaev [**30**]) considered differential inequalities with Volterra operators $\phi(t,u,v_t)$.

THEOREM 1. *Let the operator $\phi(t,u,v_t)$ be continuous in all variables and non-decreasing in the third argument. Let $v(t) \in C_T$ (or $v(t) \in S_T$) and satisfy*

(20.1) $$v'(t) < \phi(t,v(t),v_t) \quad \text{on } (0,T], \quad v(0) \le u_0,$$

and if $v(0) = u_0$, then

(20.2) $$v'(0) < \phi(0,u_0,\{u\}_0).$$

Then

(20.1′) $$v(t) < u(t) \quad \text{on } (0,T],$$

where $u(t)$ is the lower solution of the problem

(20.3) $$u'(t) = \phi(t,u(t),u_t), \quad u(0) = u_0,$$

defined on $[0,T]$.

PROOF: It is clear that from (20.2) that (20.1′) holds on some small interval $(0,t_0)$ if $v(0) = u_0$, and if $v(0) < u_0$, the same conclusion follows from continuity.

If (20.1′) does not hold on $(0,T]$, let t^* be the first t for which $u(t)=v(t)$. Then using (15) and (18) we get

$$\begin{aligned} v'(t^*) &< \phi(t^*, v(t^*), v_{t^*}) = \phi(t^*, u(t^*), v_{t^*}) \\ &\le \phi(t^*, u(t^*), u_{t^*}) = u'(t^*). \end{aligned}$$

But then $v(t) > u(t)$ to the left of t^*. This contradiction proves that (20.1′) holds on $(0,T]$.

REMARK 1: The theorem is valid if all the inequalities are reversed and $u(t)$ is the upper solution of (20.3).

REMARK 2: For $\phi(t,u,v_t) = \phi\left(t, u, \int_0^t f(t,s,v(s))\,ds\right)$, we get Theorem 1 in 15. of Chapter XI.

21. We say that the operator $\phi(t,u,v_t)$ satisfies the conditon (μ) if the problem

$$W'(t) = \phi(t, W(t), W_t) + \delta, \quad W(0) = u_0 + \delta$$

has a solution in C_T (or in S_T) for all $\delta \in [0,\mu]$.

Using Theorem 1 of 20., we can prove the following theorem:

THEOREM 1. *Let the operator $\phi(t,u,v_t)$ be continuous in all variables, nondecreasing in the third argument, and satisfy condition (μ) for some small μ. If $v(t) \in C_T$ (or in S_T) and*

$$v'(t) \le \phi(t, v(t), v_t) \quad \text{on } (0,T], \quad v(0) \le u_0, \tag{21.1}$$

then

$$v(t) \le u(t) \qquad \text{on } [0,T] \tag{21.1'}$$

where $u(t)$ is the upper solution of (20.3).

THEOREM 2. *Let the operator $\phi(t,u,v_t)$ be continuous in all variables and nondecreasing in the third argument. If the function $v(t) \in S_T$ satisfies (21.1), then (21.1′) holds where $u(t)$ is the upper solution of (20.3) defined on $[0,T]$.*

PROOF: Define the sequence of functions $u^{(n)}(t)$ by the following scheme. Let $u^{(1)}(t)$ be the solution of the problem

$$u'(t) = \phi(t, u(t), \{\gamma\}_t), \quad u(0) = u_0, \tag{21.2}$$

and recursively $u^{(n+1)}(t)$ is defined as the solution of the problem

$$u'(t) = \phi\left(t, u(t), u_t^{(n)}\right), \quad u(0) = u_0. \tag{21.3}$$

We shall prove that the sequence $\{u^{(n)}(t)\}$ is nonincreasing. From (21.3) we find that $du^{(2)}/dt(t) < \phi\left(t, u^{(2)}(t), \{\gamma\}_t\right)$, with $u^{(2)}(0) = u_0$. By the comparison theorem for differential inequalities of first order (see Theorem 2 from 2. in Chapter XI) using (21.2) we get $u^{(2)}(t) \leq u^{(1)}(t)$ on $[0,T]$. Suppose now that $u^{(n)}(t) \leq u^{(n-1)}(t)$ on $[0,T]$. Then from (21.3) we have

$$\frac{du^{(n+1)}}{dt}(t) \leq \phi\left(t, u^{(n+1)}(t), u_t^{(n-1)}\right), \quad u^{(n+1)}(0) = u_0.$$

Again using the comparison theorem we get $u^{(n+1)}(t) \leq u^{(n)}(t)$ on $[0,T]$ and the sequence in nondecreasing.

Next we prove that

$$v(t) \leq u^{(n)}(t) \quad \text{on } [0,T], \quad n = 1, 2, \ldots. \tag{21.4}$$

From (21.1) we get

$$v'(t) \leq \phi\left(t, v(t), v_t\right) \leq \phi\left(t, v(t), \{\gamma\}_t\right), \quad v(0) \leq u_0.$$

From the comparison theorem we get $v(t) \leq u^{(1)}(t)$ on $[0,T]$. Again by way of induction we assume that (21.4) holds for a certain n. Then from $v'(t) \leq \phi\left(t, v(t), u_t^{(n)}\right)$, $v(0) \leq u_0$, and (21.3) and the comparison theorem we get $v(t) \leq u^{(n+1)}(t)$ on $[0,T]$, proving (21.4) for all n.

Since the sequence $u^{(n)}(t)$ is monotonic and bounded below by $v(t)$, the $\lim_n u^{(n)}(t)$ exists on $[0,T]$, call it $\tilde{u}(t)$. From the integral equation

$$u^{(n)}(t) = u_0 + \int_0^t \phi\left(s, u^{(n)}(s), u_s^{(n-1)}\right) ds,$$

we get

$$\tilde{u}(t) = u_0 + \int_0^t \phi\left(s, \tilde{u}(s), \tilde{u}_s\right) ds,$$

so that $\tilde{u}(t)$ is a solution of (20.3). In fact, $\tilde{u}(t)$ is the upper solution of (20.3) so that it is $u(t)$. Clearly (21.1′) follows from (21.4) and the theorem is proven.

THEOREM 3. *Let the operator $\phi(t,u,v_t)$ be continuous in all variables and nondecreasing in the third argument. Let the function $v(t) \in C_T$ (or S_T) and satisfy*

$$D_-v(t) < \phi(t,v(t),v_t) \quad \text{on } (0,T], \quad v(0) < u_0$$

and in the case $v(0) = u_0$, then $D^+v(0) < \phi(0,u_0,\{u\}_0)$. Then $v(t) < u(t)$ on $(0,T]$ where $u(t)$ is the lower solution of the problem (20.3) defined on $[0,T]$.

THEOREM 4. *Let the operator $\phi(t,u,v_t)$ be continuous in all variables, nondecreasing in the third argument, and satisfy the condition (μ) for some small μ. Let the function $v(t) \in C_T$ (or S_T) satisfy the inequality*

$$D_-v(t) \le \phi(t,v(t),v_t) \quad \text{on } (0,T], \quad v(0) \le u_0,$$

or

$$D_+v(t) \le \phi(t,v(t),v_t) \quad \text{on } (0,T], \quad v(0) \le u_0.$$

Then $v(t) \le u(t)$ on $[0,T]$ where $u(t)$ is the upper solution of (20.3) on $[0,T]$.

THEOREM 5. *Let the operator $\phi(t,u,v_t)$ be continuous in all variables and nondecreasing in the third argument. Suppose that $v(t) \in S_T$ and satisfies*

$$D_-v(t) \le \phi(t,v(t),v_t) \quad \text{on } (0,T], \quad v(0) \le u_0,$$

or

$$D_+v(t) \le \phi(t,v(t),v_t) \quad \text{on } (0,T], \quad v(0) \le u_0.$$

Then

$$v(t) \le u(t) \quad \text{on } [0,T],$$

where $u(t)$ is the upper solution of (20.3) on $[0,T]$.

22. Generalizations of differential and integral inequalities for abstract operators were considered for example in:

Walter [**32**], Bertolino [**33**], Mamedov, Aširov, and Atdaev [**30**], and Kurpel and Shuvar [**34**],(see also the references given there).

Here we shall first give generalizations of the linear Gronwall inequality (see Hille [**35**, pp. 18–20], and [**36**, pp. 364–376]).

Let X be a complete metric space with metric d. A mapping $T\colon X \to X$ is called a contraction if there is a constant $q \in (0,1)$ such that $d(Tx,Ty) \le qd(x,y)$ for all x,y in X. Banach's fixed point theorem (see Hille [**35**, pp. 12]) asserts that if X is complete and if for some $m > 1$, T^m is a contraction, then T has a unique fixed point x. Moreover, for every $x \in X$, $x = \lim_n T^n x$.

If X is partially ordered by a relation $\le$, then $T\colon X \to X$ is said to be order preserving (or monotonic) if $Tx \le Ty$ whenever $x \le y$.

THEOREM 1. *Let X be a complete metric space which is partially ordered in such a way that if $x_n < x_{n+1}$ for $n \geq 1$ and $\lim_n x_n = x_0$, exists then $x_n \leq x_0$ for all n. Let T be order preserving and suppose that T is a contraction for some $m \geq 1$. If x is the unique fixed point of T then*

$$x \leq Tx \quad \textit{implies that } x \leq x_0. \tag{22.1}$$

Moreover, (22.1) remains valid for $\leq$ is replaced by $<$ or $\geq$ or $>$ in both places.

PROOF: We prove the result for $<$. Since T preserves order, we have

$$x < Tx \leq T^2 x \leq \cdots \leq T^n x. \tag{22.2}$$

Now $T^{nm}x \leq x$ by hypotheses (22.1) and $\lim_n T^{nm}x = x_0$. This implies that $T^{nm}x \leq x_0$, and by (22.2) this can be extended to $T^n x \leq x$. Thus $x < Tx \leq x_0$.

When $\leq$ is a partial order relation in a linear space X we always will assume that $\leq$ is compatible with the algebraic operations in X. That is, we assume that

$$\begin{gathered} x \leq y \quad \text{implies that } x + z \leq y + z \\ \text{and } ax \leq ay \text{ if } a > 0. \end{gathered} \tag{22.3}$$

Note that this also means that

$$\begin{gathered} x \geq 0,\ y \geq 0 \quad \text{imply that } x + y \geq 0, \text{ and} \\ x \leq y \text{ is equivalent to } y - x \geq 0 \text{ or is equivalent to } -y \leq -x. \end{gathered} \tag{22.4}$$

If X is a partially ordered Banach space with partial order $\leq$, then $X^+ = \{x \in X \mid x \geq 0\}$ is called the positive cone of X. An operator $S\colon X \to X$ is called positive if $S(X^+) \subset X$. If S is linear it is clear that S is positive if and only if it is order preserving.

For our next theorem we require the following lemma (see Hille **[35]**,**[36]**) called Volterra's fixed point theorem:

LEMMA 1. *Let $S\colon X \to X$ be a bounded linear operator on a Banach space X such that*

$$\sum_{n=1}^{\infty} \|S^n\| < \infty. \tag{22.5}$$

For each $z \in X$, the operator $T_z\colon X \to X$ defined by $T_z x = z + Sx$ has a unique fixed point $x_z \in X$ given by

$$x_z = \sum_{n=0}^{\infty} S^n z. \tag{22.6}$$

PROOF: For each $z \in X$ it follows from (22.5) that

$$\|\sum_{k=m}^{n} S^k z\| \leq \|z\| \sum_{k=m}^{n} \|S^k\|,$$

and the completeness of X that the series in (22.6) converges. It is easy to see that the sum defines a fixed point of T. On the other hand, if $x \in X$ is a fixed point of T_z, then

$$x = z + S(z + Sx) = \cdots = \sum_{k=0}^{n-1} S^k z + S^n x.$$

Since $\|S^n x\| \leq \|S^n\| \|x\|$ and $\lim_n \|S^n\| = 0$, it follows that $x = x_z$.

We note that the series in (22.6) is ofter called the Neumann series for the solution of $x = z + Sx$.

THEOREM 2. *Let X be a partially ordered Banach space with positive cone X^+ that is closed. Let $S\colon X \to X$ be a positive bounded linear operator satisfying (22.5). Let $z \in X$ and x_z be the unique solution of $x = z + Sx$. Then*

(22.7) $$y \leq z + Sy \quad \textit{implies that } y \leq x_z.$$

Moreover, (22.7) is valid if $\leq$ is replaced by $<$ or $\geq$ or $>$ in both places. Finally, if $z \in X^+$, then $x_z \in X^+$.

PROOF: Since S is order preserving,

(22.8) $$y \leq \sum_{k=0}^{n-1} S^k z + S^n y = y_n, \quad n \geq 1,$$

follows from $y \leq z + Sy$ as in the proof of the previous theorem. Hence $y_n - y \in X^+$ and since X^+ is closed

$$x_z - y = \lim_n (y_n - y) \in X^+.$$

That is $x_z - y \geq 0$. Now if $y < z + Sy$ then (22.8) becomes $y < z + Sy < y_n$ for $n \geq 1$ and in turn $y \leq x_z$.

If $z \in X^+$ then $S^k x \in X^+$ for all $k \geq 0$ and so by (22.4) $\sum_{k=0}^{n} S^k z \in X^+$, for $n \geq 0$. Since X^+ is closed it follows from (22.6) that $x_z \in X^+$.

Theorem 2 is the abstract version of the generalized Gronwall inequality. Indeed (see for example Beesack [**5**, pp. 109–110]), let $J = [\alpha, \beta]$ and $X = C(J)$ (the space of continuous functions with sup norm) and for $x, y \in C(J)$ define $x \leq y$ by $x(t) \leq y(t)$ for all $t \in J$. Then $X^+ = C^+(J) = \{x \in C(J) \,|\, x(t) \geq 0 \text{ on } J\}$ is clearly closed. If $b, k \in C^+(J)$, define $S\colon C(J) \to C(J)$ by

$$(Sx)(t) = b(t) \int_\alpha^t k(s)x(s)ds, \qquad t \in J.$$

It is easy to verify that

$$|(S^n x)(t)| \leq \|x\| \left(\|b\| \, \|k\|\right)^n (t-\alpha)^n/n!, \quad t \in J, \quad n \geq 1.$$

Hence if $M = \|b\| \, \|k\|(\beta - \alpha)$, then $\|S^n x\| \leq \|x\| M^n/n!$ for all $x \in C(J)$ whence $\|S^n\| < M^n/n!$, $n \geq 1$. Thus (22.5) is satisfied so it follows from Theorem 2 that if $x, a \in C(J)$ and

$$x(t) \leq a(t) + b(t) \int_\alpha^t k(s)x(s)ds, \quad t \in J.$$

then $x(t) \leq x_a(t)$, for $t \in J$, where $y = x_a$ is the unique solution of the equation

$$y(t) = a(t) + b(t) \int_\alpha^t k(s)y(s)ds.$$

Setting $U(t) = \int_\alpha^t ky \, ds$ this equation is equivalent to the initial value problem

$$U' - kbU = ka, \quad U(\alpha) = 0,$$

with solution $U(t) = \int_\alpha^t k(s)a(s) \exp\left(\int_s^t kb \, dr\right) ds$. Hence

$$x(t) \leq x_a(t) = a(t) + b(t) \int_\alpha^t k(s)a(s) \exp\left(\int_s^t kb \, dr\right) ds, \quad t \in J.$$

23. We observe that Theorem 2 in 22. also includes Theorem 1 from 7. of Chapter XII and Theorem 1 from 6.

For another corollary of Theorem 2 in 22. we take $X = \mathcal{M}_n C(J)$, the space of all $n \times n$ matrices $x = (x_{ij})$ with each element $x_{ij} \in C(J)$, and norm defined by

$$\|x\| = \sup_{t \in J} \sup_{1 \le i \le n} \sum_{j=1}^{n} |x_{ij}(t)|,$$

and with the partial order relation $x \le y$ if and only if $x_{ij}(t) \le y_{ij}(t)$ for all i, j and $t \in J$. The positive cone $X^+ = \mathcal{M}_n C^+(J)$ consists of all matrices all of whose entries are nonnegative functions and thus is clearly closed. For $k \in X^+$ define $S \colon X \to X$ by

$$(Sx)(t) = k \int_\alpha^t x ds = \left(\sum_{n=1}^{n} k_{ir} \int_\alpha^t x_{rj}(s) ds \right), \quad t \in J, \quad x \in \mathcal{M}_n C(J).$$

S is clearly a positive operator and for each i, j and $t \in J$

$$\begin{aligned} |(Sx)_{ij}(t)| &\le \sum_{r=1}^{n} |k_{ir}(t)|(t-\alpha) \sup_{s \in J} |x_{rj}(s)| \\ &\le \sum_{r=1}^{n} |k_{ir}(t)|(t-\alpha)\|x\| \le \|k\| \, \|x\|(t-\alpha). \end{aligned}$$

By induction we get for each i, j and $t \in J$

$$|(S^m x)_{ij}(t)| \le \|x\| \, \|k\|^m (t-\alpha)^m / m!,$$

so that

$$\|S^m x\| \le n\|x\| \, \|k\|^m (\beta-\alpha)^m / m!, \quad m \ge 1, x \in X$$

and

$$\|S^m\| \le n\|k\|^m (\beta-\alpha)^m / m!, \quad m \ge 1.$$

Therefore S satisfies condition (22.5), and we have proved:

COROLLARY 1. *Let $a = (a_{ij})$ and $k = (k_{ij})$ be constant matrices with $k \ge 0$. If $x - (x_{ij}) \subset \mathcal{M}_n C(J)$ and*

$$x(t) \le a + k \int_\alpha^t x(s) ds, \quad t \in J,$$

then $x(t) \le (\exp[k(t-\alpha)])\, a$, $t \in J$.

24. Generalizations of Theorem 2 of 22. involving nonlinear operators have been given, for example, in Chandra and Fleishman [**37**],[**38**], and Losonczi [**39**].

THEOREM 1. (Losonczi [**39**, Th. 1]) *Let X be a Banach space with a partial order $\leq$ such that the positive cone X^+ is closed. Let $A, B\colon X \to X$ be two operators such that*

(i) *$x, y \in X$, and $x \leq y$ imply that $Ax \leq By$; and*
(ii) *the equations $x = g + Ax$, $y = h + By$ have unique solutions x_g, y_h for arbitrary $g, h \in X$, and these solutions can be obtained as the limits of the sequence of successive approximations beginning with $x_1 = g$ and $y_1 = h$.*

Then

$$u - Au \leq v - Bv \quad \text{implies that } u \leq v. \tag{24.1}$$

PROOF: Set $g = u - Au$ and $h = v - Bv$ and define the sequences of successive approximations $\{u_n\}$ and $\{v_n\}$ to u and v by

$$u_1 = g,\ u_{n+1} = g + Au_n,\ v_1 = h,\ v_{n+1} = h + Bv_n, \quad n \geq 1.$$

By hypotheses $u_1 \leq v_1$. Moreover if $u_n \leq v_n$ then by (i) we have

$$u_{n+1} = g + Au_n \leq g + Bv_n \leq h + Bv_n = v_{n+1}.$$

By induction $(v_n - u_n) \in X^+$ for all $n \geq 1$. Since X is closed, $v - u = \lim_n (v_n - u_n) \in X^+$ and the proof is complete.

REMARK 1: Taking $Bx = Ax = z + Sx$ where S is a positive bounded linear operator on X satisfying (22.5), we see that the conditions (i) and (ii) are satisfied. Now (24.1) gives

$$u \leq z + Su \quad \text{implies that } u \leq x_z \ (x_z = z + Sx_z),$$

so that Theorem 1 includes Theorem 2 of 22.

REMARK 2: Condition (i) is satisfied if $Ax \leq Bx$ for all $x \in X$ and either A or B is order preserving.

REMARK 3: Condition (ii) is satisfied if both A and B are contractions (not necessarily linear) or if A and B are both linear operators such that A^n and B^m are contractions for some integers n and m.

25. More generally, according to a theorem of D. Boyd and J.S.W. Wong [40], condition (ii) is satisfied (say for A) if there is a continuous function w on R such that $w(r) < r$ for $r > 0$ and

$$\|Ax - Ay\| \leq w(\|x - y\|) \quad \text{for } x, y \in X. \tag{25.1}$$

For other conditions which assure condition (ii) see Geraghty [**41**] and Kannan [**42**].

If A is an order preserving operator and satisfies (25.1) then

(25.2) $$u \leq q + Au \quad \text{implies that } u \leq x_q \ (x_q = q + Ax_q).$$

26. The condition that $w(r) < r$ for every $r > 0$ in (25.1) is weakened in Chandra and Fleishman [**37**, Th. 3]:

THEOREM 1. *Let the partially ordered Banach space X have a closed positive cone X^+, and let the operator $A\colon X \to X$ be an order preserving compact (completely continuous) operator which satisfies conditon (25.1) where w is a nonnegative nondecreasing function on R and there is an r^* such that*

(26.1) $$w(r) + \|q\| + \|A\theta\| < r \quad \text{if } r > r^*.$$

Suppose that

(26.2) $$u \leq q + Au.$$

Then the equation $v = q + Av$ has a solution given by $v = \lim_n v_n$, where

(26.3) $$v_1 = u, \quad v_{n+1} = q + Av_n, \quad n \geq 1.$$

Moreover, $u \leq v$. In particular if $v = q + Av$ has a maximal solution ϕ, then $u \leq \phi$.

27. P.R. Beesack [**5**, p. 115] noted that the conditions involving w can be replaced by the following:

There exist numbers r^* and R^* such that $0 < r^* < R^*$ and a nonnegative nondecreasing function w on $[0, R^*]$ such that

(27.1) $$\|Ax\| \leq w(\|x\|) \quad \text{for } \|x\| \leq R^*. \qquad \text{(See (25.1))}$$

(27.2) $$w(r) + \|q\| \leq r \quad \text{if } r^* \leq r \leq R^*. \qquad \text{(See (26.1))}$$

Then the conclusion remains valid for $u \in X$ with $\|u\| \leq R^*$.

REMARK 1: In Theorem 1 of 26. we understand that it was implicitly assumed that the equation $v = q + Av$ has a maximal solution.

28. Finally we shall give one result from Chandra and Fleishman [**37**, Th.4].

Let $X = S_nC(J)$ be the class of all real $n \times n$ symmetric matrices $x = (x_{ij})$ with $x_{ij} \in C(J)$, J a compact interval. X is a Banach space with the sup norm as before and with the positive cone S^+ given by the nonnegative semi-definite symmetric matrices. This cone is closed since the quadratic form is a continuous function. Let S_n denote the constant matrices in $S_nC(J)$ and note that $\|x\| = |x| = \sup_i \sum_{j=1}^{n} |x|$ if $x \in S_n$.

THEOREM 1. *Let $a = (a_{ij}) \in S_nC(J)$ and let $G\colon S_n \to S_n$ be order preserving and satisfy a Lipschitz condition*

$$|G(x) - G(y)| \leq k|x - y| \text{ for } x, y \in S_n.$$

If $x \in S_nC(J)$ and

$$x(t) \leq a(t) + \int_{\alpha}^{t} G(x(s))\,ds, \quad t \in J,$$

then $x(t) \leq y(t)$, $t \in J$, where $y(t)$ is the unique solution of the corresponding equation on J.

29. The natural conception for integral inequalities of interval functions is theory of interval analysis, initiated by R. E. Moore in 1979. Formally, we have substitution of sign $\leq$ by $\subseteq$, i.e. connection between intervals and classical inequalities are given by:

$$[a, b] \subseteq [c, d] \quad \text{if and only if } c \leq a \leq b \leq d.$$

A short exposition of theory of integral inequalities by using interval functions is given in the book [**43**] by A.A. Martynyuk, V. Lakshmikanthan and S. Leela. This is a new theory in mathematical literature, and it started by the papers of R.E. Moore [**44**]–[**46**], O. Caprani, K. Madsen and L.L. Rall [**47**], and G. Alefeld and J. Herzberger [**48**].

30. Further references for this Chapter may be found in [**49**]–[**118**].

REFERENCES

1. RABCZUK, R., "*Elemnty nierównósci vóżniczkowych,*" Warszawa, 1976.
2. GRACE, S.R., B.S. LALLI and C.C. YEH, *Comparison theorems for difference inequalities*, J. Math. Anal. Appl. **113** (1986), 468–472.
3. PACHPATTE, B.G. and V.M. SINGARE, *Comparison theorems on difference equations of the Volterra type*, Math. Education Sect. A. **7** (1973), 45–48.
4. REDHEFFER, R. and W. WALTER, *A comparison theorem for difference inequalities*, J. Differential Equations **44** (1982), 111–117.
5. BEESACK, P.R., *Comparison theorems and integral inequalities for Volterra integral equations*, Proc. Amer. Math. Soc. **20** (1969), 61–66.
6. JONES, G., *Fundamental inequalities for discrete and discontinuous functional equations*, J. Soc. Ind. Appl. Math. **12** (1964), 43–57.

7. HULL, T. and W. LUXEMBURG, *Numerical methods and existence theorems for ordinary differential equations*, Numer. Math. **2** (1960), 30–41.
8. FILATOV, A. and L. ŠAROVA, "Integral'nye neravenstva i teorija nelineinyh kolebanii," Moskva, 1976.
9. WILLETT, D. and J. WONG, *On the discrete analogues of some generalizations of Gronwall's inequality*, Monatsh. Math. **69** (1965), 362–367.
10. LEES, M., *Approximate solutions of parabolic equations*, SIAM Journal **7** (1959), 167–183.
11. SINGARE, V.M. and B.G. PACHPATTE, *On some discrete generalizations of an inequality of Gronwall*, Chinese J. Math. **6** (1978), 121–135.
12. __________, *Wendroff type discrete inequalities and their applications*, J. Math. Phys. Sci. **13** (1979), 149–167.
13. YEH, C.C., *Discrete inequalities of the Gronwall-Bellman type in n independent variables I, II*, J. Math. Anal. Appl. **105** (1985), 322–332; **106** (1985), 282–285.
14. PACHPATTE, B.G., *On some fundamental finite difference inequalities*, Univ. Beograd. Publ. Elektrotehn. Fak. Ser. Mat. Fiz. No **577-598** (1977), 65–73.
15. POPENDA, J., *Finite difference inequalities*, Ann. Soc. Math. Polonae. Ser. I: Comm. math **26** (1986), 89–96.
16. YANG, E.H., *On some new discrete generalizations of Gronwall's inequality*, J. Math. Anal. Appl. **119** (1988), 505–516.
17. THANDAPANI, E. and R.P. AGARWAL, *On some new integro-differential inequalities; theory and applications*, Tamkang J. Math. **11** (1980), 169–184.
18. PACHPATTE, B.G.and V.M. SINGARE, *Discrete generalized Gronwall inequalities in three independent variables*, Pac. J. Math. **82** (1979), 197–210.
19. PACHPATTE, P.G., *On some nth order finite difference inequalities*, Proc. Nat. Acad. Sci. India **40** (A) (1970), 235–240.
20. __________, *Discrete generalizations of Gronwall's inequality*, Math. Education **8** (1979), A73–A75.
21. __________, *On discrete inequalities related to Gronwall's inequality*, Proc. Indian. Acad. Sci. **85** (A) (1977), 26–40.
22. __________, *A note on some new nth order discrete inequalities*, Tamkang J. Math. **9** (1978), 29–39.
23. __________, *On some nth order differential inequalities and their discrete analogues*, Proc. Nat. Acad. Sci. India. Sect A **49** (1979), 7–12.

24. __________, *On some fundamental nth order finite difference inequalities*, Ann. Ştinţ. Univ. "Al. I. Cuza" Iaşi Sect. Ia Mat. (N.S.) **26** (1980), 361–367.
25. AGARWAL, R.P. and E. THANDAPANI, *On some new integro-differential inequalities*, An. Ştinţ, Univ. "Al. I. Cuza" Iaşi. N.S. **28** (1982), 123–126.
26. NOVOTNÁ, J., *Variations of discrete analogues of Wirtinger's inequality*, Čas. Pěsto. Mat. **105** (1980), 278–285.
27. DAS, P. and R. SHARMA, *Some Stieltjes integral inequalities*, J. Math. Anal. Appl. **73** (1980), 423–433.
28. MINGARELLI, A.B., *On a Stieltjes version of Gronwall's inequality*, Proc. Amer. Math. Soc. **82** (1981), 249–252.
29. MAMEDOV, Ja. D. and S.A. AŠIROV, "Nelineǐnie uravnenija Volterra," Ašhabad, YLYM, 1977.
30. MAMEDOV, Ja.D., S. AŠIROV and S. ATDAEV, "Teoremy o neravenstvah," Ašhabad, YLYM, 1980.
31. MAMEDOV, Ja. D. and G.M. NOVRUZOV, *K teorii rešeniǐ differencial'nyh uravneniǐ s vol'terrovskim operatorom*, Differencial'nye uravnenija **8** (1972), No 6.
32. WALTER, W., "Differential and integral inequalities," New York-Heidelgerg-Berlin, 1970.
33. BERTOLINO, M., "Diferencijalne jednačine," Beograd, 1980.
34. KURPEL', N.S, and B.A. SHUVAR, "Dvustoronnie operatornye neravensta i 16 primenenija," Kiev, 1980.
35. HILLE, E., "Lectures on ordinary differential equations," Reading, Mass., 1969.
36. __________, "Methods in classical and functional analysis," Reading. Mass., 1972.
37. CHANDRA, F. and B. FLEISHMAN, *Bounds and maximal solutions of nonlinear functional equations*, Bull. Amer. Math. Soc. **74** (1968), 512–516.
38. __________, *On a generalization of the Gronwall-Bellman lemma in partially ordered Banach spaces*, J. Math. Anal. Appl. **31** (1970), 668–681.
39. LOSONCZI, L. *A generalization of the Gronwall-Bellman lemma and its applications*, J. Math. Anal. Appl. **44** (1973), 701- 709.
40. BOYD, D. and J. WONG, *On nonlinear contractions*, Proc. Amer. Math. Soc. **20** (1969), 458–464.
41. GERAGHTY, M., *On contractive mappings*, Proc. Amer. Math. Soc. **40** (1973), 604–608.

42. KANNAN, R., *Construction of fixed points of a class of nonlinear mappings*, J. Math. Anal. Appl. **41** (1973), 430–438.
43. MARTYNYUK, A.A., V. LAKSHMIKANTHAM, and S. LEELA, "Motional Stability: The Method of Integral Inequalities," Kiev, 1989.
44. MOORE, R.E., "Methods and applications of interval analysis," SIAM Studies in Appl. Math., Philadelphia, 1979, 250 pp.
45. ________, *Set-valued extensions of integral inequalities*, J. of Integral Equations **5** (1983), 187–198.
46. ________, *Set-valued extensions of integral inequalities*, Trends in theory and practice of nonlinear differential equations **80** (1984), 383–388.
47. CAPRANI, O., K. MADSEN, and L.L. RALL, *Integration of integral functions*, SIAM J. Math. Analysis **12** (1981), 321–341.
48. ALEFELD, G. and J. HERZBERGER, "Introduction to interval computations," New York, 1983.
49. AGARWAL, R.P. and E. THANDAPANI, *On nonlinear discrete inequalities of Gronwall type*, An. Ştinţ. Univ. "Al. I. Cuza" Iaşi. N.S. Sect. Ia (1981), 139–149.
50. ________, *On discrete generalizations of Gronwall's inequality*, Bull. Inst. Math. Acad. Sinica **9** (1981), 235-248.
51. ________, *On discrete inequalities in n independent variables*, Riv. Mat. Univ. Parma **4** (1987), 241–256.
52. ________, *Linear and nonlinear discrete inequalities in n independent variables*, General Inequalities **5** (Oberwolfach, 1986), Basel-Boston (1987), 303–318.
53. ________, *Systems of multi-dimensional discrete inequalities*, J. Math. Anal. Appl. **140** (1989), 241–250.
54. AŠIROV, S. and Ja.D. MAMEDOV, *Issledovanie rešenii nelineinyh operatornyh uravnenii Vol'terra-Fredgol'ma*, DAN SSSR, **229** (1976), No 2.
55. ATDAEV, S., *K teorii rešeniĭ nelineĭnyh mnogomernyh operatornyh uravneniĭ Vol'terra*, Izvestija AN TSSR. Ser. FTN i GN, 1976, No 6.
56. ________, *K teorii nelineĭnyh operatornyh uravneniĭ Vol'terra*, Izvestija AN TSSR. Ser. FTN i GN, 1977, No 5.
57. ATDAEV, S. and S. AŠIROV, *Nekotorye iteracionnye processy dlja rešenija nelineĭnyh operatornyh uravneniĭ na poluosi*, Izvestija AN TSSR Ser. FTN i GN, 1975, No. 6.
58. AZBELEV, N.V. and Z.B. CALJUK, *O zadače Čaplygina.* Ukrain. Mat. Ž. **19** (1958), No 1.
59. BEESACK, P.R., *Lower bounds from discrete inequalities of Gollwitzer-Langenhop type*, An. Ştinţ. Univ. "Al. I. Cuza" Iaşi Sect Ia Mat. (N.S.) **30** (1984), 25–30.

60. __________, *More generalised discrete Gronwall inequalities*, Z. Angew. Math. Mech. **65** (1985), 583–595.
61. __________, *On integral inequalities of Bihari type*, Acta Math. Acad. Sci. Hungar. **28** (1976), 81–88.
62. __________, *On Lakshmikantham's comparison methods for Gronwall inequalities*, Annal. Polon. Math. **25** (1977), 187–222.
63. BYKOV, Ja.V. and V. LINENKO, *Kačestvennoe issledovanie traektoriĭ nelineĭnyh sistem raznostnyh uravneniĭ*, Izv. Vysš. Učebn. Zaved. Matematika **56** (1967), 21–29.
64. __________, *O periodičeskom predel'nom režime dlja sistem raznostnyh uravneniĭ*, Izv. Vysš. Učebn. Zaved. Matematika **57** (1967), 9–19.
65. CALJUK, Z.B., *Funkcional'nye neravenstva Vol'terra*, Izvestija VUZov. Matematika, **1969**, No 3.
66. CHANDRA, J. and V. LAKSHMIKANTHAM, *Some pointwise estimates for solutions of a class of nonlinear functional-integral inequalities*, Ann. Mat. Pura. Appl. (4) **94** (1972), 63–74.
67. __________, *A discrete integral inequality of the convolution type*, Bul. Inst. Politehn. Iaşi. Sect. Ia **33** (37) (1987), 35–40.
68. DAS, P.C. and R.SHARMA, *Some Stieltjes integral inequalities*, J. Math. Anal. Appl. **73** (1980), 423–433.
69. DEMIDOVIČ, V., *Ob asimptotičeskom povedenii rešenii konačnoraznostnyh uravnenii. I,* Obščie položenija, Differencial'nie uravnenija **10** (1974), 2267–2278.
70. __________, *Ob odnom priznake ustoičivosti raznostnyh uravnenij*, Differencial'nye uravnenija **5** (1969), 1247–1255.
71. DRAGOMIR, S.S., *On a Volterra integral equation*, An. Univ. Timişoara Ser. St. Mat. **24** (1986), 3–18.
72. __________, *On Volterra integral equation with kernels of (L)-type*, An. Univ. Timişoara Ser. St. Mat. **25** (1987), 21–41.
73. Konstantinov, M., *Ob odnom funkcional'nom neravenstve*, Ukrain. Mat. Ž. **30** (1978), 226–231.
74. KVAPIŠ, M., *O suščestvovanii i edinstvennosti rešenija i shodimosti posledovatel'nyh približeniĭ dlja funkcional'nyh uravneniĭ tipa Vol'terra v Banahovom prostranstve*, Tr. sem. po teor. diff. uravn. s otkl. argumentom. M. Univ. druž. nar. im. P. Lumumby, **4** (1967).
75. MAKHMUDOV, A.P. and R.Z. ABBASOV, *Theorems on integral inequalities in Riemann-Stieltjes integrals and their applications* (Russian), Dokl. Akad. Nauk. SSSR **265** (1982), No. 5, 1069–1073.
76. MAMEDOV, Ja.D. and A.I. PEROV, *O mnogomernyh integral'nyh uravnenijah i neravenstvah tipa Vol'terra*, Izv. AN AzSSR. Ser. FT i MN, 1965, No. 1.

77. MAMEDOV, Ja.D. and S.M. SARDARLY, *K teorii rešeniǐ nelineǐnyh operatornyh uravneniǐ Vol'terra na poluosi*, Sib. Mat. Ž. **12** (1971), No 3.
78. MARTYNJUK, D., "Lekcii po kačestvennoi teorii raznostnyh uravnenii," Kiev, 1972.
79. McKEE, S., *Generalised discrete Gronwall lemmas*, Z. Angew. Math. Mech. **62** (1982), 429–434.
80. NUMIROW, T. and I. CHORSHANBAEV, *Some integral inequalities with singular kernels* (Russian), Dokl. AN UzSSR, 1983, No 1, 8–10.
81. PACHPATTE, B.G., *On the discrete generalizations of Gronwall's inequality*, J. Indian Math. Soc. N.S. **37** (1973), 147–156.
82. __________, *Finite difference inequalities and their applications*, Proc. Natn. Acad. Sci., India, **A43** (1973), 348–356.
83. __________, *A note on some finite difference inequalities*, Math. Student. **43** (1975), 405–416.
84. __________, *On some discrete inequalities of Bellman-Bihari type*, Indian. J. Pure. Appl. Math. **6** (1975), 1479–1487.
85. __________, *On a class of integral inequalities of Bellman-Bihari type*, Indian. J. Pure. Appl. Math. **7** (1976), 1166–1177.
86. __________, *A note on some new discrete inequalities*, Proc. Nat. Acad. Sci. India Sect. A **46** (1976), 190–198.
87. __________, *On some new discrete inequalities and their applications*, Proc. Nat. Acad. Sci. India Sect. A **46** (1976), 255–262.
88. __________, *On some nonlinear discrete inequalities of Gronwall type*, Bull. Inst. Math. Acad. Sinica **5** (1977), 305–315.
89. __________, *On some new integral inequalities and their discrete analogues*, Indian J. Pure Appl. Math. **8** (1977), 1093–1107.
90. __________, *A note on discrete inequalities of the Gronwall-Bellman type*, J. Math. Phys. Sci. **17** (1977), 115–124.
91. __________, *On some new generalizations of Gronwall's inequality and their discrete analogues*, J. of M.A.C.T. **10** (1977), 55–69.
92. __________, *A note on some fundamental discrete inequalities of the Gronwall-Bellman type*, Bull. Inst. Math. Acad. Sinica **5** (1977), 121–128.
93. __________, *On some new discrete inequalities involving partial sums*, Indian J. Pure Appl. Math. **9** (1978), 640–647.
94. __________, *Finite difference inequalities and discrete time control systems*, Indian J. Pure Appl. Math. **9** (1978), 1282–1290.
95. __________, *On some new discrete inequalities and their applications to a class of sum-difference equation*, Ann. Ştinţ. Univ. "Al I. Cuza" Iaşi Sect Ia Mat. (N.S.) **24** (1978), 315–326.

96. __________, *On some fundamental discrete inequalities in two independent variables*, Tamkang J. Math. **12** (1981), 22–33.
97. __________, *On some integral inequalities and their application to higher order differential equations*, Soochow J. Math. **14** (1988), 221- 230.
98. __________, *On a class of new discrete inequalities*, Tamkang J. Math. **20** (1989), 19–28.
99. PELCZAR, A., *On invariant points of monotone transformations in partially ordered spaces*, Ann. Polon. Math. **17** (1965), 49–53.
100. PEROV, A.I., *Ob integral'nyh neravenstvah*, Tr. sem. po Funkcional'nomu Analizu. Voronež, Vyp. **5** (1975), 87–97.
101. POPENDA, J., *Finite difference inequalities*, Fasc. Math. **13** (1981), 79–87.
102. __________, *On the discreteanalogy of Gronwall lemma*, Demonstratio Math. **16** (1983), 11–25.
103. __________, *On the discrete analogy of Gronwall-Wendroff inequality*, Demonstratio Math. **18** (1985), 1083–1103.
104. __________, *On some discrete Gronwall type inequalities*, Fasc. Math. **17** (1985), 109–114.
105. __________, *On the discrete inequalities of Gronwall-Bellman type*, An. Ştinţ. Univ. "Al. I. Cuza" Iaşi Ser. Ia Mat. **33** (1987), 37–52.
106. __________, *On the discrete analogy of Gronwall inequality*, Fasc. Math. **19** (1987), 63–78.
107. POPENDA, J and J. WERBOWSKI, *On the discrete analogy of Gronwall lemma.* Fasc. Math. **11** (1979), 143–154.
108. SINGARE, V.M. and B.G. PACHPATTE, *On some fundamental discrete inequalities of the Wendroff type*, An. Ştinţ. Univ. "Al. I. Cuza" Iaşi Sect Ia Mat. (N.S.) **26** (1980), 87–94.
109. __________, *On certain discrete inequalities of the Wendroff type*, Indian J. Pure Appl. Math. **11** (1980), 727–736.
110. SREE HARI RAO, V., *Gronwall-Bellman type inequalities to distributions*, Proc. Conf. Math. Math. in Physics (Mysore, 1978), Part B, pp. 60–64, Matscience Rep., **96**, Inst. Math. Sci., Madras, 1979.
111. __________, *Integral inequalities of Gronwall type for distributions.* J. Math. Anal. Appl. **72** (1979), 545–550.
112. THANDAPANI, E., *Discrete inequalites in n independent variables of Gronwall-Bellman type*, Appl. Anal. **39** (1988), 189–199.
113. __________, *Discrete inequalities in n independent variables of Gronwall-Bellman type*, Appl. Anal. **39** (1988), No 1-3, 189–199.
114. THIEME, H.R., *The asymptotic behavior of solutions of nonlinear integral equations*, Math. Z. **157** (1977), 141–154.

115. VIDYASAGAR, M., *Nonlinear functional inequalities of Gronwall-Bihari type*, private communication.
116. WIRTH, T., *Abschätzungen von Lösungen Voltenascher und Fredhomscher Integralungleichungen*, Diss. Dokt. Naturwiss. Fak. Math. und Naturwiss. Tehn. Univ. Hannover, 1976, 55 pages.
117. WU, J. and G.S. YANG, *On discrete Gronwall's inequalities.* Tamkang J. Math. **12** (1981), 161–170.
118. YANG, E.H., *On some new discrete inequalities of the Bellman-Bihari type.* Nonlinear Anal. **7** (1983), 1237–1246.

CHAPTER XV

INTEGRAL INEQUALITIES INVOLVING FUNCTIONS WITH BOUNDED DERIVATIVES

1. Ostrowski's inequality. A. Ostrowski [**1**] has proved the following result (see also 11., 17., and 52. from Chapter I):

THEOREM 1. *Let f be a differentiable function on (a,b) and let, on (a,b), $|f'(x)| \le M$. Then, for every $x \in (a,b)$,*

$$\left| f(x) - \frac{1}{b-a} \int_a^b f(x)dx \right| \le \left(\frac{1}{4} - \frac{\left(x - \frac{a+b}{2}\right)^2}{(b-a)^2} \right)(b-a)M. \tag{1.1}$$

G. V. Milovanović [**2**] has generalized the Theorem 1 when f is a function of several varibles:

THEOREM 2. *Let $f\colon R^m \to R$ be a differentiable function defined on $D = \{(x_1, \dots, x_m) \,|\, a_i \le x_i \le b_i \ (i = 1, \dots, m)\}$, and let $\left|\frac{\partial f}{\partial x_i}\right| \le M_i$ ($M_i > 0$; $i = 1, \dots, m$) in D. Furthermore, let function $x \mapsto p(x)$ be defined integrable and let $p(x) > 0$ for every $x \in D$. Then for every $x \in D$,*

$$\left| f(x) - \frac{\int_D p(y)f(y)\,dy}{\int_D p(y)dy} \right| \le \frac{\sum_{i=1}^m M_i \int_D p(y)|x_i - y_i|dy}{\int_D p(y)dy}. \tag{1.2}$$

Several related results are given in [**3**].

G. V. Milovanović and J. E. Pečarić [**4**] proved:

THEOREM 3. *Let $f(x)$ be n (> 1) times differentiable function such that $|f^{(n)}(x)| \le M$ for $x \in (a,b)$. Then, for every $x \in [a,b]$*

$$\left|\frac{1}{n}\left(f(x)+\sum_{k=1}^{n-1}F_k\right)-\frac{1}{b-a}\int_a^b f(y)\,dy\right| \le \frac{M}{n(n+1)!}\,\frac{(x-a)^{n+1}+(b-x)^{n+1}}{b-a}. \tag{1.3}$$

where F_k is defined by

$$\begin{aligned} F_k &= F_k(f;n;x,a,b) \\ &= \frac{n-k}{k!}\,\frac{f^{(k-1)}(a)(x-a)^k - f^{(k-1)}(b)(x-b)^k}{b-a}. \end{aligned} \tag{1.4}$$

PROOF: Integrating by use of Taylor's formula

$$f(x) = f(y) + \sum_{k=1}^{n-1}\frac{1}{k!}f^{(k)}(y)(x-y)^k + \frac{1}{n!}f^{(n)}(\xi)(x-y)^n, \tag{1.5}$$

where $\xi = y+\theta(x-y)$ $(0<\theta<1)$, we obtain

$$\begin{aligned} &f(x)(b-a) - \int_a^b f(y)dy - \sum_{k=1}^{n-1}\frac{1}{k!}\int_a^b f^{(k)}(y)(x-y)^k\,dy \\ &= \frac{1}{n!}\int_a^b f^{(n)}(\xi)(x-y)^n\,dy. \end{aligned} \tag{1.6}$$

Putting $I_k = \frac{1}{k!}\int_a^b f^{(k)}(y)(x-y)^k\,dy$ $(k\ge 1)$ and $I_0 = \int_a^b f(y)dy$, by means of partial integration on I_k, we have

$$I_k = \frac{1}{k!}\left\{f^{(k-1)}(b)(x-b)^k - f^{(k-1)}(a)(x-a)^k\right\} + I_{k-1} \quad (1\le k\le n-1),$$

i.e.

$$(n-k)(I_k - I_{k-1}) = -(b-a)F_k \quad (1\le k\le n-1),$$

where F_k is defined by (1.4).

Since

$$\sum_{k=1}^{n-1}(n-k)(I_k-I_{k-1}) = -(b-a)\sum_{k=1}^{n-1}F_k,$$

i.e.

$$I_0 + \sum_{k=1}^{n-1} I_k = nI_0 - (b-a)\sum_{k=1}^{n-1} F_k,$$

it follows, from (1.6),

$$\left| f(x)(b-a) + (b-a)\sum_{k=1}^{n-1} F_k - nI_0 \right| = \frac{1}{n!}\left| \int_a^b f^{(n)}(\xi)(x-y)^n dy \right|$$

$$\leq \frac{M}{n!}\int_a^b |x-y|^n dy$$

$$= \frac{M}{(n+1)!}\left\{(x-a)^{n+1} + (b-x)^{n+1}\right\}.$$

Thus the proof is finished.

For $n = 2$, the Theorem 3 reduces to the following corollary:

COROLLARY 1. *Let $f(x)$be twice differentiable function such that $|f''(x)| \leq M$ for $x \in (a,b)$). Then*

$$\left| \frac{1}{2}\left(f(x) + \frac{(x-a)f(a) + (b-x)(b)}{b-a} \right) - \frac{1}{b-a}\int_a^b f(x)dy \right|$$

$$\leq \frac{M(b-a)^2}{4}\left(\frac{1}{12} + \frac{\left(x - \frac{a+b}{2}\right)^2}{(b-a)^2} \right)$$

for every $x \in [a,b]$.

The condition $|f^{(n)}(x)| \leq M$ can be substituted by the weaker condition that $f^{(n-1)}$ is continuous and that (see [**5**])

$$|f^{(n-1)}(x) - f^{(n-1)}(y)| \leq M|x-y| \quad \text{for } x, y \in (a,b).$$

In [**5**] is also given a generalization in which $f^{(n-1)}$ satisfy the Lipschitz's condition of order α $(0 < \alpha \leq 1)$. Generalizations for positive linear functionals are given in [**6**]. Applications of previous results to quadrature formulas are given in the same papers.

2. A. M. Fink [**7**] has generalized the above to prove the following result.

THEOREM 1. *Let $f^{(n-1)}(t)$ be absolutely continuous on $[a,b]$ with $f^{(n)} \in L^p(a,b)$ then (see (1.3))*

$$\left| \frac{1}{n}\left(f(x) + \sum_{k=1}^{n-1} F_k(x) \right) - \frac{1}{b-a}\int_a^b f(y)dy \right| \leq K(n,p,x)\|f^{(n)}\|_p \tag{2.1}$$

where
(2.2)

$$K(n,p,x) = \frac{1}{n!}\frac{\left[(x-a)^{n+\frac{1}{p'}} + (b-x)^{n+\frac{1}{p'}}\right]^{1/p'}}{b-a} B\left((n-1)p'+1, p'+1\right)^{1/p'}$$

if $1 < p \leq \infty$ and

$$K(n,1,x) = \frac{(n-1)^{n-1}}{n^n n!}\,\frac{\max\{(x-a)^n(b-x)^n\}}{b-a} \tag{2.3}$$

where $B(x,y)$ is the beta function.

For $p > 1$ these are best possible in the sense that for each x in $[a,b]$ there is an f for which equality holds at x. For $p = 1$, (2.3) gives the best possible but equality holds in (2.1) for no function.

In particular we note that (1.1) and (1.3) are best possible results. The proof of this theorem writes the remainder in (1.5) as $\frac{1}{(n-1)!}\int_y^x (x-t)^{n-1} f^{(n)}(t)dt$ and uses Hölder's inequality.

3. Iyengar's inequality. K. S. K. Iyengar [8], by means of geometrical consideration, has proved the following theorem:

THEOREM 1. *Let f be a differentiable function on $[a,b]$ and $|f'(x)| \leq M$. Then*

$$\left| \int_a^b f(x)dx - \frac{1}{2}(b-a)\left(f(a)+f(b)\right) \right| \leq \frac{M(b-a)^2}{4} - \frac{1}{4M}\left(f(b)-f(a)\right)^2 . \tag{3.1}$$

In [9] the following generalization of Theorem 1 is proved analytically:

THEOREM 2. *Let $f(x)$ be a differentiable function defined on $[a,b]$ and $|f'(x)| \leq M$ for every $x \in (a,b)$. If $p(x)$ is an integrable function on (a,b) such that*

$$0 < c \leq p(x) \leq \lambda c \quad (\lambda \geq 1,\ x \in [a,b]),$$

the following inequality holds

$$\left|A(f;p)-\frac{1}{2}(f(a)+f(b))\right|\le\frac{M(b-a)}{2}\cdot\frac{(\lambda+q)(1-q^2)+2(\lambda-1)q}{2\lambda(1+q)-(\lambda-1)(1+q^2)}, \tag{3.2}$$

where A and q are defined by

$$A(f;p)=\frac{\int_a^b p(x)f(x)dx}{\int_a^b p(x)dx} \quad \text{and} \quad q=\frac{|f(b)-f(a)|}{M(b-a)}.$$

The following result from [**10**] is a special case of a more general result in which $f^{(n-1)}$ satisfy the Lipschitz's condition of order α.

THEOREM 3. *Let function $f\colon [a,b]\to \mathbf{R}$ have a continuous derivative of order $n-1$ and bounded derivation of the order n, i.e. $|f^{(n)}(x)|\le M$ for $x\in(a,b)$. If $f^{(k)}(a)=f^{(k)}(b)=0$ $(k=1,\dots,n-1)$, then the inequality*
(3.3)

$$\left|\frac{1}{b-a}\int_a^b f(x)dx-\frac{1}{2}(f(a)+f(b))\right|\le\frac{M(b-a)^n}{(n+1)!}\left\{\zeta^n-\frac{q}{2}(1+n(2\zeta-1))\right\},$$

holds, where ζ is the real root of the equation

$$\zeta^n-(1-\zeta)^n=q \quad \left(q=\frac{n!}{M(b-a)^n}(f(b)-f(a))\right).$$

The special case of the main results from [**10**] is also:

THEOREM 4. *Let the function $f(x)$ fulfill the conditions*

1° $|f'(x)-f'(y)|\le M|x-y|$ *for* $x,y\in[a,b]$;
2° $f'(a)=f'(b)=0$.

Then

$$\left|\frac{1}{b-a}\int_a^b f(x)dx-\frac{1}{2}(f(a)+f(b))\right|\le\frac{M(b-a)^2}{24}-\frac{1}{2}\cdot\frac{(f(b)-f(a))^2}{M(b-a)^2}.$$

This results is more rigorous than a similar results from [**9**].

4. The inequalities (3.1) and (3.3) were considered in [**7**]. Fink proves the following result.

THEOREM 1. *Let f have $f^{(n)} \in L^p(a,b)$ with $f^{(j)}(a) = f^{(j)}(b) = 0$ for $j = 1,\dots,n-1$ (for $n = 1$ there is no condition). Then for $1 \le p \le \infty$*

$$\left| \frac{1}{b-a} \int_a^b f(x)dx - \frac{f(a)+f(b)}{2} \right| \le \frac{R(n,p)}{n!} \|f^{(n)}\|_p \tag{4.1}$$

are best possible inequalities where

$$R(n,p) = \min_{q \in \pi_{n-2}} \frac{\|(b-t)^n - (a-t)^n - q(t)\|_{p'}}{2(b-a)} \tag{4.2}$$

and π_m is the polynomials of degree $\le m$.

The proof is based on the representation of the left side of (4.1) as

$$\frac{1}{2n!(b-a)} \int_a^b f^{(n)}(t)\left[(b-t)^n + (a-t)^n\right] dt$$

and the following lemma together with Hölder's inequality.

LEMMA. *Let $f^{(j)}(a) = f^{(j)}(b) = 0$, $j = 1,\dots,n-1$. Then $g = f^{(n)}$ satisfies $\int_a^b gq = 0$ for all $q \in \pi_{n-2}$. Conversly if $\int_a^b gq = 0$ for all $q \in \pi_{n-2}$, then there is an f such that $f^{(n)} = g$ and $f^{(j)}(a) = f^{(j)}(b) = 0$ for $j = 1,\dots,n-1$.*

By either solving the minimum problem (4.2) explicitly or choosing an appropriate q the following results are obtained in [7].

$$R(1,p) = \frac{(b-a)^{1-\frac{1}{p}}}{2(1+p')^{1/p'}}, \quad R(1,1) = \frac{1}{2}; \tag{4.3}$$

$$\begin{cases} R(2,\infty) = \frac{b-a)^2}{16}, \quad R(2,1) = \frac{b-a}{q}, \quad R(2,2) = \frac{(b-a)^{3/2}}{3\sqrt{5}} \\ R(2,p) \le \frac{(b-a)^{2-\frac{1}{p}}}{4(2p'+1)^{1/p'}}; \end{cases} \tag{4.4}$$

$$\begin{cases} R(3,\infty) = \frac{(b-a)^3}{32} \\ R(3,p) \le \frac{(b-a)^{3-\frac{1}{p}}2^{1/p}}{16} B\left(\frac{p}{2}+\frac{1}{2}, p'+1\right)^{1/p'}, \quad p > 1; \end{cases} \tag{4.5}$$

(4.6)

$$R(n,p) \le \frac{(b-a)^{n-\frac{1}{p}}}{2^{1/p'}} B\left(\left(\frac{n}{2}-1\right)p' + \frac{1}{2}, p'+1\right)^{1/p'}, \quad n \ge 4,\ p > 1;$$

(4.7)

$$R(n,1) \le (b-a)^{n-1}\frac{2}{n}\left(\frac{n-2}{n}\right)^{\frac{n-2}{2}}, \quad n \ge 3.$$

5. Mahajani's inequalities. Using some geometrical arguments, G. S. Mahajani [**11**] (see also [**4**, pp. 297–298]) proved the following results:

1° If f has a bounded derivative on $[a,b]$, i.e. if $|f'(x)| \le M$ $(M > 0)$ and if $\int_a^b f(x)dx = 0$, then for $x \in [a,b]$

$$\left|\int_a^x f(t)dt\right| \le \frac{M(b-a)^2}{8}; \tag{5.1}$$

2° If, besides the conditions given in 1°, $f(a) = f(b) = 0$, then

$$\left|\int_a^x f(t)dt\right| \le \frac{M(b-a)^2}{16}. \tag{5.2}$$

Analytic proofs of these results were given by P. R. Beesack [**12**].

We shall give generalizations of these results.

Let us define the two-parameter class of polynomials $P_n^{(m,k)}$ $(0 \le m \le k < n;\ m,k,n \in N)$ by means of

$$\begin{aligned} P_n^{(m,k)}(x) &\equiv P_n^{(m,k)}(x;a,b) \\ &= \frac{(-1)^{n-k}(n-m)!}{m!(k-m)!(n-k-1)!} \sum_{i=0}^{k-m} \frac{(b-a)^{m-n+i}}{n-m-i} \\ &\quad \times \binom{k-m}{i} (x-a)^m (x-b)^{n-m-i} \end{aligned}$$

where a and b are real parameters.

If the values of derivatives of function F at $x = a$ and $x = b$ are known, using polynomials $P_n^{(m,k)}$, Hermite's interpolation polynomial $S_{n,k}(x)$ can be represented in the following form ([**13**]):

$$S_{n,k}(x) \equiv \sum_{m=0}^{k-1} P_{n-1}^{(m,k-1)}(x; a, b) F^{(m)}(a) + \sum_{m=0}^{n-k-1} P_{n-1}^{(m,n-k-1)}(x; b, a) F^{(m)}(b),$$

and if $|F^{(n)}(x)| \leq M$ for $x \in (a, b)$, then

$$|F(x) - S_{n,k}(x)| \leq \frac{M}{n!} |(x-a)^k (x-b)^{n-k}|. \tag{5.3}$$

REMARK: For $k = 0$ $(k = n)$ the first (second sum does not exist, i.e. we have Taylor's formula.

Now, we shall give the following generalization of 1° ([**14**]):

THEOREM 1. *Let $f(x)$ be a n-times differentiable function such that $|f^{(n)}(x)| \leq M$ for $x \in (a, b)$ and $\int_a^b f(t)dt = 0$, then*

$$\left| \int_a^x f(t)dt - \bar{S}_{n,k}(x) \right| \leq \frac{M}{(n+1)!}(x-a)^k (b-x)^{n+1-k} \leq \frac{k^k (n-k+1)^{n-k+1}}{(n+1)^{n+1}(n+1)!} M (b-a)^{n+1} \tag{5.4}$$

where

$$\bar{S}_{n,k}(x) = \sum_{m=1}^{k-1} P_n^{(m,k-1)}(x; a, b) f^{(m-1)}(a) + \sum_{m=1}^{n-k} P_n^{(m,n-k)}(x; b, a) f^{(m-1)}(b)$$

(for $k = 1$, $(k = n)$ the first (second) sum does not exist.

PROOF: Using the substitutions $n \to n+1$, $F(x) = \int_a^x f(t)dt$, we get the first inequality in (5.4) from (5.3). For the second inequality we should only observe that the function $(x-a)^k (b-x)^{n-k+1}$ has maximum for $x = (kb + (n+1-k)a)/(n+1)$.

CORLLARY 1. *If, besides the conditions given in* Theorem 1, $f^{(i)}(a) = 0$ $(i = 0, 1, \ldots, k-2)$, $f^{(i)}(b) = 0$ $(i = 0, 1, \ldots, n-k-1)$ *(for* $k = 1$ $(k = n)$ *the first (second) condition does not exist), then*

$$\left|\int_a^x f(t)dt\right| \le \frac{M}{(n+1)!}(x-a)^k(b-x)^{n-k+1} \tag{5.5}$$
$$\le \frac{k^k(n-k+1)^{n-k+1}}{(n+1)^{n+1}(n+1)!} M(b-a)^{n+1}.$$

In a special case, if the conditions from 1° are fulfilled, then

$$\left|\int_a^x f(t)dt\right| \le \frac{M}{2}(x-a)(b-x) \le \frac{M(b-a)^2}{8}. \tag{5.6}$$

which is a refinement of (5.1).

Now, we shall prove the following generalization of 2°:

THEOREM 2. *Let* $f(x)$ *be* n*-times differentiable function such that* $|f^{(n)}(x)| \le M$ *for* $x \in (a, b)$, $\int_a^b f(t)dt = 0$ *and* $f^{(i)}(a) = f^{(i)}(b) = 0$ $(i = 0, 1, \ldots, n-1)$. *Then*

$$\left|\int_a^x f(t)dt\right| \le \frac{M(b-a)^{n+1}}{2^{n+1}n(n+1)!}. \tag{5.7}$$

PROOF: In the proof we use the following result from **[13]** which is also a consequence of (5.3):

Let $f(x)$ be n-times differentiable function such that $|f^{(n)}(x)| \le M$ for $x \in (a, b)$, $f^{(i)}(a) = 0$ $(i = 0, 1, \ldots, k-1)$ and $f^{(i)}(b) = 0$ $(i = 0, 1, \ldots, n-k-1)$. Then

$$\int_a^b |f(x)|dx \le \frac{k!(n-k)!}{n!(n+1)!} M(b-a)^{n+1}. \tag{5.8}$$

For a proof of Theorem 2 we may assume that $f(c) = 0$ for some $c \in (a, b)$. Moreover, by symmetry we may assume that $a < c \le (a+b)/2$. We may also assume that c is the largest zero of f on $(a, (a-b)/2]$. For $a \le x \le c$, (5.8) for $b = c$, $k = n-1$, implies that

$$\left|\int_a^x f(t)dt\right| \le \int_a^c |f(t)|dt \le \frac{M(b-a)^{n+1}}{2^{n+1}n(n+1)!} \quad (\equiv T).$$

If $f(x) \neq 0$ for $c < x < b$, then $|G(x)| = |\int_a^x f(t)dt|$ would be decreasing on $[c, b]$, so that $|G(x)| \leq |G(c)| \leq T$ would follow. We may thus assume that $f(c_1) = 0$ for some $c_1 \in (c, b)$, hence for some $c_1 \in [(a+b)/2, b)$. Now we may assume that c_1 is the least zero of f on this interval, and with no loss of generality suppose $f(x) > 0$ for $c < x < c_1$. Then

$$|G(x)| = \left| \int_x^b f(t)dt \right| \leq \int_{c_1}^b |f(t)|dt \leq T$$

if $c_1 \leq c \leq b$ by (5.8) in the case $a = c_1$, $k = 1$. So it only remains to consider the case $c < x < c_1$. On this interval $G'(x) = f(x) > 0$ so $G(x)$ increases on (c, c_1). It follows that

$$\max_{c \leq x \leq c_1} |G(x)| = \max\{|G(c)|, |G(c_1)|\} \leq T.$$

completing the proof of (5.7).

6. A. M. Fink [7] has given some variants of Mahajani's inequalities. He proved the following results.

THEOREM 1. *Let $f^{(n-1)}$ be absolutely continuous and $f^{(n)} \in L^p(a, b)$ with $f^{(j)}(a) = f^{(j)}(b) = 0$, $j = 0 \ldots, n-1$. Then*

$$\left| \frac{1}{n} f(x) - \frac{1}{b-a} \int_a^b f(y)dy \right| \leq \frac{M(n, p, x)}{n!} \|f^{(n)}\|_p \tag{6.1}$$

where

$$M(n, p, x) = \min_{q \in \pi_{n-1}} \frac{\|(x-t)^{n-1} k(t, x) - q(t)\|_{p'}}{b-a}, \tag{6.2}$$

and

$$k(t, x) = \begin{cases} t - a & a \leq t \leq x \leq b; \\ t - b & a \leq x < t \leq b. \end{cases} \tag{6.3}$$

If in addition $\int_a^b f(y)dy = 0$, then

$$|f(x)| \leq \frac{C(n, p, x)}{(n-1)!} \|f^{(n)}\|_p \tag{6.4}$$

where

$$C(n,p,x) = \min_{q\in\pi_n} \frac{\|(x-t)^{n-1}k(t,x) - q(t)\|p'}{b-a}. \tag{6.5}$$

All of these inequalities are best possible if $1 < p < \infty$. If $p = \infty$ (6.1) is always best possible.

COROLLARY 1. *If $\int_a^b f = 0$ and $f^{(j)}(a) = f^{(j)}(b) = 0$, $j = 0,\dots,n-1$ then for $1 \le p \le \infty$,*

$$|f(x)| \le \frac{\|f^{(n)}\|p}{(n-1)!\,[(n-1)p'+1]^{1/p'}} \min\{(x-a)^{n-1/p}, (b-x)^{n-1/p}\}.$$

COROLLARY 2. *If $1 \le p \le \infty$, $\int_a^b f = 0$ and $f^{(j)}(a) = f^{(j)}(b) = 0, j = 0,\dots,n-1$, then*

$$\left|\int_a^x f(y)dy\right| \le \frac{\|f^{(n)}\|p \min\{(x-a)^{n+1/p'}, (b-x)^{n+1/p'}\}}{(n-1)!\,[(n-1)p'+1]^{1/p'}\,(n+1/p')}.$$

Explicit estimates for $M(n,p,x)$ (see (6.1)) are given in Theorems 2 and 2′.

THEOREM 2. *For $1 \le p \le \infty$ and $f(a) = f(b) = 0$, we have the best possible inequality*

$$\left|f(x) - \frac{1}{b-a}\int_a^b f(y)dy\right| \le \frac{(b-a)^{1/p'}}{2(1+p')^{1/p'}}\|f'\|_p$$

that is, $M(1,p,x) = \frac{(b-a)^{1/p'}}{2(1+p')^{1/p'}}$, and $M(1,1,x) = \frac{1}{2}$. For $x \ge \frac{a+b}{2}$ an extremal is given by $(1 < p)$

$$p'f(t) = \begin{cases} \left(x - \frac{a+b}{2}\right)^{p'} - \left(x - t - \frac{b-a}{2}\right)^{p'}, & a \le t \le x - \frac{b-a}{2};\\ \left(x - \frac{a+b}{2}\right)^{p'} - \left(t - x + \frac{b-a}{2}\right)^{p'}, & x - \frac{b-a}{2} \le t \le x;\\ \left(x - \frac{a+b}{2}\right)^{p'} - \left(x - t + \frac{b-a}{2}\right)^{p'}, & x \le t \le b. \end{cases}$$

A similar formula holds when $x < \frac{a+b}{2}$. Any other extremal is a multiple of this one.

THEOREM 2′. *For $n \ge 4$ and $f^{(j)}(a) = f^{(j)}(b) = 0$, $j = 0,\dots,n-1$*

$$\left|\frac{1}{n}f(x) - \frac{1}{b-a}\int_a^b f(y)dy\right|$$
$$\le (b-a)^3 \frac{2}{n-2}\left(\frac{n-3}{n-2}\right)^{n-3} \max\{(x-a)^{n-3}, (b-x)^{n-3}\}\|f^{(n)}\|_1.$$

Moreover for $n = 3$

$$\left|\frac{1}{3}f(x) - \frac{1}{b-a}\int_a^b f(y)dy\right| \leq 2(b-a)\max(x-a, b-x)\|f'''\|_1$$

and for $n = 2$ *we have the best possible bound*

$$\left|\frac{1}{2}f(x) - \frac{1}{b-a}\int_a^b f(y)dy\right| \leq \frac{(b-a)}{8}\max\{(x-a)^2, (b-x)^2\}\|f''\|_1.$$

If $n \geq 3$, *then for* $p > 1$

$$M(n,p,x) \leq 2(b-a)^{-\left(1+\frac{1}{p'}\right)} B\left(1+p'(n-3), 1+p'\right)^{1/p'} \times \left[(x-a)^{p'(n-2)+1} + (b-x)^{p'(n-2+1}\right]^{1/p'}$$

For $C(n,p,x)$ (see (6.4)) the following are given in [7]:

THEOREM 3. *If* $f(a) = f(b)$ *and* $\int_a^b f = 0$ *then we have the best possible inequalities*

$$|f(x)| \leq \frac{1}{2}\|f'\|_1, \text{ and}$$
$$|f(x)| \leq (b-a)g(x)\|f'\|_\infty, \text{ where}$$

$$g(x) = \begin{cases} \frac{x-a}{b-a}, & a \leq x \leq \frac{3}{4}a + \frac{1}{4}b; \\ \left(\frac{1}{2} + \frac{\left(x - \frac{a+b}{2}\right)^2}{(b-a)^2}\right)^{1/2} - \frac{1}{2}, & \frac{3}{4}a + \frac{1}{4}b \leq x \leq \frac{1}{4}a + \frac{3}{4}b, \\ \frac{b-x}{b-a}, & \frac{1}{4}a + \frac{3}{4}b \leq x \leq b. \end{cases}$$

For $1 < p < \infty$ *we have*

$$|f(x)| \leq \frac{(x-a)^{1/p'}(b-x)^{1/p'}}{[(x-a)^{p-1} + (b-x)^{p-1}]^{1/p}}\|f'\|_p.$$

That is $C(1,1,x) = \frac{1}{2}$, $C(1,\infty,x) = (b-a)g(x)$ *and*
$C(1,p,x) \leq \frac{(x-a)^{1/p'}(b-x)^{1/p'}}{[(x-a)^{p-1}+(b-x)^{p-1}]^{1/p}}$.

THEOREM 3′. *If* $f^{(j)}(a) = f^{(j)}(b) = 0$, $j = 0, \ldots, n-1$ *and* $\int_a^b f = 0$, *then*

$$|f(x)| \leq (b-a)^n \frac{\sigma^{n-1}(1-\sigma)^{n-1}}{\left[1 - \sigma^{\frac{n-1}{n-2}}\right]^{n-2}}\|f^{(n)}\|_1 \quad n > 2,$$

where

$$\sigma = \min\left(\frac{x-a}{b-a}, \frac{b-x}{b-a}\right)$$

and

$$|f(x)| \le (x-a)(b-x)\|f''\|_1.$$

Two other results are also given that are related to Mahajani type inequalities.

THEOREM 4. *Suppose* $\int_a^b f = 0$. *Then the best possible inequalities are*

$$|f(x)| \le \|f'\|_p \frac{\left[(x-a)^{p'+1} + (b-x)^{p'+1}\right]^{1/p'}}{(b-z)(1+p')^{1/p'}}, \quad 1 < p \le \infty,$$

and

$$|f(x) \le \|f'\|_1 \min\left\{\frac{x-a}{b-a}, \frac{b-x}{b-a}\right\}.$$

Moreover, we also have best possible inequalities

$$\left|\int_a^x f(t)dt\right| \le \frac{(b-x)(x-a)}{(b-a)^{1/p}} \frac{\|f'\|_p}{(1+p')^{1/p'}}, \qquad 1 < p \le \infty,$$

and

$$\left|\int_a^x f(t)dt\right| \le \frac{b-x)(x-a)}{b-a}\|f'\|_1.$$

THEOREM 5. *Let* $\int_a^b f = 0$, $f(a) = f(b) = 0$, *and* $p = 1$, *or* ∞. *Then*

$$\left|\int_a^x f(t)dt\right| \le h_p(x)\|f'\|_p$$

are best possible inequalities where

$$h_\infty(x) = \begin{cases} \frac{(x-a)^2}{2}, & a \le x \le \frac{3}{4}a + \frac{1}{4}b \\ \frac{1}{16} - \frac{1}{2}\frac{\left(x - \frac{a+b}{2}\right)^2}{(b-a)^2}, & \frac{3}{4}a + \frac{1}{4}b \le x \le \frac{1}{4}a + \frac{3}{4}b, \\ \frac{(b-x)^2}{2}, & \frac{1}{4}a + \frac{3}{4}b \le x \le b \end{cases}$$

and $h_1(x) = \frac{1}{2}x(1-x)$.

For $1 < p < \infty$

$$\left|\int_a^x f(t)dt\right| \le \frac{1}{(1+p')^{1/p'}} \min\{(x-a)^{1+1/p'}, (b-x)^{1+1/p'}\}\|f'\|_p.$$

We note that in Theorem 4 we get

$$\left|\int_a^x f(t)dt\right| \le \frac{(b-x)(x-a)}{2}\|f'\|_\infty$$

which is superior to (5.1) except when $x = \frac{a+b}{2}$.

7. Note that inequality (5.8) is a generalization of results from [**15**],[**16**]. The following more general result obtained in [**13**] is a simple consequence of (5.3):

THEOREM 1. *Let $f(x)$ be a n-times differentiable function such that $|f^{(n)}(x)| \le M$ for $x \in (a,b)$. If $p(x)$ is an integrable function on (a,b) such that $0 < c \le p(x) \le \lambda c$ $(\lambda \ge 1,\ x \in [a,b])$, the following inequality*

(7.1)
$$M^{[r]}(|g|;p) \le \frac{MC(b-a)^n}{n!}\left(\frac{\lambda B(rk+1, r(n-k)+1)}{C^r + (\lambda-1)B(rk+1, r(n-k)+1)}\right)^{1/r}$$
$$(r > 0)$$

holds, where B is beta function, $C = k^k(n-k)^{n-k}/n^n$, $g(x) = f(x) - S_{n,k}(x)$, $S_{n,k}$ is Hermite's interpolation polynomial of function f and $M^{[r]}(f;p)$ is the integral mean of order r of function f with weighted function p.

With the same conditions as for (5.8) we get the inequality

(7.2)
$$\left(\frac{1}{b-a}\int_a^b |f(x)|^r dx\right)^{1/r} \le \frac{M(b-a)^n}{n!} B(rk+1, r(n-k)+1)^{1/r} \quad (r > 0).$$

From Theorem 1 can be obtained the following inequality

$$\left|\int_a^b f(x)dx - \sum_{k=1}^n \frac{(2m-k)!}{(2m)!}\binom{m}{k}(b-a)^k \left(f^{(k-1)}(a) - (-1)^k f^{(k-1)}(b)\right)\right| \tag{7.3}$$
$$\le \frac{M(m!)^2(b-a)^{2m+1}}{(2m)!(2m+1)!}.$$

where $|f^{(2m)}(x)| \le M$ for $x \in (a,b)$.

The following generalization of (4.3) is also given in [**13**]:

Theorem 2. *Let $I_n = \{0,1,\dots,n\}$ and let $\{P_k\}_{k\in I_n}$ be a harmonic sequence of polynomials on $[0,1]$ ($P_n'(x) = P_{n-1}(x)$). If $f(x)$ is n- times differentiable function such that $|f^{(n)}(x)| \le M$ for $x \in (a,b)$, then*

$$\left| P_0 \int_a^b f(x)dx - \sum_{k=1}^{n} (-1)^k (b-a)^k \left(P_k(0) f^{(k-1)}(a) - P_k(1) f^{(k-1)}(b) \right) \right| \le M(b-a)^{n+1} \int_0^1 |P_n(t)|dt. \tag{7.4}$$

Proof: If $h(t) = f(a + t(b-a))$ we have $\int_a^b f(x)dx = (b-a)\int_0^1 h(t)dt$, from which, applying integration by parts on the last integral, we get

$$\int_0^1 h(t)dt = h(1) - h(0) - \int_0^1 th'(t)dt. \tag{7.5}$$

Since $P_1(t) = P_0 t + P_1(0)$ ($P_0(t) = P_0$), equality (7.5) may be represented in the form

$$P_0 \int_0^1 h(t)dt = P_1(1)h(1) - P_1(0)h(0) - \int_0^1 P_2'(t)h'(t)dt.$$

By successive integration by parts of $\int_0^1 P_2'(t)h'(t)dt$ $(n-1)$-times, we obtain

$$P_0 \int_0^1 h(t)dt = \sum_{k=1}^{n} (-1)^k \left(P_k(0)h^{(k-1)}(0) - P_k(1)h^{(k-1)}(1) \right) + (-1)^n \int_0^1 P_n(t)h^{(n)}(t)dt,$$

from which (7.4) follows.

Corollary 1. *Let function $f(x)$ satisfy the conditions as in* Theorem 2 *and let $f^{(k)}(b) = (-1)^{k-1} f^{(k)}(a)$ $(k = 0,1,\dots,n-1)$. Then*

$$\left| \int_a^b f(x)dx \right| \le \frac{M(b-a)^{n+1}}{2^n (n+1)!}. \tag{7.6}$$

To prove this, take $P_n(t) = \frac{1}{n!}\left(t - \frac{1}{2}\right)^n$, in Theorem 2.

REMARK: Inequality (7.6) is obtained in **[10]** with somewhat stricter conditions for f.

8. Several various results can be obtained by applications of well-known quadrature formulas. Here we shall give only the following results from **[1]**:

If $F(-1) + F(1) = 0$, $|F^{(4)}(x)| \leq 1$, then

$$\int_{-\sqrt{3}}^{\sqrt{3}} F(x)dx \leq \frac{1}{15}\sqrt{3}. \tag{8.1}$$

This is a simple consequence of the well-known two-point Gauss-Legendre quadrature formula

$$\int_{-\sqrt{3}}^{\sqrt{3}} F(x)dx = (F(-1) + F(1)) \cdot \sqrt{3} + \frac{1}{15}\sqrt{3} \cdot F^{(4)}(\xi), \quad |\xi| < \sqrt{3}.$$

REFERENCES

1. OSTROWSKI, A., *Über die Absolutabweichung einer differentiiebaren Funktion von ihrem Integralmittelwert*, Comment. Math. Helv. **10**(1938), 226–227.
2. MILOVANOVIĆ, G.V., *On some integral inequalities*, Univ Beograd. Publ. Elektrotehn. Fak. Ser. Mat. Fiz. No. 498–541 (1975), 119–124.
3. ________, *O nekim funkcionalnim nejednakostima*, Univ. Beograd. Publ. Elektrotehn. Fak. Ser. Mat. Fiz. No. 599 (1977), 1–59.
4. MILOVANOVIĆ, G.V. and J.E. PEČARIĆ, *On generalization of the inequality of A. Ostrowski and some related applications*, Univ. Beograd. Publ. Elektrotehn. Fak. Ser. Mat. Fiz. No. 544–576 (1976), 155–158.
5. PEČARIĆ, J. and B. SAVIĆ, *O novom postupku razvijanja funkcije u red i nekim primjenama*, Zbornik radova VA KoV (Beograd) **9**(1983), 171–202.
6. ________, *O generalizacijama nekih rezultata J. Karamate i nekim primjenama*, Zbornik radova VA KoV (Beograd) **10**(1984), 245–268.
7. FINK, A.M., *Bounds on the deviation of a function from its averages*, submitted manuscript.
8. IYENGAR, K.S.K., *Note on an inequality*, Math. Student **6**(1938), 75–76.

9. VASIĆ, P.M. and G.V. MILOVANOVIĆ, *On an inequality of Iyengar*, Univ. Beograd. Publ. Elektrotehn. Fak. Ser. Mat. Fiz. No. 544–576 (1976), 18–24.
10. MILOVANOVIĆ, G.V. and J.E. PEČARIĆ, *Some considerations on Iyengar's inequality and some related applications*, Univ. Beograd. Publ. Elektrotehn. Fak. Ser. Mat. Fiz. No. 544–576 (1976), 166–170.
11. MAHAJANI, G.S., *A note on an inequality*, Math. Student **6**(1938), 125–128.
12. BEESACK, P.R., Private communications.
13. MILOVANOVIĆ, G.V. and J.E. PEČARIĆ, *On an application of Hermite's interpolation polynomial and some related results*, Symposium Num. Meth. Appr. Th. Niš. (1984), 93–98.
14. MITRINOVIĆ, D.S. and J.E. PEČARIĆ, *On some applications of Hermite's interpolation polynomial*, C. R. Math. Rep. Acad. Sci. Canada **9**(1987), 55–58.
15. ANON., *Problem E2155*, Amer. Math. Monthly **76**(1969), 188, 1142–1143.
16. ZAIDMAN, S., *Problem E2622*, Amer. Math. Monthly **83**(1976), 740–741.

CHAPTER XVI

INEQUALITIES OF BERNSTEIN-MORDELL TYPE

1. If a real polynomial $f_n(x) = a_0 + a_1x + \cdots + a_nx^n$ reaches the value of 1 anywhere on the segment $[-1, 1]$, then

$$\int_{-1}^{+1} (a_0 + a_1x + \cdots + a_nx^n)^2 dx \geq \frac{2}{(n+1)^2}. \tag{1.1}$$

Except for an obvious error, this relation has been obtained by S. N. Bernstein [**1**].

A related result is obtained by O. Bottema in [**2**]:

Let $\int_a^b (a_0 + a_1x + \cdots + a_nx^n)^2 dx = I$, where a, b, n and I are fixed numbers. The maximum value of a_n appears to be

$$m_n = \frac{(2n)!}{(n!)^2} \frac{((2n+1)I)^{1/2}}{(b-a)^{n+1/2}},$$

and

$$\lim_{n\to\infty} m_n = \begin{cases} 0 & \text{for } |b-a| > 4, \\ \sqrt{\frac{I}{2\pi}} & \text{for } |b-a| = 4, \\ \infty & \text{for } |b-a| < 4. \end{cases}$$

F. Bowman (see [**3**]) proved that if $f_n(0) = 1$, then

$$\int_0^1 (a_0 + a_1x + \cdots + a_nx^n)^2 dx \geq \frac{1}{(n+1)^2}. \tag{1.2}$$

The following inequality was posed as a problem in the book [**4**] by F. Bowman and F. A. Gerard. A simple proof was given by L. J. Mordell in [**3**].

If $a_1, \ldots, a_n$ are real numbers, then

$$\int_0^\infty e^{-x}(1 + a_1x + \cdots + a_nx^n)^2 dx \geq \frac{1}{n+1}. \tag{1.3}$$

F. Smithies [5] has proved the same result using the orthogonality properties of the Laguerre polynomials. At the same time he has derived the explicit expression for minimizing polynomial. In [6] L. J. Mordell has solved, in some cases, the problem of finding the minimum value of integrals of the form

$$\int_a^b p(x)(a_0 + a_1 x + \cdots + a_n x^n)^2 dx \equiv \int_a^b p(x) f_n(x)^2 dx \tag{1.4}$$

where $p(x) \geq 0$ is such that the integrals $\int_a^b p(x)x^r dx$ $(r \geq 0)$ exist and the coefficient a_k of the term x^k in the bracket is given as 1.

L. Mirsky [7] proved

$$\begin{aligned} &\min \int_a^b w(t)\left(t^{k_0} + \lambda_1 t^{k_1} + \cdots + \lambda_n t^{k_n}\right)^2 dt \\ &= \det_{0\leq r,s\leq n} M(k_r + k_s) / \det_{1\leq r,s\leq n} M(k_r + k_s) \end{aligned} \tag{1.5}$$

where $M(p) = \int_a^b w(t)t^p dt$, and where the minimum on the left-hand side is taken with respect to all real numbers $\lambda_1, \ldots, \lambda_n$.

2. In [8]–[10] the minimum values of (1.4) were found by using a different kind of normalization of $f_n(x)$ which is especially appropriate in solving the approximation problem in synthesis of filtering networks in communication engineering. We shall give results from [9] due to Rakovich and Vasić.

Let $w(x)$ be a nonnegative function on the interval $[a, b]$ such that all the moments $\int_a^b w(x)x^r dx$ are finite for $r \geq 0$.

Consider the integral of the form

$$I_n = \int_a^b w(x)\left(f_n(x)\right)^2 dx. \tag{2.1}$$

The problem to be solved is to determine the polynomial $f_n(x)$ of order n which minimizes the integral (2.1) under the constraint that $f_n(p) = 1$ for any given real number p. Since the integrand is nonnegative for any value of $x \in [a, b]$ such a minimum value does exist.

If $Q_0, Q_1, Q_2, \ldots$ is a set of orthogonal polynomials associated with the weight function $w(x)$ on $[a,b]$, the polynomial f_n can be expanded into a finite series of the $Q_i(x)$ so that

$$I_n = \int_a^b w(x) \left(\sum_{i=1}^n a_i Q_i(x) \right)^2 dx. \tag{2.2}$$

Using standard minimization techniques [8] one considers

$$\varphi(a_0, a_1, \ldots, a_n, \beta) = \int_a^b w(x) \left(\sum_{i=0}^n a_i Q_i(x) \right)^2 dx + \beta \left(\sum_{i=0}^n a_i Q_i(p) - 1 \right) \tag{2.3}$$

where β is the Lagrangian multiplier. The necessary conditions are

$$\begin{aligned} \frac{\partial \varphi}{\partial a_i} &\equiv 2 \int_a^b w(x) a_i Q_i(x)^2 dx + \beta Q_i(p) = 0, \\ & \sum_{i=0}^n a_i Q_i(p) = 1. \end{aligned} \tag{2.4}$$

Let

$$b_i = \int_a^b w(x) Q_i(x)^2 dx. \tag{2.5}$$

Then

$$a_i = \frac{Q_i(p)}{b_i} \frac{1}{\sum_{j=0}^n \frac{Q_j(p)^2}{b_j}} \tag{2.6}$$

so that the minimum value M of the integral (2.1) under the constraint is

$$M = \frac{1}{\sum_{i=0}^n \frac{Q_i(p)^2}{b_i}}. \tag{2.7}$$

3.

THEOREM 1. *If c is finite and $c \leq a$ or $c \geq b$, the integral*

$$I_n = \int_a^b w(x) f_n(x)^2 dx, \tag{3.1}$$

where f_n is any real polynomial of degree n such that $f_n(c) = 1$, reaches its minimum value if and only if f_0, f_1, f_2, ... form a set of orthogonal polynomials on $[a, b]$ with respect to the weight function $(x - c)w(x)$ for $c \leq a$, and with respect to the weight function $(c - x)w(x)$ for $c \geq b$.

PROOF: Suppose $c(\leq a)$ is finite. From (2.2) and (2.6) the polynomial f_n that minimizes the definite integral (2.1), subject to the condition $f_n(c) = 1$, is

$$f_n(x) = \left(\sum_{j=0}^{n} \frac{Q_j(c)^2}{b_j} \right)^{-1} \sum_{i=0}^{n} \frac{Q_i(c)^2}{b_i} Q_i(x). \tag{3.2}$$

But, from the Christoffel-Darboux theorem [**24**, p. 42], it follows that

$$\sum_{i=0}^{n} \frac{Q_i(c) Q_i(x)}{b_i} = \frac{k_n}{k_{n+1} b_n} \frac{Q_n(c) Q_{n+1}(x) - Q_{n+1}(c) Q_n(x)}{x - c}, \tag{3.3}$$

where k_n is the coefficient of x^n in $Q_n(x)$, so that the minimizing polynomial takes the form

$$f_n(x) = K \frac{Q_n(c) Q_{n+1}(x) - Q_{n+1}(c) Q_n(x)}{x - c} \quad (K = \text{const.}). \tag{3.4}$$

The Christoffel formula [**27**, p. 28] states that if Q_0, Q_1, Q_2, ... form a set of orthogonal polynomials associated with the weight function w on $[a, b]$, then the polynomials R_0, R_1, R_2, ..., where

$$R_n(x) = K \frac{Q_n(c) Q_{n+1}(x) - Q_{n+1}(c) Q_n(x)}{x - c} \quad (n = 0, 1, 2 \ldots) \tag{3.5}$$

are orthogonal on the same segment $[a, b]$ with respect to the weight function $(x - c)w(x)$. This evidently completes the proof of Theorem 1 for $c \leq a$. A similar results holds if $c(\geq b)$ is finite.

An immediate consequence of Theorem 1 is the following, more general result:

THEOREM 2. *Let g be an increasing function on $[a,b]$ and w a nonnegative weight function on the same interval such that the integral*

$$\int_a^b w(x)g(x)^r\,dx \quad (r \geq 0)$$

exists. Then the sequence of functions $f_0(g(x))$, $f_2(g(x))$, $f_2(g(x))$, ... that minimizes the integrals

$$I_n = \int_a^b w(x)f_n(g(x))^2\,dx \quad (n = 0,1,2,\dots), \tag{3.6}$$

where f_n is a real polynomial of degree n, forms an orthogonal system on $[a,c]$ with the weight function $(g(x)-c)\,w(x)$ for $c \leq g(a)$, or with $(c-g(x))\,w(x)$ for $c \geq g(b)$.

PROOF: Substituting $g(x) = t$ in (3.6) we have

$$I_n = \int_{g(a)}^{g(b)} p(t)f_n(t)^2 dt \quad (n = 0,1,2,\dots), \tag{3.7}$$

where

$$p(t) = \frac{w\left(g^{-1}(t)\right)}{g'\left(g^{-1}(t)\right)} \tag{3.8}$$

According to Theorem 1, the functions that minimize (3.6) form an orthogonal sysem on $[g(a),g(b)]$ with respect to $(t-c)p(t)$ for $c \leq g(a)$, that is

$$\int_{g(a)}^{g(b)} (t-c)p(t)f_j(t)f_k(t)dt = 0 \quad (j,k = 0,1,2,\dots;\ j \neq k). \tag{3.9}$$

This is the same as

$$\int_a^b (g(x)-c)\,w(x)f_j(g(x))\,f_k(g(x))\,dx = 0 \ (j,k = 0,1,2,\dots;\ j \neq k). \tag{3.10}$$

4. If the polynomials Q_i $(i = 0, 1, 2, \dots)$ are othogonal on $[a, b]$ with respect to w, then from Theorem 1, we have for $c \geq b$,

$$\min I_n = \int_a^b w(x) \left(\sum_{i=0}^{a} \frac{Q_i(c)M}{b_i} Q_i(x) \right)^2 dx \tag{4.1}$$

$$= \int_a^b w(x) \left(\frac{R_n(x)}{R_n(c)} \right)^2 dx,$$

where $R_0, R_1, R_2, \dots$ are the polynomials orthogonal with respect to $(c-x)w(x)$ on $[a, b]$. Since for any given nonnegative weight function w and any prescribed segment $[a, b]$ there is only one set of orthogonal polynomials which are determined to within a constant multiplier, it follows that

$$\sum_{i=1}^{n} \frac{Q_i(c)M}{b_i} Q_i(x) = \frac{R_n(x)}{R_n(c)}, \tag{4.2}$$

and by comparing the coefficients of x^n, we find

$$M = \frac{b_n k_n^R}{Q_n(c) R_n(c) k_n^Q} \tag{4.3}$$

where k_n^Q and k_n^R are the coefficients of x^n in Q_n and R_n respectively. Hence from (4.2) and (4.3)

$$R_n(x) = \sum_{i=1}^{n} \frac{Q_i(c)}{Q_n(c)} \frac{b_n}{b_i} \frac{k_n^R}{k_n^Q} Q_i(x). \tag{4.4}$$

5. Now, we shall give some special cases:

1^0 ([8]) If $\sum\limits_{i=1}^{n} a_i = 1$, $\nu > -1$, then

$$\int_{-1}^{+1} (1-x^2)^\nu \left(\sum_{i=0}^{n} a_i x^i \right)^2 dx$$

$$\geq \frac{\pi n! \Gamma(2\nu + 3)}{(2\nu+2)_n \Gamma\left(\nu + \frac{3}{2}\right)^2 (\nu + n + 1)\, 2^{2+2\nu}}.$$

For $\nu = -1/2$ we get

$$\int_{-1}^{+1} (1-x^2)^{-1/2} \left(\sum_{i=0}^{n} a_i x^i \right)^2 dx \geq \frac{\pi}{2n+1}.$$

2^0 ([9] If $f_n(1) = 1,\ u, v > -1$, then

$$\min \int_{-1}^{+1} (1-x)^u (1+x)^v \left(\sum_{i=0}^{n} a_i x^i \right)^2 dx$$
$$= \frac{2^{u+v+1} n! \Gamma(u+1)\Gamma(u+2)\Gamma(n+v+1)}{\Gamma(n+u+2)\Gamma(n+u+v+2)}.$$

The last result is obtained for $f_n(1) = 1$ but it still holds if the polynomial f_n reaches the value of 1 at any point in the interval $[-1, +1]$.

3^0 In [9] was proved that (1.2) can be obtained from (1.1), i.e. the following results are obtained from (1.1):

$$\int_{c}^{1} (a_0 + a_1 x + \cdots + a_n x^n)^2 \, dx \geq \frac{1-c}{(n+1)^2},$$

and

$$\int_{-1}^{c} (a_0 + a_1 x + \cdots + a_n x^n)^2 \, dx \geq \frac{1+c}{(n+1)^2}.$$

where $-1 \leq c \leq 1$ and $f_n(c) = 1$.

4^0 In the design of electric filters the polynomial f_n is constrained to be an even function of frequency and if, in addition a monotonic magnitude response is required, the error integral has the form

$$I = \int_{0}^{1} (1-x^2)^{p-q} x^{2q-1} \left(\sum_{i=0}^{n} a_i x^{2i} \right)^2 dx \quad (p-q > -1,\ c > 0).$$

The following results are valid ([**9**]):

If $f_n(1) = 1$, then: $\min I = \frac{n!\Gamma(n+q)\Gamma(p-q+1)\Gamma(p-q+2)}{2\Gamma(n+p+1)\Gamma(n+p-q+2)}$, and if $f(0) = 1$, then: $\min I = \frac{n!\Gamma(n+p-q+1)\Gamma(q)\Gamma(q+1)}{2\Gamma(n+p+1)\Gamma(n+q+1)}$

5⁰ Theorems 1 and 2 still hold if one of the integration limits is infinite. For example with $w(x) = x^\nu e^{-x}$, $a = 0$, $b = +\infty$ we have from (4.3)

$$\min \int_0^\infty x^\nu e^{-x} (1 + a_1 x + \cdots + a_n x^n)^2 \, dx = \Gamma(\nu + 1) / \binom{n + \nu + 1}{n}.$$

6⁰ Let n and k be integers with $0 \le k \le n$. W. Janous [**11**] determined the minimum value $M_{n,k}$ of

$$\int_0^1 (P_{n,k}(x))^2 \, dx$$

where $P_{n,k}$ runs through the set of all polynomials with real coefficients and degree at most n such that the coefficient of x^k is 1. Janous proved that

$$M_{n,k} = \left((2k+1) \binom{n+k+1}{n-k}^2 \binom{2k}{k}^2 \right)^{-1}.$$

This is a generalization of (1.2). This appears as a problem in the American Mathematical Monthly [**11**]. See the answer [**12**] for one solution.

As an application of this result he proved the following result:

Let m, n be nonnegative integers. Furthermore let $f(x)$ be n-times continuously differentiable and $f^{(k)}(0) = f^{(k)}(1)$, $k = 0, \ldots, n-1$ and $\int_0^1 x^j f(x) dx = 0$, $j = 1, \ldots, m$. Then ([**12**])

$$\left(\int_0^1 f(x) dx \right)^2 \le (2n+1) \left(\frac{n!m!}{(2n+m+1)!} \right)^2 \int_0^1 \left(f^{(n)}(x) \right)^2 dx.$$

6. Finally, we shall give some related results which were obtained as applications of quadrature formulae. The following text was communicated to us by D. Acu, [**13**]–[**16**]:

If $[a,b]$ is a finite interval we shall call $L[a,b]$ the class of functions $f(x)$ Lebesgue-integrable (summable) in $[a,b]$ and $AC^k[a,b]$ the class of functions $f(x)$ whose k-th derivative $f^{(k)}(x)$ is absolutely continuous in $[a,b]$ $(k = 0,1,2,\dots)$.

We shall call quadrature formula relative to the function $f \in L[a,b]$ and to the nodes $x_1, x_2, \dots, x_n$, any formula of the type

$$\int_a^b f(x)dx = \sum_{i=1}^{n} A_i f(x_i) + R_n(f). \tag{6.1}$$

The constants A_i are the coefficients and $R_n(f)$ is the remainder of the quadrature formula (6.1).

The number $\gamma \in N$ with the property that $R_n(f) = 0$ for all $f \in P_r$ and there exists a $g \in P_{r+1}$ so that $R(g) \neq 0$ is named the degree of exactness of the quadrature formula.

The problem which arises regarding to a quadrature formula is to determine the parameters A_i, $i = 1,\dots,m$, and x_i, $i = 1,\dots,k$, in some given conditions and to study the remainder term $R(f)$ fo the obtained values of A_i and x_i.

For a broader information on quadrature formulae the works **[17]**–**[22]** can be consulted.

Further on we shall present two procedures for using the quadrature formulae in obtaining inequalities.

1. If in a quadrature formula (6.1) all coefficients are nonnegative, then for $f \in L[a,b]$, removing the quadrature sums, we obtain inequalities of the type

$$\int_a^b f(x)dx \geq R_n(f) \quad \text{if } f(x) \geq 0 \text{ for an } x \in [a,b]; \tag{6.2}$$

$$\int_a^b f(x)dx \leq R_n(f) \quad \text{if } f(x) \leq 0 \text{ for any } x \in [a,b]. \tag{6.3}$$

Equality holds for functions of the form

$$f(x) = h(x)\prod_{i=1}^{n}(x - x_i)^{2k}, \ k \in N, \ h \in L[a,b] \text{ with } h(x) \geq 0$$

for $x \in [a,b]$ in the case of the inequalities (6.2) and $h(x) \leq 0$ for $x \in [a,b]$ in the case of the inequalities (6.3).

This method has been used by F. Locher [23] to obtain certain inequalities with polynomials, inequalities used in solving some extremal problems for the quadrature formulae.

Consider Gauss-Jacobi's quadrature formula [20]:

$$(6.4)\qquad \int_{-1}^{1}(1-x)^{\beta}(1+x)^{\beta}f(x)dx = \sum_{i=1}^{m} A_{mi}^{(\alpha,\beta)} f\left(x_{mi}^{(\alpha,\beta)}\right) + R_{2m-1}(f),$$

$\alpha > -1$, $\beta > -1$, in which the nodes $x_{mi}^{(\alpha,\beta)}$, $i = 1,\ldots,m$, are the zeros of Jacobi's polynomial $P_m^{(\alpha,\beta)}(x)$, and the coefficients $A_{mi}^{(\alpha,\beta)}$, $i = 1,\ldots,m$ are positive.

For a function $f \in AC^{2m-1}[-1,1]$ the remainder is written in the form

$$R_{2m-1}(f) = \frac{2^{\alpha+\beta+2m+1}m!\Gamma(1+m+1)\Gamma(\beta+m+1)\Gamma(\alpha+\beta+m+1)}{(\alpha+\beta+2m+1)\left(\Gamma(\alpha+\beta+2m+1)\right)^2}\,\frac{f^{(2m)}(\xi)}{(2m)!},$$

where $\xi \in (-1,1)$. For $\alpha = \beta = 0$ we obtain Gauss quadrature formula.

If in (6.4) we take as f and arbitrary polynomial $p_{2m}(x)$ of the degree $2m$, $p_{2m}(x) \geq 0$ on $[-1,1]$ and with leading coefficient equal to 1, then the inequality

$$\int_{-1}^{1}(1-x)^{\alpha}(1+x)^{\beta}p_{2m}(x)dx \geq R_{2m-1}(p_{2m}).$$

Thus we have obtained the inequality given by F. Locher in [23].

PROPOSITION 1. *For any polynomial $p_{2m}(x) \geq 0$, $x \in [-1,1]$, with leading coefficient equal to 1, the inequality*

$$\int_{-1}^{1}(1-x)^{\alpha}(1+x)^{\beta}p_{2m}(x)dx$$
$$\geq \frac{2^{\alpha+\beta+2m+1}m!\Gamma(\alpha+m+1)\Gamma(\beta+m+1)\Gamma(\alpha+\beta+m+1)}{(\alpha+\beta+2m+1)\left(\Gamma(\alpha+\beta+2m+1)\right)^2}$$

is true, in which the equality is attained only if

$$P_{2m}(x) = \left(\frac{2^m m!\Gamma(\alpha+\beta+m+1)}{\Gamma(\alpha+\beta+2m+1)}\right)^2\left(p_m^{(\alpha,\beta)}(x)\right)^2.$$

For $\alpha = \beta = 0$ we obtain the inequality

$$\int_{-1}^{1} p_{2m}(x)dx \geq \frac{2}{2m+1}\left(\frac{2^m(m!)^2}{(2m)!}\right)^2,$$

with the equality for

$$p_{2m}(x) = \left(\frac{2^m(m!)^2}{(2m)!}\right)^2 (p_m(x))^2 .$$

F. Locher [23] also demonstrates that:

PROPOSITION 2. *For any polynomial $p_{2m+1}(x) \geq 0$, $x \in [-1,1]$ of the degree $2m+1$ and with the leading coeffient equal to 1, the inequality*

$$\int_{-1}^{1} (1-x)^\alpha (1+x)^\beta p_{2m+1}(x)dx$$

$$\geq \frac{2^{\alpha+\beta+2m+2} m!\Gamma(\alpha+m+1)\Gamma(\beta+m+2)\Gamma(\alpha+\beta+m+2)}{(\alpha+\beta+2m+2)((\Gamma(\alpha+\beta+2m+2))^2}$$

id valid, with the equality for

$$p_{2m+1}(x) = \left(\frac{2^m m!\Gamma(\alpha+\beta+m+2)}{\Gamma(\alpha+\beta+2m+2)}\right)^2 (x+1)\left(p_m^{(\alpha\beta+1)}(x)\right)^2 .$$

PROPOSITION 3. *For any polynomial $p_{2m+1}(x) \leq 0$, $x \in [-1,1]$, of the degree $2m+1$ and with the leading coefficient equal to 1, the inequality*

$$\int_{-1}^{1} (1-x)^\alpha (1+x)^\beta p_{2m+1}(x)dx$$

$$\leq \frac{2^{\alpha+\beta+2m+2} m!\Gamma(\alpha+m+2)\Gamma(\beta+m+1)\Gamma(\alpha+\beta+m+2)}{(\alpha+\beta+2m+2)\left(\Gamma(\alpha+\beta+2m+2)\right)^2}$$

is true, with the equality for

$$p_{2m+1}(x) = \left(\frac{2^m m!\Gamma(\alpha+\beta+m+2)}{\Gamma(\alpha+\beta+2m+2)}\right)^2 (x-1)\left(p_m^{(\alpha+1,\beta)}(x)\right) .$$

PROPOSITION 4. *For any polynomial $p_{2m}(x) \leq 0$, $x \in [-1,1]$, of the degree $2m$ and with the leading coefficient equal to 1, we have the equality*

$$\int_{-1}^{1} (1-x)^\alpha (1+x)^\beta p_{2m}(x)dx$$

$$\leq -\frac{2^{\alpha+\beta+2m+1}(m-1)!\Gamma(\alpha+m+1)\Gamma(\beta+m+1)\Gamma(\alpha+\beta+m+2)}{(\alpha+\beta+2m+1)\left(\Gamma(\alpha+\beta+2m+1)\right)^2},$$

with equality only if

$$p_{2m}(x) = \left(\frac{2^{m-1}(m-1)!\Gamma(\alpha+\beta+m+2)}{\Gamma(\alpha+\beta+2m+1)}\right)^2 (x^2-1)\left(p_{m-1}^{(\alpha+1,\beta+1)}(x)\right)^2 .$$

In [15] it is demonstrated that:

PROPOSITION 5. *For each polynomial $p_{2m}(x) \geq 0$ of degree $2m$ and with the leading coefficient equal to 1, the inequality*

$$\int_0^\infty x^\alpha e^{-x} p_{2m}(x)dx \geq m!\Gamma(\alpha+m+1), \quad \alpha > -1,$$

is valid, with equality only if

$$p_{2m}(x) = (m!)^2 \left(p_m^{(\alpha)}(x)\right)^2,$$

where $p_m^{(\alpha)}(x)$ is the Laguerre polynomial.

PROPOSITON 6. *For each polynomial $p_{2m+1}(x) \geq 0$, $x \in (0,\infty)$, with the leading coefficient equal to 1, the inequality*

$$\int_0^\infty x^\alpha e^{-x} p_{2m+1}(x)dx \geq m!\Gamma(\alpha+m+1),$$

is valid, with equality only if

$$p_{2m+1}(x) = (m!)^2 x \left(p_m^{(\alpha+1)}(x)\right)^2,$$

where $p_m^{(\alpha+1)}(x)$ is the polynomial of degree m out of the system of orthogonal polynomials on the interval $(0,\infty)$ refering to the weight $x^{\alpha+1}e^{-x}$.

The proof of Proposition 5 is obtained from the Gauss-Laguerre quadrature formula [8]

$$\int_0^\infty x^\alpha e^{-x} f(x)dx = \sum_{i=1}^m A_i f(x_i) + R_{2m-1}(f)$$

in which the coefficients A_i, $i = 1,\ldots,m$, are positive and the remainder is given by

$$R_{2m-1}(f) = \frac{m!\Gamma(\alpha+m+1)}{(2m)!} f^{(2m)}(\xi), \quad \xi \in (0,\infty).$$

The proof of Proposition 6 is based on the Gauss-Radau quadrature formula [24]

$$\int_0^\infty x^\alpha e^{-x} f(x)dx = Bf(0) + \sum_{i=1}^m A_i f(x_i) + R_{2m}(f)$$

where

$$A_i > 0, \quad i = 1, \ldots, m, \quad B > 0$$

and

$$R_{2m}(f) = \frac{m!\Gamma(\alpha + m + 2)}{(2m+1)!} f^{(2m+1)}(\xi).$$

The generalizations for the results from the Propositions 1–6 were presented in [**16**].

7. Examples given in the previous text of D. Acu are special cases of general results which follow from Gauss quadrature formula (see [**25**, pp. 86–91]):

$$\int_a^b w(x)f(x)dx = \sum_{i=1}^{n} A_i f(x_i) + R_n, \tag{7.1}$$

where $w(x) \geq 0$ is an integable function and

$$R_n = \frac{f^{(2m)}(\xi)}{(2m)!} \int_a^b w(x) \left(\pi_n(x)\right)^2 dx \quad (a < \xi < b), \tag{7.2}$$

$$\pi_n(x) = (x - x_1) \ldots (x - x_n), \tag{7.3}$$

and $x_1, \ldots, x_n$ are such that

$$\int_a^b w(x)\pi_n(x)x^k dx = 0 \quad (k = 0, 1, \ldots, n-1). \tag{7.4}$$

In this case we have $A_i \geq 0$ $(i = 1, \ldots, n)$, so for a positive function $f(x)$ we get

$$\int_a^b w(x)f(x)dx \geq \frac{f^{(2n)}(\xi)}{(2n)!} \int_a^b w(x) \left(\pi_n(x)\right)^2 dx. \tag{7.5}$$

For an arbitrary polynomial p_{2n} of the degree $2n$, $p_{2n}(x) \geq 0$ and with the dominant coefficient equal to 1, we get from (7.5):

$$\int_a^b w(x)p_{2n}(x)dx \geq \int_a^b w(x) \left(\pi_n(x)\right)^2 dx, \tag{7.6}$$

with equality for $p_{2n}(x) = \left(\pi_n(x)\right)^2$.

For various classes of orthogonal polynomials we can obtain results like Proposition 1,2,5,6 from Section 6. Of course for negative functions we have opposite results.

References

1. BERNSTEIN, S.N., "*Leçons sur les propriétés extrémales et la meilleure approximation des fonctions analytiques d'une variable réelle*," Paris, 1926.
2. BOTTEMA, O., *Problem 170*, Wisk Opgaven **14** (1929) 342-344.
3. MORDELL, L. J., *The minimum value of a definite integral*, Math. Gaz. **52** (1968), 135–136.
4. BOWMAN, F. and F. A. GERARD, "Higher Calculus," Cambridge 1967 p. 327.
5. SMITHIES, F., *Two remark on note by Mordell*, Math. Gaz. **54** (1970), 260–261.
6. MORDELL, L.J., *The minimum value of a definite integral, II*, Aequationes Math. **2** (1969), 327–331.
7. MIRSKY, L., *A footnote to a minimum problem of Mordell*, Math. Gaz. **57** (1973), 51–56.
8. VASIĆ, P.M. and B.D. RAKOVICH, *A note on the minimum value of a definite integral*, Univ. Beograd. Publ. Elektrotehn. Fak. Ser. Mat. Fiz. No. **544–576** (1976), 13–17.
9. RAKOVICH, B.D. and P.M. VASIĆ, *Some extremal properties of orthogonal polynomials*, Ibid. No. **634–677** (1979), 25–32.
10. JANOUS, W., *A Minimum Problem for a Class of Polynomials*, to appear.
11. American Mathematical Monthly **97** (1989), 655–656.
12. Aufgabe 933, Elem Math. **40** (1985).
13. ACU, D., *Extremal problems in the numerical integration of the functions*, doctor's thesis, Cluj-Napoca, 1980 (Romania).
14. __________, *The Use of Quadrature Formulae in Obtaining Inequalitis*, The Scientific Paper Session "Mathematical Inequalities", 13-15 November, 1981, Sibiu, Romania.
15. __________, *V-Optimal formulas of quadrature with weight function of the Laguerre type*, Proceedings of the second symposium of Mathematics and Its Applications, 30-31 October, 1987, Timişoara, Romania (1988), 51–54.
16. __________, *Extremal problems with polynomials*, Symposium on "Numerical Methods with Applications to Sciences and Technique", 13-14 November, 1987.
17. BAHVALOR, S.V., *Méthodes numériques.* Moscou, 1976.
18. BEREZIN, I.S. and N.I. ZHIDKOV, *Computational Methods* (in Russian), V. 1, "Nauka", Moskow, 1966.
19. DAVIS, I. and B. RABINOWITZ, "Numerical Integraltion," London, 1967.
20. GHIZZETTI, A. and A. OSSICINI, "Quadrature Formulae," Berlin 1970.
21. IONESCU, D.V., "Numerical Quadratures" (in Romain), 1957 Romania.

22. NIKOL'SKIĬ, S. M., "Quadrature Formulas" (in Russian), Moskow, 1974.
23. LOCHER, P., *Pozitivitat bei Quadratur formulas; Habilitationnchrift*, im Fachbereich Mathematik der Eberhard-Karls Universität zu Tübingen, 1972.
24. STANCU, D.D. and H.A. STROUD, *Quadrature formulas with simple Gaussian nodes and multiple fixed nodes*, Math. Comput. **17**, Nr. **84** (1983), 384–394.
25. TOŠIĆ, D., "Uvod u numeričku analizu," Beograd, 1987.
26. MITRINOVIĆ, D.S., "Analytic Inequalities," Berlin-Heidelberg-New York, 1970.
27. SZEGÖ, G., *Orthogonal Polynomials*, Amer. Math. Soc. Colloquium Publ. **23**, New York, 1939.

CHAPTER XVII

METHODS OF PROOFS FOR INTEGRAL INEQUALITIES

1. In this Chapter we give a number of different methods for proofs of various integro-differential inequalities, with emphasis on the more elementary methods. The simpler inequalities of the Chapter are of the form

$$\int_a^b F(x,u,u')dx \geq 0, \qquad u \in U, \tag{1.1}$$

where U is an apropriate class of functions. We illustrate most of the methods to be used in the Chapter by applying them to inequality (2.1) from Chapter II, i.e.

$$\int_0^\pi u^2 dx \leq \int_0^\pi u'^2 dx \quad \text{for} \quad u(0) = u(\pi) = 0, \tag{1.2}$$

provided $u \in AC[0,\pi]$ with $u' \in L_2[0,\pi]$. Moreover, equality holds in (1.2) precisely for $u(x) \equiv C \sin x$, for some constant C.

The first six methods we discuss are from the unpublished manuscript [**1**], with few editorial changes.

2. Use of calculus of variations. The inequality (1.2) is the special case of (1.1) with $F(x,y,y') = y'^2 - y^2$. The Euler equation in this case is

$$\frac{d}{dx}F_{y'}(x,y,y') - F_y(x,y,y') \equiv 2y'' + 2y = 0.$$

From the general theory, a necessary condition that $u(x)$ be a solution to the problem of minimizing the integral in (1.1), subject to fixed end-point conditions such as $u(a) = u(b) = 0$, is that $u = y$, where y is a solution of the corresponding Euler equation which satisfies the boundary conditions. (See, for example, [**2**, p. 34] or [**3**, pp. 5, 21].) Hence the *extremals* for (1.2) are the solutions $y = c\sin x$ of $y'' + y = 0$, $y(0) = y(\pi) = 0$.

The problem of showing that a given extremal function y_0 *does* provide a minimum value of the integral

$$J(u) = \int_a^b F(x, u, u')dx \qquad (u(a) = A,\ u(b) = B),$$

is, of course, the main problem of the calculus of variations. Sufficient conditions depend on the existence of an appropriate *field* (or *flow*) of solutions of the Euler equation

$$\frac{d}{dx}F_{y'} - F_y = 0. \tag{2.1}$$

If y_0 is an extremal of $J(u)$, then by a *field $\mathcal{F}$ containing* y_0 we mean a one-parameter family $\{\phi(x,\alpha)\}$ of solutions of equation (2.1) defined on a rectangle $[a,b]\times(\alpha_1,\alpha_2)$, having the property that $y_0(x) \equiv \varphi(x,\alpha_0)$ for some $\alpha_0 \in (\alpha_1,\alpha_2)$, and which *simply* covers the domain

$$S = \{(x,y)\colon a \le x \le b,\ \varphi(x,\alpha) \le y \le \varphi(x,\alpha')\ \textit{for some}\ \alpha,\alpha' \in (\alpha_1,\alpha_2)\}.$$

That is we assume that through each point of S there passes a *unique* curve $y = \varphi(x,\alpha)$. In addition it is assumed that the equation $y - \varphi(x,\alpha) = 0$ defines $\alpha = \alpha(x,y)$ uniquely for $(x,y) \in S$ as a twice-continuously differentiable function. The *slope-function* $p\colon S \to \mathbf{R}$ of the field $\mathcal{F}$ is defined by $p(x,y) \equiv \varphi_x[x,\alpha(x,y)]$.

The *Weierstrass E-function* (E for excess) for J is defined by

$$E(x,y,p,q) \equiv F(x,y,q) - F(x,y,p) - (q-p)F_p(x,y,p). \tag{2.2}$$

Under appropriate smoothness conditions on F (cf. [**2**, pp. 55–59], [**3**, pp. 26–27, 30]) a sufficient condition for $J(y_0)$ to be a *minimum* value of J is that y_0 be an extremal (so $y = y_0$ is a solution of (2.1) and $y_0(a) = A$, $y_0(b) = B$) embedded in a field $\mathcal{F}$ as above, such that

$$E\{x, u(x), p(x,u(x)), u'(x)\} \ge 0,\ a \le x \le b, \tag{2.3}$$

holds for all *admissible functions* u (so $u(a) = A$, $u(b) = B$ in particular) whose graphs lie in S. In this case the conclusion is only that $J(y_0) \le J(u)$ for all such u. If strict inequality holds in (2.3) provided $u'(x) \not\equiv p(x,u(x))$, the conclusion is that $J(y_0) < J(u)$ for all such u.

We will not be more precise concerning the class of admissible functions for $J(u)$ except to note that in the classical theory it is usually assumed that u is piecewise smooth on $[a,b]$. (In [**2**, Ch. 9] the theory is extended to include

Lebesgue integrals and absolutely continuous functions.) The point we wish to make just now is that *the above theory does not apply to the simple minimum problem* (1.2) *even if one restricts the admissible function to those* $u \in C^1[a,b]$. For, no field $\mathcal{F}$ can exist in this case. To see this consider any one extremal function $y_0 = c\sin x$. The two-parameter family of *all* solutions of the Euler equation is $y = \beta\sin t + \gamma\cos t$, $0 \le t \le \pi$, $(\beta, \gamma \in \mathbf{R})$. If $\gamma \neq 0$, then y has opposite signs at $t = 0$, $t = \pi$ so the graph of y necessarily intersects the graph of y_0. Hence the only possible members of a field containing y_0 are the solutions $y = \beta\sin t$, and *all* of these intersect at the points $(t, y) = (0,0), (\pi, 0)$.

On the other hand, the derivation of the sufficient condition (2.3) in the general theory essentially depends on a certain integral identity involving the so-called *Hilbert invariant integral*

$$(2.4) \qquad J^*_{x_1,x_2}(u) = \int_{x_1}^{x_2} \{F(t,u,p(t,u)) - [u' - p(t,u)]\, F_p(t,u,p(t,u))\}\, dt$$

which can be written as a line integral, and *is independent of the path* in S. That is, if $a \le x_1 < x_2 \le b$ and u_1, u_2 are two piecewise smooth functions satisfying $u_1(x_1) = u_2(x_1)$, $u_1(x_2) = u_2(x_2)$ and whose graphs lie in S, then

$$J^*_{x_1,x_2}(u_1) = J^*_{x_1,x_2}(u_2).$$

The integral identity in question now is

$$(2.5) \qquad \int_{x_1}^{x_2} |F(t,u,u') - F(t,y_0,y_0')|dt = \int_{x_1}^{x_2} E\left[t, u(t), p\left[t, u(t)\right], u'(t)\right] dt,$$

valid for all piecewise smooth functions u whose graphs lie in S and which satisfy $u(x_i) = y_0(x_i)$ for $i = 1, 2$. The proof of (2.5) depends on the fact that the left side of (2.5) is equal to $J^*_{x_1,x_2}(u) - J^*_{x_1,x_2}(y_0)$, using the invariance of J^*.

By taking $x_1 = a$, $x_2 = b$ the proof of the sufficiency of the Weirstrass condition (2.3) follows directly from (2.5) – *when the embedding field $\mathcal{F}$ exists.* However, for $a < x_1 < x_2 < b$, the identity (2.5) remains valid provided only the one-parameter family $y = \varphi(x, \alpha)$ of solutions of (2.1) simply covers the subset S_1 of S for which $x_1 \le x \le x_2$. In our example, with

$$F(x,y,y') = y'^2 - y^2, \quad E(x,y,p,q) = (q-p)^2, \quad y_0 = c\sin x,$$
$$\varphi(x,\alpha) = \alpha\sin x, \quad \alpha(x,y) = y/\sin x, \quad p(x,y) = y\cot x,$$

and $0 < x_1 < x_2 < \pi$, $u(x_i) = c\sin x_i$, the identity (2.5) becomes

$$(2.6) \qquad \int_{x_1}^{x_2} (u'^2 - u^2)dt = \int_{x_1}^{x_2} (u' - u\cot t)^2 dt + c^2 \int_{x_1}^{x_2} \cos 2t\, dt,$$

valid for piecewise smooth functions u on $[x_1, x_2]$ for which $u(x_i) = c\sin x_i$ $(i = 1, 2)$. Letting $x_1 \to 0^+$ and $x_2 \to \pi^-$, this suggests, but does not prove, that the inequality (1.2) is valid for all piecewise smooth functions u on $[0, \pi]$ having $u(0) = u(\pi) = 0$.

Instead of trying to justify this procedure, we return to the identity (2.6), let $x_1 \to 0^+$, $x_2 \to \pi^-$ to "obtain" the conjectured identity

$$\int_0^\pi (u'^2 - u^2)dt = \int_0^\pi (u' - u\cot t)^2 dt. \tag{2.7}$$

We now show *directly* that (2.7) is valid for all $u \in AC[0, \pi]$ with $u' \in L_2[0, \pi]$ and $u(0) = u(\pi) = 0$. In fact, if $0 < \alpha < \beta < \pi$, then

$$\begin{aligned}\int_\alpha^\beta (u' - u\cot t)^2 dt &= \int_\alpha^\beta u'^2 dt - \int_\alpha^\beta (\cot t)(u^2)' dt + \int_\alpha^\beta u^2 \cot^2 t dt \\ &= \int_\alpha^\beta u'^2 dt - \int_\alpha^\beta u^2 dt - u^2(t)\cot t\Big|_\alpha^\beta .\end{aligned} \tag{2.8}$$

Now $u(t) = \int_0^t u' ds = -\int_t^\pi u' ds$, so that

$$0 \le u^2(t) \le t \int_0^t u'^2 ds, \qquad 0 \le u^2(t) \le (\pi - t)\int_t^\pi u'^2 ds$$

follows from the Cauchy-Schwarz inequality. It now follows that $u^2(\beta)\cot\beta \to 0$ as $\beta \to \pi^-$, and $u^2(\alpha)\cot\alpha \to 0$ as $\alpha \to 0^+$, so that (2.7) follows. This proves (1.2), and by (2.7) we see that equality holds in (1.2) precisely when $u' = u\cot t$ a.e. on $[0, \pi]$, that is precisely when $u = c\sin t$ for some constant c.

We have dealt with this example in some detail to show that although the classical calculus of variations may not, by itself, completely handle an integral inequality such as (1.1), nevertheless an appropriate use of the Euler equation (2.1) and the Hilbert identity (2.5) may lead to a direct solution. For further discussion and examples of this approach, see also Hardy, Littlewood, and Pólya [4, Ch. 7].

Before leaving this subsection, we note that the Hilbert identity (2.6) can be used to obtain *other* integral inequalities to which the standard theory of the calculus of variations *does* apply. For example, taking $0 < x_1 < \pi/2 < x_2 = \pi - x_1 < \pi$, and $u(x_1) = u(\pi - x_1) = u_0$, where $u_0 = c\sin x_1$, (2.6) becomes

$$\int_{x_1}^{\pi - x_1} (u'^2 - u^2)dt = \int_{x_1}^{\pi - x_1} (u' - u\cot t)^2 dt - 2u_0^2 \cot x_1,$$

so that

$$\int_{x_1}^{\pi - x_1} u^2 dt \leq 2u_0^2 \cot x_1 + \int_{x_1}^{\pi - x_1} u'^2 dt$$

holds for all such u, with equality if and only if $u = u_0 \sin t/\sin x_1$. A simple change of variable gives: *if $a < b$ with $b - a < \pi$, and $v(a) = v(b) = u_0$, then*

$$\int_a^b v^2 ds \leq 2u_0^2 \tan\left(\frac{b-a}{2}\right) + \int_a^b v'^2 ds \tag{2.9}$$

holds for all piecewise smooth functions v on $[a, b]$, with equality if and only if $v = u_0 \sec\left(\frac{b-a}{2}\right)\cos\left(\frac{a+b}{2} - t\right)$. Although this result requires only the classical theory of the calculus of variations, a direct proof which shows that (2.9) *is valid for all $v \in AC\ [a, b]$ with $v' \in L_2[a, b]$ and $v(a) = v(b) = u_0$*, can also be given by expanding the integral

$$\int_a^b \left[v' - v\tan\left(\frac{a+b}{2} - t\right)\right]^2 dt.$$

3. Use of a differential identity. The integral identity (2.8) is actually equivalent to the differential identity

$$(v' - v\cot t)^2 = v'^2 - v^2 - (v^2 \cot t)', \quad 0 < t < \pi, \tag{3.1}$$

as one readily verifies. However, (3.1) is valid for all $v \in AC(0, \pi)$ provided equality is interpreted as being *almost everywhere* (a.e.). [Note that here, and later, when I is an open interval and we write $v \in AC(I)$ we mean that v *is absolutely continuous on each compact subinterval of I*.] Suppose in addition that $v' \in L_2[0, \pi]$. Then $v' \in L_1[0, \pi]$ and as in the analysis following (2.8) if $0 < \varepsilon < x < \pi$,

$$|v(x) - v(\varepsilon)| \leq \int_\varepsilon^x |v'|dt \leq K\sqrt{x - \varepsilon}.$$

This proves that $\lim_{x\to 0^+} v(x) = A$ exists, and similarly $\lim_{x\to\pi^-} v(x) = B$ exists. Since $v(x) - v(\varepsilon) = \int_\varepsilon^x v'dt$ and $v' \in L_1$, on letting $\varepsilon \to 0^+$ we obtain

$$v(x) = A + \int_0^x v'dt, \quad 0 < x < \pi.$$

Similarly, $v(x) = B - \int_x^\pi v'dt$, $0 < x < \pi$. Thus if we redefine $v(0) = A$, $v(\pi) = B$ if necessary, it follows that $v \in AC[0,\pi]$. We have proved the following elementary lemma.

LEMMA 1. *Let $-\infty < a < b < \infty$ and suppose that $v \in AC(a,b)$ and $v' \in L_2[a,b]$. Then $\lim_{x\to a^+} v(x) = A$, $\lim_{x\to b^-} v(x) = B$ both exist and redefining $v(a) = A$, $v(b) = B$ if necessary, we have $v \in AC[a,b]$.*

We remark that the conclusion is false if one only assumes that $v \in AC(a,b)$, even if one assumes that $\lim_{x\to a} v(x)$ and $\lim_{x\to b} v(x)$ both exist. On the other hand, one may clearly replace the hypothesis $v' \in L_2[a,b]$ by the weaker hypothesis that $v' \in L_1[a,b]$.

This lemma shows that in dealing with *bounded* intervals such as in (3.1), we may as well assume that $v \in AC[0,\pi]$. Now integrate (3.1) over $[\alpha,\beta]$ where $0 < \alpha < \beta < \pi$, to obtain the integral identity (2.8) with $u = v$. Note that the only *possible* boundary values for $v(0)$, $v(\pi)$ which can lead to an integral inequality are $v(0) = v(\pi) = 0$. The rest of the proof of the inequality (1.2) under these hypotheses has already been given (after (2.8)), including the conditions for equality.

The identity (3.1) is a special case of an identity used by B. Florkiewicz and A. Rybarski [5].

4. Integral identities and Riccati equations. The preceding method is essentially the same as one used by Beesack [6] somewhat earlier. The function $y_0 = c\sin x$ ($c \neq 0$) is a solution of the (Euler) differential equation $y_0'' + y_0 = 0$. If we set

$$h = y_0'/y_0,$$

then h is a solution of the associated Riccati equation

$$h' = -h^2 - 1, \quad 0 < x < \pi.$$

Hence for all $v \in AC(0,\pi)$,

$$(v' - hv)^2 = v'^2 - h(v^2)' + h^2v^2 = v'^2 - v^2 - \{h'v^2 + h(v^2)'\}$$
$$= v'^2 - v^2 - (v^2h)', \qquad \text{for } x \in (0,\pi).$$

This is the identity (3.1) and leads to the integral identity

$$\int_\alpha^\beta (v' - hv)^2 = \int_\alpha^\beta v'^2 dt - \int_\alpha^\beta v^2 dt - v^2(t)h(t)\Big|_\alpha^\beta$$

obtained before.

5. Integral identities and Jacobi multipliers. Let $y_0 = \sin t$ and for $v \in AC(0,\pi)$ define z by

$$v = zy_0 = z\sin t. \tag{5.1}$$

Then $v' = z'\sin t + z\cos t$,

$$(v')^2 = z'^2 \sin^2 t + 2zz'\sin t\cos t + z^2\cos^2 t,$$

so

$$\int_\alpha^\beta v'^2 dt = \int_\alpha^\beta z'^2 \sin^2 t\, dt + \int_\alpha^\beta z^2\cos^2 t\, dt + z^2 \sin t\cos t\Big|_\alpha^\beta$$
$$- \int_\alpha^\beta z^2(\cos^2 t - \sin^2 t)dt,$$

or using (5.1),

$$\int_\alpha^\beta v'^2 dt = \int_\alpha^\beta v^2 dt + \int_\alpha^\beta z'^2 \sin^2 t\, dt + v^2(t)\cos t\Big|_\alpha^\beta \tag{5.2}$$

valid for $0 < \alpha < \beta < \pi$. Thus

$$\int_\alpha^\beta v'^2 dt \geq \int_\alpha^\beta v^2 dt + v^2(t)\cot t\Big|_\alpha^\beta$$

and equality holds if and only if $z' = 0$ or $v = c\sin t$ (a.e) for some constant c. The proof of (1.2) now follows by the same argument used following (2.8), including the conditions for equality.

We note that this method—which can be regarded as replacing the constant c in the family $y = c\sin t$ of extremals by the multiplier function z—was used as long ago as 1837 by C. G. Jacobi. (cf. the discussion given in [**7**, pp. 709].) It is clearly related to Lagrange's method of "variation of parameters" but we shall call it the method of Jacobi multipliers.

6. Spectral theory. Consider the eigenvalue problem

$$y'' + \lambda y = 0, \qquad y(0) = y(\pi) = 0. \tag{6.1}$$

This problem has the simple eigenvalues $\lambda_n = n^2$ $(n = 1, 2, \dots)$ and the corresponding eigenfunctions $y_n = \sin nx$. By known results from the spectral theory of linear differential equations ([**8**, p. 268] or [**9**, p. 273]) we have

$$1 = \lambda_1 = \inf\left\{\int_0^\pi u'^2 dt \Big/ \int_0^\pi u^2 dt : u \in \mathcal{D}_1\right\}, \tag{6.2}$$

where $\mathcal{D}_1 = \{u \in AC[0,\pi] : u(0) = u(\pi) = 0, \quad u' \in L_2[0,\pi]\}$. Moreover we also have for $n \geq 2$,

$$n^2 = \lambda_n = \inf\left\{\int_0^\pi u'^2 dt \Big/ \int_0^\pi u^2 dt : u \in \mathcal{D}_n\right\} \tag{6.3}$$

where $\mathcal{D}_n = \left\{u \in \mathcal{D}_1 : \int_0^\pi u y_k dt = 0,\ 1 \leq k \leq n-1\right\}$. The quotient appearing in (6.2), (6.3) is usually called the *Rayleigh quotient*, and the infimum is attained in (6.3) precisely for $u = cy_n$. The equality (1.2), together with the conditions for equality, therefore follows from (6.2). Similarly from (6.3) we obtain

$$\int_0^\pi u^2 dt < \frac{1}{n^2}\int_0^\pi u'^2 dt \quad \text{unless } u = c\sin n\,t, \tag{6.4}$$

for all $u \in AC[0,\pi]$, *with* $u' \in L_2[0,\pi]$, $u(0) = u(\pi) = 0$, *and* $\int_0^\pi u(t)\sin k\,t\,dt = 0$ *for* $1 \leq k \leq n-1$.

It is clear that this method is not so elementary as the last three methods , depending as it does on spectral theory for differential equations. See [**14**] or

[15] for proofs of the minimum properties of the eigenvalues of Sturm-Liouville boundary value problems, including a justification of (6.2), (6.3).

7. Use of Fourier series. If $u \in AC[0,\pi]$ with $u(0) = u(\pi) = 0$ and $u' \in L_2[0,\pi]$ then we have the Fourier series expansions (for the odd periodic extension of u to $[-\pi,\pi]$)

$$u(x) \sim \sum_1^\infty b_k \sin k\,x, \quad u'(x) \sim \sum_1^\infty k\,b_k \cos k\,x, \tag{7.1}$$

where $b_k = 2\pi^{-1} \int_0^\pi u(t) \sin k\,t\,dt$, as well as the Parseval relations

$$\int_0^\pi u^2 dt = \frac{\pi}{2} \sum_1^\infty b_k^2, \quad \int_0^\pi u'^2 dt = \frac{\pi}{2} \sum_1^\infty k^2 b_k^2. \tag{7.2}$$

From this it follows at once that

$$\int_0^\pi u^2 dx \leq \int_0^\pi u'^2 dx$$

and that equality holds precisely when $b_k = 0$ for $k \geq 2$, that is for $u = b_1 \sin x$ for some constant b_1. Similary, if u satisfies the above conditions and, in addition $\int_0^\pi u(t) \sin k\,t\,dt = 0$ for $1 \leq k \leq n-1$, it follows that

$$n^2 \int_0^\pi u^2 dx \leq \int_0^\pi u'^2 dx \tag{7.3}$$

with equality precisely for $u = b_n \sin n\,x$. This is just the inequality (6.4) obtained in the last subsection.

The same technique leads to the following inequality, usually referred to as *Wirtinger's inequality: if $u \in AC[0,2\pi]$ with $u' \in L_2[0,2\pi]$ and $u(0) = u(2\pi)$, $\int_0^{2\pi} u\,dt = 0$, then*

$$\int_0^{2\pi} u^2 dt \leq \int_0^{2\pi} u'^2 dt \tag{7.4}$$

with equality if and only if $u = A\cos x + B\sin x$. Indeed, if

$$(7.5)\qquad \begin{aligned} u(x) &\sim \frac{1}{2}a_0 + \sum_1^\infty (a_k \cos kx + b_k \sin kx) \\ u'(x) &\sim \frac{1}{2}A_0 + \sum_1^\infty (A_k \cos kx + B_k \sin kx), \end{aligned}$$

then $a_0 = \pi^{-1}\int_0^{2\pi} u\,dt = 0$, and

$$\begin{aligned} A_k &= \frac{1}{\pi}\int_0^{2\pi} u'(t)\cos kt\,dt \\ &= \frac{1}{\pi}\left\{ u(t)\cos kt\big|_0^{2\pi} + k\int_0^{2\pi} u(t)\sin kt\,dt \right\} \\ &= kb_k \quad \textit{for } k \geq 0, \end{aligned}$$

and similarly, $B_k = -kA_k$ for $k \geq 1$. Hence (7.5) becomes

$$(7.6)\qquad \begin{aligned} u(x) &\sim +\sum_1^\infty (a_k \cos k\,x + b_k \sin k\,x), \\ u'(x) &\sim +\sum_1^\infty k\,(b_k \cos k\,x - a_k \sin k\,x), \end{aligned}$$

and the corresponding Parseval relations become

$$(7.7)\qquad \begin{aligned} \int_0^{2\pi} u^2 dx &= \pi\sum_1^\infty (a_k^2 + b_k^2), \\ \int_0^{2\pi} u'^2 dt &= \pi\sum_1^\infty k^2\,(b_k^2 + a_k^2). \end{aligned}$$

The inequality (7.4) follows at once, together with the conditions for equality. For further details of Wirtinger's inequality and related results, see Chapter II.

As in (7.3) *if to the hypotheses for* (7.4) *we add the orthogonality conditions*

$$\int_0^{2\pi} u(x)\cos kx\,dx = \int_0^{2\pi} u(x)\sin kx\,dx = 0, \quad 1 \leq k \leq n-1,$$

then

$$n^2 \int_0^{2\pi} u^2 dx \le \int_0^{2\pi} u'^2 dx \tag{7.8}$$

with equality precisely for $u = A\cos nx + B\sin nx$. This extension of Wirtinger's inequality was given by Everitt [**10**].

There are other extensions of (7.4) which follow by the arguments used above. For example, we may delete the condition $\int_0^{2\pi} u\,dt = 0$ and obtain the following: *if* $u \in AC[0, 2\pi]$ *with* $u(0) = u(2\pi)$ *and* $u' \in L_2[0, 2\pi]$, *then*

$$\int_0^{2\pi} u^2 dx \le \frac{1}{2\pi}\left(\int_0^{2\pi} u\,dx\right)^2 + \int_0^{2\pi} u'^2\,dx, \tag{7.9}$$

with equality precisely for $u = C + A\cos x + B\sin x$. This follows from the (full) Parseval relation

$$\int_0^{2\pi} u^2\,dx = \pi\left\{\frac{1}{2}a_0^2 + \sum_1^{\infty}(a_k^2 + b_k^2)\right\}.$$

Similarly, *if* $u' \in AC[0, 2\pi]$ *with* $u'' \in L_2[0, 2\pi]$ *and* $u(0) = u(2\pi)$, $u'(0) = u'(2\pi)$, *then*

$$\int_0^{2\pi} u^2 dx \le \frac{1}{2\pi}\left(\int_0^{2\pi} u\,dx\right)^2 + \int_0^{2\pi} u''^2\,dx \tag{7.10}$$

with equality precisely for $u = C + A\cos x + B\sin x$. The inequlaity (7.10) was proved as long ago as 1915 by N. M. Krylov [**11**], who proved a number of inequalities of this kind using the Fourier series method. We note that this same method can be applied to any complete orthonormal sequence of functions using the corresponding Parseval relations.

8. Minimum properties of eigenvalues of Sturm-Louiville problems. P. R. Beesack [**12**] considered the Sturm-Louiville eigenvalue problem

$$(ry')' + (\lambda p - q)y = 0, \tag{8.1}$$

$$\beta_a y(a) - \alpha_a r(a)y'(a) = 0,\quad \beta_b y(b) + \alpha_b r(b)y'(b) = 0, \tag{8.1''}$$

where we assume that $p, q, r, \in C[a, b]$ with $r > 0$, $p \geq 0$ (but $p \not\equiv 0$ on any subinterval of $[a, b]$), and $\beta_a^2 + \alpha_a^2 > 0$, $\beta_b^2 + \beta_b^2 > 0$. Under these it is known that there exists a real sequence $\{\lambda_n\}$ of eigenvalues of (8.1) such that for $\lambda = \lambda_n$ (and only for such values), (8.1) has nontrivial solutions $y = y_n$ $(n = 0, 1, 2, \dots)$. Moreover, if $\lambda_0 < \lambda_1 < \dots$, then $\lambda_n \to +\infty$ and y_n has exactly n zeros on (a, b) (see [**9**, p. 251],[**13**, p. 337], [**14**, p. 212], [**8**, p. 277]). We note that solutions of (8.1) are functions $y \in C^2[a, b]$ such that $(ry') \in C^1[a, b]$, which satisfy (1') on $[a, b]$.

The next Theorem 1 was proved in [**12**] (proof is also given in [**1**]).

We shall use the following notations:

$$(8.2) \qquad A(y) = \begin{cases} 0, & \text{if } \alpha_a = 0, \\ \frac{\beta_a}{\alpha_a} y^2(a), & \text{if } \alpha_a \neq 0, \end{cases} \quad B(y) = \begin{cases} 0 & \text{if } \alpha_b = 0, \\ \frac{\beta_b}{\alpha_b} y^2(b), & \text{if } \alpha_b \neq 0. \end{cases}$$

THEOREM 1. *Under the preceding hypotheses on* (8.1), *set*

$$\mathcal{D}_0 = \Big\{u \in AC[a, b]\colon u(a) = 0 \text{ if } \alpha_a = 0,\ u(b) = 0 \text{ if } \alpha_b = 0; \\ \int_a^b ru'^2 dx < \infty,\ \int_a^b pu^2 dx \neq 0\Big\}.$$

(If $\alpha_a \neq 0$, $\mathcal{D}_0$ includes no boundary condition at a, and similarly for b; only "essential" boundary condition of (8.1′) remain.)

For $n \geq 1$, set

$$\mathcal{D}_n = \left\{ u \in \mathcal{D}_0 \colon \int_a^b p y_k u \, dx = 0,\ 0 \leq k \leq n-1 \right\}.$$

Then for all $n \geq 0$,

$$(8.3) \qquad \lambda_n = \min_{u \in \mathcal{D}_n} \left(\int_a^b \left(ru'^2 + qu^2 \right) dx + A(u) + B(u) \right) \Big/ \int_a^b pu^2 dx,$$

where A, B are defined by (8.2). *The minimum in* (8.3) *is attained if and only if $u = cy_n$ for some constant $c \neq 0$.*

We shall give some applications of this theorem given in [**1**].

First, we note that the above theorem remains valid, with only minor modifications to the proof, if it is only supposed that the nth eigenfunction y_n has $k_n \leq n$ zeros on (a, b). Moreove, corresponding results hold for certain singular systems, that is cases where a or b or both are infinite, or where one or more of the coefficients p, q, r are singular at the end-points, or $r(x) = 0$ for $x = a$ or b. See [**12**] for a general theorem of this type. Examples of singular cases given in [**12**] are listed below:

$$(8.4) \qquad n(n+1)\int_{-1}^{1} u^2\,dx \le \int_{-1}^{1}(1-x^2)u'^2\,dx \quad n = 0,1,2,\ldots)$$

for all $u \in AC[-1,1]$ *such that* $\int_{-1}^{1}(1-x^2)u'^2\,dx < \infty$, *and* $\int_{-1}^{1} P_j(x)u(x)dx = 0$ *for* $0 \le j \le n-1$ *there* P_k *is the Legendre polynomial of degree* k. *Equality holds in* (8.4) *if and only if* $u = cP_n$ *for some constant* c.

$$(8.5) \qquad 2n\int_{-\infty}^{+\infty} e^{-x^2}u^2\,dx \le \int_{-\infty}^{+\infty} e^{-x^2}u'^2\,dx \quad (n = 0,1,2,\ldots)$$

for all $u \in AC(\mathbf{R}))$ *such that* $u(x) = 0\left(|x|^k\right)$ *as* $|x| \to \infty$ *for some* $k > 0$, *the right side of* (8.5) *is finite, and* $\int_{\mathbf{R}} e^{-x^2}H_j(x)u(x)dx = 0$ *for* $0 \le j \le n-1$ *where* H_k *is the Hermite polynomial of degree* k. *Equality holds in* (8.5) *if and only if* $u = cH_n$ *for some constant* c.

$$(8.6) \qquad k_n^2\int_0^1 u^2\,dx \le \int_0^1 u'^2\,dx + \left(m^2 - \frac{1}{4}\right)\int_0^1 x^{-2}u^2\,dx \quad (n = 1,2,\ldots)$$

for $m \ge \frac{1}{2}$, *and all* $u \in AC[0,1]$ *such that* $u(0) = 0$, *both integrals on the right side of* (8.6) *are finite, and* $\int_0^1 x^{1/2}J_m(k_jx)u(x)dx = 0$ *for* $1 \le j \le n-1$, *where* k_n *is the* n^{th} *positive zero of the Bessel function* $J_m(x)$. *Equality holds in* (8.6) *if and only if* $u = cx^{1/2}J_m(k_nx)$ *for some constant* c.

For further examples of singular eigenvalue problems, see [**15**, pp. 324–331, 400, 415]. In the proof of Theorem 1 the following lemma is used:

LEMMA. *Let* $n \ge 1$, *so that the* n*-th eigenfunction* y_n *has consecutive zeros at* $x_k \in (a,b)$, $1 \le k \le n$, *where*

$$x_0 = a < x_1 < x_2 < \cdots < x_n < b = x_{n+1}.$$

If $u \in D_0$ satisfies the conditions $u(x_k) = 0$ $(1 \le k \le n)$, then

$$\lambda_n \int_a^b pu^2 dx \le \int_a^b \left(ru'^2 + qu^2\right) dx + A(u) + B(u). \tag{8.7}$$

Moreover, equality holds in (8.7) if and only if $u(x) = c_k y_n(x)$ on $[x_{k-1}, x_k]$ $(1 \le k \le n_1)$, for some constants c_k.

We can use the Lemma and Theorem 1 to obtain certain maximum-minimum characterizations of the eigenvalues of the Sturm-Liouville problem (8.1) under the hypotheses given there.

THEOREM 2. *For each $n = 0, 1, \dots$, let*

$$\begin{aligned}
&\mathcal{H}_n = \{u \colon u(x_k) = 0 \text{ for at most } n \text{ distinct points } x_k \in (a,b)\},\\
&\mathcal{H}_n(u) = \{v \in \mathcal{D}_0 \colon v = 0 \text{ if } u = 0\}, \quad u \in \mathcal{H}_n,\\
&d_n(u) = \inf\{R(v) \colon v \in \mathcal{H}_n(u)\},
\end{aligned}$$

where R is the Rayleigh quotient

$$R(v) \equiv \left[\int_a^b (rv'^2 + qv^2) dx + A(v) + B(v)\right] \Big/ \int_a^b pv^2 dx. \tag{8.8}$$

Then

$$\lambda_n = \sup\{d_n(u) \colon u \in \mathcal{H}_n\}. \tag{8.9}$$

PROOF: Since y_n has n zeros on (a, b), it follows from Lemma 2 that

$$\lambda_n = d_n(y_n) \le \sup\{d_n(u) \colon u \in \mathcal{H}_n\}.$$

To prove the opposite inequality, take any $u \in \mathcal{H}_n$ and suppose the zeros of u on (a, b) are $x_1, \dots, x_k$ where $k \le n$. We now construct a function $v \in \mathcal{H}_n(u)$ such that $R(v) \le \lambda_n$. This will prove $d_n(u) \le \lambda_n$ for all $u \in \mathcal{H}_n$ whence (8.9) will follow. In fact it suffices to take $v = \sum_0^n c_i y_i$. We have $v \in \mathcal{H}_n(u)$ if the c_i (not all 0) can be chosen so that

$$\sum_{i=1}^n c_i y_i(x_j) = 0, \quad 1 \le j \le k \ (\le n).$$

Such a nontrivial solution always exists for this homogeneous system. But then, as in the proof of Theorem 1 (see [**12**]), we have:

$$\int_a^b (rv'^2 + qv^2)dx + A(v) + B(v) = \sum_0^n c_i^2 \lambda_i \int_a^b py_i^2\, dx,$$

$$\lambda_n \int_a^b pv^2\, dx = \lambda_n \sum_0^n c_i^2 \int_a^b py_i^2\, dx.$$

Thus, $\lambda_n \int_a^b pv^2\, dx \geq \int_a^b (rv'^2 + qv^2)dx + A(v) + B(v)$ is clear so that $R(v) \leq \lambda_n$ holds, completing the proof of (8.9).

The above result is given in Courant-Hilbert [**15**, p. 463] and attributed there to K. Hohenemser. A second maximum-minimum characterization of λ_n is due to R. Courant [**15**, p. 406] and can be formulated as follows.

THEOREM 3. *Let C_n be the class of all n-tuples $(v_0, \ldots, v_{n-1})$ of functions such that $0 < \int_a^b pv_i^2 dx < \infty$ for $0 \leq i \leq n-1$, and set*

$$C_n(v_0, \ldots, v_{n-1}) = \left\{ u \in \mathcal{D}_0 : \int_a^b puv_i\, dx = 0,\ 0 \leq i \leq n-1 \right\},$$

$$d_n(v_0, \ldots, v_{n-1}) = \inf \{R(u) \colon u \in C_n(v_0, \ldots, v_{n-1}\}.$$

Then

$$\lambda_n = \sup\{d_n(v_0, \ldots, v_{n-1}) \colon (v_0, \ldots, v_{n-1}) \in C_n\}. \tag{8.10}$$

PROOF: By Theorem 1 we have $d_n(y_0, \ldots, y_{n-1}) = \lambda_n$ so that

$$\lambda_n \leq \sup\{d_n(v_0, \ldots, v_{n-1}) \colon (v_0, \ldots, v_{n-1}) \in C_n\}.$$

Now for any $(v_0, \ldots, v_{n-1}) \in C_n$ define the function

$$u(x) = \sum_0^n c_i y_i(x)$$

where the c_i are any nontrivial solution of the n homogeneous equations (in $n+1$ unknowns)

$$\int_a^b puv_j\, dx = \sum \int_a^b pv_j y_i\, dx = 0 \quad (0 \leq j \leq n-1).$$

Then $u \in C_n(v_0,\ldots,v_{n-1})$ and, as in the proof of Theorem 2, $\lambda_n \geq R(u) \geq d_n(v_0,\ldots,v_{n-1})$, completing the proof of (8.10).

9. Benson's Method. D. C. Benson [16] has given an elementary method for proving inequalities and using its deduced large number of known classical inequalities. This method is contained in the following:

1° Start from a suitable chosen algebraic inequality and use it to show that an expression which contains a function together with its derivatives is nonnegative.

2° To both sides of the inequality add something which is, for example, an exact derivative.

3° Integrate both sides of the inequality to obtain the wanted inequality.

Benson started from the following algebraic inequality

$$x^{2n} - 2nx + 2n - 1 \geq 0, \tag{9.1}$$

which is true for every real x and $n = 1,2,\ldots$, equality holds if and only if $x = 1$, and he got the following:

THEOREM 1. *Let $u(x)$, $P(u,x)$, $G(u,x)$ be continuously differentiable functions and let $P(u,x) > 0$. Then, the following inequality holds*

$$\begin{aligned}\int_a^b \left(Pu'^{2n} + (2n-1)P^{-1/(2n-1)}G_u^{2n/(2n-1)} + 2nG_x\right) dx \\ \geq 2n\left(G\left(u(b),b\right) - G\left(u(a),a\right)\right),\end{aligned} \tag{9.2}$$

where $G_u = \frac{\partial}{\partial u}G(u,x)$, $G_x = \frac{\partial}{\partial x}G(u,x)$.

Equality holds if and only if $u' = (G_u/P)^{1/(2n-1)}$.

Benson also proved:

THEOREM 2. *Let u be twice continuously differentiable, $P(u',u,x)$, $G(u',u,x)$ continuously differentiable functions and $P(u',u,x) > 0$. Then the following inequality holds*

$$\begin{aligned}\int_a^b (pu''^2 + G_{u'}^2p^{-1} + 2u'G_u + 2G_x)dx \\ \geq 2G\left(u'(b),u(b),b\right) - 2G\left(u'(a),u(a),a\right),\end{aligned} \tag{9.3}$$

where the same abbreviations are used as in Theorem 1.

Equality holds if and only if $u'' = G_{u'}P^{-1}$.

Inequalities (9.2) and (9.3) contain a number of known inequalities. We shall show how Wirtinger's inequality can be obtained from (9.2).

Let u be a continuously differentiable periodic function of period 2π and let $m = \inf u(x)$, $M = \sup u(x)$. Further, let $m = u_1(x_1)$, $M = u_2(x_2)$, $(0 \leq x_1 - x_2 \leq 2\pi)$. Putting in (9.2) $P(u,x) = 1$, $n = 1$ and assuming that $x_2 < x_1$,

$$G(u,x) = -\int_M^u ((M-u)(u-m))^{1/2}\,du \qquad (x_2 \leq x \leq x_1),$$

$$= \int_M^u ((M-u)(u-m))^{1/2}\,du \qquad (x > x_1 \text{ or } x < x_2),$$

we get the following two inequalities:

$$\int_{x_2}^{x_1} \Big(u'^2 + (M-u)(u-m)\Big)\,dx \geq -2\int_M^m ((M-u)(u-m))^{1/2}\,du = \frac{\pi(M-m)^2}{4}$$

and

$$\int_{x_1}^{x_2+2\pi} \Big(u'^2 + (M-u)(u-m)\Big)\,dx \geq -2\int_M^m ((M-u)(u-m))^{1/2}\,du$$
$$= \frac{\pi(M-m)^2}{4}.$$

From those inequalities we obtain

$$\int_0^{2\pi} \left(u'^2 - \left(u - \frac{M+m}{2}\right)^2\right) dx \geq 0.$$

If additionally we assume $\int_0^{2\pi} u\,dx = 0$,, then we have

$$\int_0^{2\pi} \left(u'^2 - u^2\right) dx \geq \frac{\pi(M+m)^2}{2}.$$

Since the right-hand side of this inequality is always positive, as a special case we obtain Wirtinger's inequality (see Chapter II; for a generalization of the above result see especially 12. from Chapter II).

Let us now state a useful special case of (9.2) which can be applied to prove Weyl's inequality ((6.8) from Chapter I.).

Let u and g be continuously differentiable. Then

$$\int_a^b \left(u'^2 + \left(g'(x) + g(x)^2\right) u^2\right) dx \geq u(b)^2 g(b) - u(a)^2 g(a). \tag{9.4}$$

Equality holds if and only if $u' = ug(x)$.

Inequality (9.4) can be directly obtained from (9.2) by putting $n = 1$, $P = 1$ and $G(u, x) = \frac{1}{2}u^2 g(x)$.

Weyl's inequality can be proved in the following way. Replace $g(x)$ in (9.4) by $\lambda g(x)$. Then (9.4) becomes

$$\lambda^2 \int_a^b g(x)^2 u^2 dx - \lambda \left(u(b)^2 g(b) - u(a)^2 g(a) - \int_a^b g'(x) u^2 dx\right) + \int_a^b u'^2 dx \geq 0.$$

Since this inequality holds for every real λ, the discriminant of the above quadratic polynomial in λ must be nonpositive, i.e.

$$\left(\int_a^b g' u^2 dx + u(a)^2 g(a) - u(b)^2 g(b)\right)^2 \leq 4 \int_a^b g^2 u^2 dx \int_a^b u'^2\, dx.$$

Putting in the last inequality $g(x) = x$, $a = 0$ and letting b tend to infinity, we get Weyl's inequality

$$\left(\int_0^\infty u^2 dx\right)^2 \leq 4 \int_0^\infty x^2 u^2 dx \int_0^\infty u'^2 dx,$$

provided that the above integrals exist.

D. C. Benson [**16**] has cited a number of inequalities which are special cases of Theorems 1 and 2. Some of them are listed below:

1° Let u be a differentiable periodic function of period L and let $m = \inf u(x)$, $M = \sup u(x)$. Then

$$\int_0^L \left(u'^2 + (M-m)^{2b}(m-u)^{2a}\right) dx \geq 4(M-m)^{a+b+1} \frac{\Gamma(a+1)\Gamma(b+1)}{\Gamma(a+b+2)},$$

where $a > -1/2$ and $b > -1/2$.

2° Let u be a differentiable periodic function of period L. Let m and M be as in 1°. Then

$$\int_0^L \left(u'^{2n} + (2n-1)(u-m)^n(M-u)^n\right) dx \geq \frac{\pi n(2n)!}{(n!)^2 2^{2n-2}}(M-m)^{2n},$$

where n is a natural number.

Benson's method is given in monograph [**17**] with the following generalization:

THEOREM 3. *Let u be an n-times continuously differentiable function, let $P\left(u^{(n-1)},\dots,u,x\right)$ and $G\left(u^{(n-1)},\dots,u,x\right)$ be continuously differentiable functions and let $P > 0$. Then*

$$\int_a^b (P(u^{(n)})^{2m} + (2m-1)P^{-1/(2m-1)}G_{u^{(n-1)}}^{2m/(2m-1)} + 2mu^{(m-1)}G_{u^{(n-2)}} + \cdots +$$

$$+ 2mG_x)dx \geq 2\left(G(u^{(n-1)}(b),\dots,u(b),b) - G(u^{(n-1)}(a),\dots,u(a),a)\right)$$

with equality if and only if $u^{(n)} = \left(P^{-1}G_{u^{(n-1)}}\right)^{1/(2m-1)}$.

The following sentence is from [**17**]:

"It would be desirable if the elegant Benson method were further developed, since the very first applications indicate its fruitfulness and give nice results."

Indeed, D. T. T. Shum has written his doctoral disertation [**18**] in which he developes and applies Benson's method.

Shum unifies Benson's method and Beesack's method from his papers [**6**], [**12**], [**19**] [**20**].

Instead of (9.1), Shum used the following elementary algebraic inequalities [**21**] and [**4**, Th. 41]:

$$s^p + (p-1)t^p - pst^{p-1} \geq 0 \quad (p > 1 \text{ or } p < 0); \tag{9.5}$$

$$s^p + (p-1)t^p - pst^{p-1} \leq 0 \quad (0 < p < 1). \tag{9.6}$$

Here, s and t are nonnegative (positive if $p < 0$), and in both cases strict inequality holds unless $s = t$. (Note also that when $p = 0$ or $p = 1$, the left sides of both (9.5) and (9.6) become identically zero for all s and t.)

THEOREM 4. *Let $v(x)$ be absolutely continuous on $[\alpha,\beta]$ with $v'(x) \geq 0$ a.e. Also, suppose that $Q(x)$ is nonnegative a.e. and measurable on $[\alpha,\beta]$, and $G(v,x)$*

is continuously differentiable for $x \in [\alpha, \beta]$ *and* v *in the range of the function* $v(x)$, *with* $G_v(v, x) \geq 0$ *(or* $G_v(v, x) > 0$ *in case* $p < 1$*). Then if the integrals exist*

$$\int_\alpha^\beta \left\{Qv'^p + (p-1)(G_v)^{p/(p-1)}Q^{-1/(p-1)} + pG_x\right\} dx \tag{9.7}$$
$$\geq p\{G(v(\beta), \beta) - G(v(\alpha)\alpha)\} \qquad (p > 1 \text{ or } p < 0),$$

$$\int_\alpha^\beta \left\{Qv'^p + (p-1)(G_v)^{p/(p-1)}Q^{-1/(p-1)} + pG_x\right\} dx \tag{9.8}$$
$$\leq p\{G(v(\beta), \beta) - G(v(\alpha), \alpha)\} \qquad (0 < p < 1),$$

with the same abbreviations as in Theorem 1. *Equality in both* (9.7) *and* (9.8) *holds if and only if the differential equation*

$$v' = (G_v/Q)^{1/(p-1)} \tag{9.9}$$

is satisfied almost everywhere.

PROOF: (The following proof illustrates how Benson's method works.) By taking $s = v'Q^{1/p}$, $t = (G_v)^{1/(p-1)}Q^{-1/(p(p-1))}$ in (9.5) we have, almost everywhere

$$Qv'^p + (p-1)(G_v)^{p/(p-1)}Q^{-1/(p-1)} \geq pv'G_v.$$

That is,

$$Qv'^p + (p-1)(G_v)^{p/(p-1)}Q^{-1/(p-1)} + pG_x \geq p\frac{d}{dx}G(v, x),$$

proving (9.7) by integrating both sides of the above inequality from α to β.

The proof of (9.8) follows from the above argument, but using (9.6) instead of (9.5).

Using these results, Shum obtained various generalizations of Opial's inequalities (for some of his generalizations see Chapter III. and generalizations of Bomieri's inequality (27.2) from Chapter IV.

D. C. Benson **[22]** extended a method of his own to comparison theory. So, he proved:

THEOREM 5. *Let* $p(x)$, $p_1(x)$, $q(x)$ *and* $q_1(x)$ *be continuous on* $[c, d]$ *with* $p > 0$, $p_1 > 0$. *On* $[c, d]$, *let* $y(x)$ *and* $y_1(x)$ *be solutions of*

$$y'^2 - 2py = -pq,$$
$$y_1'^2 - 2p_1y_1 = -p_1q_1,$$

respectively, with $y_1' > 0$ *on* (c,d). *Taking* C *to be a positive constant, let* $x_1(t)$ *be the inverse function of*

$$F(x) = C \exp \int \frac{p_1(x)}{y_1'(x)} dx$$

and let a *and* b *satisfy* $x_1(a) = c$, $x_1(b) = d$, *with* $a, b > 0$. *Then the following inequality holds:*

$$\frac{1}{2} \int_a^b \left(\left(\frac{dx_1}{dt} \right)^2 (p_1(x_1) - p(x_1)) + t^{-2} (q(x_1) - q_1(x_1)) \right) dt$$
$$\leq \frac{y_1(d) - y(d)}{b} - \frac{y_1(c) - y(c)}{a}.$$

Equality holds here only if $p = p_1$ *and* $q = q_1$.

REFERENCES

1. BEESACK, P.R., D.S. MITRINOVIĆ and P.M. VASIĆ, *Integral inequalities*, (the manuscript not achieved and not published).
2. EWING, G.M., "Calculus of Variations with Applications," New York, 1969.
3. YOUNG, L.C., "Calculus of Variations and Optimal Control Theory," Philadelphia-London-Toronto, 1969.
4. HARDY, G.H., J.E. LITTLEWOOD, and G. PÓLYA, "Inequalities," Cambridge, 1934, 1952.
5. FLORKIEWICZ, B. and A. RYBARSKI, *Some integral inequalities of Sturm-Liouville type*, Colloq. Math. **36** (1976), 127–141.
6. BEESACK, P.R., *Integral inequalities of Wirtinger type*, Duke Math. J. **25** (1958), 477–498.
7. ―――――, *Integral inequalities involving a function and its derivative*, Amer. Math. Monthly **78** (1971), 705–741.
8. KAMKE, E., "Differentialgleichungen, Lösungsmethoden und Lösungen," 3rd ed., Berlin, 1944.
9. REID, W.T., "Ordinary Differential Equations," New York, 1971.
10. EVERITT, W.N., *Spectral theory of the Wirtinger inequality*, in Lecture Notes in Mathematics **564**, Berlin, 1976, pp. 93–105.

11. KRYLOV, N.M., *On some inequalities which will be established in the exposition of the Schwartz-Poincaré-Steklov method and are also encountered in solving many minimization problems*, Zapiski Gornago Institute **2** 6 (1915), 10–14 (Russian).
12. BEESACK, P.R., *Elementary proofs of the extremal properties of the eigenvalues of the Sturm-Liouville equation*, Canad. Math. Bull. **3** (1960), 59–77.
13. HARTMAN, P., "Ordinary Differential Equations," Baltimore, 1973.
14. CODDINGTON, P. and N. LEVINSON, "Theory of Ordinary Differential Equations," New York, 1955.
15. COURANT, R. and D. HILBERT, "Methods of Mathematical Physics," Vol. 1, New York, 1953.
16. BENSON, D. C., *Inequalities involving integrals of functions and their derivatives*, J. Math. Anal. Appl. **17** (1967), 292–308.
17. MITRINOVIĆ, D.S., "Analytic Inequalities," Berlin-Heidelberg-New York, 1970.
18. SHUM, D.T.T., *Integral inequalities using Benson's method*, Doctoral dissertation, Carleton University, Ottawa, 1971, 212 pages.
19. BEESACK, P.R., *Extensions of Wirtinger's inequality*, Trans. Roy. Soc. Canada **53** 3 (1959), 21–30.
20. __________, *Hardy's inequality and its extensions*, Pacific J. Math. **11** (1961), 39–61.
21. BECKENBACH, E.F. and R. BELLMAN, "Inequalities," Berlin, 1965.
22. BENSON, D.C., *Comparison theory for* ${y'}^2 - 2p(x)y = -p(x)q(x)$, SIAM J. Math. **21** (1971), 279–286.

CHAPTER XVIII

PARTICULAR INEQUALITIES

1. This Chapter is devoted to various unconnected results which do not easily relate to the types we have presented in the previous chapters. A derivative or integral of a function of one or two variables appear in each inequality of this Chapter.

2. If $f(x) \in L(-\infty, +\infty)$ and for a positive b the inequality

$$\left| \frac{1}{2\pi} \int_{-\infty}^{+\infty} f(t) e^{-ixt} dt \right| \leq \frac{e^{-bx^2}}{1 + 4\pi^2 x^2}$$

holds on the real x-axis, then for all positive integers k the inequality

$$\left| \frac{1}{2\pi} \int_{-\infty}^{+\infty} f(t)^k e^{-ixt} dt \right| \leq \frac{e^{-bx^2/k}}{1 + 4\pi^2 x^2}$$

holds on the whole real axis. This is a result of Turán [**1**].

3. Let $f(x)$ be a real-valued integral function on $[0, \pi]$ which is subject to conditions:

1^0 $0 \leq f(x) \leq 1$,

2^0 $\int_0^\pi f(x) \cos x \, dx = 0$ and

3^0 there exists a number λ, with $0 \leq \lambda \leq 2\pi/3$, such that

$$f(x) = 0 \quad \text{for } \lambda \leq x \leq \lambda + \frac{\pi}{3}.$$

Then we have

$$\int_0^\pi f(x) \sin x \, dx \leq 1,$$

equality holding if and only if

$$
\begin{aligned}
f(x) &= 0 \quad \text{for } \lambda \le x \le \lambda + \frac{\pi}{3}, \\
f(x) &= 0 \quad \text{for } \left|\lambda + \frac{\pi}{6} - x\right| \ge \frac{\pi}{2}, \\
f(x) &= 1 \quad \text{otherwise}
\end{aligned}
$$

(or any function that differs from it on a set of measure zero.)

REMARK: The above statement is due to H. Rademacher, [**2**], and two different proofs of this statement were given by M. Kraft and G. Szegö.

4. Let $f(t)$ be a real-valued integrable and 2π- periodic function. Consider the n-th Fourier sum

$$f_n(t) = \frac{1}{\pi} \int_{-\pi/2}^{\pi/2} f(t+2z) \frac{\sin(2n+1)z}{\sin z}\, dz.$$

Let, further, $M[g]$ denote upper bound of $|g|$ and A, B, C, A_1, B_1 absolute constants.

Then we have the following Lebesgue inequality [**3**]:

$$M[f_n] \le A \cdot M[f] + B \log n \cdot M\left[f\left(t + \frac{1}{n}\right) - f(t)\right]. \tag{4.1}$$

Assume now that the functions P and Q_1, defined by

$$
\begin{aligned}
P(x) &= \int_{-1}^{x} \frac{p(u)}{\sqrt{1-u^2}}\, du \quad \text{with } p(u) \ge 0, \\
Q_1(x) &= \int_{-1}^{x} \frac{dQ(u)}{\sqrt{1-u^2}} \quad \text{with } dQ(u) \ge 0,
\end{aligned}
$$

satisfy the following equalities

$$m_k = \int_{-1}^{1} u^k \frac{p(u)}{\sqrt{1-u^2}} du = \int_{-1}^{1} u^k \frac{dQ(u)}{\sqrt{1-u^2}},$$

where $k = 0, 1, \ldots, 2n-1$.

Put $f(t) = p(\cos t)$. Then, for $k = 0, 1, \ldots, 2n-1$,

$$\int_{-1}^{1} u^k \frac{p(u)du - f_n(t)du + M[f_n]du}{\sqrt{1-u^2}} = M[f_n] \int_{-1}^{1} u^k \frac{du}{\sqrt{1-u^2}}, \tag{4.2}$$

$$\int_{-1}^{1} u^k \frac{dQ(u) - f_n(t)du + M[f_n]du}{\sqrt{1-u^2}} = M[f_n] \int_{-1}^{1} u^k \frac{du}{\sqrt{1-u^2}}. \tag{4.3}$$

From (4.2) and (4.3) we obtain, for $-1 \leq x \leq 1$,

$$\left| \int_{-1}^{x} \frac{p(u)du - f_n(t)du}{\sqrt{1-u^2}} \right| < \frac{C}{2} \frac{M[f_n]}{n}, \tag{4.4}$$

$$\left| \int_{-1}^{1} \frac{dQ(u) - f_n(t)du}{\sqrt{1-u^2}} \right| < \frac{C}{2} \frac{M[f_n]}{n}. \tag{4.5}$$

Inequalities (4.4) and (4.5) yield

$$\left| \int_{-1}^{x} \frac{dQ(u)}{\sqrt{1-u^2}} - \int_{-1}^{x} \frac{p(u)du}{\sqrt{1-u^2}} \right|$$
$$< CM[f_n] = \frac{A_1}{n} M[p(x)] + \frac{B_1 \log n}{n} M\left[p\left(x + \frac{1}{n}\right) - p(x) \right].$$

REMARK: The Lebesgue inequality (4.1) can be found in [**3**], pp. 19–23. The above results are due to M. Krawtchouk [**4**].

5. Another inequality involving $\sin t/t$ is given by P. Masani, I. Schoenberg and F. Steutel [**5**].

$$\frac{2}{\pi} \int_{x}^{xp} \left(\frac{\sin t}{t} \right)^2 dt \leq 1 - \frac{1}{p} \quad \text{if } x \geq 0,\ p > 1.$$

6. Let $f(t)$ be a real-valued differentiable function on $u \leq t \leq v$. Suppose $f'(t)$ is monotonic and that $|f'(t)| > \lambda > 0$ for $u \leq t \leq v$. Then

$$\left| \int_{u}^{v} \exp[i f(t)]dt \right| < \frac{1}{\lambda}.$$

This lemma of Van der Corput is used in [6] in the proof of the following result:

Let $a_1 < a_2 < \cdots < a_n$ be fixed nonnegative real numbers and let $b_1, \ldots, b_n$ be real numbers. Then $\left|\int_{-\infty}^{+\infty} \exp\{i(b_1[x]^{a_1} + \cdots + b_n[x]^{a_n}\}\frac{dx}{x}\right| \le K(a_1, \ldots, a_n)$, where K does not depend on $b_1, \ldots, b_n$, and $[x]$ is the integer part of x.

7. A result of A. M. Fink [7]:

If $f(x) = \int_0^x (x-t)^n d\mu(t)$ for some non-negative measure μ, then

$$\frac{f(t)^{n+1}}{(n+1)^n} \le \int_0^t y(u_1) \int_0^{u_1} y(u_2) \ldots \int_0^{u_{n-2}} y(u_{n-1}) \int_0^{u_{n-1}} y(u_n) y^{(n)}(u_n) d\mu_n \ldots d\mu_1$$
$$\le f(t)^{n+1}.$$

8. Some inequalities are of the Tauberian type, for example a result of V. Avakumović and J. Karamata [8].

Let $f(t)$ be a nonnegative nondecreasing function for $t \ge 0$. Further let the integral

$$J(\theta) = \int_0^{+\infty} e^{-\theta t} f(t) dt \quad \text{be convergent for } \theta > 0.$$

Then the inequalities

$$mx^k e^x \le J(x) \le Mx^k e^x,$$

where k, m, M are positive constants, imply

$$\exp\left(2\sqrt{t} - a^3\sqrt{t}\sqrt{\log t}\right) < f(t) < Mt^{(k-1)/2} \exp\left(2\sqrt{t}\right)$$

for $a > 2$ and $t > t(a)$.

These inequalities are best possible.

9. Let $P_n(t)$ be the normalized Legendre polynomial of order n. Then

$$\left|\int_{-1}^{x} P_n(t) dt\right| \le Cn^{-1}, \quad -1 \le x \le 1.$$

This inequality is due to L. Fejér [**9**]. A generalization of this inequality was given by S. Z. Rafal'son [**10**].

10. Inequalities for the incomplete gamma function are given by H. R. van der Vaart [**11**, pp. 442–443].

Consider the incomplete gamma function $\gamma(a,x) = \int_0^x e^{-t}t^{a-1}\,dt$. As $e^{-a}a^a > \int_a^{a+1} e^{-t}t^a\,dt$, we obtain

$$a\int_0^a e^{-t}t^{a-1}\,dt = e^{-a}a^a + \int_0^a e^{-t}t^a\,dt > \int_0^{a+1} e^{-t}t^a\,dt,$$

whence

$$a\gamma(a,a) > \gamma(a+1,a+1).$$

Therefore, for $a > 0$ we have

$$\frac{\gamma(a,a)}{\Gamma(a)} > \frac{\gamma(a+1,a+1)}{\Gamma(a+1)},$$

where Γ is the gamma function.

The following inequality

$$\frac{\gamma(a,a)}{\Gamma(a)} > \frac{1}{2}$$

is valid for $a > 0$.

11. Consider $f(x) = e^{-x^2}P(x)$, where P is a real polynomial of degree n. If $|f(x)| \le L$ (L is a positive constant) for $-\infty < x < +\infty$, then

$$|f'(x)| < (1+\varepsilon(x))\cdot 4e|x|L \quad \text{for } -\infty < x < +\infty,$$

where $\lim_{|x|\to+\infty} \varepsilon(x) = 0$.

This statement is due to S. Bernstein [**12**]. W. E. Milne [**13**] proved the following result:

$$|f'(x)| < (1+\varepsilon(n))\cdot 1.0951\sqrt{n}\,L, \tag{11.1}$$

with $\lim_{n\to+\infty} \varepsilon(n) = 0$ and L defined above.

The proof of (11.1) is based upon certain properties of solutions of Weber's differential equation

$$u''(x) + \left(m + \frac{1}{2} - \frac{x^2}{4}\right) u(x) = 0.$$

12. Here is a probem posed in the SIAM Review, April 1973. The proof is by the proposer, Robert Shafer.

If a, b, c, m, n denote real numbers with $a, b > 0$, $1 \geq m/n > 0$, show that

$$\int_0^\infty z^{c-1} \exp\{-az - bz^{-m/n}\} dz$$

$$\geq 2p^c \left\{\frac{m}{2n-m}\right\}^{q/2} e^{(c-q)n/m} \cdot K_q \left\{\frac{ap}{n^2}(m+n)\sqrt{n(2n-m)}\right\},$$

where

$$q = c + ap\left(1 - m^2/n^2\right), \qquad ap = \{a^m b^n m^n / n^n\}^{1/(m+n)},$$

and K_q denotes a modified Bessel function.

PROOF: Let $\alpha = m/n$. Note that for $z > 0$ and $0 < \alpha \leq 1$, the maximum of $\exp\{-az - bz^{-\alpha}\}$ occurs at $z = p$ where p satisfies $bp^{-\alpha} = ap/\alpha$. We shall show that for $z > 0$ and $0 < \alpha \leq 1$,

$$(12.1) \qquad \begin{aligned} z^{-\alpha} \leq p^{-\alpha} \{&(1-\alpha^2) - \alpha(1-\alpha^2)\log(z/p) \\ &+ \frac{1}{2}\alpha^2(1-\alpha)z/p + \frac{1}{2}\alpha^2\,(1+\alpha)p/z\}, \end{aligned}$$

from which the stated result will readily follow. Letting $u = \log(z/p)$, we transform inequality (12.1) to the equivalent inequality

$$(12.2) \qquad \exp(-\alpha u) \leq (1-\alpha^2) - \alpha(1-\alpha^2)u + \alpha^2\cosh(u) - \alpha^3\sinh(u).$$

For $-\infty < u \le 0$, (12.2) can be verified by expanding both sides in powers of u and using the fact that $0 < \alpha \le 1$. For $0 < u < \infty$, the inequality is somewhat more difficult to prove. First, we note that for all real s,

$$\exp(-\alpha s) \le \cosh(s) - \alpha \sinh(s),$$

or equivalently,

$$\exp(-\alpha s) \le \left(\frac{1-\alpha}{2}\right)\exp(s) + \left(\frac{1+\alpha}{2}\right)\exp(-s),$$

which follows because the exponential function is convex. It then follows that

$$\int_0^u dt \int_0^t ds \exp(-\alpha x) \le \int_0^u dt \int_0^t ds\,(\cosh(s) - \alpha\sinh(s)),$$

from which we obtain (12.2).

From (12.1) we obtain the inequality

$$\begin{aligned}&\int_0^\infty z^{c-1}\exp\{-az - bz^{-a}\}dz \\ &\geq p^{c-q}\exp\{(c-q)/\alpha\}\int_0^\infty z^{q-1}\exp\{-Az - B/z\}dz,\end{aligned} \tag{12.3}$$

where $A = a(2-\alpha)(1+\alpha)/2$, $B = ap^2\alpha(1+\alpha)/2$, and $q = c + ap(1-\alpha^2)$. From an integral representation of the modified Bessel function K_q, [14], we obtain

$$\int_0^\infty z^{q-1}\exp\{-Az - B/z\}dz = 2(B/A)^{q/2}K_q\left(2\sqrt{AB}\right). \tag{12.4}$$

Combining (12.3) and (12.4) with $A = a(2-m/n)(1+m/n)/2$ and $B = ap^2(m/2)(1+m/n)/2$, we obtain the stated result.

13. If p is a positive integer and if we put

$$S_p = \int_0^1 e^{-x}\frac{x^{p-1}(x-1)^p}{(p-1)!}dx,$$

then

$$|S_p| \le \frac{1}{2^{2p-2}(p-1)!}, \tag{13.1}$$

$$|S_p| \le \frac{1}{p\binom{2p}{p}(p-1)!} < \frac{2+(1/p)}{2^{2p}(p-1)!}. \tag{13.2}$$

REMARK: In order to prove (13.2), consider $\int_0^1 x^m(1-x)^n\,dx$, where m and n are natural numbers. See Ostrowski [**15**, p. 96].

14. If $a > 0$ and $b > 0$, then with $t > 0$

$$0 < \frac{\sqrt{\pi}}{2} - \sqrt{bt}\int_0^a \exp\left(-btx^2\right)dx < \frac{\exp\left(-a^2bt\right)}{\sqrt{bt}}.$$

See [**15**, p. 303].

15. If f is a nonnegative and nondecreasing function in $I = (a, b)$ where $-\infty \le a < b \le +\infty$, and if g and k have integrable derivatives in I with

$$g(u) \ge 0, \quad g'(u) \ge 0, \quad \lim_{u\to b} k(u) = 0, \quad k'(u) \le 0, \tag{15.1}$$

then

$$\int_a^b f(u)g'(u)k(u)du \le -\int_a^b \left(\min_{u\le q<b} f(q)g(q)\right) k'(u)du.$$

PROOF: For each pair of numbers $q, u \in I$ such that $q \ge u$, we have

$$f(q)g(q) \ge f(q)\int_a^q g'(t)dt \ge \int_a^q f(t)g'(t)dt \ge \int_a^u f(t)g'(t)dt,$$

and

$$\min_{u\le q<b} f(q)g(q) \ge \int_a^u f(t)g'(t)dt.$$

Hence,

$$-\int_a^b \left(\min_{u\le q<b} f(q)g(q)\right) k'(u)du \ge -\int_a^b k'(u)du \int_a^u f(t)g'(t)dt$$

$$= -\int_a^b f(t)g'(t)dt \int_t^b k'(u)du = \int_a^b f(t)g'(t)k(t)dt.$$

REMARK: If f is a nonnegative and nonincreasing function in I, and if condition $\lim_{u\to b} k(u) = 0$ in (15.1) is replaced by $\lim_{u\to a} k(u) = 0$, then

$$-\int_a^b \left(\max_{a<q\le u} f(q)g(q)\right) k'(u)du \le \int_a^b f(u)g'(u)k(u)du.$$

The above results are due to J. G. van der Corput [16]. Paper [16] contains also a number of similar inequalities.

16. Suppose g is continuously differentiable positive function defined in some interval $[t_0, +\infty]$. Let a, b, c, x, y, z and w be numbers with

$$0 < a \le b, \quad t_0 \le x \le y \le z \le w$$

such that

$$\int_y^z \frac{dt}{g(t)} = a, \quad \int_x^w \frac{dt}{g(t)} \le b,$$

and

$$|g'(t)| \le c \quad \text{for } t \in [t_0, +\infty).$$

If h is defined by

$$h(a,b,c) = \left\{\frac{a}{c}\left(\frac{1}{a}+c\right)^2\left(\exp(bc-ca)-\frac{1}{a^2}\right)\right\}^2\left(\frac{1}{a}+c\right)a,$$

then we have

$$\int_x^w g(t)dt \leq h(a,b,c) \int_y^z g(t)dt.$$

See J. Walter [**17**].

17. We give an inequality that is often used. A proof can be found in K. Tatarkiewicz [**18**].

If A_1 and A_2 are two disjoint measurable sets with $A = A_1 \cup A_2$, if f_1, f_2 and g are functions such that f_1, f_2, gf_1 and gf_2 are integrable on A, the inequalities $f_1(p_1) \leq f_2(p_1)$, $f_2(p_2) \leq f_1(p_2)$, $g(p_1) \leq g(p_2)$ hold for all $p_1 \in A_1$ and $p_2 \in A_2$ and $\int_A f_2 = \int_A f_1$, then

$$\int_A f_2 g \leq \int_A f_1 g.$$

18. Let functions $f_i(x)$ $(i = 1,2)$ have integral representations

$$f_i(x) = \int_0^\infty K_i(x,t)d\sigma(t),$$

where $K_i(x,t) \geq 0$, $\sigma(t)$ is an arbitrary nondecreasing function, $x \in R^+$, $t \in R^+$. Then for nondecreasing functions g and h

$$\int_0^\infty f_1(x)dg(x) \leq \int_0^\infty f_2(x)dh(x)$$

iff for every $t \in (0,\infty)$ we have

$$\int_0^\infty K_1(x,t)dg(x) \leq \int_0^\infty K_2(x,,t)dh(x).$$

See V. N. Volkov [**19**].

19.

1. If a twice differentiable real function f satisfies the inequalitity $f(x)f''(x) < 0$ for $x \in (0,1)$, then f satisfies at least one of the following inequalities

$$|f(x)| > px \quad \text{or} \quad |f(x)| > p(1-x)$$

where $p = |f(1^-) - f(0^+)|$.

2. If f is a differentiable real nonzero function in $(0,1)$, and if at least one of the following inequalities $|f'(x)| > px^r$ or $|f'(x)| > p(1-x)^r$, with $p, r > 0$ is satisfied in $(0,1)$, then at least one of the following inequalities

$$|f(x)| > \frac{p}{r+1}x^{r+1} \quad \text{or} \quad |f(x)| > \frac{p}{r+1}(1-x)^{r+1}$$

also holds on $(0,1)$.

3. If f is an n-times differentiable function in (a,b), if $\lim_{x\to a} f(x) = \lim_{x\to b} f(x) = 0$, and if

$$|f^{(n)}(x)| \geq m|f(x)| > 0 \quad (m \text{ is a positive constant}),$$

then

$$b - a < Am^{-1/n},$$

where A does not depend on n.

See I. G. Mikusiński [**20**].

20. Let a function $f(x)$ be defined on $[a,b]$. We shall say that f belongs to the class $Lip(\alpha, p)$ if

$$\left(\int_a^b |f(x+h) - f(x)|^p dx\right)^{1/p} \leq Mh^\alpha,$$

where $h > 0$ is an arbitrary number, M, p, α are fixed numbed independent of h, such that $p \geq 1$, $0 < \alpha \leq 1$. (In $R \setminus [a,b]$ we suppose $f(x) \equiv 0$).

The following results are obtained by G. H. Hardy and J. E. Littlewood [**21**]: If $f(x)$ is from $Lip(\alpha, p)$ then :

1) for $\alpha p > 1$, $f(x)$ is equivalent to a function which belongs to the class of Lipschitz functions of order $\alpha - 1/p$;
2) for $\alpha p \leq 1$, $f(x) \in Lip\left(\alpha - \frac{1}{p} + \frac{1}{q}; q\right)$, where q is an arbitrary number such that $p \leq q < p/(1-\alpha p)$.

A new proof of this result is given by V. P. Il'in [**22**].

21. A result of A. M. Ostrowski [**23**]:

The inequality

$$\int_0^1\int_0^1 \left|\frac{f(x)-f(y)}{x-y}\right|^\lambda dx dy \leq (\log 4)\int_0^1 |f'(x)|^\lambda dx \quad (\lambda \geq 1),$$

is valid if the righthand integral exists as a Lebesgue integral, and the constant $\log 4$ cannot be improved for $\lambda = 1$.

22. A. F. Timan [**24**] determines explicitly the best bound for the difference between a continuous function of period 2π and its Poisson integral

$$f(r,h) = \frac{1}{2\pi}\int_{-\pi}^{\pi} f(x+t)\frac{1-r^2}{1-2r\cos t + r^2}dt,$$

when

$$|f(x+h) - f(x)| \leq M|h|$$

or when $f(x)$ has a pth derivative with a prescribed bound. In the first case the result is, for $0 \leq r < 1$,

$$|f(x) - f(r,h)| \leq \frac{2M}{\pi}\left((1-r)\log\frac{1}{1-r} + \int_0^{1-r}\left(\frac{1}{1-t}\log\frac{2-t}{t^t}+1\right)dt\right).$$

Some related results are given by I. P. Natanson [**25**].

23. Let $S(k,n)$ denote the class of functions f which are continuous on $[0,1]$ such that:

1^0 f is $k-1$ times differentiable (continuous for $k=1$);

2^0 there is a subdivision of $[0,1]$ into at most n intervals I_j $(j=1,\dots,m;$ $m \neq n)$; $\bigcup_{j=1}^{m} I_j = [0,1]$, so that the restriction of f to I_j is a polynomial of degree at most k.

For $k=0$, $f \in S(0,n)$ is a step function taking a constant value at each interval I_j $(j=1,\dots,m;\ m \leq n)$.

For an arbitrary continuous function g, let

$$\varepsilon_n^{(k)}(g) = \min_{f \in S(k,n)} \|f-g\|_{C[a,b]}.$$

Starting from these assumptions the following results was proved by G. Freud and V. Popov [**26**].

Let $f \in C[0,1]$ be absolutely continuous, and for a function $F_n \in S(k,n)$ let

$$\int_0^1 |f'(x) - F_n(x)|dx < \eta_n;$$

then there exists an integer $r(k)$ depending neither on n nor on f so that

$$\varepsilon_{rn}^{(k+1)}(f) \leq \frac{\eta_n}{n}.$$

24. Let $f_1, f_2, \dots$ be an infinite sequence of real measurable functions on $[0,1]$, such that

$$0 \leq f_n(x) \leq 1 \qquad (x \in [0,1];\ n = 1,2,\dots).$$

Let a be an accumulation point of the numbers $\int_0^1 f_n(x)dx$ $(n=1,2,\dots)$. Then for any $\varepsilon > 0$, there exists two indices p and q $(p \neq q)$ such that

$$\int_0^1 |f_p(x) - f_q(x)|dx < 2a(1-a) + \varepsilon.$$

Let $f_1, f_2 \ldots$ be an infinite sequence of real measurable functions on $[0,1]$ with the properties

$$0 \leq f_n(x) \leq 1 \qquad (x \in [0,1];\ n = 1,,2,\ldots),$$

and

$$\int_0^1 f_n(x)dx \geq a \qquad (n = 1,2,\ldots).$$

Then given $\varepsilon > 0$ and a natural number m, there exists an infinite subsequence $f_{n_1}, f_{n_2}, \ldots,$ with the property that for any $r = 1,2,\ldots m$

$$\int_0^1 f_{n_{k_1}}(x)\ldots f_{n_{k_r}}(x)dx > a^r - \varepsilon$$

for arbitrary indices $k_1, \ldots, k_r$ no two of which are equal.

These two results appear in C. Visser [**27**].

25. Let $f(x)$ and $f'(x)$ be continuous functions on $[a,b]$ and let f'' exist and be absolutely bounded, i.e. $|f''(x)| \leq m$. Then, for any $x \in [a,b]$ we have

$$\frac{f(b)-f(a)}{b-a} = f'(x) + \frac{m\theta}{2}(b-a) \quad \text{with } |\theta| \leq 1,$$

and therefore, in particular,

$$\min_{a \leq x \leq b} f'(x) + \frac{b-a}{2} \geq \frac{f(b)-f(a)}{b-a}$$
$$\geq \max_{a \leq x \leq b} f'(x) - \frac{b-a}{2}m.$$

See Ostrowski [**28**].

26. Let $f(x)$ be a differentiable function on $[a,b]$. Let σ denote the oscillation of f on $[a,b]$, and let, for a positive constant m,

$$\left|\frac{f'(x)-f'(y)}{x-y}\right| \leq m \quad \text{if } x, y \in [a,b],\ x \neq y.$$

Then, for $x \in [a,b]$,

$$|f'(x)| \le \frac{\sigma}{b-a} + m\frac{b-a}{2}.$$

Under the same assumption we have

$$\max_{x \ge y} (f(x) - f(y)) \ge (b-a) \max_{a \le x \le b} f(x) - m\frac{b-a}{2},$$

when $a \le y \le x \le b$.

If $b - a \ge \sqrt{2\sigma/m}$, then

$$|f'(x)| \le \sqrt{2m\sigma}.$$

See A. M. Ostrowski [**29**].

27. Here is a result of J. G. van der Corput [**30**].

If the first and second derivatives of the function f exist on $[a,b]$, with $f'(x) \neq 0$ in that interval, and if $f''(x)/f'(x)$ is a nonincreasing function, then

$$|f(x) - f(b)| \ge |a-b|\sqrt{f'(a)f'(b)}.$$

28. Let $f(x)$ be a real 2π-periodic function, absolutely continuous in $[0,2\pi]$ and let $f' \in L^2(0,2\pi)$. Then for all x and y,

$$|f(x)^2 - f(y)^2| \le 2\pi \int_0^{2\pi} f(t)^2 dt \int_0^{2\pi} f'(t)^2 dt. \tag{28.1}$$

This inequality is due to S. E. Warschawski [**31**], p. 18, and is used and proved also in [**32**], pp. 68–69.

Proof: We may take $x, y \in [0, 2\pi]$ and $y < x$. Since $f(x)^2$ is also absolutely continuous, we have

$$f(x)^2 - f(y)^2 = 2\int_y^x f(t)f'(t)dt = -2\int_x^{y+2\pi} f(t)f'(t)dt.$$

This implies that

$$|f(x)^2 - f(y)^2| \le 2 \int_y^x |f(t)||f'(t)|dt \tag{28.2}$$

and

$$|f(x)^2 - f(y)^2| \le 2 \int_x^{y+2\pi} |f(t)||f'(t)|dt. \tag{28.3}$$

By using the Schwarz inequality, we obtain from (28.2) and (28.3)

$$\begin{aligned} |f(x)^2 - f(y)^2| &\le \frac{1}{2}\left(2\int_y^x |f(t)||f'(t)|dt + 2\int_x^{y+2\pi} |f(t)||f'(t)|dt\right) \\ &= \int_0^{2\pi} |f(t)||f'(t)|dt \\ &\le \left(\int_0^{2\pi} f(t)^2 dt\right)^{1/2} \left(\int_0^{2\pi} f'(t)^2 dt\right)^{1/2}. \end{aligned}$$

This finishes the proof of (28.1).

Using the Cauchy-Schwarz inequality directly to (28.2) one gets

$$\left|f(x)^2 - f(y)^2\right|^2 \le 4 \int_0^{2\pi} f(t)^2 dt \int_0^{2\pi} f'(t)^2 dt, \qquad [\mathbf{15}, \text{p. } 118].$$

REMARK: Let f be a real-valued function 2π-periodic, absolutely continuous on $[0, 2\pi]$ and $f' \in L^q[0, 2\pi]$, with $q > 1$. Suppose that $f(y) = 0$ for a value $y \in [0, 2\pi]$. Let further $1/p + 1/q = 1$ and let $a > 1$ be arbitrary. Then as above

$$\begin{aligned} |f(x)|^a &\le \frac{a}{2} \int_0^{2\pi} |f(t)|^{a-1} |f'(t)| dt \\ &\le \frac{a}{2} \left(\int_0^{2\pi} |f(t)|^{(a-1)p} dt\right)^{1\varphi} \left(\int_0^{2\pi} |f'(t)|^q dt\right)^{1/q}. \end{aligned}$$

For $a = p = q = 2$ we have

$$|f(x)|^2 \le 2\pi \int_0^{2\pi} |f(t)|^2 dt \int_0^{2\pi} |f'(t)|^2 dt. \tag{28.4}$$

The constant 2π in (28.4) is best possible. In order to prove this fact consider

$$f(x) = e^{-c|x|} - \frac{1}{\pi c}\left(1 - e^{-\pi c}\right) \quad (c \text{ is a real constant})$$

and let $c \to +\infty$.

29. If f, a non-negative Lebesgue measurable function on $[0, \infty)$, and $p \ge 0$ are such that $\int_0^T t^{-p} f(t)dt = \infty$ for all $T > 0$ then for each $q > p$ there is a sequence $t_n > 0$ converging to zero with

$$t_n f(t_n) \le q \int_0^{t_n} f(t)dt.$$

See D. Bindschadler [**33**].

30. (a) Let Ψ be positive and monotone increasing and let w be positive on $(0, A)$ where $0 < A \le +\infty$ with $w(x) \le hx$ for $0 < x < A$ and some constant $h > 0$. If $0 < q < p < \infty$, then there exists an absolute constant $M = M(h, p, q) > 0$ such that for all positive, measurable,, subadditive functions ϕ on $(0, A)$ we have

$$\left(\int_0^A \left(\frac{\phi}{\Psi}\right)^p \frac{dx}{w}\right)^{1/p} \le M \left(\int_0^A \left(\frac{\phi}{\Psi}\right)^q \frac{dx}{w}\right)^{1/q}. \tag{30.1}$$

(b) Moreover, if Ψ is positive and monotone decreasing on $(0, A)$ and there are constants c, k such that $0 < c < 1/2$, $k > 0$, and $\Psi(cx) \le k\Psi(x)$ for $0 < x < A$, then the inequality (30.1) is still valid for some $M = M(c, k, h, p, q) > 0$.

A special case of these results is:

Let w be positive on $(0, A)$, where $0 < A < \infty$ and satisfy $w(x) \leq hx^b$ for some constants $h > 0$, $b > 1$. Let $0 < q < p < \infty$ and Ψ, ϕ satisfy the previous hypotheses. Then the inequality (30.1) holds with

$$M = \left(2^q h A^{b-1} / \log 2\right)^{1/q-1/p} \qquad \text{in case (a),}$$
$$M = \left((2k)^q h A^{b-1} / \log\left(2(1-c)\right)\right)^{1/q-1/p} \qquad \text{in case (b).}$$

These results are generalizations of some results of R. P. Gosselin and F. C. Hsiang, [**34**]–[**38**], and they are given by P. R. Beesack in [**39**], [**40**].

31. Let M be a real continuously differentiable function of $[a, b]$ with $M' > 0$. If z is an absolutely continuous real function on $[a, b]$ with $z(a) = z(b) = 0$ and $z' \in L^2(a, b)$, then

$$\int_a^b M' z^2 dx \leq 2 \int_a^b |M||zz'| dx \leq 4 \int_a^b \frac{M^2 (z')^2}{M'} dx.$$

This result was obtained by D. B. Hinton and R. T. Lewis [**41**].

32. If $a < b$ are real numbers and f is a continuous complex valued function on $[a, b]$, with continuous first derivative in (a, b), then, for any u in $[a, b]$,

$$f(u) = \frac{1}{b-a} \int_a^b f(x) dx + \int_u^b \frac{x-b}{b-a} f'(x) dx + \int_a^u \frac{x-a}{b-a} f'(x) dx;$$

hence

$$|f(u)| \leq \frac{1}{b-a} \int_a^b |f(x)| dx + \int_a^b |f'(x)| dx.$$

See H. Davenport,[**42**].

33. If $f^{(n)}(x) \geq A > 0$ on an interval I of length $2h$ $(h > 0)$, and if $M = \max_{x \in I} |f(x)|$, then

$$M \geq \frac{2A(h/2)^n}{n!}.$$

See Boas, [**43**].

34. Let f be a 2π-periodic function which is continuous and absolutely continuous over a period with $f' \in L^p[0, 2\pi]$, where $1 < p \leq 2$. If there exists an integer $m > 1$ such that the Fourier coefficients a_k and b_k of f satisfy

$$a_k = 0 \text{ for } k = 0, 1, \ldots, m-1 \quad \text{and } b_k = 0 \text{ for } k = 1, \ldots, m-1,$$

then

$$|f(x)| \leq \left(\frac{q}{\pi}\right)^{1/p} \|f'\|_p m^{1/q} \quad \left(\frac{1}{p} + \frac{1}{q} = 1\right). \tag{34.1}$$

This result is due to W. A. J. Luxemburg [**44**].

REMARK: Inequality (34.1) was improved by A. A. Jagers. Namely he proved that

$$|f(x)| \leq 2^{-1/p} \|f'\|_p m^{-1/q}, \tag{34.2}$$

with the condition that the assumptions made above for (34.1) remain valid for (34.2).

35. Let $f(t) = e^{it}/(e^{it} - a)$,, with t real and a complex ($|a| \neq 1$). Then f is infinitely differentiable in t and we have

$$|f^{(n)}(t)| \leq 2^n n! |e^{it} - a|^{-n} \quad \text{for } |e^{it} - a| < 1,$$
$$|f^{(n)}(t)| \leq 2^n n! \quad \text{for } |e^{it} - a| \geq 1,$$

where $n = 0, 1, 2, \ldots$. See K. Yabuta, [**45**, p. 93].

36. Let $f(x)$ and $g(x)$ be real-valued functions which have derivatives of n-th order for $x > a$. Let $g(x) \neq 0$ for $x > a$, and let $|f^{(n)}(x)| \leq |g^{(n)}(x)|$ for $x > a$.

Suppose that there is a sequence (y_n) with $y_n \to +\infty$ as $n \to +\infty$, and also an integer $m < n$ such that the following limits

$$\lim_{x \to +\infty} \frac{f(y_\lambda)}{y_\lambda^m} = A \quad \text{and} \quad \lim_{\lambda \to +\infty} \frac{g(y_\lambda)}{y_\lambda^m} = B$$

exist.

Consider further the expressions

$$F(x) = \prod_{k=0}^{n} (x - x_k) \quad \text{and} \quad G(f; x_0, x_1, \ldots, x_n) = \sum_{k=0}^{n} \frac{f(x_k)}{F'(x_k)}.$$

Then the following inequality

$$|G(f;\ x_0, x_1, \ldots, x_m) - A| \le |G(g;\ x_0, x_1, \ldots, x_m) - B|$$

holds for any numbers $x_0, x_1, \ldots, x_m$ greater than a. These results appear in N. Obrechkoff, [**46**].

37.

$$\int_x^{x+1} \sin t^2 dt < 1/x \quad \text{for } x > 0.$$

The bound $1/x$ is the best possible, see M. Landau, J. Gillis and M. Shimshoni, [**47**]. For generalizations of this result see E. K. Godunova and V. I. Levin, [**48**].

38. Let $f(x)$ be a function admitting continuous derivatives up to the order n in the closed interval $[a, b]$, $a < b$, which may be finite or infinite with $a = 0$ and $b = \infty$, and $\lim_{x \to \infty} f(x) = f(\infty)$. Let $(\phi_n(x))$ be a normalized orthogonal system of functions on $[a, b]$ and let us suppose that $\phi_n(x)$ can be expressed in the form $\phi_n(x) = \Psi(x) \frac{d^n}{dx^n} F_n(x)$, where $F_n(x)$ and $\Psi(x)$ satisfy the following conditions:

(i) $F_n^{(p)}(x) = 0$, $p = 0, 1, 2, , \ldots, n-1$, (for $x = a$ and $x = b$),

(ii) $F_n(x)$ does not change sign in $[a, b]$,

(iii) $f(x)/\Psi(x)$ is continuous in $[a, b]$.

If λ_n denote the lower bound of $|f^{(n)}(x)|$ in $[a, b]$, then

$$\lambda_n \le \left(\int_a^b \frac{f(x)^2}{\Psi(x)^2} dx \right)^{1/2} \Big/ \left| \int_a^b F_n(x) dx \right|.$$

For example, using Čebyšev polynomials we get

$$\lambda_n \le \frac{(2n)!}{2^{n+1/2}\Gamma(n+1/2)} \left(\int_{-1}^{1} (1-x^2)^{-1/2} f(x)^2 dx \right)^{1/2} ;$$

using Legendre's polynomial we get the following inequality of J. Soula

$$\lambda_n \le \frac{(2n)!\,((2n+1)/2)^{1/2}}{2^n n!} \left(\int_{-1}^{1} f(x)^2 dx \right)^{1/2} ;$$

while using Laguerre polynomial we get

$$\lambda_n \le \left(\int_0^\infty e^{-x} f(x)^2 dx \right)^{1/2} .$$

These results were obtained by H. Sircar, [**49**].

39. Let $0 \le \lambda \le 1$ and $0 \le y \le x$. Suppose that $f(t)$ is integrable for $0 \le t \le y$. Then, as M. Riesz [**50**] has shown,

$$\left| \int_0^y (x-t)^{\lambda-1} f(t) dt \right| \le \max_{0 \le u \le y} \left| \int_0^u (u-t)^{\lambda-1} f(t) dt \right| .$$

L. S. Bosanquet [**51**] has replaced integrals by sums and has proved the following:

Let $A_\nu^\sigma = (\sigma+\nu)!/(\sigma!\nu!)$. If $0 \le \lambda \le 1$ and $0 \le m \le n$ (m and n are integers), then

$$\left| \sum_{\nu=0}^{m} A_{n-\nu}^{\lambda-1} a_\nu \right| \le \max_{0 \le \mu \le m} \left| \sum_{\nu=0}^{\mu} A_{\nu-\nu}^{\lambda-1} a_\nu \right|$$

for each real or complex sequence $a_1, \ldots, a_m$.

Bosanquet [**51**] has shown by examples that the last estimate can fail in the case when $\lambda > 1$ or when $\lambda < 0$.

40. Let $f(x)$ be a real nonnegative function defined in R such that

$$\int_0^\infty f(t)dt = 1, \quad \int_0^\infty tf(t)dt = 0, \text{ and } \int_0^\infty t^2 f(t)dt = a^2 < \infty.$$

If $0 < p \le q$ we have

$$\int_{-p}^{q} f(t)dt \ge \frac{p^2}{a^2 + p^2} \quad \text{for } p(q-p) \ge 2a^2 \tag{40.1}$$

and

$$\int_{-p}^{q} f(t)dt \ge \frac{4(pq - a^2)}{(p+q)^2} \quad \text{for } p(q-p) \le 2a^2. \tag{40.2}$$

When $p = q$, inequality (40.2) reduces to the following Čebyšev inequality

$$\int_{-p}^{p} f(t)dt \ge 1 - \frac{a^2}{p^2}.$$

For the above results, see [**52**]. In [**53**] the following is proved.

If there exists x_0 such that the function f, given above, is nondecreasing for $x < x_0$ and nonincreasing for $x > x_0$, then

$$\int_{-p}^{p} f(t)dt \ge 1 - \theta\frac{a^2}{p^2} > 1 - \frac{3}{5}\,\frac{a^2}{p^2},$$

where θ is a zero between 0 and 1 of the equation $\theta^3 - 9\theta^2 + 3\theta + 1 = 0$, i.e. $0.56 < \theta < 0.57$.

41. We give a solution by O. P. Lossers, of a problem of D. J. Newman, see [**54**]. If $f \in C^2$, then

$$\int_0^\infty \{|f(x)| + |f''(x)|\}dx \geq \sqrt{2}|f(0)|$$

and $\sqrt{2}$ is the best possible constant.

PROOF: First one argues that if the integrals are finite then $\lim_{x\to\infty} f'(x) = 0 = \lim_{x\to\infty} f(x)$. If $f(0) = 0$ there is nothing to prove, so one may assume $f(0) = 1$. Then $g(x) = \frac{f(x)-1}{x}$ has a negative minimum $-\alpha$ at ξ. It follows that $f(x) \geq 1 - \alpha x$ and $f'(\theta\xi) = -\alpha$ for some $0 < \theta < 1$, by the mean value theorem. Then $\int_0^\infty |f(x)|dx \geq \int_0^{1/\alpha} |f(x)|dx \geq \frac{1}{2\alpha}$ and $\int_0^\infty |f''(x)|dx \geq |\int_{\theta\xi}^\infty f''(x)dx| = \alpha$. So $\int_0^\infty \{|f(x)| + |f''(x)|\}dx \geq \frac{1}{2\alpha} + \alpha \geq \sqrt{2}$. But $f(x) = 1 - \frac{x\sqrt{2}}{4}$ on $[0, 2\sqrt{2}]$, $f(x) = 0$ else, gives equality. This function can be approximated by a C^2 function.

42. Let f, g be two integrable functions on $[0,1]$ such that $0 \leq g(x) \leq 1$ for all $x \in [0,1]$ and $\int_0^1 f(x)dx = 0$. Then

$$\int_0^1 f(x)^2 dx \geq 4\left(\int_0^1 g(x)f(x)dx\right)^2.$$

This improvement of the Cauchy-Schwarz estimate is due to H. Hadwiger and E. Heil [**55**].

43. If positive continuous functions $Q(x)$ and $R(x)$ have in (a,b) continuous derivatives $Q'(x)$, $R'(x)$ and $R''(x)$ and if

(a) $R'(x_0) > 0,\ x_0 \in (a,b)$;

(b) $(Q(x)/R(x))' < 0$;

(c) $(R'(x)/R(x))' - (R'(x)/R(x))^2 \geq 0$,

then

$$\int_{x_0}^{x} R(t)Q(t)dt \le \frac{R(x_0)Q(x_0)}{R'(x_0)}(R(x) - R(x_0)) \quad (a \le x_0 < x \le b).$$

See [**56**].

44. A moment inequality is given by G. Aumann [**57**]. Let $\lambda > 0$, $\int_0^\infty W(t)dt = 1$ and $\int_0^\infty W(t)^2 dt = 1$. Then

$$\frac{(2\lambda+2)^\lambda}{(2\lambda+1)^{\lambda+1}} \le \int_0^\infty t^\lambda W(t)dt$$

with equality holding if

$$W(x) = \frac{(2\lambda+1)^{\lambda+1}}{2\lambda(2\lambda+2)^\lambda}\left(\frac{(2\lambda+2)^\lambda}{(2\lambda+1)^\lambda} - x^\lambda\right) \quad \text{for } 0 < x \le \frac{2\lambda+2}{2\lambda+1},$$

or if

$$W(x) = 0 \quad \text{for } x > \frac{2\lambda+2}{2\lambda+1}.$$

As a corollary we get

$$\left(\int_0^\infty W(t)^2 dt\right)\left(\int_0^\infty \left(t - \int_0^\infty tW(t)dt\right)^2 W(t)dt\right)^{1/2} \ge \frac{3}{5\sqrt{5}}$$

$$\text{for } \int_0^\infty W(t)dt = 1.$$

45. Let the function f be defined and integrable on each compact subinterval of $[0, +\infty)$, with $f(x) \to 0$ as $x \to +\infty$. Let g be a continuous function for $x \ge 0$ and $h(x) = \int_0^x g(t)dt \ge 0$ with h bounded. Then $J_1 = \int_0^\infty f(x)f(x)dx$ exists and $J_1 \ge 0$.

More generally, let $f(x_1,\dots,x_n)$ be defined and integrable on each compact subset of the domain $\{(x_1,\dots,x_n)\,|\,x_1 \ge 0\dots,x_n \ge 0\}$ and let $f \to 0$ as any $x_i \to +\infty$ $(i = 1,\dots,n)$. Let $(-1)^n\partial^n f/\partial x_1 \dots \partial x_n \ge 0$. Let $g_1(x),\dots,g_n(x)$ be continuous on $x \ge 0$ and let $h_1(x) = \int_0^x g_1(t)dt$, $\dots$, $h_n(x) = \int_0^x g_n(t)dt$ be bounded and nonnegative. Then the repeated integral

$$J_n = \int_0^\infty \dots \int_0^\infty f(x_1,\dots,x_n)g(x_1)\dots g(x_n)dx_1\dots dx_n$$

exists and $J_n \ge 0$. See L. J. Mordell [**58**].

46. Let $f(x)$ and $g(x)$ be complex-valued functions of the real variable x. Consider the following two conditions.

A. $|f(x)| \le |g(x)|$;

B. For any real number t the function $\arg\left(e^{it}g(x)\cdot f(x)\right)$ is nondecreasing in some neighbourhood of x.

To N. Meǐman [**59**] are due the following results:

1. If f and g satisfy A and B on $[a,b]$, then

$$\theta(x) = \arg g(x)$$

is strictly increasing in any subinterval of (a,b), where $f(x)/g(x)$ is not constant, and if

$$\theta(x_2) - \theta(x_1) \le \pi \quad \text{for } a \le x_1 < x_2 \le b,$$

then

$$|h(x_2) - h(x_1)| \le 2\sin\frac{1}{2}\left(\theta(x_2) - \theta(x_1)\right),$$

where $h(x) = f(x)/|g(x)|$;

2. If f and g satisfy A and B, and if f' and g' exist at x, then

$$|f'(x)| \le |g'(x)|;$$

3. If f and g satisfy A and B at x, if f' and g' exist at x and if $f'(x)/f(x)$ is real, then

$$(46.1)\qquad \begin{aligned} &|f'(x)| \le |g'(x)| \cos(\beta - \alpha) \quad \text{or} \\ &|f'(x)| \le -|g'(x)| \cos(\beta + \alpha), \end{aligned}$$

according as $f'(x)/f(x)$ is positive or negative at x.

In (46.1) α $(0 \le \alpha \le \pi/2)$ and β $(0 \le \beta \le \pi)$ are defined by

$$\cos\alpha = \frac{|f(x)|}{|g(x)|} \quad \text{and} \quad \cos\beta = \frac{|g(x)|'}{|g'(x)|};$$

4. If f is a real-valued function on $[a, b]$, where the conditions A and B are satisfied, then

$$|h(x_2) - h(x_1)| \le \sin\left(\theta(x_2) - \theta(x_1)\right)$$

for $\theta(x_2) - \theta(x_1) \le \pi/2$, where $a \le x_1 < x_2 \le b$.

REMARK: The above interesting results of N. Meĭman [59] extend Bernstein's theorem on the derivative of an entire function of exponential type.

47. We give a problem of D. Boyd [**60**].

Let f be a real-valued continuously differentiable function on $[0, 1]$ for which $f(0) = f(1) = 0$. Then

$$(47.1)\qquad \max_{0 \le x \le 1} |f(x)|^2 \le \left(\frac{1}{2}\tanh\frac{1}{2}\right) \int_0^1 \left(|f(x)|^2 + |f'(x)|^2\right) dx$$

and the constant $\frac{1}{2}\tanh\frac{1}{2}$ is the best possible.

PROOF: (W. Nuij). Since $f(0) = f(1) = 0$, we have

$$\int_0^1 \left(|f(x)|^2 + |f'(x)|^2\right) dx = \int_0^1 |f(x) - f'(x)|^2 dx.$$

Applying Schwarz' inequality to the right-hand side of the equality

$$e^{-x} f(x) = \int_0^x (f'(t) - f(t))\, e^{-t} dt,$$

we get

$$\frac{e^{-x}|f(x)|^2}{\sinh x} \leq \int_0^x |f'(t) - f(t)|^2 dt, \tag{47.2}$$

and similarly

$$\frac{e^{1-x}|f(x)|^2}{\sinh(1-x)} \leq \int_x^1 (|f'(t) - f(t)|^2)\, dt. \tag{47.3}$$

If we add (47.2) and (47.3) we obtain

$$|f(x)|^2 \leq \frac{\sinh x \sinh(1-x)}{\sinh 1} \int_0^1 |f(t) - f'(t)|^2 dt. \tag{47.4}$$

The function $\sinh x \sinh(1-x)/\sinh 1$ takes its maximal value $\frac{1}{2}\tanh\frac{1}{2}$ at $x = \frac{1}{2}$. This proves (47.1).

W. Nuij proved, by an interesting procedure, that inequality (47.1) is sharp.

48. Let the function $f(x, y)$ vanish on the boundary of the square $S = \{(x, y) \mid 0 \leq x \leq 1 \text{ and } 0 \leq y \leq 1\}$ and let $|\partial^4 f(x, y)/\partial x^2 \partial y^2| \leq A$ (A is a positive constant). Then

$$\left| \iint_B f(x, y) dx dy \right| \leq \frac{1}{144} A.$$

REMARK: H. Flanders has signed himself in this problem as "Anon, Erewhon-upon-Wabash" [**61**]. The authors of this book have learned this pseudonym from Flanders himself.

49. If $a_n = -\sum(-1)^k\binom{n}{k}k^r$ for $0 < r < 1$ and $n = 1, 2, \ldots$, then $a_n > 0$ $(n = 1, 2, \ldots)$ and $\lim\limits_{n\to+\infty} a_n = 0$.

PROOF: From

$$I_k = \int_0^\infty \left(1 - e^{-yk}\right) y^{-r-1}\,dy = k^r I_1$$

follows that

$$a_n I_1 = -\sum(-1)^k\binom{n}{k} I_k = \int_0^\infty \left(1 - e^{-y}\right)^n y^{-r-1}\,dy.$$

Hence

$$0 < a_n I_1 < \int_0^\infty y^{n-r-1}\,dy + \left(1 - e^{-M}\right)^n \int_1^M y^{-r-1}\,dy$$
$$+ \int_M^\infty y^{-r-1}\,dy,$$

with $M > 1$.

The right side of the above inequality becomes smaller than any $\varepsilon > 0$ if we take first M and then n large enough. It therefore follows that $a_n > 0$ and $\lim\limits_{n\to+\infty} a_n = 0$.

50. Let $u_n = \int\limits_0^\infty t^n e^{-t} e^{-at^k}\,dt$, $n = 0, 1, 2, \ldots$, and k is a fixed natural number, and let $z_n = u_n/u_{n+1}$.

For $k = 2$ we have

$$\frac{1}{n+1} + \frac{2a}{(n+1)(1/(n+2) + 2a(n+3)/(n+2))}$$
$$< z_n < \frac{1}{n+1} + \frac{2a(n+2)}{n+1}, \quad n = 0, 1, \ldots$$

and for $k = 3$:

$$\frac{1}{n+1} < z_n < \frac{1}{n+1} + \frac{3a(n+2)(n+3)}{n+1}, \qquad n = 0, 1, 2, \ldots.$$

See R. Bellman [**62**].

51. Let $f(t)$ be a nonnegative summable function on $[0, a]$ such that it is either nondecreasing or twice differentiable with $f'^2(t) - f(t)f''(t) \geq 0$ for every $t \in (0, a)$. Then the sequence

$$\left\{ \frac{1}{k} \frac{\int_0^a t^k f(t)dt}{\int_0^a t^{k-1} f(t)dt} \right\} \quad (k = 1, 2, 3, \dot{)}$$

is nonincreasing.

This result is due to W. Sciamplicotti [**63**]. Some generalizations are obtained by D. S. Mitrinović and J. E. Pečarić [**64**].

52. A necessary and sufficient condition for the existence of three linearly independent periodic solutions of

$$x''' - \frac{p'}{p}x'' + (1 + p^2)x' - \frac{p'}{p}x = 0, \quad p(t + \omega) \equiv p(t), \ p(t) \not\equiv 0,$$

is that

$$\int_0^\omega \sqrt{1 + p^2}\, dt > 2\pi.$$

See H. Guggenheimer [**65**].

53. For all real $n \geq 1$ and $[x]$ the integer part of x,

$$\int_0^n \frac{n^{[x]}}{[x]!} dx \geq e^{n-1}.$$

See M. Schultz and G. A. Heuer, [**66**].

54. Azencott [**67**] has given the following revised proof.

Let u and v be positive increasing functions defined for $t \in [1, +\infty)$, such that

$$\frac{u''(t)}{u(t)} \geq \frac{v''(t)}{v(t)}. \tag{54.1}$$

Then

$$\frac{u'(t)}{u(t)} \geq \frac{v'(t)}{v(t)} - c, \quad \text{with} \quad c = \left|\frac{u'(1)}{u(1)} - \frac{v'(1)}{v(1)}\right|. \tag{54.2}$$

PROOF: Put $V(t) = v'(t)/v(t)$, $U(t) = u'(t)/u(t)$, $w(t) = V(t) - U(t)$. Then $U(t) > 0$ and $V(t) > 0$, and in view of (54.1) also

$$w'(t) \leq -w(t)\left(U(t) + V(t)\right). \tag{54.3}$$

From (54.3) follows that w is a decreasing function in all its points of positivity.

1° Let $w(1) > 0$. Then $w(t) \leq w(1)$ for all $t \geq 1$ such that $w(t) > 0$. If $w(t) \leq 0$, then a fortiori $w(t) < w(1)$.

2° Let $w(1) \leq 0$. Then $w(t) \leq 0$ for all $t \geq 1$, and therefore $w(t) \leq |w(1)|$ for all $t \in [1, +\infty)$. Namely, in the opposite case for some $b > 1$ we would have $w(b) > 0$, and the zeros of w on $[1, b]$ would form a nonempty set. Let a be the supremum of that set. Since the function w is nonnegative on $[a, b]$, it is nonincreasing on that interval. Hence $w(b) \leq w(a)$, and thus $w(b) \leq 0$, which contradicts the assumption.

55. If $\sum a_k e^{ikx}$ and $\sum A_k e^{ikx}$ are trigonometric polynomials and $|a_k| \leq A_k$ for all k, then for all even integers p

$$\int_0^{2\pi} |\sum a_k e^{ikx}|^p dx \leq \int_0^{2\pi} |\sum A_k e^{ikx}|^p dx.$$

A reference is G. H. Hardy and J. E. Littlewood [68].

The result is generalized in papers of A. Garsia and Ch. A. Greenhall [69].

56. A lemma of G. van der Corput:

Let $f(t)$ be a real-valued differentiable function on $[u, v]$. Suppose f' is monotone and that

$$|f'(t)| > c > 0 \quad \text{for } u \le t \le v.$$

Then

$$\left|\int_u^v \exp(if(t))\right| dt < \frac{1}{c}.$$

For the proof of this lemma see [**70**, p. 197].

57. Rahman [**71**] proved:

Let $f(z)$ be a rational function which is the quotient of two real-valued polynomials of degrees m and n, respectively. If f has neither zeros nor poles inside the unit circle and if $|f(x)| \le 1$ for $-1 \le x \le 1$, then for $0 < c < 1$ and $-1 + c \le x \le 1 - c$, we have

$$|f'(x)| < \frac{1}{c}\,\frac{m+n}{\sqrt{m+n-1}} \quad \text{when } m+n > 1.$$

If $m + n = 1$, then, under the above conditions,

$$|f'(x)| \le \max\left(\frac{1}{2}, \frac{1}{4c}\right).$$

58. We give another result of Ostrowski [**15**].

Consider

$$f_1(x) = \frac{x^{p-1}(x^2-2)^p}{(p-1)!} \quad \text{and} \quad f_2(x) = \frac{x^{p+1}(x^2-2)^{p-1}}{(p-1)!}$$

where p is a positive integer.

For $k = 1$ and $k = 2$ we have

$$\left|\int_0^{\pm\sqrt{2}} e^{-x} f_k(x)dx\right| < \frac{2^{5p/2}e^{\sqrt{2}}}{(p+2)(p+4)\dots 3p} < \frac{\left(\sqrt{8}\right)^p e^{\sqrt{2}}}{p!}.$$

59. Let $y(x)$ be a solution of the initial value problem

$$y''(x) - p(x)y(x) = 0, \quad y(0) = 0, \quad y'(0) = a \neq 0,$$

where p is a continuous nonnegative real function for $0 \leq x < +\infty$. Then

$$\frac{x^2}{R(x) + x} < \frac{y(x)}{y'(x)} < \frac{x^2}{R(x) + x - \int\limits_0^x \frac{1}{t^2} R(t)^2 dt}, \tag{59.1}$$

where $R(x) = \int\limits_0^x t^2 p(t) dt$.

The lower bound in (59.1) is valid for all $x \geq 0$, and the upper bound is valid only for $0 \leq x < x_1$, where x_1 is the first zero (different from the origin) of the function $R(x) + x - \int\limits_0^x \frac{1}{t^2} R(t)^2 dt$.

For example, if $p(x) \equiv 1$, inequalities (59.1) become

$$\frac{3x}{3 + x^2} < \tanh x < \frac{3x}{3 + x^2 - \frac{x^4}{15}},$$

where the upper bound is valid for $0 \leq x \left(\frac{1}{2}\left(15 + \sqrt{405}\right)\right)^{1/2}$. This is a result of A. Ronveaux [**72**].

60. Let $x(t) = (x_1(t), \ldots, x_n(t))$ be a vector, whose components $x_1, \ldots, x_n$ are real-valued and continuous functions on $[a, b]$. Let $A(t) = (a_{ij}(t))$ be a square $n \times n$ matrix whose entries are also real-valued and continuous functions on $[a, b]$. Let further

$$|x| = \left(\sum_{k=1}^{n} |x_k(t)|^p\right)^{1/p} \quad (p > 1) \text{ and } |A(t)| = \max_{|x|>0} \frac{|Ax|}{|x|},$$

where $Ax = (A_1, \ldots, A_n)$ with $A_i = \sum\limits_{j=1}^{n} a_{ij} x_j$.

Then, any solution $x(t)$ of the system of linear differential equations

$$\frac{dx_i(t)}{dt} = \sum_{j=1}^{n} a_{ij}(t) x_j(t) \quad (i = 1, \ldots, n)$$

satisfies the following inequalities

$$\exp\left(-\int_a^b |A(t)|dt\right) \le \left|\frac{x(b)}{x(a)}\right| \le \exp\left(\int_a^b |A(t)|dt\right),$$

where equalities hold if and only if

$$x(t) = x(a)\exp\left(\pm\int_a^b |A(t)|dt\right).$$

A very short and elegant proof of the above result [73], due to H. Tôyama, was given by E. Kamke [74].

61. We give a result of G. W. Veldkamp and J. Brands [75].

Let $f(x)$ and $g(x)$ be independent real solutions of the differential equation

$$\frac{d}{dx}\left(p(x)\frac{dy}{dx}\right) + q(x)y = 0, \tag{61.1}$$

where p and q are real continuous functions for $x > 0$ such that

$$p(x) > 0 \quad \text{and} \quad \frac{d}{dx}(p(x)q(x)) = 0,$$

Then, if $f(x_1) = f(x_2) = 0$ for $x_2 > x_1 > 0$, we have

$$|g(x_1)| \ge |g(x_2)| > 0. \tag{61.2}$$

PROOF: By Abel's formula, we obtain

$$p(x)(f(x)g'(x) - f'(x)g(x)) = C \text{ (constant} \neq 0).$$

Thus it follows that the neither $g(x)$ and $g'(x)$ can vanish at x_1 or x_2. Whence

$$|g(x_1)| = \frac{|C|}{p(x_1)|f'(x_1)|}, \quad |g(x_2)| = \frac{|C|}{p(x_2)|f'(x_2)|}. \tag{61.3}$$

Multiplying (61.1) (we take $y \equiv f$) by $p(x)f(x)$ and integrating between x_1 and x_2, we get

$$(p(x_2)f'(x_2))^2 - (p(x_1)f'(x_1))^2 = \int_{x_1}^{x_2} f(x)^2 \frac{d}{dx}(p(x)q(x))\,dx > 0.$$

Hence,

(61.4) $$p(x_2)|f'(x_2)| > p(x_1)|f'(x_1)|.$$

Inequality (61.2) immediately follows from (61.3) and (61.4).

62. Let (f_n) be a sequence of continuous orthonormal functions in a bounded interval (a, b) and let c be any number such that $a \leq c \leq b$. Then

$$\sum_{n=1}^{+\infty} \int_a^b \left| \int_c^x f_n(t)dt \right|^2 dx \leq \frac{1}{2}\left((a-c)^2 + (b-c)^2\right),$$

with equality if and only if (f_n) is closed in $L^2(a, b)$.

This inequality is due to R. E. Graves who in the papers [76], [77] proved a more general result.

63. Let $a_0, a_1, \ldots, a_n$ be nonnegative real numbers and let $f(t)$ be an increasing function for $t \geq 0$ such that $f(0) = 0$. If

$$\sum_{i=k}^{n} a_i \leq f^{-1}(a_k^{1-r}) \qquad (k = 0, 1, \ldots, n)$$

for some r, $0 < r < 1$, then

$$\sum_{i=0}^{n} a_i^k \leq \int_0^b \frac{1}{f(t)}dt,$$

where $b = \sum_{i=0}^{n} a_i$. See Yu. G. Ol'hovskiĭ [78].

64. Let $P(x)$ be a homogeneous polynomial of degree d in variables $\{x_{ij}\}$ with nonnegative coefficients. Further, let $x = \{x_{ij}\}$ be any point of the domain D, defined by

$$\left\{ x_{ij} \mid x_{ij} \geq 0, \sum_{j=1}^{q_i} x_{ij} = 1 \text{ with } i = 1, \ldots, p;\ j = 1, \ldots, q_i \right\}.$$

For $x = \{x_{ij}\} \in D$ let $J(x) = J\{x_{ij}\}$ denote the point of D, whose (i,j)-coordinate is

$$J(x)_{ij} = \frac{x_{ij} \left.\frac{\partial P}{\partial x_{ij}}\right|_{(x)}}{\left.\sum_{j=1}^{q_i} x_{ij} \frac{\partial P}{\partial x_{ij}}\right|_{(x)}}$$

Then the following inequality is valid

$$P(J(x)) > P(x)$$

unless $J(x) \equiv x$. This is proved in L. E. Baum and J. A. Eagon [79].

65. Let $f(x,y,z)$ be a real-valued continuous function on $\{(x,y,z) \mid 0 \leq x,y,z \leq \pi\}$ and let f have the following expansion

$$f(x,y,z) \sim \sum_{\ell,m,n=1}^{+\infty} a_{\ell mn} \sin \ell x \sin my \sin nz.$$

Then

$$\max_{\substack{x,y,z \\ 0 \leq t < +\infty}} \left| \sum_{\ell,m,n=1}^{+\infty} a_{\ell mn} \exp\{-(\ell^2 + m^2 + n^2)t\} \sin \ell x \sin my \sin nz \right|$$
$$\leq \max_{x,y,z} |f(x,y,z)|.$$

This result is connected with the problem of determining the majorants for solutions of the equation

$$u_{xx} + u_{yy} + u_{zz} - u_t = 0.$$

66. For every trigonometric polynomial

$$S_n(x) = \frac{1}{2}a_0 + \sum_{k=1}^{n}(a_k \cos kx + b_k \sin kx)$$

we have

$$\left(\int_{-\pi}^{\pi} |S_n'(x)|^p dx\right)^{1/p} \le n \left(\int_{-\pi}^{\pi} |S_n(x)|^p dx\right)^{1/p}$$

with $p \ge 1$.

When $p \to +\infty$, the last inequality reduces to

$$\max_{-\pi\le x\le\pi} |S_n'(x)| \le n \max_{-\pi\le x\le\pi} |S_n(x)|.$$

See A. Zygmund [**80**] and R. P. Boas [**81**].

67. A function f is called cyclically monotone on (a,b), if f and its derivatives are each of constant sign on (a,b) and

$$f^{(k)}(x)f^{(k+2)}(x) \le 0 \text{ for } a \le x \le b \text{ and for every } k \ge 0.$$

For a cyclically monotone function the following inequality, for any $r = 0, 1, \ldots,$

$$\left|f^{(r)}\left(\frac{a+b}{2}\right)\right| \le \left(\frac{2}{b-a}\right)^r \left|f\left(\frac{a+b}{2}\right)\right|,$$

is valid.

This result, due to Bernstein [**82**], is dated from 1928.

68. Let f be defined and bounded on the rectangle $D = [a,b] \times [c,d]$. Then the lower and upper iterated and double Riemann integrals of f over D satisfy the inequalities

$$\underline{\int}\underline{\int} f(x,y)dA \le \underline{\int}\underline{\int} f(x,y)dx\,dy \le \begin{Bmatrix} \int \overline{\int} f(x,y)dx\,dy \\ \overline{\int} \underline{\int} f(x,y)dx\,dy \end{Bmatrix} \tag{68.1}$$

$$\le \overline{\int}\overline{\int} f(x,y)dx\,dy \le \overline{\int}\overline{\int} f(x,y)dA.$$

The same inequalities hold with the opposite order of integration for the iterated integrals. A proof of this is given in T. Apostol [**83**, p. 260] together with examples where strict inequality holds. See also the examples (and partial proof) given in Hobson [**84**, pp. 514–518]. As noted in Hobson, the existence of both the Riemann integrals $\int_a^b f(x,y)dx$ for all $y \in [c,d]$ and $\int_c^d f(x,y)dy$ for all $x \in [a,b]$ implies the existence and equality of the two iterated integrals. However, examples [**84**, Ex. 4, p. 518], [**83**, Ex. 10-9, p. 299] show that this does not imply the existence of the double Riemann integral.

If f is *not* bounded on D, W. H. Young [**85**] has shown that the inequalities (68.1) may not remain valid. In this case the integrals are improper lower and upper integrals in the sense of de la Vallée Poussin, defined as follows. For unbounded functions f on an interval I, one sets

$$f_{N,M}(x) = \begin{cases} f(x), & \text{if } -N < f(x) < M, \\ -N, & \text{if } f(x) \leq -N, \\ M, & \text{if } f(x) \geq M. \end{cases}$$

Then by definition,

$$\underline{\int} f\,dx = \lim_{N,M\to\infty} \underline{\int} f_{N,M}\,dx, \quad \overline{\int} f\,dx = \lim_{N,M\to\infty} \overline{\int} f_{N,M}\,dx,$$

provided these limits exist. The function f is said to be *imporperly integrable* precisely when its lower and upper integrals exist and are equal. In this case, it is clear that the common value is just the Lebesgue integral of f. Similar definitions apply to unbounded functions f defined on the rectangle D.

For such improper integrals, Young [**85**, Th. 3] proves that the first inequality of (68.1) remains valid if f is bounded below on D, while the last inequality of (68.1) remains valid if f is bounded above on D. Moreover, in the first case equality holds if f is lower semicontinuous on D, and in the second case if f is upper semicontinuous on D. An explicit example is given of a function f which is positive but unbounded on D (although continuous in the extended sense that $\lim f(x,y) = +\infty$ at points where f is infinite) for which

$$\underline{\int}\left(\int f(x,y)dx\right)dy = \iint f(x,y)dA < \overline{\int}\left(\int f(x,y)dx\right)dy.$$

Here, the inner integrals and the double integral are imporperly integrable. Young also gives two examples of bounded functions f, g for which

$$\underline{\int}\overline{\int} f(x,y)dx\,dy < \overline{\int}\underline{\int} f(x,y)dx\,dy,$$
$$\overline{\int}\underline{\int} g(x,y)dx\,dy < \underline{\int}\overline{\int} g(x,y)dx\,dy,$$

which shows that no general inequality can hold for these repeated integrals.

69. Here is a problem posed by Richard Kellner and solved by O. P. Lossers [**54**].

If

$$\phi(f) = 20\int_0^\pi f(\theta)^2 \sin\theta d\theta + 2\int_0^\pi f'(\theta)^2 \sin\theta d\theta$$
$$- 10\left\{\int_0^\pi f(\theta)\sin\theta d\theta\right\}^2 - 9\left\{\int_0^\pi f(\theta)\sin\theta d\theta\right\}^2,$$

then $\phi(f) \geq 0$ with equality if and only if

$$f = A + B\cos\theta \text{ for } A, B \text{ constants.}$$

PROOF: For f given, define $g = f - A - B\cos\theta$ with A and B selected so that $\int_0^\pi g(\theta)\sin\theta d\theta = \int_0^\pi g(\theta)\sin 2\theta d\theta = 0$. Then

$$\phi(f) = \phi(g) = 20\int_0^\pi g(\theta)^2 \sin\theta d\theta + 2\int_0^\pi g'(\theta)^2 \sin 2\theta d\theta$$

from which all the results follow.

70. If f and g are such that all the integrals exist, then

$$\int_{-\infty}^{\infty} \sup_t |g(x-t)f(t)|dx \geq \sup_x \int_{-\infty}^{\infty} g(x,t)f(t)dt.$$

PROOF:

$$\begin{aligned}
\int_{-\infty}^{\infty} \sup_t \left[g(x-t)f(t)\right] dx &\geq \sup_t \int_{-\infty}^{\infty} g(x-t)f(t)dx \\
&= \sup \left[f(t) \int_{-\infty}^{\infty} g(x-t)dx\right] = \sup_t \left[f(t) \int_{-\infty}^{\infty} g(x)dx\right] \\
&= \left(\sup_x f(x)\right) \int_{-\infty}^{\infty} g(t)dt = \int_{-\infty}^{\infty} \left[\sup_x f(x)\right] g(t)dt \\
&= \int_{-\infty}^{\infty} \left[\sup_x f(x-t)\right] g(t)dt = \int_{-\infty}^{\infty} \sup_x \left[f(x-t)g(t)\right] dt \\
&\geq \sup_x \int_{-\infty}^{\infty} f(x-t)g(t)dt = \sup_x \int_{-\infty}^{\infty} g(x-t)f(t)dt.
\end{aligned}$$

This was proposed by R. P. Boas and solved by O. P. Lossers [**54**].

71. Let ε_k, $k = 0, \ldots, n$, be a prescribed sequence of signs $+1$ or -1. Let f be infinitely differentiable on $[a, b]$. Define

$$G_r(x) = \varepsilon_k \begin{vmatrix} f^{(r)}(x) & \ldots & f^{(r+k)}(x) \\ f^{(r+1)}(x) & \ldots & f^{(r+k+1)}(x) \\ \vdots & & \vdots \\ f^{(r+k)}(x) & \ldots & f^{(r+2k)}(x) \end{vmatrix}.$$

If $G_r(0) > 0$ for $k = 0, \ldots, n$ and $r = 0, 1, 2 \ldots$, then

$$G_r(x) > 0 \quad \text{on } [a, b] \text{ for } k = 0, \ldots, n \text{ and } r = 0, 1, 2 \ldots .$$

See S. Karlin and C. Loewner [**86**, p. 179].

72. Tagamlicki gave related results in three papers [**87**], [**88**], and [**89**]. We shall give only the following two results from [**87**] and [**89**] respectively.

(i) Let $f(x)$ have derivatives of all order for $x < a$ and let

$$|f^{(k)}(x)| \le Ae^x \quad (k = 0, 1, 2, \dots).$$

Then

$$f(x) = Be^x,$$

where B is constant such that $|B| \le A$.

(ii) If $f(x)$ satisfies

$$|f^{(k)}(x)| \le (-1)^k \frac{d^k}{dx^k}(x_0 - x)e^{-x} = (x_0 + k - x)e^{-x},$$

for $a < x \le x_0 + k$, $k = 0, 1, \dots$, then

$$f(x) = C(x_0 - x)e^{-x}$$

where C is a constant such that $|C| \le 1$.

73. Let $f(x)$ be continuous, $g(x)$ differentiable on $I = [a, b]$, $g(a) = 0$, and $\lambda \ne 0$ a constant. If

$$|g(x)f(x) + \lambda g'(x)| \le |g(x)| \quad \text{on } I$$

then $g(x) \equiv 0$ on I.

This is a Gronwall type lemma. It implies the usual Gronwall lemma since if $g(a) = 0$ and $|g'(x)| \le |g(x)|$ on I then $g \equiv 0$ (the case $f \equiv 0$ and $\lambda = 1$). See [**90**].

74. We consider a vector field $\underline{u}(x)$, continuously differentiable on the closure $R + B$ of a bounded domain R with boundary B in two or three dimensions.

Let x_i be coordinates in a given rectangular Cartesian coordinate system, and denote by u_i the components of u in this system. Define the functionals D and S by

$$D(\underline{u}) = \int_R u_{i,j}u_{i,j}dV, \tag{74.1}$$

$$S(\underline{u}) = \frac{1}{4}\int_R (u_{i,j} + u_{j,i})(u_{i,j} + u_{j,i})\,dV, \tag{74.2}$$

where Latin subscripts have the range 1 to n, and n is the number of dimensions. Summation over repeated subscripts is implied, and subscripts preceded by a comma indicate differentiation with respect to the corresponding coordinate.

A. Korn [**91**], [**92**] proved the existence of a number $K_1 > 0$, depending only on the domain R, such that

$$D(\underline{u}) \leq K_1 S(\underline{u}) \tag{74.3}$$

for all vector fields $\underline{u}$ satisfying certain side conditions. Since equality holds in (74.3) for any K_1 if $\underline{u}$ is a constant vector, we will agree henceforth to identify vector fields differing only by a constant.

The necessity for imposing some side conditions on $\underline{u}$ in order that (74.3) should hold can be seen from the fact that if $\underline{u}$ is a pure rotation, then $S(u) \equiv 0$, $D(u) > 0$.

Let

$$R(\underline{u}) = \frac{1}{4}\int_R (u_{i,j} - u_{j,i})(u_{i,j} - u_{j,i})\,dV.$$

Then

$$D(\underline{u}) = S(\underline{u}) + R(\underline{u}),$$

so that K_1 cannot be less than unity, and the inequality (74.3) is equivalent to either of the following inequalities:

$$R(\underline{u}) \leq (1 - 1/K_1)D(\underline{u}), \qquad R(\underline{u}) \leq (K_1 - 1)S(\underline{u}). \tag{74.4}$$

K. O. Friedrichs [**93**] proved Korn's inequality in the form of the second of (74.4) under each of the following three side conditions on $\underline{u}$:

First Case: $\underline{u} = 0$ on boudnary B,

Second Case: $\int_R (u_{i,j} - u_{j,i})dV = 0,$

Main Case: $\begin{cases} \int_R (u_{i,j} - u_{j,i})dV = 0, \\ \underline{u} \in C^2(R), \quad u_{i,jj} + u_{j,ij} = 0 \text{ in } R. \end{cases}$

Either of the conditions in First and Second cases serves to eliminate pure rotations. The class of admissible domains R is specified in detail in [**93**] (see also [**94**]).

Extension of Main case is considered in [**94**] and [**95**]. For other references about this and related problems see [**94**], [**95**].

75. In 1965, D. Ž. Djoković [**96**] formulated the following interesting proposition which has intrigued mathematicians:

Let $x_0 < x_1 < \cdots < x_n$ be real numbers and

(75.1) $f(x; x_0, x_1, \ldots, x_n) = (x - x_0)(x - x_1) \ldots (x - x_n), \quad (= f(x)),$

(75.2) $$M = \max_{x_0 < x < x_n} |f(x; x_0, x_1, \ldots, x_n)|,$$

$$\phi(x_0, x_1, \ldots, x_n) = \frac{1}{M} \int_{x_0}^{x_n} f(x; x_0, x_1, \ldots, x_n) dx,$$

prove (or disprove) the inequality:

(75.3) $$(-1)^{k+1} (\partial\phi/\partial x_k) > 0.$$

In 1966, the editor of the Amer. Math. Monthly pointed out [**97**]: From a reexamination of the original proposal there is reason to think that the inequality has been stated as

(75.4) $$(-1)^{n+1-k} (\partial\phi/\partial x_k) > 0.$$

For $n = 1$ $(k = 0, 1)$, D. A. Hejhol showed that the inequality (75.3) is untrue, but (75.4) is true.

In [**AI**] (75.4) was called Djoković's conjecture. H. Guanchu and T. Jiankang [**98**] disproved this conjecture. Namely, the following results are obtained:

THEOREM 1. *Let $x_0 < x_1 < x_2$ be real numbers, f, M, $\emptyset$, defined above. Let the numbers a_1 and a_2 be defined by*

$$a_1 = \left(8 + \left(-109 + \sqrt{109^2 + 512}\right)^{1/3} - \left(109 + \sqrt{109^2 + 512}\right)^{1/3}\right)/18$$
$$\simeq 0.1824879,$$
$$a_2 = 1 - a_1 \simeq 0.8175121,$$

and $n = 2$. *Then the inequality (75.4) is true for $a_1(x_2 - x_0) < x_1 - x_0 < a_2(x_2 - x_0)$, and it is untrue for $x_1 - x_0 < a_1(x_2 - x_0)$* or *$x_1 - x_0 > a_2(x_2 - x_0)$. Similarly, for $n > 2$, the validity of the inequality (75.4) depends on the distribution of $x_0 < x_1 < \cdots < x_n$.*

THEOREM 2. *Let $x_0 < x_1 < \cdots < x_n$ be $n + 1$ uniformly spaced points of the closed interval $[x_0, x_n]$, then for $n \leq 6$, form (75.4) is true, and for large n (75.4) is untrue.*

THEOREM 3. *Let $a < x_0 < x_1 < \cdots < x_n < b$ be the roots of the $(n + 1)$-th orthogonal polynomial $P_{n+1}(x)$, associated with the weight function $p(x)$ in the interval $[a, b]$, where*

$$\phi(x_0, x_1, \ldots, x_n) = \frac{1}{M} \int_a^b p(x) f(x) dx,$$

then (75.4) is true.

76. Montgomery [**99**] discusses certain results of Gallagher [**100**] and improves upon them.

Lemma 1. *Let f be a continuous (complex valued) function on $[a,b]$ with a continuous first derivative in (a,b). Then*

$$\left|f\left(\frac{a+b}{2}\right)\right| \leq (b-a)^{-1}\int_a^b |f(t)|dt + \frac{1}{2}\int_a^b |f'(t)|dt \tag{76.1}$$

and for any $x \in [a,b]$,

$$|f(x)| \leq (b-a)^{-1}\int_a^b |f(t)|dt + \int_a^b |f'(t)|dt. \tag{76.2}$$

The result (76.1) is in [**100**] while the second is a special case of a general Sobolev inequality.

Proof: (Montgomery) By an integration by parts

$$f(x) = (b-a)^{-1}\int_a^b f(x) + \int_x^b \frac{t-b}{b-a}f'(t)dt + \int_a^x \frac{t-a}{b-a}f'(t)dt.$$

The fractions $\frac{t-b}{b-a}$ and $\frac{t-a}{b-a}$ are at most 1 so (76.2) follows. If $x = \frac{a+b}{2}$, these fractions are at most $1/2$ so (76.1) is obtained.

The inequality (76.2) may be iterated.

Lemma 2. (Montgomery) *Let T_0, $T \geq \delta > 0$, and let P be a partition of $\left[T_0 + \frac{\delta}{2},\ T_0 + T - \frac{\delta}{2}\right]$ and define*

$$N_\eta(x) = \sum_{\substack{t\in P\\ |t-x|<\eta}} 1.$$

If $f \in C^1(T_0, T_0+1)$ then

$$\sum_{t\in P} N_\delta(x)^{-1}|f(x)| \leq \delta^{-1}\int_{T_0}^{T_0+T} |f(t)|dt + \frac{1}{2}\int_{T_0}^{T_0+T} |f'(t)|dt. \tag{76.3}$$

PROOF: By Lemma 1

$$|f(x)| \leq \delta^{-1} \int_{x-\frac{\delta}{2}}^{x+\frac{\delta}{2}} |f(t)dt + \frac{1}{2} \int_{x-\frac{\delta}{2}}^{x+\frac{\delta}{2}} |f'(t)|dt.$$

Thus

$$\sum_{x \in P} N_\delta(x)^{-1}|f(x)| \leq \delta^{-1} \int_{T_0}^{T_0+T} |f(t)|w(t)dt + \frac{1}{2} \int_{T_0}^{T_0+T} |f'(t)|w(t)dt \tag{76.4}$$

where $w(t) = \sum_{\substack{s \in P \\ |s-t|<\delta/2}} N_\delta(s)^{-1}$. Now $N_\delta(x)$ is the number of $t \in P$ such that $|x-t| < \delta$. Consequently, if $|x-t| < \delta/2$, then $N_\delta(x) \geq N_{\delta/2}(s)$ if $|x-s| < \delta/2$. Thus $w(t) \leq N_{\delta/2}(t)^{-1} \sum_{\substack{s \in P \\ |s-t|<\delta/2}} 1 = 1$. Using $w(t) \leq 1$ in (76.4) yields (76.3).

Two results of Gallagher [**100**] follow from this.

LEMMA 3. *With the hypotheses of Lemma 2, let* $|x_1 - x_2| \geq \delta$ *for any two points* $x_1, x_2 \in P$. *Then*

$$\sum_{x \in P} |f(x)| \leq \delta^{-1} \int_{T_0}^{T_0+T} |f(t)|dt + \frac{1}{2} \int_{T_0}^{T_0+T} |f'(t)|dt.$$

LEMMA 4. *With the hypotheses of Lemma 2,*

$$\sum_{x \in P} N_\delta(x)^{-1}|f(x)|^2 \leq \delta^{-1} \int_{T_0}^{T_0+T} |f(t)|^2 + \left(\int_{T_0}^{T_0+T} |f(t)|^2 dt \right)^{1/2} \left(\int_{T_0}^{T_0+T} |S'(t)|^2 dt \right)^{1/2}.$$

PROOFS: Lemma 3 follows from Lemma 2 by observing that $N_\delta(x) \equiv 1$ for each $x \in P$. To get Lemma 4 we apply Lemma 3 to f^2 and use the Cauchy-Schwarz inequality.

77. G. Freud and G. P. Névai [**101**, pp. 68–69] proved the following lemma. If $x > 0$, $\alpha \leq 0$, $\beta \geq 1$ and for $j = 0, 1, 2, \ldots$, then

$$I_j(x) = \int_x^\infty (y-x)^j y^\alpha e^{-y^\beta} dy \leq \frac{j!}{(\beta x^{\beta-1})^{j+1}} x^\alpha e^{-x^\beta}.$$

The proof is by induction.

$$I_0(x) = \int_x^y y^\alpha e^{-y^\beta} dy \le \frac{x^\alpha}{\beta x^{\beta-1}} \int_x^\infty \beta y^{\beta-1} e^{-y^\beta} dy$$
$$= \frac{1}{\beta x^{\beta-1}} x^\alpha e^{-x^\beta}$$

so that for $j \ge 1$,

$$I_j(x) = j \int_x^\infty (y-x)^{j-1} I_0(y) dy$$
$$\le j \int_x^\infty (y-x)^{j-1} \frac{y^\alpha e^{-y^\beta}}{\beta y^{\beta-1}} dy \le \frac{j}{\beta x^{\beta-1}} I_{j-1}(x).$$

78. W. N. Everitt and A. P. Guinand [**102**] gave a classical proof of a result of M. K. Kwong and A. Zettl [**103**]: if $w > 0$ and increasing with $\lim_{x \to -\infty} w(x) > 0$, then

$$\left(\int_{-\infty}^\infty w(x) f'(x)^2 dx \right)^2 \le 4 \int_{-\infty}^\infty w(x) f(x)^2 dx \int_{-\infty}^\infty w(x) f''(x)^2 dx.$$

The constant 4 is best possible. They also show that

$$\left(\int_0^\infty x f'(x)^2 dx \right)^2 \le k \int_0^\infty x f(x)^2 dx \int_0^\infty x f''(x)^2 dx.$$

where $2.35070 < k < 2.35075$.

REFERENCES

1. TURÁN, P. *Problem 392*, Elem. Math. **17**(1960), 19–20.
2. RADEMACHER, H., *Problem 26*, Jber. Deutsch. Math.-Verein. **35** (1926), 50–52.
3. JACKSON, D., "The Theory of Approximation," New York, 1930.
4. KRAWTCHOUK, M., *Sur quelques inégalités dans le problème des moments*, C. R. Acad. Sci. Paris **200** (1935), 1567–1569.
5. MASANI P. R., I. J. SCHOENBERG, and F. W. STEUTEL, *Problem No. 134*, Nieuw Arch. Wisk. (3) **15** (1967), 172–173.
6. STEIN, E. M. and S. WAINGER, *The estimation of an integral arising in multiplier transformation*, Studia Math. **35** (1970), 101–104.
7. FINK, A. M., *Pointwise inequalities and continuation of solutions of an nth order equation*, Tôhoku Math. J. (2) **32** (1980), 169–175.
8. AVAKUMOVIĆ, V. and J. KARAMATA, *Über einige Taubersche Sätze, deren Asymptotik von Exponentialcharakter ist*, I. Math. Z. **41** (1936), 346–356.
9. FEJÉR, L., *Abschätzungen für die Legendreschen und verwandte Polynome*, Math. Z. **24** (1925), 285–298.
10. RAFAL'SON, S. Z., *Ob odnom obobščenii neravenstva Fejera*, Izv. Vyš. Uč. Zav. Matematika **2** (57) (1967), 57–63.
11. VAN DER VAART, H. R., *Some extensions of the idea of Bias*, Ann. Math. Statist. **32** (1961), 436–447.
12. BERNSTEIN, S., "Leçons sur les propriétés extrémales et la meilleure approximation des fonctions analytiques d'une variable réelle," Paris, 1969 — See especially p. 77.
13. MILNE, W. E., *On the maximum absolute value of the derivative of* $e^{-x^2}P(x)$, Trans. Amer. Math. Soc. **33** (1931), 143–146.
14. WATSON, G. N., "A Treatise on the Theory of Bessel Functions," 2nd ed., Cambridge, 1944, §6.22.
15. OSTROWSKI, A., "Aufgabensammlung zur Infinitesimalrechnung, III," Basel-Stuttgart, 1977, Problem 31, pp. 118 and 350.
16. VAN DER CORPUT, J. G., *Einige Ungleichungen bei bestimmten Integralen*, Proc. Roy. Acad. Amsterdam **35** (1932), 334–346.
17. WALTER, J., *Absolute continuity of the essential spectrum of* $-d^2/dt^2 + q(t)$ *with monotony of* q, Math. Z. **129**, 83–94 (1972).
18. TATARKIEWICZ, K., *Sur une inégalité*, Demonstration Math. **2** (1970), 159–163.
19. VOLKOV, V. N., *Dokazatel'stvo odnogo neravenstva Littl'vuda*, Moskov. Gos. Ped. Inst. Učen. Zap. **460** (1972), 106–109.

20. MIKUSIŃSKI, G. I., *Sur l'inégalité différentielle* $|f^{(n)}(x)| \geq mf(x)$, C. R. Acad. Sci. Paris **222** (1946), 359–361.
21. HARDY, G. H. and J. E. LITTLEWOOD, *A convergence criterion for Fourier series*, Math. Z. **28** (1928), 122-147.
22. IL'IN, V. P., *Ob odnoĭ teoreme F. H. Hardi i Dž. E. Littl'vuda*. Akad. Nauk. SSSR, Trudy Mat. In-ta im. V. A. Steklova **8** (1959), 128–144.
23. OSTROWSKI, A. M., *On some integral inequalities II*, Basel Mathematical Notes **24** (1968), 1–11.
24. TIMAN, A. F., *Točnaja ocenka ostatka pri približenii periodičeskih differenciruemyh funkciĭ integralami Puassona*, Doklady Akad. Nauk. SSSR **74** (1950), 17–20.
25. NATANSON, I. P., *O porjadke približenija nepreryvnoĭ* 2π*-periodičeskoĭ funkcii pri pomošči ee integrals Puassona*, Doklady Akad. Nauk SSSR **72** (1950), 11-14.
26. FREUD, G. and V. POPOV, *On approximations by spline functions*, Proc. Budapest, p. 163.
27. VISSER, C., *On certain infinite sequences*, Nederl. Akad. Wetensch. Proc. **40** (1937), 358–367.
28. OSTROWSKI, A., "Aufgabensammlung zur Infinitesmalrechnung 2A," Basel-Stuttgart, 1972, p. 87.
29. _________, "Aufgabensammlung zur Infinitesmalrechnung 2A," Basel-Stuttgart, 1972, p. 88.
30. VAN CORPUT, J. G., *Problem*, Wisk Opgaven **16** (1935), 241–243.
31. WARSCHAWSKI, S. E., *On Theodorsen's method of conformal mapping of nearly circular regions*, Quart. Appl. Math. **3** (1945), 12–28.
32. GAIER, D., *Konstructive Methoden der konformen Abbildung*, (Springer Tracts in Natural Philosophy, vol 3). Berlin/Göttingen/Heidelberg 1964.
33. BINDERSCHADLER, D. *An integral inequality*, Real Analysis Exchange **4** (1978–79), 16–163.
34. GOSSELIN, R. P., *Some integral inequalities*, Proc. Amer. Math. Soc. **13** (1962), 378–384.
35. _________, *Integral norms of subaddative functions*, Bull. Amer. Math. Soc. **69** (1963), 255–259.
36. _________, *A maximal theorem for subadditive functions*, Acta Math. **112** (1964), 163–180.
37. _________, "General Subadditive Functions. Inequalities." New York, 1967, pp. 127–135.
38. HSIANG, F. C., *An integral inequality*, Portugaliae Math. **28** (1969), 63–66.

39. BEESACK, P. R. *Equivalent L_p-norms of subadditive functions*, Univ. Beograd. Publ. Elektrotehn. Fak. Ser. Mat. Fiz. No. 544–576 (1976), 39–42.
40. __________, *Comparable L_p-norms of subadditive functions*, Univ. Beograd. Publ. Elektrotehn. Fak. Ser. Mat. Fiz. No. 602–633 (1978), 47–49.
41. HINTON, D. B. and R. T. LEWIS, *Discrete spectra criteria for singular differential operators with middle terms*, Math. Proc. Camb. Phil. Soc. **77** (1975), 337–341.
42. DAVENPORT, H., *The zeros of trigonometrical polynomials*, Mathematica **19** (1972), 88–90.
43. BOAS, R. P., *Signs of derivatives and analytic behavior*, Amer. Math. Monthly **78** (1971), 1085–1093.
44. LUXEMBURG, W.A.J., *Problem 319*, Nieuw Arch. Wisk. (3) **21** (1973), 108–109.
45. YABUTA, K., *On the distribution of values of functions in some function classes in the abstract Hardy space theory*, Tôhoku Math. J. **25** (1973), 89–102.
46. OBRECHKOFF, N., *Quelques inégalités nouvelles sur les dérivées des fonctions*, C. R. Acad. Sci. Paris **222** (1946), 531–533.
47. LANDAU, M., J. GILLIS and M. SHIMSHONI, *Problem E2211*, Amer. Math. Monthly **77** (1970), 79 and 1107–1108.
48. GODUNOVA, E. K. and V. I. LEVIN, *Nekotorye zadači na neravenstva*, Moskov. Gos. Ped. Inst. Učen. Zap. **460** (1972), 21–27.
49. SIRCAR, H., *On the inequality satisfied by the derivative of order n of a function*, Bull. Calcutta Math. Soc. **37** (1947), 21–23.
50. RIESZ, M., Acta Univ. Hung. Szeged **1** (1923) 114–126.
51. BOSANQUET, L. S., *A mean value theorem.* J. London Math. Soc., **16** (1941), 146–148.
52. SELBERG, H. L., *Zwei ungleichungen zur ergänzung des Tschebycheffschen lemmas*, Skand. Aktuarietidskr. **23** (1940), 121–125.
53. __________, *Über eine Ungleichung der mathematischen Statistik*, Skand. Aktuarietidskr, **23** (1940), 114–120.
54. LOSSERS, O.P., private communications.
55. HADWIGER, H. and E. HEIL, *Aufgaben 705*, Elem. Math. **28** (1973), 155–156 and **29** (1974), 150–151.
56. MAUKEEV, B. I. and M. AĬTŽANOV, *Ob odnom itegral'nom neravenstve*, Mat. Meh. (Sb. St. asp. i soisk.) Alma-Ata, 1969, 82–89.
57. AUMANN, G., *Über eine Ungleichung der Wahrscheinlichkeitsrechnung*, Acta Univ. Szeged. Sect. Sci. Math. **13** (1950), 163–168.
58. MORDELL, L. J. *On the sign of a multiple integral*, J. Austral. Math. Soc. **3** (1963) 129–133.

59. MEĬMAN, N., *On conditions under which the derivative of a majorant function is a majorant for the derivative of the function* (Russian), Doklady Akad. Nauk. SSSR (N.S.) **71** (1950), 609–612.
60. BOYD, D., *Problem 304*, Nieuw Arch. Wisk. (3) **20** (1972), 260–262.
61. FLANDERS, H., *Problem E2207*, Amer. Math. Montly **76** (1969), 1138.
62. BELLMAN, R., *On a generalization of continued fractions and the estimation of some integrals*, J. Math. Anal. Appl. **33** (1971), 16–19.
63. SCIAMPLICOTTI, W., *Studio di una particolare successione collegata ai momenti di una funzione*, Istituto di Matematica Finanziaria dell'Università di Torino, 1967, Serie II, No. 22 3–6.
64. MITRINOVIĆ, D. S. and J. E. PEČARIĆ, *An application of the Čebyšev inequality*, To appear.
65. GUGGENHEIMER, H., *Aufgabe 606*, Elem. Math. **26** (1971), p. 14.
66. SCHULTZ, M. and G. A. HEUER, *Problem E 2117*, Amer. Math. Monthly **75** (1968), 898 and **76** (1969), 829–830.
67. AZENCOTT, R., *Behavior of diffusion semi-groups at infinity*, Bull. Soc. Math. France **103** (1974), 193–240. See in particular, p. 232.
68. HARDY, G. H. and J. E. LITTLEWOOD, *A problem concerning majorants of Fourier series*, Quarterly J. Math. (Oxford Series) **6** (1935), 305–315.
69. GARSIA, A. M. and Ch. A. GREENHALL, *Positive orthogonal systems*, Acta Math. Acad. Sci. Hung. **31** (1970), 47–63.
70. ZYGMUND, A., "Trigonometric Series," Vol. 1, 2nd ed., Cambridge, 1959.
71. RAHMAN, Q. I., *On extremal properties of the derivatives of polynomials and rationals functions*, J. London Math. Soc. **35** (1960), 334–336.
72. RONVEAUX, A., *Bounds on the logarithmic derivative of solutions of second order differential equations*, Math. Mag. **41** (1968), 231–234.
73. TÔYAMA, H., *Some inequalities in the theory of linear differential equations*, Tôhoku Math. J. **47** (1940), 210–216.
74. KAMKE, E., Review in Zbl **26** (1942), 121. See also Review in MR **2** (1974), 288 by W. M. Whyburn.
75. VELDKAMP, G. W. and J. J. M. BRANDS, *Problem 64–17*, SIAM Review **8** (1966), 236–237.
76. GRAVES, R.E., *A closure criterion for orthogonal functions*, Proc. Int. Congress Math. 1950, Cambridge (Mass.). Amer. Mah. Soc., vol. 1, (1952), 415.
77. __________, *A closure criterion for orthogonal functions*, Canadian J. Math. **4** (1952), 198–203.

78. OL'HOVSKIĬ, Yu. G., *On applications of Loyasevič's inequality to the theory of relaxation processes*, Sibirsk. Mat. Ž. **14**, 1134–1138. See in particular, p. 1136.
79. BAUM, L. E. and J. A. EAGON, *An inequality with applications to statistical estimation for probabilistic functions of Markov processes and to a model for ecology*, Bull. Amer. Math. Soc. **73** (1967), 360–363.
80. ZYGMUND, A., "Trigonometric Series," vol. 2, Cambridge, 1968, p. 11.
81. BOAS, R. P. *Inequalities for the derivatives of polynomials*, Math. Mag. **42** (1969), 165–174.
82. BERNSTEIN, S. N., "Sobranie sočinenii," vol. 1, Akad. Nauk. SSSR, Moscow 1952, 350–360.
83. APOSTOL, T., "Mathematical Analysis," Reading Mass., 1957.
84. HOBSON, E. W., "The theory of functions of a real variable and the theory of Fourier series," Vol. 1, 3rd ed., New York, 1957.
85. YOUNG, W. H., *On the inequalities connecting the double and repeated upper and lower integrals of a function of two variables.* Proc. London Math. Soc. (2) **6** (1908), 240–254.
86. KARLIN, S. and C. LOEWNER, *On some classes of functions associated with exponential polynomials*, Studies in Math. Anal. and Related Topics (H. Chernoff et al. eds), Stanford (1962), 175–182.
87. TAGAMLICKI, Ja., *Funkcii, koito udovletvorjavat izvestni neravenstva v'rhu realnata os*, Ann. Univ. Sofia. Fac. Phys.-Math. **42** (1945-1946), 239–266.
88. __________, *Recherches sur une classes de fonctions*, Ibid. **44** (1947–1948), 317–356 (Bulgarian).
89. __________, *O funkcijah, proizvodnye kotoryh udovletvrjajut nekotorym neravenstvam*, Doklady Akad. Nauk SSSR **75** (1950), 337–340.
90. *Problem 5483*, Amer. Math. Monthly **74** (1967), 446 and **75** (1968).
91. KORN, A., *Die Eigenschwingungen eines elestischen Körpers mit ruhender Oberfläche*, Akad. der Wissensch., München, Math.-phys. Kl. Berichte **36** (1906), 351–401.
92. __________, *Über eininge Ungleichungen, welche in der Theorie der elastischen und elektrischen Schwingungen eine Rollespielen*, Bulletin international de l'Académie de Cracovie (1909), 705–724.
93. FRIEDRICHS, K.O., *On the boundary-value problems of the theory of elasticity and Korn's inequality*, Annals of Mathematics **48** (1947), 441–471.
94. HORGAN, C.O. and J.K. KNOWLES, *Eigenvalue problems associated with Korn's inequalities*, Arc. Ration. Mech. Anal. **40** (1971), 384–402.

95. HORGAN, C.O., *Inequalities of Korn and Friedrichs in elasticity and potential theory*, Zeits. angew. Math. Phys. (J. App. Math. Phys.) **26** (1975), 155–164.
96. DJOKOVIĆ, D.Ž., *Problem 5311*, Amer. Math. Monthly **72** (1965), 794.
97. _________, Amer. Math. Monthly **73** (1966), 788.
98. GUANCHU, H. and T. JIANKANG, *The Djoković's conjecture on an integral inequality*, J. Math. Research and Exposition **10** (1990), 271–278.
99. MONTGOMERY, H.L., *Topics in multiplicative number theory*, Lecture Notes in Mathematics **227** (1971), 178 pp.
100. GALLAGHER, P.X., *The large sieve*, Mathematika **14** (1967), 14–20.
101. FREUD, G. and G. P NÉVAI, *Über einseitige Approximation durch Polynome III*, Acta. Sci. Math. **35** (1973), 65–79.
102. EVERITT, W. N. and A. P. GUINAND, *On a Hardy-Littlewood type integral inequality with a monotonic weight function*, in "General Inequalities," **5**, ed. W. Walter, Basel, 1987, pp. 29–63.
103. KWONG, M. K. and A. ZETTL, *Norm inequalities of product form in weighted L^p spaces*, Proc. Roy. Soc. Edinburgh A **89** (1981), 293–307.

INDEX

MIX
Papier aus verantwortungsvollen Quellen
Paper from responsible sources
FSC® C105338

If you have any concerns about our products,
you can contact us on
ProductSafety@springernature.com

In case Publisher is established outside the EU,
the EU authorized representative is:
Springer Nature Customer Service Center GmbH
Europaplatz 3, 69115 Heidelberg, Germany

Printed by Libri Plureos GmbH
in Hamburg, Germany